Environmental Science and Engineering
Subseries: Environmental Science

Series Editors: R. Allan • U. Förstner • W. Salomons

For further volumes:
http://www.springer.com/series/7487

Lars Håkanson · Henrik Ragnarsson Stabo ·
Andreas C. Bryhn

The Fish Production Potential of the Baltic Sea

A New General Approach for Optimizing Fish Quota Including a Holistic Management Plan Based on Ecosystem Modeling

Springer

Lars Håkanson
Department of Earth Sciences
Uppsala University
Villavägen 16
SE-752 36 Uppsala
Sweden
lars.hakanson@geo.uu.se
lars.hakanson@natgeog.uu.se

Henrik Ragnarsson Stabo
Department of Earth Sciences
Uppsala University
Villavägen 16
SE-752 36 Uppsala
Sweden

Andreas C. Bryhn
Department of Earth Sciences
Uppsala University
Villavägen 16
SE-752 36 Uppsala
Sweden
andreas.bryhn@geo.uu.se

ISBN 978-3-642-11561-5 e-ISBN 978-3-642-11562-2
DOI 10.1007/978-3-642-11562-2
Springer Heidelberg Dordrecht London New York

Library of Congress Control Number: 2010923407

Cover design: Integra Software Services Pvt. Ltd., Pondicherry

Printed on acid-free paper

Springer is part of Springer Science+Business Media (www.springer.com)

What is wanted is not the will to believe, but the will to find out, which is the exact opposite

Bertrand Russell

Contents

Prologue

Two books related to coastal studies and management strategies with a focus on the conditions in the Baltic Sea were published in 2008 by our group at Uppsala University, "Tools and criteria for sustainable coastal ecosystem management – examples from the Baltic Sea and other aquatic systems" and "Eutrophication in the Baltic Sea – present situation, nutrient transport processes, remedial strategies". We will complete the series with this book, and discuss and model the bioproduction potential of the Baltic Sea system with a focus on the production and biomass of prey and predatory fish. Our ambition has been that a reader of this book should be able to understand the motivation and structuring of the model used (CoastWeb) without consulting other publications, therefore there is a slight overlap between these books.

Our aim is not to write a literature review on aspects already covered by a plethora of other publications; instead we will provide key references to the particular problems discussed in this book. Our aim with this book has been to do what has not been done before and we will also explain and motivate the particular perspective on ecosystem modeling taken in this work. This is a book on "how it works" with a focus on the conditions in the Baltic Sea at the ecosystem scale and related to monthly quantifications of production and biomasses of ten key functional groups of organisms. The actual work carried out by functional groups of organisms rather than by individual species will be in focus. For example, predatory fish does the work of eating prey fish. The second target group in this book, prey fish, has a much more varied diet ranging from zoobenthos in the sediments to zooplankton in the water. The ten functional groups are: (1) predatory fish, (2) prey fish, (3) zoobenthos, (4) herbivorous zooplankton, (5) predatory zoopankton, (6) jellyfish, (7) phytoplankton, (8) bacterioplankton, (9) benthic algae and (10) macrophytes.

This books deals with the quantification of what these functional groups of organisms eat, how much they eat, when they eat, and how this relates to basic abiotic properties of the system, such as salinity, water temperature, oxygen conditions and nutrient concentrations, and how those things relate to fundamental transport process, such as water fluxes, stratification/mixing, sedimentation of suspended particulate matter, particulate and dissolved forms of nutrients, resuspension, burial, biouptake of nutrients, etc. In spite of the fact that so much has been written on eutrophication and ecosystem conditions in the Baltic Sea and on strategies to "save" the Baltic Sea, there are no previous studies on the holistic ecosystem modeling addressed in this book, except by members of our team at Uppsala University. Pluralism benefits the progress of science, and we believe that the science presented in this work increases our understanding of how natural aquatic ecosystems work. However, we do not disregard other alternatives and complementary approaches to Baltic Sea studies.

The ecosystem represents an important scale in aquatic science and management, e.g., in contexts of impact assessment of water pollutants and when remedial measures are discussed. Few people are interested in the content of a sampling bottle and most people in science and management are interested in what this content may actually represent. There is, however, no contradiction between work at this larger ecosystem scale and sampling and work at smaller scales, since, e.g., the mean or median values characterizing ecosystem conditions of necessity must emanate from sampling at individual sites.

Most of the data discussed in this book come from comprehensive "data-mining" of public sources available via the Internet. Several persons in our group at Uppsala University have participated in the data-mining and the work discussed in this book, especially Dan Lindgren.

Chapter 1
Introduction, Background and Aim

During the last decades there has been a steadily increasing flow of reports on declining and even collapsed fisheries in many parts of the world (e.g., Myers and Worm, 2005), from the well known cod disaster along the Canadian Atlantic coast, the well documented fishery problems in the Black Sea (Zaitsev and Mamaev, 1997) to the overfishing in the Baltic Sea and the Kattegat/Skagerrak (see, e.g., FAO, 2000; IBSFC, 2003). These problems have also been discussed intensively in the media, at environmental authorities at local, regional and national levels, and certainly not least among the general public. An excellent book on these matters has been published in Swedish by Isabella Lövdin (2007; Tyst hav; Silent Sea). Another example of the interest in this problem is that 50 million euro have been donated by an industrialist (Björn Carlsson; see http://www.balticsea2020.com/) and a trust has been established to use this money to try to "save" the Baltic Sea. It is evident that the general public, most politicians, many environmental managers and scientists are convinced that many parts of the Baltic Sea are in a deplorable state, with increasing extension of "dead bottom areas" (Jonsson, 1992), major regime shifts (Swedish Environmental Advisory Council, 2005; Wulff, 2006), increasing blooms of toxic algae (see the HELCOM website) and increasing anthropogenic eutrophication and that these problems add to the problems with the intensive fishing. Our results in this book are meant to give an honest quantitative picture of the situation in the Baltic Sea related to the fish production potential. This should help to produce a more solid scientific base for discussions on how to set fish quota and adjust those to changes in salinity, temperature, nutrient loading and especially related to different remedial actions and strategies and what to expect from various actions.

The Swedish Environmental Protection Agency has discussed the main threats to the Baltic Sea and ranked eutrophication as the worst threat (17 points) followed by overfishing (14 points), organic toxins (13 points) and invasion of alien species (13 points). The reason why we devote relatively much space to eutrophication in this book is not only because eutrophication is ranked as the worst threat to the Baltic Sea but because the nutrient loading will regulate the primary production of the system, which will regulate the secondary production, including the fish production. Changes in salinity and water temperature will also affect the fish production, but it is easier for man – i.e., the governments of the Baltic Sea states – to influence

L. Håkanson et al., *The Fish Production Potential of the Baltic Sea*, Environmental Science and Engineering, DOI 10.1007/978-3-642-11562-2_1,

the nutrient loading to the system than the salinity and water temperature of the system. Overfishing is partly an administrative issue, but the criteria for setting the fish quota should be based on science – i.e., on the production potential of fish in the given system – and this is what this book addresses from a new angle. We have not carried out any fishing expeditions and we have not looked into fishnets or fish stomachs during this study, but we have done extensive data-mining to obtain the best possible empirical framework for our model calculations. It should already at the outset be stressed that we would have liked to have access to much more extensive and reliable data on the biomasses of the ten functional groups discussed in this book, and how the biomasses vary within the system, from the Bothnian Bay in the north to the southern part of the Baltic Proper, from smaller coastal areas and archipelagos to the open water areas.

Fisheries in general, and certainly also in the Baltic Sea, are highly dependent on government subsidies (Hatcher, 2000) and on policy decisions. Invasions of alien species largely depend on shipping policies (Leppäkoski et al., 2002). Decreasing the levels of organic toxins require reliable, general, validated predictive models, of which there are none for the target organics in the Baltic Sea (such as dioxins and PCBs) that can relate abatement action to environmental effects.

As already stressed, there are numerous reports on declining fish stocks of several species from many parts of the world (e.g., cod in the Baltic Sea, the North Sea and along the Atlantic coast of Canada; Hutchings, 1996; FAO, 2000; Anonymous, 2000). The traditional method to set fish quota using fish catch statistics is highly uncertain (uncertainties up to a factor of four have been reported) and the quota are also often violated (Karagiannikos, 1996; Anonymous, 2002). Many persons (about 1,600 according to informal sources) are also directly and indirectly involved in setting fish quotas in Northern Europe for different species and areas. For example, the permitted quota for cod in the Baltic Sea in 2002 was 75 kt (75,000 t) and for 2008 40 kt. Scientists, experts and journalists have claimed that the overfishing may be 50–100% of the legal quota. Note that overfishing could mean both more fishing than the legal quota and more fishing than the system could sustain. The fish production evidently depends on the survival of the roe, the supply of food for the young and adult fish, the food supply for the organisms consumed by fish all the way along the food chain, i.e., the habitat conditions from "the cradle to the grave" including predation and fishing by man, birds and seals. This includes how the abiotic conditions (e.g., oxygen, salinity, nutrient concentrations and temperature) vary and set the framework for the fish production (Busch and Sly, 1992; Håkanson and Boulion, 2002a). Many processes and compensatory effects are involved. This means that models are essential tools to handle these complicated relationships in a structured manner so that rational and sustainable fish quota can be set and adjusted to prevailing and anticipated environmental changes. There are also many different types of models addressing different target variables. This work is not meant as a literature review on models in fishery science, but in Chap. 3, we will give a brief discussion on models mainly to provide a background to the new modeling approach presented in this book.

Illegal overfishing exists since there are evident short-term economic benefits to the fishermen, and in many countries with poor legal controls, the risks of reprisals are small (see Corten, 1996), so there should be major benefits if a new approach could be developed to complement existing approaches to set fish quotas used by, e.g., ICES (see http://www.ices.dk/indexfla.asp).

It is important to emphasize the problems related to extensive fishing, eutrophication, toxic contamination and high levels of toxins in fish and other environmental changes (e.g., in salinity, redox conditions, water exchange and sediment conditions) cannot and should not be separated. They are all pieces of the same jigsaw puzzle. For the Baltic Sea, which is the focus of this work, these threats have been discussed in many contexts (Voipio, 1981; Ambio, 1990, 2000; Wulff et al., 2001; Håkanson et al., 2002; Håkanson and Gyllenhammar, 2005):

- Eutrophication. The nutrient concentrations (mainly N and P) continue to stay at high levels; sediment anoxia have been high in the open water system for several decades (Jonsson, 1992; Jonsson et al., 2003). The nutrient loading from different sources and different Baltic countries are rather uncertain but known in terms of order of magnitude values (Stålnacke et al., 1999; Håkanson and Bryhn, 2008b).
- Toxic contamination. Most organic toxins, but not all, decrease in fish (the ecological half-life is about 6 years for organic toxins like PCBs and DDT in the Baltic Sea; Håkanson, 1999). There are EU recommendations on restrictions concerning consumption and sale of fat fish (e.g., salmon, whitefish and herring) due to high levels of dioxins in fish muscle. This is a key concern for the Baltic Sea fishery.
- Extensive fishing. There have been drastic reductions in catches of cod but also of other species (like pike and perch from Baltic coastal areas; see Söderberg and Gårdmark, 2003).

These threats are certainly not unique to the Baltic Sea but appear in many systems, e.g., the Black Sea (Zaitsev and Mamaev, 1997) and many other coastal areas (Aertbjerg, 2001).

Scaling is a central problem in communications among scientists, and also a reason for much misunderstanding in contexts of modeling. This is evident in practically all matters dealing with, e.g., fish modeling. Very many species, groups and size classes are involved, and differently in different systems. To sort this out in great detail at the scale dealing with hourly or daily changes for individual species is a very difficult task indeed for one system, and to try to do this in a general, predictive manner for many systems is even more difficult. The CoastWeb-model discussed in this book does not address the finer details of trophic interactions or the changes from one sample site to the next, or from one day to the next, or from one species to the next, in a given well-studied system. The CoastWeb-model is based on ordinary differential equations (i.e., box or compartment modeling) and this means that in this book we focus on entire defined sub-basins in the Baltic Sea, i.e., the Bothnian Bay (BB), the Bothnian Sea (BS), the Baltic Proper (BP), the Gulf of Finland (GF)

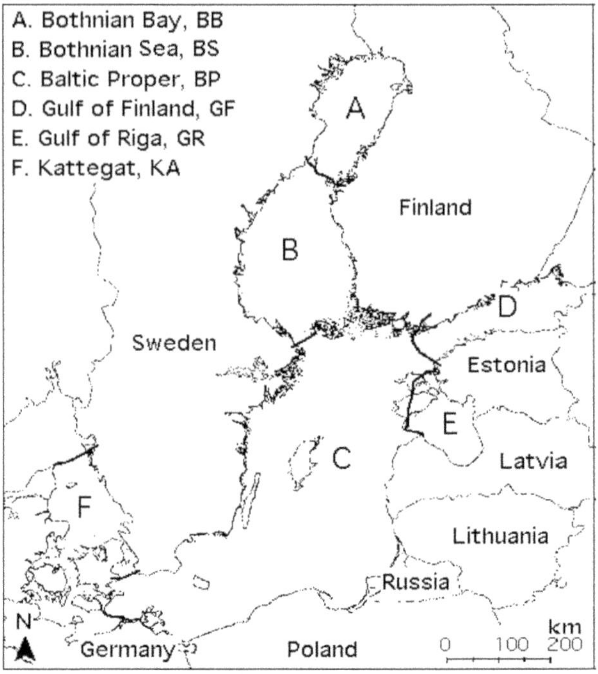

Fig. 1.1 Location map of the Baltic Sea

and the Gulf of Riga (GR). Figure 1.1 gives a location map of the area. These sub-basins are separated into 2 or 3 vertical layers, the surface-water (SW) layer above the theoretical wave base (see Chap. 2 for definition), the middle-water layer (MW) in the two deepest sub-basins, the Baltic Proper and the Gulf of Finland, and the deep-water layer (DW) below either the SW-layer or the MW-layer. When there are three layers, the DW-layer represents the water mass below the average depth of the halocline. The five sub-basins of the Baltic Sea covered in this book, including their bottom types are shown in Fig. 1.2. Thick black lines in this figure shows the limits between the different sub-basins.

Production of predatory fish is a target variable in the CoastWeb-model. This value represents an integration over space and time in the sense that the prey fish eaten by the predatory fish grow, live and eat within a much wider area than the area where the predatory fish are caught. And the organisms eaten by the prey fish, i.e., zooplankton and zoobenthos also move on their own accord in their search for food and suitable habitats for reproduction. Zooplankton are also distributed and transported by water currents. The latter aspects are important and one can note that a typical theoretical water turnover time for smaller Baltic Sea coastal areas is about 2–6 days (Håkanson, 2000), that the coastal current (the Coriolis driven coastal jet) has a typical velocity of 25 cm/s or 150 km/week (FRP, 1978) and that during a month winds can blow from many directions and with different speeds. All this creates a strong internal mixing in spite of the fact that the theoretical water turnover

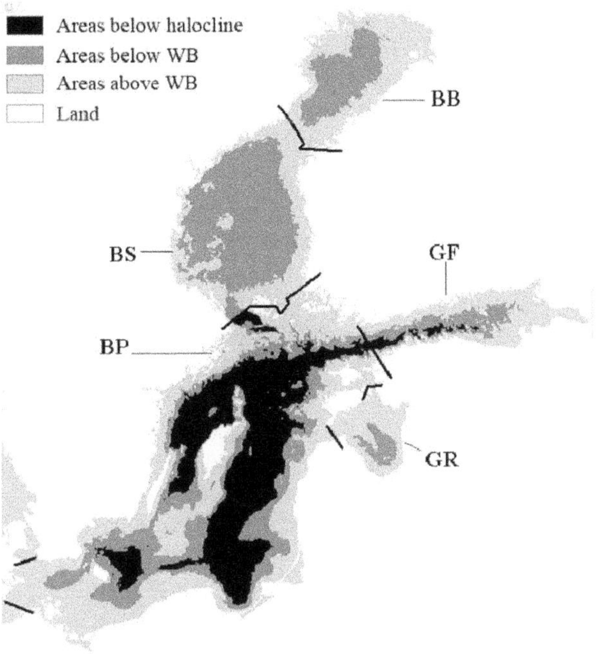

Fig. 1.2 The five sub-basins of the Baltic Sea covered in this book, including their *bottom* types. WB = theoretical wave base. *Thick black lines* indicate sub-basin limits

time for the Baltic Sea water is very long (about 20 years; see Håkanson and Bryhn, 2008b and Chap. 2). The mixing also influences the distribution of the food eaten by zooplankton, i.e., phytoplankton and bacterioplankton. Thus, to understand and predict the factors regulating fish production one needs to look at the production potential of an entire system, i.e., to take an ecosystem approach and integrate many processes over a wide area and a relatively long period of time. To illustrate this, one can first estimate phytoplankton biomass from measurements of primary production given in mg C per m^3 per day; multiplication with 30 gives the monthly production; multiplication with the depth of the photic zone (D) and the size of the area (A) gives, after dimensional adjustments, the phytoplankton biomass expressed in kg ww a given month as a function of D and A – the larger the area, the larger the phytoplankton biomass if all else is constant. Then, a given phytoplankton biomass (BM_{PH}) can sustain a biomass of herbivorous zooplankton (BM_{ZH}) that would be a factor of 5–20 times lower than BM_{PH}. Then, a given biomass of herbivorous zooplankton would sustain a biomass of predatory zooplankton and prey fish biomass that would be a factor of 5–20 lower than BM_{ZH}. Finally, a given prey fish biomass would sustain a predatory fish biomass that would be even lower. So, in order to calculate a given production and biomass of predatory fish, the production potential of a larger water body has to be considered, as given by A and D in this example.

The CoastWeb-model is structured to handle such processes and, to the best of our knowledge, no other ecosystem model does so in the same way, except the original LakeWeb-model (from Håkanson and Boulion, 2002a), which is the basis for the CoastWeb-model discussed in this work. CoastWeb is meant to incorporate all key aspects in a general and mechanistic manner – from the environmental factors regulating primary production to the processes regulating secondary production, including fish production. The focus is on the ecosystem scale also partly because society is not interested in the contents of a sampling bottle or a particular fishing site, but in a much larger entity, the ecosystem (Odum, 1986).

The basic LakeWeb-model has been critically tested with positive results against very comprehensive empirical data for lakes mainly from Eastern and Western Europe. The model has also been tested by means of extensive sensitivity and uncertainty analyses (using Monte Carlo techniques) for both uniform and characteristic coefficients of variation for all model variables to quantify and rank the uncertainties influencing the model predictions (Håkanson and Boulion, 2002a). An important feature of the model, and a pre-requisite for its practical applicability in contexts of water management, is that it can be run from relatively few driving variables readily accessible from standard maps and monitoring programs. It also includes a sub-model for toxic substances in fish, which is of special interest in the Baltic Sea where fat fish have so high levels of dioxins that there are restrictions on consumption and sale, but the latter aspects will not be discussed in this book.

The basic structure of this work and some of the main features of the CoastMab-model are illustrated in Fig. 1.3.

First (at Level 1), we will use CoastMab, i.e., the coastal mass-balance model for salt. This approach is explained in detail in Håkanson and Bryhn (2008b) for the Baltic Sea basins. CoastMab for salt quantifies the water fluxes to, within and from all the sub-basins and vertical layers in the Baltic Sea, including mixing and diffusion. The main results will be given in Chap. 2. It should be stressed that the CoastMab-modeling has been tested in many coastal areas and lakes and also discussed in Håkanson and Bryhn (2008b). This model will calculate the water fluxes needed to explain the measured salinities. This means that data on salinities in the inflowing water from the Kattegat to the Baltic Proper are needed to run the model and in most of the following simulations, we will use the same data from the period 1997 to 2006 as was done by Håkanson and Bryhn (2008b). This modeling also needs morphometric data (mean depth, form factor, dynamic ratio, etc.), hypsographic curves and volume curves for all the sub-basins and also those data emanate from Håkanson and Bryhn (2008b).

At Level 2, we use CoastMab for phosphorus. This version of the CoastMab-model for phosphorus is modified as compared to the modeling presented by Håkanson and Bryhn (2008b). The main reason for this is that in this work, we will not use a regression predicting the concentrations of suspended particulate matter (SPM) from modeled concentrations of total phosphorus (TP). Instead, we will use the dynamic CoastMab-model for SPM (Layer 3). One should note that many of the algorithms to quantify the transport processes for phosphorus are also valid for salt and SPM, e.g., inflow from the Kattegat to the Baltic Proper, sedimentation of particulate phosphorus and SPM, mixing, diffusion of salt and dissolved phosphorus, resuspension of particulate phosphorus and SPM, TP and SPM added from land uplift and burial of phosphorus and SPM. There are specific transport processes for

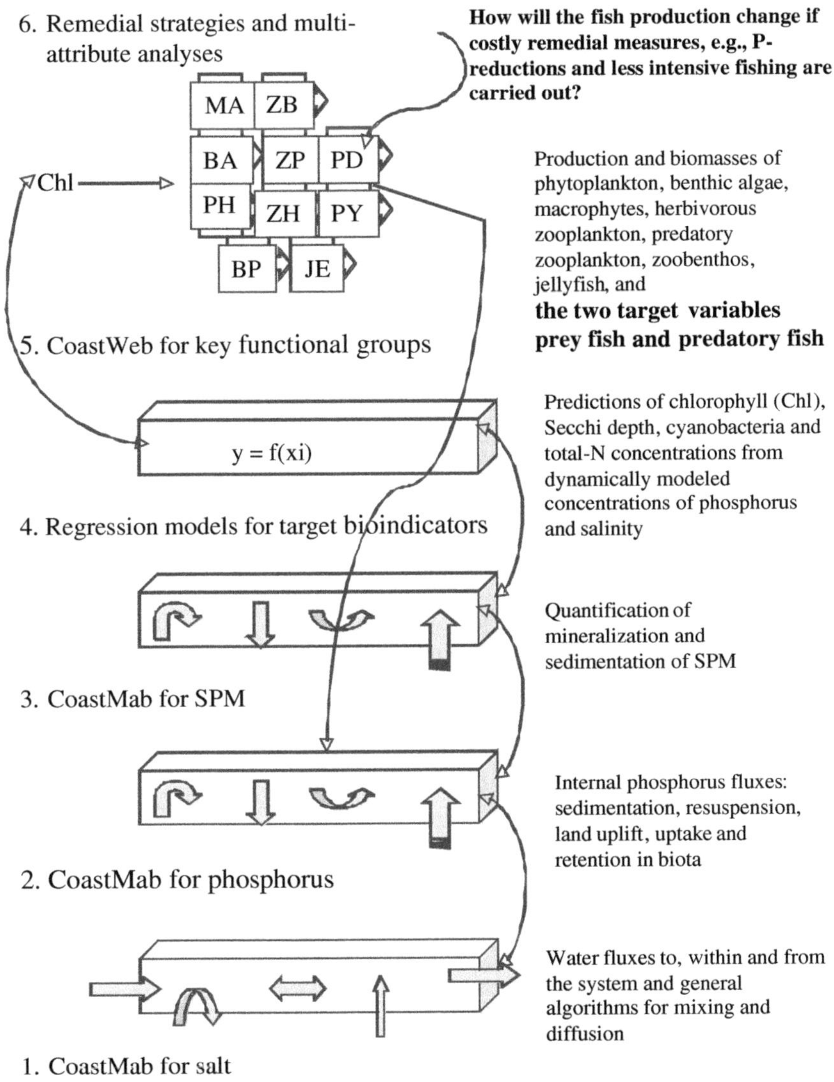

Fig. 1.3 A general outline of the structure of the CoastWeb-model and the modeling carried out in this work

nitrogen such as atmospheric deposition, gas transport (nitrogen also appears in a gaseous phase), N_2 fixation and denitrification. Nitrogen modeling is not included in this work, but it has been done for the Kattegat (Håkanson, 2009) with the CoastMab-model. The reasons why we have not used CoastMab for nitrogen will be explained in more detail in Chap. 2.

At Level 3, we will use CoastMab to model the inflow, production, sedimentation, burial and mineralization of suspended particulate matter (SPM; see Håkanson,

2006). Sedimentation is important for the oxygen consumption and oxygen status of the system, and especially for the oxygen conditions in the deep-water layer and for the diffusion of phosphorus from sediments to water.

At Level 4, we will use both regional and general regression models to predict how key bioindicators such as the Secchi depth (a standard measure of water clarity and the depth of the photic zone) and the concentration of chlorophyll-a (a standard measure of both primary phytoplankton production and biomass and the key driving variable for the foodweb model, CoastWeb) and cyanobacteria (= bluegreen algae; an important measure of harmful algae in the Baltic Sea) would change in relation to changing phosphorus concentrations, salinities, SPM-values, temperatures and light conditions. This will also be explained in detail in the Chap. 2.

At Level 5, which is the target level including the production and biomasses of prey and predatory fish, we will ask questions like: How will the fish production change if, e.g., costly remedial measures are taken to reduce eutrophication in the Baltic Sea and if less intensive fishing would be carried out? In Chaps. 4 and 5, we will give several scenarios addressing questions related to "optimal fishing rates", the influence of mussels or jellyfish on Baltic Sea fisheries, the impact of salt-water intrusions on Baltic Sea fisheries, and assess how increasing or decreasing the nutrient loading would likely influence the fish production in the Baltic Sea.

Figure 1.4 stresses the fact that this work has a focus not on individual species of fish but on key functional groups which are present in most/all aquatic systems;

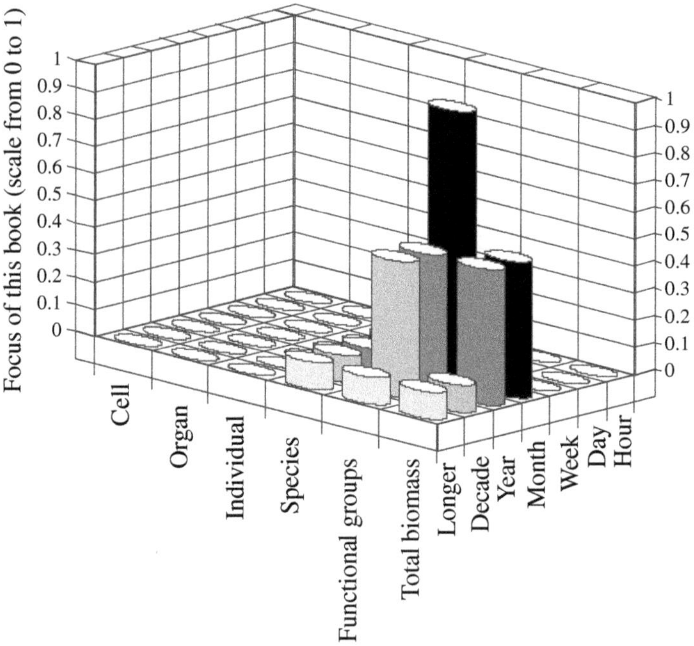

Fig. 1.4 Focus of the CoastWeb-model, i.e., on entire basins, functional groups of organisms and monthly predictions

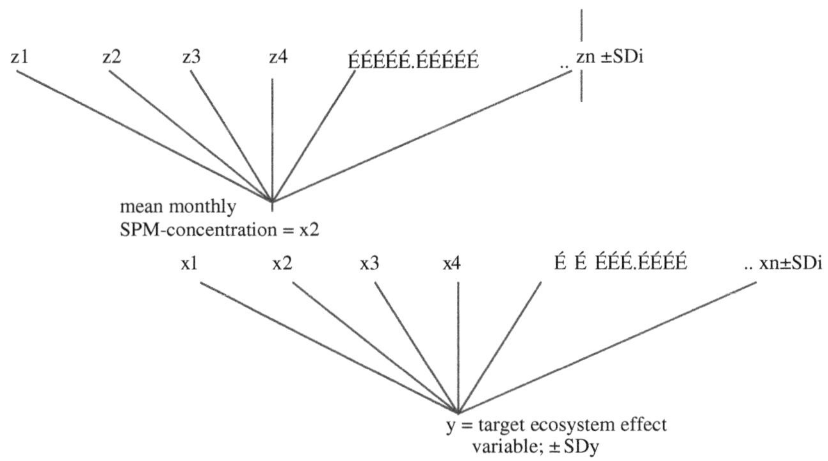

Fig. 1.5 Illustration of how a target variable in geosciences, such as the concentration of suspended particulate matter (SPM), may be used as an x-variable in predictive models in water management, where the y-variables may express key functional aspects of ecosystem status, such as water clarity, chlorophyll-a concentrations and/or biomasses of key functional groups of organisms. In this work, the focus is set on the production and biomasses of prey and predatory fish

the focus in not on the conditions at individual sites in the Baltic Sea, but on entire defined sub-basins (the ecosystem scale); not on short-term (hourly, daily, weekly) predictions but on monthly and annual predictions.

Figure 1.5 has been included here to highlight that many of the variables included in this modeling as "x-variables", e.g., the SPM-concentration, may be regarded a target y-variable in other contexts, e.g., in studies related to sedimentation. However, in water management, aquatic ecology and fishery science, the main focus is not on SPM but rather on how different abiotic factors (such as temperature, nutrient concentrations, SPM and salinity) may influence the structure and function of aquatic ecosystems. In such cases, SPM can be regarded as one of many important x-variables influencing the target y-variables. In this work, the target y-variables are the monthly production values and biomasses of prey and predatory fish.

In this introduction, we would also like to stress that a dynamic mass-balance approach to the problem of understanding the role of the salinity, SPM or nutrients in aquatic systems, may start with a calculation of the inflow to the system. This can be accomplished by using data on water discharge (Q in m^3/time) and the salinity, SPM or nutrient concentration (g dw/m^3; dw = dry weight) of the inflowing water (see Fig. 1.6a). Differential equations are often used to quantify fluxes (e.g., g X/time), amounts (g X) and concentrations (g X/m^3) of all types of materials (such as SPM, organic matter, toxins and nutrients), but not generally ecosystem effect variables such as the standard bioindicators included in this work (the Secchi depth or the concentration of chlorophyll-a). Regressions based on empirical data are often necessary to relate concentrations of chemicals in water or sediments to

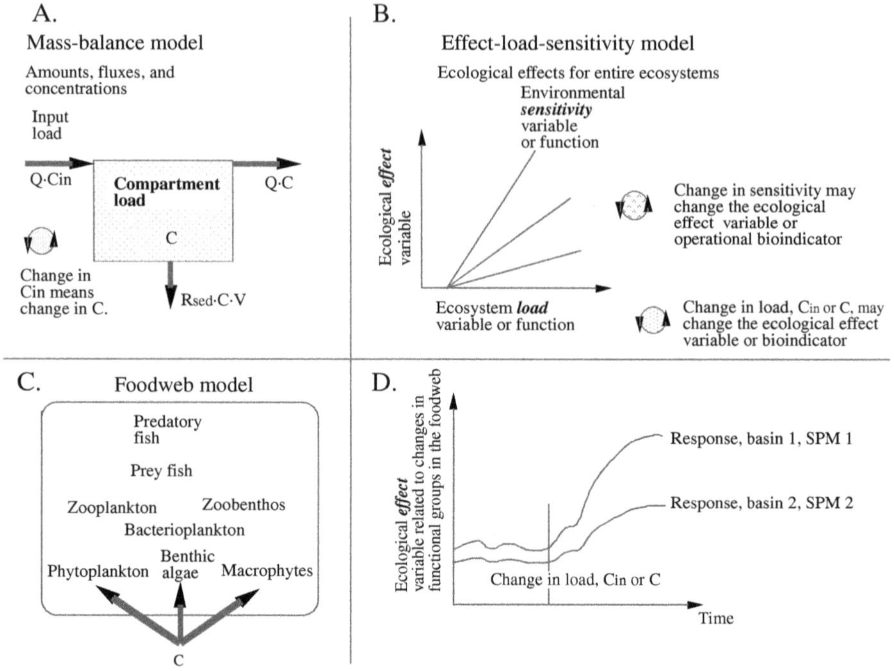

Fig. 1.6 Illustration of the fundamental difference between dynamic mass-balance models (**a**) and effect-load-sensitivity models (ELS) based on regressions (**b**) and ELS-models related to dynamic foodweb models (**c**) and (**d**) how changes in the load at a given time may cause different responses in the foodweb in aquatic systems with different SPM-concentrations (basin 1 compared to basin 2). The wheels indicate that by means of remedial measures one may reduce the load variable in dynamic models and the load and the sensitivity variables in ELS-models

target bioindicators. In theory, both these model approaches (see Fig. 1.6a, b) may be used for the effect-load-sensitivity analyses (ELS; see Håkanson, 1999) provided that at least one operationally defined ecological effect variable relevant for the load variables(s) in question is included in the model. Ideally, the effect variable should express the production or biomass of defined functional organisms (preferably fish at the top trophic level, see Fig. 1.6c), which characterize a given system. Figure 1.6d illustrates schematically that two systems, e.g., two sub-basins in the Baltic Sea, with different salinities, SPM or nutrient concentrations are likely to react differently to a change in the load of toxins and/or nutrients to the system. The classical approach to carry out environmental consequence analyses is to use, (1) dynamic mass-balance models to predict concentrations of pollutants and (2) empirical models (like regressions) to link these concentrations to measured data on the operational bioindicators. So, there are good reasons why these two modelling approaches are addressed in this book.

Figure 1.7 illustrates a central argument in this work. The fish productivity of water bodies is connected to primary production by many intermediate trophic links

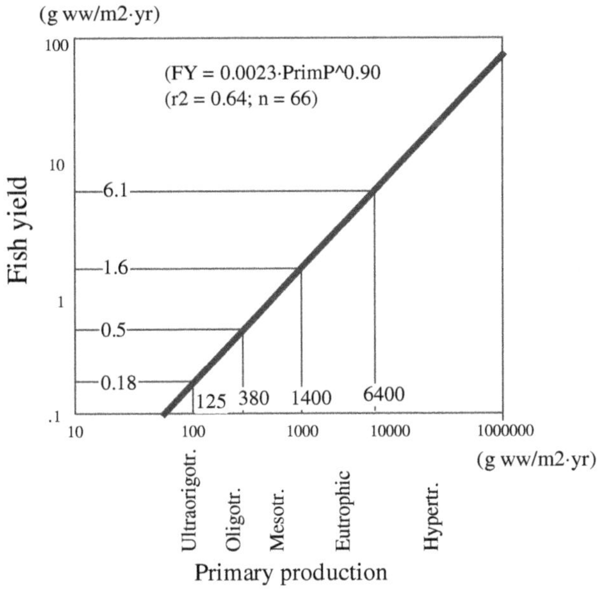

Fig. 1.7 The relationship between fish yield (FY) and phytoplankton primary production (PrimP). The figure also gives the corresponding values related to the different trophic classes based on chlorophyll (from Håkanson and Boulion, 2002a)

(see Håkanson and Boulion, 2002a). This question has, evidently, many practical implications. It has been discussed for a long time (see, e.g., Oglesby, 1977; Morgan, 1980; Boulion and Winberg, 1981). Generally, the annual phytoplankton primary production (PrimP) is compared to the fish yield (FY). Note that it is relatively easy to calculate PrimP from chlorophyll (see Håkanson and Boulion, 2002a), but that it is often very difficult to obtain reliable empirical values on fish biomasses and fish production (= fish yield). This means that the empirical data on fish production are generally quite uncertain, and this will restrict the predictive power of models for fish production, including the CoastWeb-model used in this work.

There are also at least three different measures referred to as "fish yield", (1) total production of fish (FY_1), (2) total catches of fish (FY_2) and (3) commercial catches of fish (FY_3). Håkanson and Boulion (2002a) have argued that for lakes the order of magnitude differences between these three expressions for fish yield generally would be about, $FY_1 \approx 3{\cdot}FY_2 \approx 10{\cdot}FY_3$. It is evident that there are great uncertainties among aquatic systems in this respect.

A regression between primary production and fish yield (FY_1) is given in Fig. 1.7 based on data from lakes, reservoirs and marine areas (see Håkanson and Boulion, 2002a). There is, in fact, a close correlation between FY_1 (the data range from 0.13 to 693 g ww/m^2·year) and PrimP (from 170 up to 14,000 g ww/m^2·year):

$$FY_1 = 0.0023 \cdot PrimP^{0.90} \tag{1.1}$$

$$(r^2 = 0.64; n = 66)$$

This equation predicts that the average fish yield is about 0.1–0.2% of the primary production. A deviation from this general relationship, e.g., as a result of excessive fishing, is likely to create changes in ecosystem structure and function. It is important to stress that the ratio FY_1/PrimP is rather independent of the absolute level of the primary production since the exponent in Eq. (1.1) is close to 1.

These results imply that one can quite confidently predict order-of-magnitude values, e.g., that fish catches should not exceed about 0.3% of the primary production. On the other hand, fish catches lower than 0.05% of PrimP would indicate that the fishing is sub-optimal. This will be elaborated more closely in the following text by using the CoastWeb-model.

Figure 1.8 is meant to give an introductory illustration of some of the key questions raised in this work. We have made preliminary simulations to illustrate how the total annual fishing of predatory (PD) fish (Fig. 1.8a; this includes all types of fishing by man, seals, bird, etc.) and the total biomass (BM) of prey fish (PY; Fig. 1.8b)

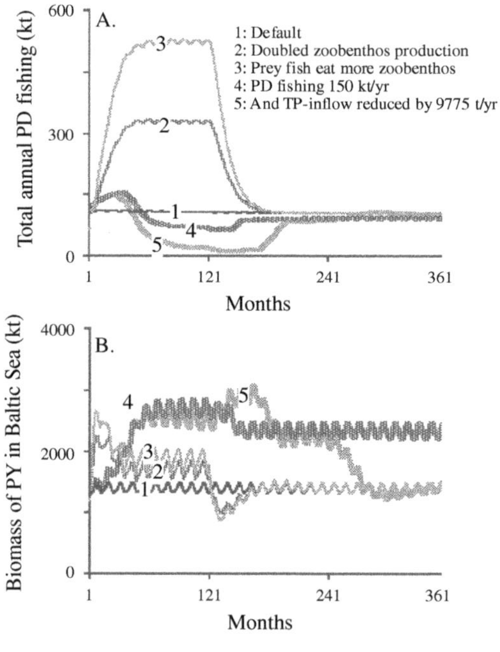

Fig. 1.8 An initial example illustrating key questions discussed in this work. *Curves 1* give default steady-state conditions for (**a**) total fishing of predatory fish (PD) and (**b**) total biomass of prey fish in the entire Baltic Sea; *Curves 2* illustrates changes from default conditions when the production of zoobenthos has been increased by a factor of 2 between months 1 and 121; *Curves 3* give changes when also the distribution coefficient regulating the feeding habits of prey fish between zooplankton and zoobenthos has been changed so that more zoobenthos are consumed; *Curves 4* illustrate results when the total predatory fishing is increased from 100 to 150 kt/year from month 1 to month 121; *Curves 5* illustrate results when also the total phosphorus loading to the Baltic Sea system has been reduced (compared to the results in *Curve 4*) by 9,775 t/year from month 1 to month 121

could be in the entire Baltic Sea (i.e., in the Bothnian Bay, the Bothnian Sea, the Baltic Proper, the Gulf of Finland and the Gulf of Riga).

- Curves 2 have been obtained when the zoobenthos production has been increased by a factor of 2 between months 1 and 121, e.g., as a response to large intrusions of oxygenated saltwater from the Kattegat. Such intrusions are well documented (see Fig. 1.9) and they have been discussed in many papers (see, e.g., Gustafsson and Stigebrandt, 2007; Eilola et al., 2009). Increasing the oxygen concentrations in the system would increase the survival, production and biomasses of zoobenthos (see, e.g., Pearson and Rosenberg, 1976) and this would mean more food for bentivorous prey fish, which would increase in biomass. This would also mean more food for predatory fish eating prey fish and increase the possibilities for fishing predatory fish. But an increasing biomass of predatory fish would also imply increased predation pressure on the prey fish eaten by the predatory fish. This is easy to state, but the idea with this work is to quantify such interactions and explain such relationships not only by words but by equations and calculations. As already stressed, there have been drastic variations in the biomasses (and catches) of cod, sprat and herring in recent decades (see Fig. 1.10). Total biomasses have varied from a peak value of about 1,000 kt cod (both larger predatory cod and smaller bentivorous cod) to values as low as 50 kt

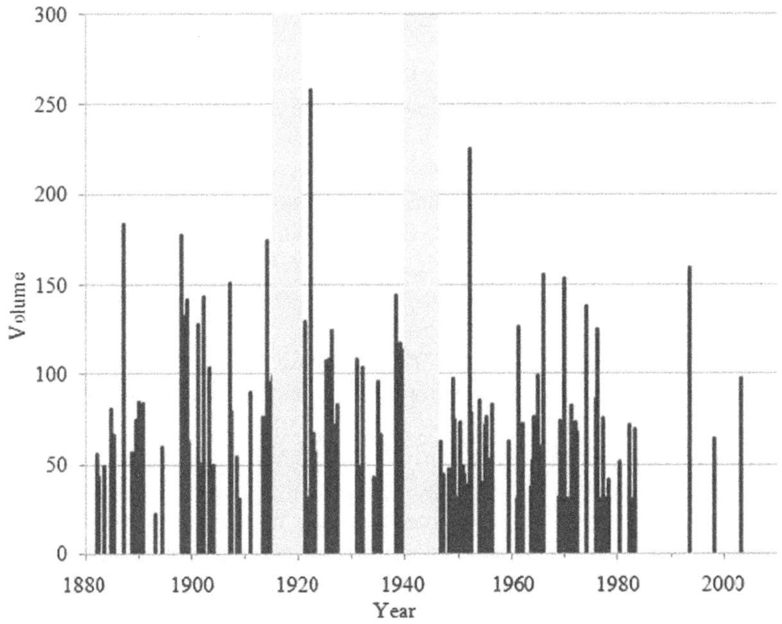

Fig. 1.9 Short-term intrusions of oxygen-rich saltwater from Kattegat (based on data from Fischer and Matthäus, 1996). Water volumes in km^3. *Grey* areas indicate periods (with wars) with insufficient data

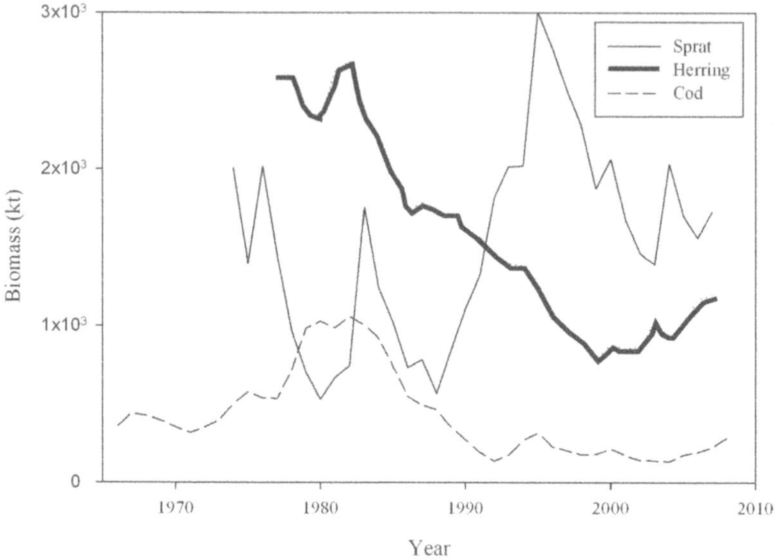

Fig. 1.10 The total biomass of important fish species in the Baltic Sea, ICES subdivisions 25–32. Data from ICES (http://www.ices.dk/). Note that the underlying data are based on information from sampling stations which do not provide an even area cover of the basins in the Baltic Sea

(see IBSFC, 2003). How can such variations be explained and understood? This work aims at shredding light onto that important question by quantifying the fish production potential of all major sub-basins in the Baltic Sea. One can note from Fig. 1.8 that the given hypothetical increase in zoobenthos biomass (Curves 2) may not be sufficient (in this introductory example) to produce more than about 300 kt of predatory fish (see Sect. 2.1). So, what else could explain the very high biomasses of cod shown in Fig. 1.10? A relatively high percentage of the predatory fish caught in the Baltic Sea around 1980 was large cod and a high fraction of the prey fish was herring, sprat and small cod (see, e.g., Elmgren, 1984; Hjerne and Hansson, 2002). We will present compilations of literature data on the fish species making up the major parts of the predatory and prey fish groups in Chap. 2.

- Curves 3 in Fig. 1.8 have been derived by changing the feeding habits of the prey fish. The biomass of zoobenthos in the Baltic Sea is high compared to the biomass of zooplankton. If prey fish eat a higher fraction of zoobenthos, the prey fish production will increase, and so will the production of predatory fish. Such a change in feeding habits would increase the predatory fish biomass. And it should be stressed again that a significant amount of the prey fish is small cod feeding on mainly zoobenthos (see Chap. 2).
- Curves 4 in Fig. 1.8 illustrate how changes in fishing would influence the two variables on the y-axes. It is reasonable to assume that in situations when there

are more fish than usual, this would imply increased recreational and professional fishing and more fish to feed seals, other animals and birds. For the results given by Curves 4, the total fishing of predatory fish has been increased by a factor of 1.5 (to 150 kt/year). More intensive fishing on predatory fish would initially increase the annual fishing but after a few years this could imply significant reductions in the biomass of predatory fish, which in turn could lead to lower predations pressure on the prey fish, which would increase in biomass. This is also well documented (see Fig. 1.10).

- Curves 5 in Fig. 1.8 have been included to stress the link between nutrient reductions and how the Baltic Sea ecosystem (in this case prey and predatory fish) would likely react to such remedial actions in contexts of eutrophication abatement. In these simulations, we have reduced (months 1–121) the total annual inflow of phosphorus via the tributaries to the Baltic Sea by 9,775 t (as discussed by Håkanson and Bryhn, 2008b) and studied the effects when the total predatory fishing is calculated according to Curves 4 (i.e., a total fishing of predatory fish of 150 kt/year). Such heavy fishing under increasingly more oligotrophic conditions could imply that the Baltic Sea could reach the verge of predatory fish extinction. The prey fish biomass would, on the other hand, increase very much due to an even lower predation pressure from the predatory fish.

It should be noted again that the results given in Fig. 1.8 are only included here to stress key questions for Chaps. 4 and 5.

At the outset, it should also be pointed out that there are many simplifications in the CoastWeb-model. They are necessary for several reasons, (1) to keep the model as small as possible (it is still fairly extensive), (2) to keep the driving variable as few and as accessible as possible, (3) to be able to critically test the model using empirical data or empirical regressions. The idea has not been to include everything but to focus on the key functional groups of organisms and on key abiotic/biotic relationships. So, CoastWeb does not address the finer details related to individual species. At any modeling scale, the complexity of natural ecosystems always exceeds the complexity and size of any model. So, simplifications are always needed, and the ultimate goal in achieving predictive power and general validity for a model is to find the most appropriate simplifications (Peters, 1991; Monte, 1995, 1996; Monte et al., 1999).

It should be evident that the aim of this work is to model general, basin-typical conditions, since there are only relatively few obligatory driving variables in the CoastWeb-model. It should also be emphasized that CoastWeb is primarily intended to handle feedbacks among the functional groups. Because biotic/abiotic feedbacks are also of great importance, such interrelationships are also included in CoastWeb, e.g., the influence of material produced in the system itself (autochthonous materials) on the depth of the photic zone, reductions in TP-concentrations related to biouptake of phosphorus, changes in suspended particulate matter (= SPM = seston) from changes in primary phytoplankton production; SPM is also a key regulatory factor for bacterial production, sedimentation and other processes influencing the oxygen conditions.

There will be situations where the model will not predict well. Since the default set-up of the model should reflect normal interactions among the most important factors influencing a given y-variable in the Baltic Sea for the period 1997–2006, a deviation between modeled and measured values could indicate, e.g., that some factors not accounted for in the model would be of importance. This is an important basis for a correct analysis of the dynamics of the system and can of course always lead to improvements of the model.

We would like to stress that many of the algorithms in CoastWeb are the same as in LakeWeb and have been thoroughly tested and found to work well. It should also be stressed that it is much more difficult to find the appropriate data for testing the model for coastal areas than for lakes, but this may, hopefully, change in the future, if this modeling approach becomes more widely used in coastal management and science.

In the following, we will present results and background information in the following order:

First, Chap. 2 will give basic information on the conditions in the Baltic Sea, on:

- The morphometry, i.e., depth, area, volume, form characteristics of the five main basins and the criteria to define the limit for the surface-water layer from the theoretical wave base.
- The water transport processes among and within the sub-basins and vertical water layers, since this is important for the concentrations and fluxes of salt, phosphorus, SPM and for the water transport of organisms among the sub-basins.
- The results related to the water balance and the mass-balance for phosphorus for the Baltic Sea have been published before (see Håkanson and Bryhn, 2008b) and the main results of relevance for the CoastWeb-modeling will be summarized in Chap. 2. This work also presents the dynamic mass-balance model for suspended particulate matter for the entire Baltic Sea. Since those results have not been published before, we will give a more thorough account on this (CoastMab for SPM).
- In Chap. 2, we will also present the approaches to predict chlorophyll-a concentrations and Secchi depths from dynamically modeled values of phosphorus and SPM-concentrations and salinity (and monthly light conditions), since these approaches are of fundamental importance for the CoastWeb-modeling. CoastWeb is driven by chlorophyll-a concentrations and the Secchi depths represent a measure of the depth of the photic layer, an integral part of the model.
- As a background to understand present conditions, Chap. 2 will also give rather comprehensive trend analyses based on empirical data to demonstrate how the in the Baltic Sea have changed since 1960 concerning nutrient concentrations, oxygen and chlorophyll-a concentrations.

- Finally, Chap. 2 will give important background information and data based on a comprehensive literature survey on the most dominating species within each functional group present in the Baltic Sea and what they eat. The aim is to get an overview of the organisms making up the main biomasses of each functional group. This section will also discuss differences between the five sub-basins in the Baltic Sea, and also between the smaller coastal areas and the open basins.

Chapter 3 will present the basic structures of the CoastWeb-model and a short comparison between this modeling approach and a well known and much used modeling approach in contexts of marine fish production, Ecopath/Ecosim (Walters et al., 1997, 2000; Sandberg et al., 2000; Christensen et al., 2000; Harvey et al., 2003). This is done to highlight the specific features of the CoastWeb-approach and to stress that it actually provides a new dimension to understand and quantitatively simulate the factors regulating fish production and biomass. It should be stressed that CoastWeb is based on general principles and processes that apply for most aquatic systems. The LakeWeb-model has been presented in a book (Håkanson and Boulion, 2002a) and the different parts of the model and of the CoastMab and CoastWeb model have been discussed in several papers in scientific journals over the last years (see Table 1.1). In Chap. 3, we will also present changes in the model related to the fact that the previous version of the CoastWeb-model (see Håkanson and Bryhn, 2008d and Håkanson and Lindgren, 2008a) was for smaller coastal areas and this version has been applied for the entire Baltic Sea with its five main sub-basins.

Table 1.1 The LakeWeb/CoastWeb-model has been presented in the following papers

Model descriptions	
The complete LakeWeb-model	Håkanson and Boulion (2002a)
Uncertainty analyses of ecosystem models	Håkanson (2003a)
The fish sub-model	Håkanson and Boulion (2004a)
The zoobenthos sub-model	Håkanson and Boulion (2003d)
The zooplankton sub-model	Håkanson and Boulion (2003a)
The phytoplankton sub-model	Håkanson and Boulion (2001a, 2003c)
The bacterioplankton sub-model	Boulion and Håkanson (2003)
The benthic algae sub-model	Håkanson and Boulion (2004b)
The macrophyte sub-model	Håkanson and Boulion (2002b)
The water clarity sub-model	Håkanson and Boulion (2003b)
The mass-balance sub-model for phosphorus (LakeMab)	Håkanson and Boulion (2003b)
The sub-model for the duration of the growing season	Håkanson and Boulion (2001b)
Practical applications using the LakeWeb-model	
Effects of land-use changes on aquatic foodwebs	Håkanson (2002)
Effects of extensive fishing on aquatic foodwebs	Håkanson et al. (2003a)
Effects of eutrophication and climate changes on aquatic foodwebs	Håkanson et al. (2003b)
Effects of acid rain on aquatic foodwebs	Håkanson (2003b)

Table 1.1 (continued)

BaltWeb (a preliminary first version of the CoastWeb-model)	Håkanson and Gyllenhammar (2005)
The CoastWeb-model	
Applied for the Ringkobing Fjord, Denmark	Håkanson and Bryhn (2008d)
Model description	Håkanson and Lindgren (2008a)
The CoastMab-model (the mass-balance model)	
Applied and tested to Baltic Sea coastal areas	Håkanson and Eklund (2007)
Applied for the Ringkobing Fjord, Denmark	Håkanson and Bryhn (2008e)
Applied for radionuclides in the Black Sea	Håkanson and Lindgren (2009)
Applied for water fluxes in the Baltic Sea	Håkanson and Lindgren (2008b)
Applied for SPM in lakes, rivers and marine areas	Håkanson (2006)
Applied for several marine coastal areas	Håkanson and Bryhn (2008b)
Applied for the entire Baltic Sea	Håkanson and Bryhn (2008a)
Applied for the Kattegat (including a mass-balance for nitrogen)	Håkanson (2009)

Chapter 4 presents all sub-models for all functional groups in detail and also comparisons between dynamically modeled values and empirically-based values (norm-values) for all functional groups in three scenarios which are meant to provide a framework for Chap. 5. Chapter 4 discusses details for each functional group which were not included in the general overview of the CoastWeb-modeling in Chap. 3 and presents results from three scenarios to pave the way for Chap. 5, which will discuss the use of the CoastWeb-model in contexts of management scenarios. The three scenarios in Chap. 4 are:

(1) TP-concentrations will be varied in four 10-year steps by reducing the TP-inflow to the Baltic Proper. Note that this scenario is of special interest in contexts related to Baltic Sea management plans since man can actually implement – costly (see Chap. 5) – remedial measures and significantly influence the nutrient fluxes from land to water.
(2) We will also change the salt-water inflow from the Kattegat in steps and study how this would influence the system.
(3) Water temperatures will also be increased in steps.

In Chap. 5, we will give several scenarios to illustrate the practical use of the CoastWeb-approach by providing results demonstrating the potentials of the model in contexts of settling fish quotas adjusted to changing environmental conditions, the recovery after periods of overfishing, the possible influences of invasions of jellyfish and cultivation of mussels and the expected consequences for the fish production potential of reducing the anthropogenic nutrient loading. An important scenario for Baltic Sea management given in Chap. 5 combines the effects of nutrient reductions, the consequences for the fisheries and a cost-benefit analysis of different remedial strategies. This chapter also includes a scenario where we have accounted for aquaculture and studied the effects of increasing fish farming in the Baltic Sea and discussed criteria for where to place the fish farm cages and how large fish farms could be recommended and for what reasons. Finally, we will try to put all of this together and present a holistic management plan for the Baltic Sea including a cost-benefit analysis.

Chapter 2
Basic Information on the Baltic Sea

2.1 Introduction and Aim

The size and form of a given aquatic system, i.e., the morphometry, influences the way in which the system functions, since the depth-characteristics influence resuspension and internal loading of nutrients, the nutrient concentrations regulate the primary production, which in turn regulates the secondary production, including zooplankton and fish (see Håkanson and Boulion, 2002a). Such relationships will be addressed in this chapter, which is structured in the following manner:

- First, we will present morphometric data for the Baltic Sea and its main sub-basins and highlight why and how the given morphometrical parameters are important for the fish production of the system. This has been discussed in more detail for lakes by Håkanson (2004) and the idea here is to provide a background illustrating how morphometric parameters are used in the CoastWeb-model.
- Then, we will give brief compilations of water, salt and phosphorus fluxes. The water fluxes are accounted for in quantifying the transport of phytoplankton, bacterioplankton and zooplankton between the different sub-basins in the Baltic Sea, and are evidently very important for the salinities in the various basins and for the phosphorus concentrations influencing chlorophyll-a concentrations and Secchi depths.
- We will also present the dynamic mass-balance model for suspended particulate matter (CoastMab for SPM), which is a new integral part of the CoastWeb-model quantifying sedimentation. SPM causes scattering of light in the water and influences the Secchi depth and hence the depth of the photic zone; SPM also influences the bacterial decomposition of organic matter included in SPM, and hence also the oxygen situation and the conditions for zoobenthos, by definition an important food source for benthivorous prey fish.
- As a background to the present conditions in the Baltic Sea, we will also give trend analyses concerning phosphorus, nitrogen and oxygen concentrations, water temperatures and ice-cover based on very comprehensive datasets from mainly ICES (2008a, b, c). This information is also used in our discussions in Chap. 5 connected to the mass-balance model for oxygen and the discussions

L. Håkanson et al., *The Fish Production Potential of the Baltic Sea*, Environmental Science and Engineering, DOI 10.1007/978-3-642-11562-2_2,

on useless and sub-optimal methods to combat eutrophication in the Baltic Sea. The data on the ratio between phosphate and total phosphorus is used to test our algorithm for the dissolved fraction in the deep-water compartment of the Baltic Proper. The trend analyses for nitrogen for the Baltic Proper provide information of importance for interpretations of the effects of the past rather extensive and costly nitrogen reductions. The empirical phosphorus concentrations are very important for testing the model predictions using CoastMab for phosphorus.
- We will also present the methods used to predict chlorophyll-a concentrations, which actually drive the CoastWeb-model, and the Secchi depths, which define the photic zones of the systems.
- Finally, we will give results based on a comprehensive literature review related to biomasses for all major species of fish and for all functional groups of organisms in the CoastWeb-model, including information on what these organisms eat.

So, extensive databases on the conditions in the Baltic Sea and other aquatic systems have been "data-mined" and in this chapter we will present the data used in the CoastWeb-modeling. An important aspect of this concerns the use of hypsographic curves (i.e., depth/area-curves for defined basins) to calculate the necessary volumes of water of the defined vertical layers. Depth/area-curves and volume curves (from Håkanson and Bryhn, 2008b) for the Baltic Sea and its five major sub-basins will be used. These curves have been derived using the best available public dataset on the bathymetry of the Baltic Sea (from Seifert et al., 2001). This information is essential in the mass-balance modeling for salt, phosphorus and SPM discussed in this chapter. If there are errors in the defined volumes, there will also be errors in the calculated concentrations since, by definition, the concentration is the mass of the substance in a given volume of water.

This chapter also presents an approach to define and differentiate between surface-water and deep-water layers. Traditionally, this is done by water temperature data, which define the thermocline, or by salinity data, which define the halocline. Our approach is based on the water depth separating areas where sediment resuspension of fine particles occurs from bottom areas where periods of sedimentation and resuspension of fine cohesive newly deposited material are likely to happen (the erosion and transportation areas, the ET-areas). The depth separating areas with discontinuous sedimentation (the T-areas) from areas with more continuous sediment accumulation (the A-areas) of fine materials is called the theoretical wave base. This is an important concept in mass-balance modeling of aquatic systems (see Håkanson, 1977, 1999, 2000). The theoretical wave base will also be used to define algorithms (1) to calculate concentrations of matter in the given volumes/compartments, (2) to quantify sedimentation by accounting for the mean depths of these compartments, (3) to quantify internal loading via advection/resuspension as well as diffusion (the vertical water transport related to concentration gradients of dissolved substances in the water), (4) to quantify upward and downward mixing between the given compartments and (5) to calculate in- and outflow of substances from the given compartments.

In this work, the Baltic Sea has first been divided into its five traditional main sub-basins, the Bothnian Bay (BB), the Bothnian Sea (BS), the Baltic Proper (BP), the Gulf of Finland (GF) and the Gulf of Riga (GR). Empirical monthly values of the salinity for the period 1997–2006 (this period is often referred to as the default period in this book) have been used to calibrate the CoastMab-model for salt and those calculations provide data of great importance for the mass-balances for phosphorus and SPM, namely:

(1) The fluxes of water to and from the defined compartments.
(2) The monthly mixing of water between layers within the given basins.
(3) The basic algorithm for diffusion of dissolved substances in water in each compartment.
(4) The water retention rates influencing the turbulence in each compartment, and hence also
(5) The sedimentation of particulate phosphorus and SPM in the given compartments.

So, this chapter will provide and discuss the data necessary to run the CoastMab-model within the CoastWeb-model. It should be stressed that in this work we use the mass-balance model for salt and the results presented by Håkanson and Bryhn (2008b) and this means that we will only briefly summarize those results here. There are a few changes in the mass-balance for phosphorus, which will be explained in Sect. 2.4.

2.2 Morphometric Data and Criteria for the Vertical Layers

The basin-specific data used are compiled in Table 2.1 and will be briefly explained in this section. This table gives data on, e.g., total area, volume, mean depth, maximum depth and the depth of the theoretical wave base (D_{wb} in m), the fraction of bottoms areas dominated by fine sediment erosion and transport (ET-areas) above the theoretical wave base, the water discharge to the given sub-basins (from literature sources; see Håkanson and Bryhn, 2008b), the catchment area, latitude and mean annual precipitation for each basin.

The data from the Baltic Proper emanate from samplings all seasons of the year between latitudes 53.9 and 60.2 (°N) and longitudes 12.2 and 23.3 (°W) (data mainly from ICES, 2006a and ICES, 2009). Most parts of the Baltic Proper are covered by these data. This is the basin with the most extensive and reliable data. More than 40,000 measurements on water temperature, salinity, TN- and TP-concentrations and over 12,000 data on chlorophyll-a concentrations have been used in this work. The available empirical data on the functional organisms included in the CoastWeb-model are much more scarce and scattered (see the final section in this chapter).

Table 2.1 Basic data (and abbreviations) for the five main sub-basins in the Baltic Sea. These concepts are explained in the text

	Gulf of Finland (GF)	Gulf of Riga (GR)	Bothnian Bay (BB)	Bothnian Sea (BS)	Baltic Proper (BP)
Land uplift 1 (LU_1) (mm/year)	1.2	0.55	8.0	6.5	1.75
Land uplift 2 (LU_2) (mm/year)	2.0	0.75	9.0	8.0	2.75
Mean land uplift (LU) (mm/year)	1.6	0.625	8.5	7.25	2.25
Area (A) (km^2)	29,600	16,700	36,300	79,300	211,100
Wave base (WB) (m)	43.8	39.2	41.1	42.5	43.8
Area above WB (ET) (km^2)	18,650	13,190	23,000	32,510	87,600
Volume "clay" (km^3/year)	0.03	0.008	0.21	0.24	0.19
ET-areas (ET) (%)	63	79	63	41	47
Area below WB ($Area_{WB}$) (km^2)	10,950	3,510	13,300	46,790	123,500
Depth, E-areas (D_E) (m)	25.4	24.0	25.8	27.1	28.3
Erosion (E)-areas (km^2)	12,020	7,810	18,050	25,240	55,630
Max. depth (D_{Max}) (m)	105	56	148	301	459
Volume (V) (km^3)	1,073.3	409.4	1,500.0	4,889.0	13,055
Mean depth (D_{MV}) (m)	36.3	24.5	41.3	61.7	61.8
Form factor (V_d) (–)	1.04	1.31	0.84	0.61	0.40
Dynamic ratio (DR) (–)	4.74	5.27	4.61	4.56	7.43
Halocline depth (D_{hc}) (m)	75	–	–	–	75
Water discharge (Q) (km^3/year)	29.0	33.2	100	95	250
Catchment area (ADA) (km^2)	421,000	167,000	269,500	229,700	568,973
Latitude (Lat) (°N)	60	57.7	64	62	58
Precipitation (Prec) (mm/year)	593	590	650	700	750

The theoretical wave base is defined from the ETA-diagram (see Fig. 2.1; erosion-transport-accumulation; from Håkanson, 1977), which gives the relationship between the effective fetch, as an indicator of the free water surface over which the winds can influence the wave characteristics (speed, height, length and orbital velocity). The theoretical wave base separates the transportation areas (T), with discontinuous sedimentation of fine materials, from the accumulation areas (A), with more continuous sedimentation of fine suspended particles. The theoretical wave base (D_{wb} in m) is, e.g., at a water depth of 43.8 m in the Baltic Proper. This is calculated from Eq. (2.1) (Area = area in km^2):

$$D_{wb} = \left(45.7 \cdot \sqrt{Area}\right) / \left(\sqrt{Area} + 21.4\right) \tag{2.1}$$

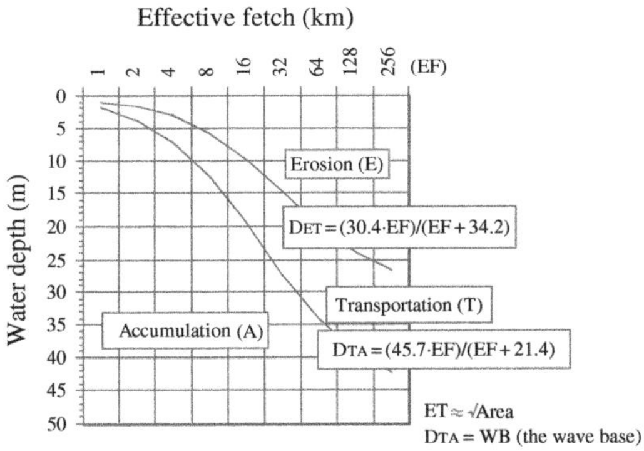

Fig. 2.1 The ETA-diagram (erosion-transportation-accumulation; redrawn from Håkanson, 1977) illustrating the relationship between effective fetch, water depth and potential bottom dynamic conditions. The theoretical wave base ($D_{wb} = D_{TA}$; 43.8 m in the Baltic Proper) may be used as a general criterion in mass-balance modeling to differentiate between the surface-water layer with wind/wave induced resuspension and deeper areas without wind-induced resuspension of fine materials. The depth separating E-areas with predominately coarse sediments from T-areas with mixed sediments is at 28.5 m in the Baltic Proper

It should be stressed that this approach to separate the surface-water layer from the deep-water layer has been used and motivated in many previous contexts for both lakes (Håkanson et al., 2004), smaller coastal areas in the Baltic Sea (Håkanson and Eklund, 2007) and for the sub-basins discussed in this work (Håkanson and Bryhn, 2008b). This approach gives one value for the theoretical wave base related to the area of the system.

Figure 2.2 illustrates empirical data on salinities and TP-concentrations from the Baltic Proper from 100 randomly selected verticals from months 5 to 9 for the default period at stations with water depths larger than 100 m. The idea is to show how these variables vary during the growing season and to illustrate the relevance of the depth intervals used in this modeling. These empirical data support the validity of the theoretical wave base also for large systems and this has also been more thoroughly demonstrated by Håkanson and Bryhn (2008b). So, in this modeling, the Baltic Proper (BP) and the Gulf of Finland (GF) have been divided into three depth intervals:

(1) The surface-water layer (SW), i.e., the water above the theoretical wave base.
(2) The middle-water layer (MW), as defined by the volume between the theoretical wave base and the average depth of the halocline.
(3) The deep-water layer (DW) is defined as the volume of water beneath the average halocline.

The Bothnian Bay (BB), the Bothnian Sea (BS) and the Gulf of Riga (GR) have been divided into two layers, the SW- and the DW-layers separated by the theoretical wave base. Note that the maximum depth of the Gulf of Riga is just 56 m. From

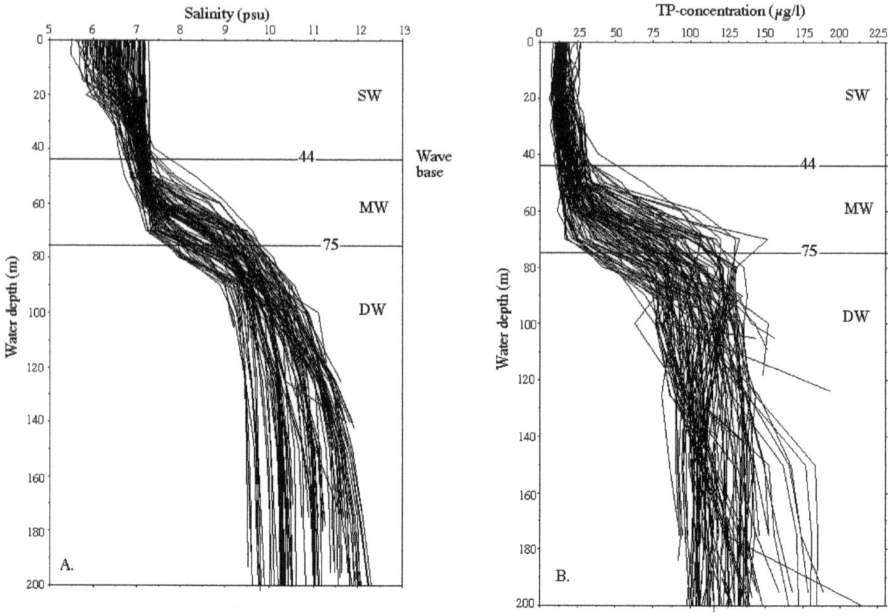

Fig. 2.2 One hundred daily verticals selected at random from stations deeper than 100 m from the Baltic Proper collected months 5–9 between 1997 and 2005: (**a**) salinity and (**b**) TP-concentrations; and *lines* indicating surface-water areas (SW), middle-water areas (MW) and deep-water areas (DW) (redrawn from Håkanson and Bryhn, 2008b)

Table 2.1, one can see that the theoretical wave base is at 43.8 m in GF, 39.2 m in GR, 41.1 m in BB, 42.5 m in BS and 43.8 m in BP. The areas below this depth vary from 3,510 km^2 in the Gulf of Riga to 123,500 km^2 in the Baltic Proper.

It should be stressed that both the theoretical wave base and the depth of the halocline describe average conditions. It is clear from Fig. 2.2a that the halocline varies considerably around 75 m. The actual wave base also varies around 43.8 m in the Baltic Proper; during storm events, the wave base will be at greater water depths (Jönsson, 2005) and during calm periods at shallower depths. The actual wave base also varies spatially within the studied areas. From Fig. 2.2, it is evident that the two boundary depths describe the average conditions in the Baltic Sea very well.

There are clear differences in the salinity profiles in the five basins (see Table 2.2). The aim of the salinity modeling is to predict the monthly water fluxes so that the modeled salinities are as close as possible to the empirical salinity data.

2.2.1 Size and Form Characteristics of the Sub-basins

Figure 2.3a exemplifies the hypsographic curve (a) and the volume curve (b) for the Baltic Proper and how the areas above and below the theoretical wave base are defined, and also how the SW-volume and the DW-volume are defined. Håkanson

Table 2.2 Data on volumes and areas (below the given depths; e.g., 10,900 km^2 is the area below the theoretical wave base, which defines the upper limit for the MW-layer in the Gulf of Finland) and salinities (mean values, medians, standard deviations and number of data; data from ICES, 2006a)

Basin	Level	Volume (km^3)	Area (km^2)	Salinity (mean)	Salinity (median)	Salinity (SD)	Salinity (CV)	Number of data (n)
Gulf of Finland	SW	851	29,600	6.18	6.11	1.09	0.17	676
	MW	202	10,900	7[a]	7[a]	–	–	0
	DW	20.0	2,400	10.2[a]	10.2[a]	–	–	0
Gulf of Riga	SW	392	16,700	5.67	5.72	0.25	0.044	260
	DW	17.5	3,500	7.5[a]	7.5[a]	–	–	0
Bothnian Bay	SW	1,067	36,300	3.33	3.38	0.38	0.11	355
	DW	433	13,327	3.58	3.61	0.25	0.069	200
Bothnian Sea	SW	2,779	79,300	5.40	5.41	0.12	0.022	216
	DW	2,110	46,703	6.18	6.12	0.38	0.061	215
Baltic Proper	SW	7,315	211,100	7.04	7.05	0.90	0.13	12,374
	MW	3,050	123,500	7.72	8.92	2.64	0.34	3,989
	DW	2,690	73,000	10.28	10.66	1.75	0.17	6,289

[a] Missing data, assumed values.

and Bryhn (2008b) gave a compilation of the hyposographic curves for all five sub-basins, as derived using GIS (Geographical Information System) and bathymetric data from Seifert et al. (2001) and those data have also been used in this work. Figure 2.3b shows the corresponding volume curve for the Baltic Proper. One can note that the area below the theoretical wave base (D_{wb}) at 43.8 m in BP is $123.5 \cdot 10^3$ km^2 and the area below the average depth of the halocline (D_{hc}) at 75 m is $73 \cdot 10^3$ km^2. The volume of the SW-, MW- and DW-layers in the Baltic Proper (BP) are 7,315, 3,050 and 2,690 km^3 and the entire volume is 13,055 km^3. Limitations in the resolution of the bathymetric dataset imply that areas and volumes in shallow regions are slightly underestimated and hence the GIS-calculated data have been harmonized with the data provided by HELCOM (1990). For the values of the maximum depths, data from SMHI (2003) were used.

Among the morphometric parameters characterizing the studied sub-basins, three main groups can be identified (see Håkanson, 2004):

- Size parameters; different parameters in length units, such as the maximum depth; parameters expressed in area units, such as water surface area; and parameters expressed in volume units, such as water volume and SW-volume.
- Form parameters based on size parameters, such as mean depth and the form factor.
- Special parameters, e.g., the dynamic ratio and the effective fetch.

The CoastMab-model uses several of these variables. They are listed in Table 2.1 and will be defined in the following text.

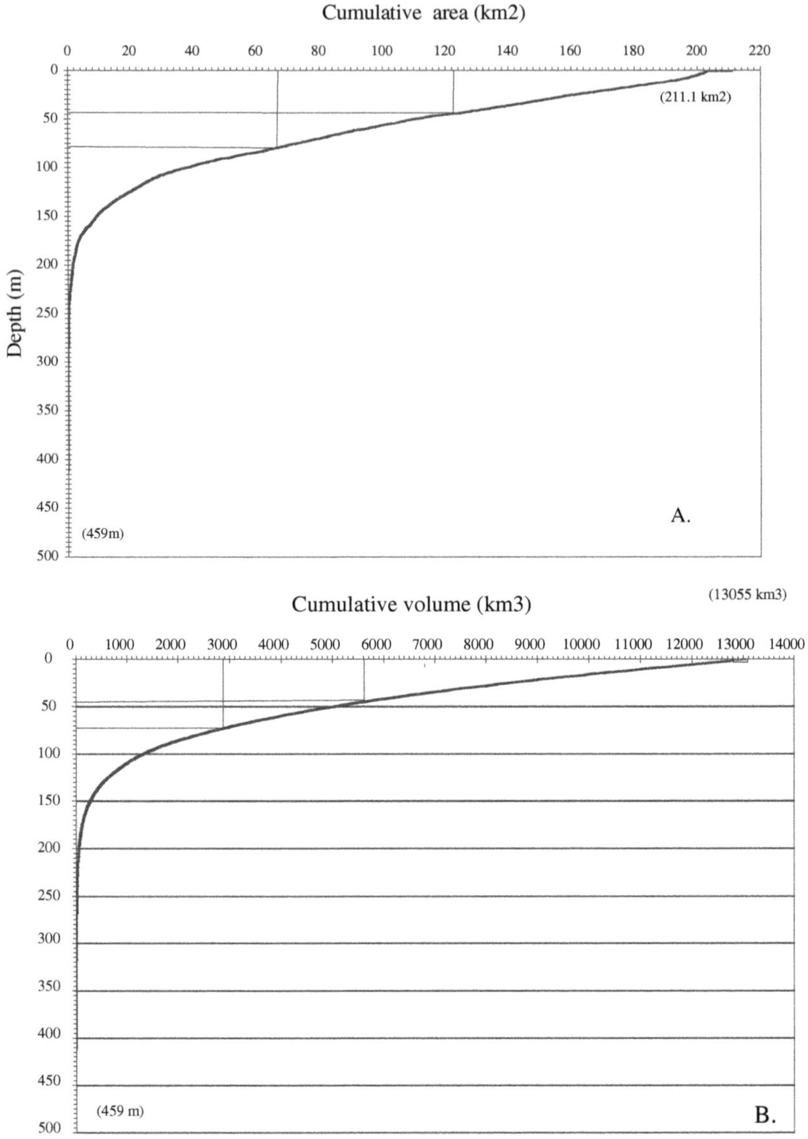

Fig. 2.3 Hypsographic curve (**a**) and volume curve (**b**) for the Baltic Proper (redrawn from Håkanson and Bryhn, 2008b)

Traditionally, the mean depth (D_{MV} in m) is defined as the ratio between the water volume (V in m^3) and the area (A in m^2), or $D_{MV} = V/A$. The mean depth is a most informative and useful parameter in aquatic sciences and it is an integral part of the CoastWeb-model.

The volume development, also often called the form factor (V_d, dimensionless) is defined as the ratio between the water volume and the volume of a cone, with a base equal to the water surface area (A in km^2) and with a height equal to the maximum depth (D_{Max} in m):

$$V_d = (A \cdot D_{MV} \cdot 0.001)/(A \cdot D_{Max} \cdot 0.001 \cdot 1/3) = 3 \cdot D_{MV}/D_{Max} \quad (2.2)$$

Where D_{MV} is the mean depth (m). The form factor describes the form of the basin (see Fig. 2.4). The form of the basins is very important, e.g., for internal sedimentological processes and Fig. 2.4 illustrates relative hypsographic curves for basins with different forms and hence also V_d-values. In basins of similar area but with different form factors, one can presuppose that the system with the smallest form factor would have a larger area above the theoretical wave base, and more of the resuspended matter transported to the surface-water compartment than to the deep-water compartment below the theoretical wave base compared to a system with a higher form factor. This is also the way in which the form factor is used in the CoastMab-model.

The dynamic ratio (DR; see Håkanson, 1982) is defined by the ratio between the square-root of the water surface area (in km^2 not in m^2) and the mean depth, D_{MV} (in m; DR = $\sqrt{Area}/D_{MV}$). DR is a standard morphometric parameter in contexts of resuspension and turbulence in entire basins. ET-areas above the theoretical wave base (i.e., areas where fine sediment erosion and transport processes prevail)

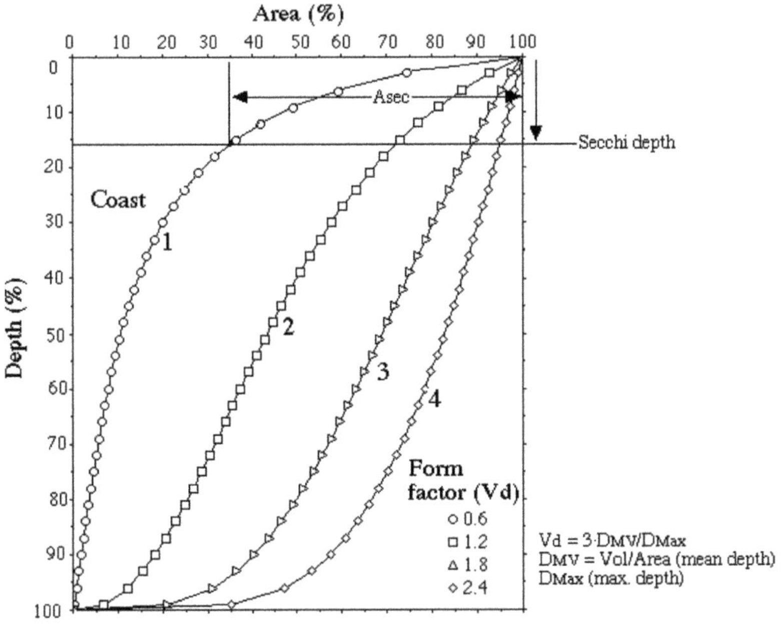

Fig. 2.4 Illustration of relative hypsographic curves (= depth-area curves) for four coasts with different forms (and form factor = volume development = V_d). The form influences, e.g., the areas above the Secchi depth (A_{sec})

are likely to dominate the bottom dynamic conditions in basins with dynamic ratios higher than 3.8. Slope processes are known (see Håkanson and Jansson, 1983) to dominate the bottom dynamic conditions on slopes greater than about 4–5%. Slope-induced ET-areas are likely to dominate basins with DR values lower than 0.052. One should also expect that in all basins there is a shallow shoreline zone where wind-induced waves will create ET-areas, and it is likely that most basins have at least 15% ET-areas. If a basin has a DR of 0.26, one can expect that in this basin the ET-areas would occupy 15% of the area. If DR is higher or lower than 0.26, the percentage of ET-areas is likely to increase. Basins with high DR-values, i.e., large and shallow system are also likely to be more turbulent than small and deep basins. This will influence sedimentation. During windy periods with intensive water turbulence, sedimentation of suspended fine particles in the water will be lower than under calm conditions. This is accounted for in the CoastMab-model and the dynamic ratio is used as a proxy for the potential turbulence in the monthly calculations of the transport processes.

Among the sub-basins in the Baltic Sea, the Baltic Proper has the highest DR (7.43) and the Bothnian Sea the lowest (4.56).

It should be stressed that the form factor and the dynamic ratio provide different and complementary aspects of how the form may influence the function of aquatic systems.

The effective fetch (see the ETA-diagram in Fig. 2.1) is often defined according to a method introduced by the Beach Erosion Board (1972). The effective fetch (L_{ef} in km) gives a more representative measure of how winds govern waves (wave length, wave height, etc.) than the effective length, since several wind directions are taken into account. Using traditional methods, it is relatively easy to estimate the effective fetch by means of a map and a special transparent paper (see Håkanson, 1977). The central radial of this transparent paper is put in the main wind direction or, if the maximum effective fetch is requested, in the direction which gives the highest L_{ef}-value. Then the distance (x_i in km) from the given station to land (or to islands) is measured for every deviation angle a_i, where a_i is $\pm$ 6, 12, 18, 24, 30, 36 and 42°. L_{ef} may then be calculated from:

$$L_{ef} = \sum x_i \cdot \cos(a_i) \Big/ \left(\sum \cos(a_i)\right) \cdot SC' \qquad (2.3)$$

$\Sigma\cos(a_i) = 13.5$, a calculation constant.

SC' = the scale constant; if the calculations are done on a map in scale 1:250,000, then $SC' = 2.5$.

The effective fetch attains the highest values close to the shoreline and the minimum values in the central part of a basin. This relationship is important in, e.g., contexts of shore erosion and morphology, for bottom dynamic conditions (erosion-transportation-accumulation), and hence also for internal processes, mass-balance calculations, sediment sampling and sediment pollution.

For entire basins, the mean effective fetch may be estimated as $\sqrt{\text{Area}}$ (see Fig. 2.1). In a round basin, the requested value should be somewhat lower than the diameter ($d = 2 \cdot r$; $r =$ the radius); the area is $\pi \cdot r^2$ and hence $d = 1.13 \cdot \sqrt{Area}$ and the mean fetch approximately $\sqrt{Area}$.

2.2.2 Sediments and Bottom Dynamic Conditions

As stressed in Fig. 2.1, the wave base may also be determined from the ETA-diagram. This approach focuses on the behavior of the cohesive fine materials settling according to Stokes' law in laboratory vessels:

- Areas of erosion (E) prevail in shallow areas or on slopes where there is no apparent deposition of fine materials but rather a removal of such materials; E-areas are generally hard and consist of sand, consolidated clays and/or rocks with low concentrations of nutrients and other substances.
- Areas of transportation (T) prevail where fine materials (such as the carrier particles for water pollutants) are deposited periodically (areas of mixed sediments). This bottom type generally dominates where wind/wave action regulates the bottom dynamic conditions. It is sometimes difficult in practice to separate areas of erosion from areas of transportation. The water depth separating transportation areas from accumulation areas, the theoretical wave base, is, as stressed, a fundamental component in mass-balance calculations.
- Areas of accumulation (A) prevail where the fine materials (and particulate forms of water pollutants) are deposited continuously (soft bottom areas). It is in these areas (the "end stations") where high concentrations of pollutants are most likely to appear.

So, the generally hard or sandy sediments within the areas of erosion and transport (ET) often have a relatively low water content, low organic content and low concentrations of nutrients and pollutants (see Håkanson and Jansson, 1983). In connection with a storm, the material on the ET-area may be resuspended and transported up and away, generally in the direction towards the accumulation areas in the deeper parts, where continuous deposition occurs. It should also be stressed that fine materials are rarely deposited as a result of simple vertical settling in natural aquatic environments. The horizontal velocity is generally at least 10 times larger, sometimes up to 10,000 times larger, than the vertical component for fine materials or flocs that settle according to Stokes′ law (Bloesch and Burns, 1980; Bloesch and Uehlinger, 1986).

An evident boundary condition for this approach to calculate the ET-areas is that if the depth of the theoretical wave base, $D_{wb} > D_{Max}$, then $D_{wb} = D_{Max}$.

In CoastMab, there are also two boundary conditions for ET (= the fraction of ET-areas in the basin):

If $ET > 0.99$ then $ET = 0.99$ and if $ET < 0.15$ then $ET = 0.15$.

ET-areas are generally larger than 15% ($ET = 0.15$) of the total area since there is always a shore zone dominated by wind/wave activities. For practical and functional reasons, one can also generally find sheltered areas, macrophyte beds and deep holes with more or less continuous sedimentation, that is, areas which actually function as A-areas, so the upper boundary limit for ET may be set at ET = 0.99 rather than at ET = 1.

The value for the ET-areas is used as a distribution coefficient in the CoastMab-model. It regulates whether sedimentation of the particulate fraction of the substance (here phosphorus or SPM) goes to the DW- or MW-areas or to ET-areas.

Håkanson and Bryhn (2008b) gave a compilation of sediment data from different basins and sites in the Baltic Sea and from this, one can note that:

1. Most TP-values from the upper decimeter of Baltic Sea sediments vary in the range from 0.36 to 2 mg TP/g dw, see also Emelyanov (2001) who have presented a detailed map of TP-concentrations in sediments from many parts pf the Baltic Proper. This range will be used as a reference in this work – if modeled TP-concentrations in accumulation area sediments are higher than 2 or lower than 0.36 mg TP/g dw, it indicates that the TP-fluxes to (i.e., sedimentation of particulate phosphorus) and from (i.e., burial of phosphorus) these sediment compartments may be wrong.
2. Due to substrate decomposition by bacteria and compaction from overlying sediments, the TP-concentration and the organic content (loss on ignition, IG) generally decrease with sediment depth (see Håkanson and Jansson, 1983).
3. The TN-concentrations are generally a factor of 3–10 higher than the TP-concentrations.
4. The bulk density (d in g/cm^3 ww) is between 1.2 and 1.3 in the upper 10 cm sediments in the Baltic Sea.
5. The water content (W in % ww) may be set to 75% for the upper 10 cm accumulation areas sediments in the Baltic Sea basins.
6. The organic content (= loss on ignition, IG in % dw) may be set to 12% for the upper 10 cm accumulation areas sediments in the Baltic Sea basins. The IG-value in Baltic Sea glacial clays is around 5% dw.

2.2.3 The Role of Land Uplift

The amount of suspended particulate matter (SPM) normally depends on two main causes: Allochthonous inflow (from rivers entering the given system) and autochthonous production (in the given system). In the Baltic Sea , however, there is also another dominating source, land uplift (see Fig. 2.5 and Voipio, 1981; Jonsson et al., 1990; Svensson, 2006). Thousand-year-old sediments influence the Baltic Sea ecosystem today. When the old bottom areas rise after being depressed by the glacial ice, they will be influenced by the waves, which will resuspend the sediments. The land uplift in the Baltic Sea (measured in relation to the sea surface) varies from about 9 mm/year in the northern part of the Bothnian Bay to about 0 for the south-central part of the Baltic Proper. It has been shown that land uplift may contribute

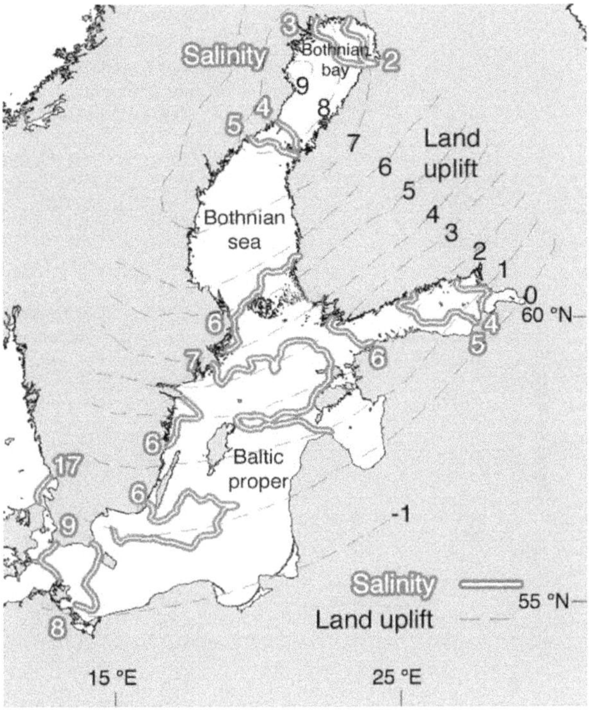

Fig. 2.5 Geographical overview of the Baltic Sea (from Håkanson and Gyllenhammar, 2005). Land uplift in mm/year. Salinity in psu. The scales give latitudes and longitudes

with 50–80% of the materials settling below the wave base in the open Baltic Proper (Jonsson et al., 1990; Jonsson, 1992; Blomqvist and Larsson, 1994; Eckhéll et al., 2000). This means that the primary allochthonous and autochthonous sources only contribute with less than half of the sediments, clay particles and iron supplied to the Baltic Sea (see also Blomqvist et al., 2004; Håkanson and Bryhn, 2008b). Actually, land uplift influences the entire system in many profound ways (see Håkanson and Bryhn, 2008b). There are at least two maps illustrating land uplift, the one in Fig. 2.5 and the map presented by Svensson (2006). Generally, Svensson's map give values which are about 0.5–1 mm/year higher than the map shown in Fig. 2.5, and in all the following calculations we have used the mean values derived from the areal mean values from the two maps to quantify how land uplift contributes with nutrients and sediments to the Baltic Sea.

When there is land uplift, the new supply of matter eroded from the sediments exposed to wind-generated waves does not emanate just from the newly raised areas related to the wave base but also from increased erosion of previously raised areas. This is schematically illustrated in Fig. 2.6.

It is assumed that the water content of the more compacted sediments from land uplift is 15% lower than the recently deposited sediments close to the theoretical wave base and that the bulk density (d in g/cm^3) is 0.2 units higher than in the recently deposited sediments. The bulk density (d) is calculated from a standard

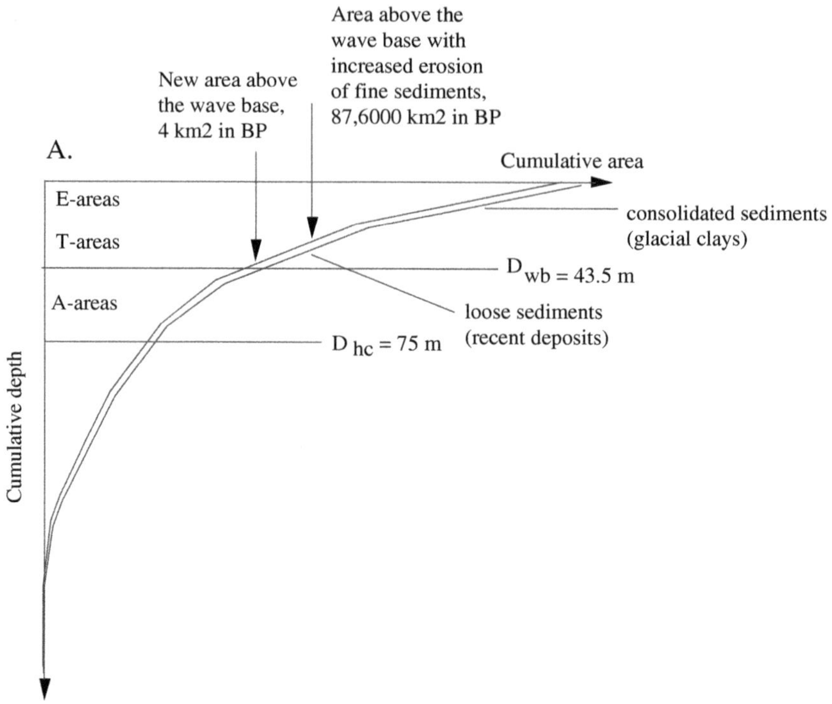

Fig. 2.6 Illustration of how land uplift influences the area above the theoretical wave base. If there is no land uplift materials deposited above the theoretical wave base, on areas of fine sediment erosion and transport, will only stay on these bottoms until the next resuspension event, often related to increase wind/wave activity. There is by definition no net deposition on the areas of fine sediment erosion and transport (the ET-areas) when there is no land uplift. Land uplift provides a net input of materials to the surface-water layer. The sediments within the areas of fine sediment erosion (i.e., the older more compacted glacial clays) are relatively consolidated, whereas the more recently deposited sediments close to the theoretical wave base are less consolidated with higher water content, organic content and contents of nutrients and iron

equation (see Håkanson and Jansson, 1983). The TP-concentration in the material added to the Baltic Sea system from land uplift will be calculated (see Håkanson and Bryhn, 2008b) from the reference value for the TP-concentration in glacial clays ($TP_{clay} = 0.36$ mg TP/g dw) and the fraction of the E-areas above the theoretical wave base ($Area_E/Area_{ET}$) and the value calculated by the CoastMab-model for the TP-concentration in the A-sediments beneath the theoretical wave base (TP_{AMWsed} in basins with three layers or TP_{ADWsed} in basins with two vertical layers).

The areas of erosion ($Area_E$) are calculated from the hyposgraphic curves and the corresponding depth given by the ETA-diagram (Fig. 2.1). This means that the depth separating E-areas from T-areas is given by:

$$D_{ET} = \left(30.4 \cdot \sqrt{}Area\right) / \left(\sqrt{}Area + 34.2\right) \qquad (2.4)$$

Note that the area is given in km^2 in Eq. (2.4) to get the depth in m.

The material added from land uplift does not just contain phosphorus, nitrogen and clay particles but also iron, manganese and many other substances.

2.3 Water Exchange and Water Transport in the Baltic Sea

2.3.1 Background on Mass-Balances for Salt and the Role of Salinity

The cod roe has a certain density and will stay floating at a water depth were the density of the water matches the density of the roe. The density of the water depends on water temperature and salinity. For the cod roe to survive in the Baltic Sea, the oxygen concentration must be higher than about 2 mg/l; the salinity must be about 10.5 psu. Increasing nutrient concentrations increase primary and secondary production and the amount of organic matter, and hence also the oxygen consumption from bacterial degradation of dead organic matter. Increased oxygen consumption means lower oxygen concentrations. This implies that the chances for the cod roe to survive are smaller if the salinity is low and the nutrient concentrations in the system high. Then, the cod roe will appear at greater water depths where the oxygen concentrations are likely relatively low (see Ambio, 1999, 2000 and Fig. 2.30 later).

The salinity is of vital importance for the biology of coastal areas influencing, e.g., the number of species in a system (see Fig. 2.7, from Remane, 1934), and also the reproductive success, food intake and growth of fish (Rubio et al., 2005; Nissling et al., 2006). Furthermore, a higher salinity increases the flocculation and aggregation of particles (see Håkanson, 2006) and hence affects the rate of sedimentation, which is of particular interest in understanding variations in water clarity within and among coastal areas. The salter the water, the greater the flocculation of suspended particles. This does not only influence the concentration of particulate matter, but also the concentration of any substance with a substantial particulate phase such as phosphorus. The salinity also affects the relationship between total phosphorus and primary production/biomass (chlorophyll-a; Håkanson, 2008). This relationship is shown in Fig. 2.8 and it is used in this work to calculate chlorophyll-a concentrations from dynamically modeled salinities in the different sub-basins, from dynamically modeled phosphorus concentrations and from information on the number of hours with daylight (the daylight data given in Table 2.3). The salinity is easy to measure and the availability of salinity data for the Baltic Sea is better than for most other water variables.

Figure 2.8 is important in the CoastWeb-modeling and it illustrates the role of salinity in relation to the Chl/TP-ratio. The figure gives the number of data in each salinity class, the box-and-whisker plots give the medians, quartiles, percentiles and outliers, and the table below the diagram provides information on the median Chl/TP-values, the coefficients of variation (CV = SD/MV; SD = standard deviation; MV = mean value) and the number of systems included in each class (n).

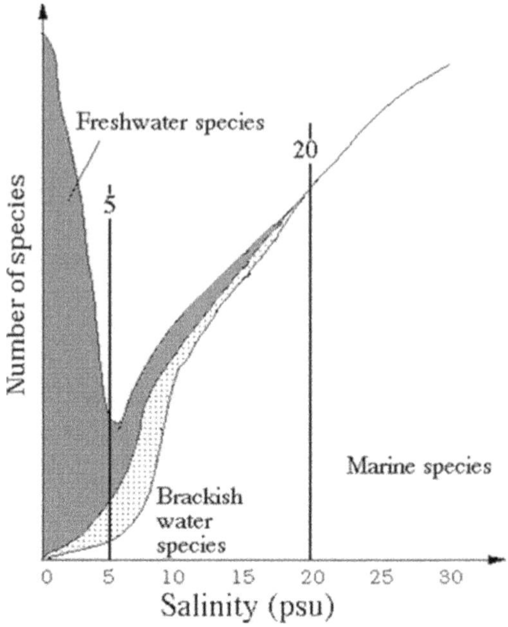

Fig. 2.7 The relationship between salinity and number of species (redrawn from Remane, 1934)

These results are evidently based on many data from systems covering a wide salinity gradient. An interesting aspect concerns the pattern shown in the figure. One can note:

- The salinity clearly influences the Chl/TP-ratio.
- The median value for lakes is 0.29, which is almost identical to the slope coefficient for the key reference model for lakes (0.28 in the OECD-model; see OECD, 1982).
- The Chl/TP-ratio changes in a wave-like fashion when the salinity increases. It is evident that there is a minimum in the Chl/TP-ratio in the salinity range between 2 and 5 psu. Subsequently, there is an increase up to the salinity range of 10–15, and then a continuous decrease in the Chl/TP-range until a minimum value of about 0.012 is reached in the hypersaline systems.

The salinity gradient in the Baltic Sea and the transient areas of the Kattegat and the Danish straits are clearly seen in Fig. 2.9. The salinity in the open water areas outside the coastal zone varies from about 2–4 psu in estuaries such as the the Bothnian Bay, to 6–8 psu in the Baltic Sea and to values in the range 20–30 psu in the Kattegat, the Skagerrak and the open ocean. Beneath the halocline, one finds denser water with salinities higher than in the surface water.

The water exchange in the Baltic Sea is calculated using the CoastMab-model for salt. This section presents monthly budgets for water and salt in the Baltic Sea and its five main sub-basins.

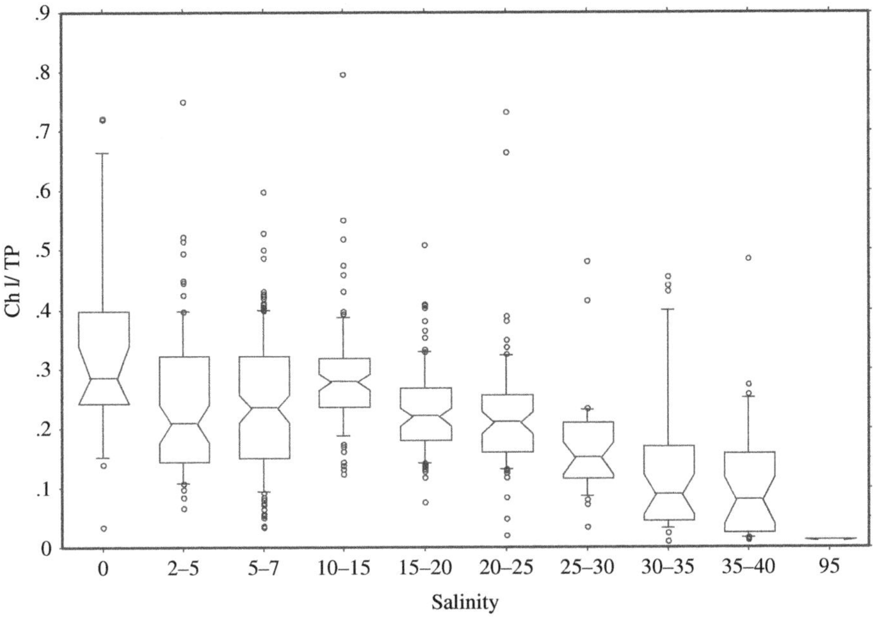

Median:	0.29	0.21	0.24	0.28	0.22	0.21	0.15	0.11	0.079	0.012
CV:	0.54	0.52	0.48	0.32	0.29	0.42	0.53	0.89	0.98	0.16
n:	20	74	140	82	79	61	30	28	24	3

Fig. 2.8 Box-and-whisker plot (showing medians, quartiles, percentiles and outliers) illustrating the Chl/TP-ratio for 10 salinity classes. The statistics give the median values, the coefficients of variation (CV) and the number of data in each class (redrawn from Håkanson, 2008)

Table 2.3 Daylight table giving average number of hours with daylight different months at different latitudes on the northern hemisphere

Month	Equator	10°	20°	30°	40°	50°	60°	70°	80°	Pole
Jan	12:07	11:35	11:02	10:24	9:37	8:30	6:38	0:00	0:00	0:00
Feb	12:07	11:49	11:21	11:10	10:42	10:07	9:11	7:20	0:00	0:00
Mar	12:07	12:04	12:00	11:57	11:53	11:48	11:41	11:28	10:52	0:00
Apr	12:07	12:21	12:36	12:53	13:14	13:44	14:31	16:06	24:00	24:00
May	12:07	12:34	13:04	14:22	15:22	17:04	22:13	24:00	24:00	24:00
Jun	12:07	12:42	13:20	14:04	15:00	16:21	18:49	24:00	24:00	24:00
Jul	12:07	12:40	13:16	13:56	14:49	15:38	17:31	24:00	24:00	24:00
Aug	12:07	12:28	12:50	13:16	13:48	14:33	15:46	18:26	24:00	24:00
Sep	12:07	12:12	12:17	12:23	12:31	12:42	13:00	13:34	15:16	24:00
Oct	12:07	11:55	11:42	11:28	11:10	10:47	10:11	9:03	5:10	0:00
Nov	12:07	11:40	11:12	10:40	10:01	9:06	7:37	3:06	0:00	0:00
Dec	12:07	11:32	10:56	10:14	9:20	8:05	5:54	0:00	0:00	0:00

From http://encarta.msn.com/media_701500905/Hours_of_Daylight_by_Latitude.html

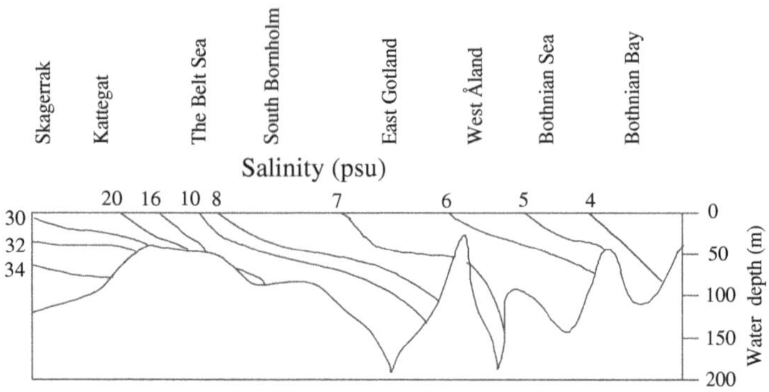

Fig. 2.9 Overview of characteristic salinities in the Baltic Sea (from Håkanson, 1991)

Mass-balance models have long been used as a tool to study lake eutrophication (Vollenweider, 1968; OECD, 1982) and also used in different coastal applications (see Håkanson and Eklund, 2007; Håkanson and Bryhn, 2008a, b). Mass-balance modeling makes it possible to predict what will likely happen to a system if the conditions change, e.g., a reduced discharge of a pollutant related to a remedial measure. Mass-balance modeling can be performed at different scales depending on the purpose of the study. A large number of coastal models do exist, all with their pros and cons. For example, the 1D-nutrient model described by Vichi et al. (2004) requires meteorological input data with a high temporal resolution, which makes forecasting for longer time periods than one week ahead problematic. The 3D-model used by Schernewski and Neumann (2005) has a temporal resolution of 1 min and a spatial resolution of 3 nm (nautical miles), which means that it is difficult to find reliable empirical data to run and validate the model. Several water balance studies have also been carried out in the Baltic Sea, see, e.g., Jacobsen (1980), HELCOM (1986, 1990), Bergström and Carlsson (1994), Omstedt and Rutgersson (2000), Stigebrandt (2001), Rutgersson et al. (2002), Omstedt and Axell (2003), Omstedt et al. (2004) and Savchuk (2005). The result of such mass-balance calculations for salt or for other substances, depend very much on how the system is defined and the model structured.

Within the BALTEX program (BALTEX, 2006; BACC, 2008), the water and heat balances are major research topics and estimates on the individual terms in the water balance are frequently being revised (e.g., Bergström and Carlsson, 1993, 1994; Omstedt and Rutgersson, 2000; Rutgersson et al., 2002). The major water balance components in the Baltic Sea are the in- and outflows at the entrance area, river runoff and net precipitation (Omstedt et al., 2004). Change in water storage needs also to be considered at least for shorter time periods. The different results depend on the time period studied and the length of the period. Several studies have also divided the Baltic Sea into sub-basins and from the water and salt balances estimated the flows (e.g., Omstedt and Axell, 2003; Savchuk, 2005).

The necessary empirical data on salinity for this modeling have been obtained from the HELCOM datasets available at the ICES website (ICES, 2006a) and data from the period 1997 to 2006 have been used in this work.

There are inter-annual and seasonal variations in both net precipitation and riverine water input to the Baltic Sea (HELCOM, 1986; Bergström and Carlsson, 1993, 1994; Winsor et al., 2001) as well as in the exchange of water with the Kattegat and the salinity of this water (Samuelsson, 1996). As a result, the mean salinities in the basins of the Baltic Sea also vary over time (Samuelsson, 1996; Winsor et al., 2001). We have focused on a period where we have access to comprehensive data for the mass-balances for salt and phosphorus, but also for this period there are inherent uncertainties in the data (see Håkanson and Bryhn, 2008b).

The fluxes and retention rates for the different sub-basins and compartments of the Baltic Sea, as defined in this mass-balance modeling for salt, will be used in the following mass-balance modeling for phosphorus and SPM. The basic structuring of this model (CoastMab) should enable extensions not just to other substances than salt, but also to other systems than the Baltic Sea.

2.3.2 Water Fluxes

Figure 2.10 illustrates the basic structure of the model with its twelve compartments (SW, MW and DW in BP and GF, SW and DW in BS, BB and GR) and also results of the modeling. Note that this modeling is done on a monthly basis to achieve seasonal variations, which is important in the mass-balance models for phosphorus

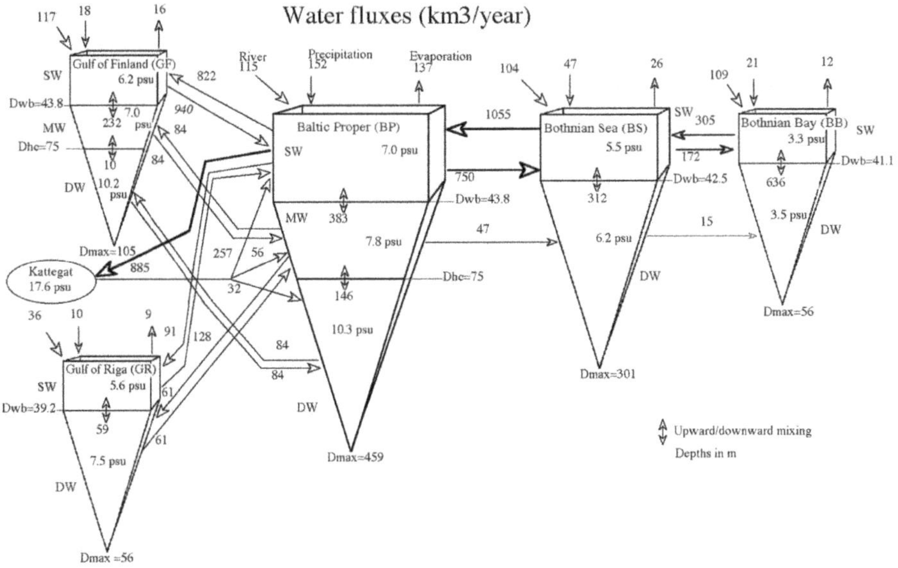

Fig. 2.10 Annual water fluxes to, from and within the five major sub-basin of the Baltic Sea (from Håkanson and Bryhn, 2008b)

and SPM and for the foodweb modeling. All the water fluxes in Fig. 2.10 are given in km^3/year to get an overview. This figure also shows water fluxes from rivers, precipitation and evaporation. For the tributary fluxes to BB, BS and BP data from Omstedt and Axell (2003) for the period 1981–1998 have been used and these data also largely agree with data used by Monitor (1988). The annual fresh-water flux to the GF is the average value from Savchuk (2000; 3.552 m^3/s), Myrberg (1998; 3,615 m^3/s) and Stålnacke et al. (1999; 3,875 m^3/s); the annual tributary flux to GR is the average value from Laznik et al. (1998; 1,149 m^3/s), Ostmann et al. (2001; 1,046 m^3/s) and Savchuk and Swaney (2000; 1,202 m^3/s). The monthly data on water discharge have been calculated from the annual data using the dimensionless moderator for this purpose (from Abrahamsson and Håkanson, 1998). This moderator is based on data on catchment area, mean annual precipitation and latitude (see Table 2.1 for data). Since we do not have access to reliable monthly data on water discharge for the default period (1997–2006), it should be stressed that we model the average, characteristic conditions on a monthly bases and not the actual sequence of months.

The median salinity in the inflowing water from Kattegat was calculated to 17.6 psu (see Håkanson and Bryhn, 2008b).

The model quantifies the fluxes needed to achieve steady-state concentrations for the salinity that correspond as closely as possible to the empirical monthly salinities in the twelve compartments shown in Fig. 2.10. All equations have been given by Håkanson and Bryhn (2008b) and will not be elaborated here.

One can note from Fig. 2.10 that the greatest water fluxes are the surface-water (SW) fluxes from the Bothnian Sea (BS) to the Baltic Proper (BP; 1,055 km^3/year), the SW-flux from the Gulf of Finland (GF) to BP (940 km^3/year) and the return SW-flow from BP to Kattegat (KA, 885 km^3/year).

From the fluxes of water and of salt, one can also define the associated retention times (T) and retention rates (1/T). The retention rates for water may be used in mass-balance models for, e.g., nutrients since these rates indicate the potential turbulence in the given compartment, and the turbulence regulates the settling velocity for suspended particles – the higher the potential turbulence, the lower the settling velocity for particulate phosphorus (Håkanson and Bryhn, 2008b). The retention time for water in each compartment is defined from the total inflow of water (m^3/year) and the volume of the compartment (m^3). Table 2.4 gives the monthly theoretical retention times for water (T_Q) in the twelve compartments. One can see that the mean T_Q-value for the DW-compartment in BS is 154 months and the corresponding value for the DW in the Gulf of Riga (GR), which has a small deep-water volume, is only 1.9 months. GR is the most dynamic sub-basin with the shortest T_Q-values.

Empirical salinity data are compared to modeled values in Fig. 2.11, which gives a regression based on all annual data. The results based on empirical median annual values gives a slope of 1.0 when empirical data are compared to modeled values and a coefficient of determination (r^2) of 0.997, and it is not possible to obtain better results because of the uncertainties in the empirical data. The inherent empirical uncertainties in the mean values (the SD-values) are between 0.12 and 2.6 (see

Table 2.4 Theoretical water retention times (in months) for the 12 compartments in the Baltic Sea

Month	DWBB	DWBP	DWBS	DWGF	DWGR	MWBP	MWGF	SWBB	SWBP	SWBS	SWGF	SWGR
1	5.74	88.71	52.87	2.53	1.94	33.63	7.59	10.74	26.78	19.93	8.61	24.97
2	5.36	93.04	53.00	2.58	1.93	35.38	7.82	10.18	26.86	20.10	8.64	24.29
3	5.76	90.7	52.82	2.54	1.85	34.14	7.81	10.77	26.87	19.99	8.67	24.4
4	5.06	78.24	52.67	2.52	1.73	28.39	7.50	9.74	25.65	19.95	8.59	23.39
5	4.98	103.45	36.68	2.67	1.43	39.07	6.64	9.62	26.85	18.99	8.14	19.58
6	8.33	240.76	251.59	2.77	1.39	129.61	6.76	14.06	30.84	29.66	8.11	18.69
7	73.28	245.57	312.98	2.48	1.39	141.22	6.09	35.77	32.58	27.42	8.03	20.43
8	85.09	247.28	353.89	2.39	1.47	145.14	5.48	36.79	33.08	27.00	7.80	21.76
9	50.92	241.05	322.37	2.72	3.04	139.67	19.03	32.91	33.24	26.53	10.31	34.27
10	9.05	225.15	276.55	2.72	3.06	124.13	20.44	14.87	32.96	26.21	10.4	34.31
11	6.41	103.34	42.58	2.44	1.98	41.07	7.12	11.66	28.51	18.52	8.51	26.79
12	7.26	87.92	37.1	2.44	1.91	33.51	6.35	12.77	26.97	17.73	8.2	25.4
Mean	22.3	153.8	153.8	2.57	1.9	77.1	9.1	17.5	29.3	22.7	8.7	24.9
Median	6.8	103.4	52.9	2.54	1.9	40.1	7.3	12.2	27.7	20.0	8.6	24.3
Min	5.0	78.2	36.7	2.39	1.4	28.4	5.5	9.6	25.7	17.7	7.8	18.7
Max	85.1	247.3	353.9	2.77	3.1	145.1	20.4	36.8	33.2	29.7	10.4	34.3
SD	29.6	76.6	134.5	0.13	0.6	52.3	5.0	10.8	3.0	4.3	0.8	5.0

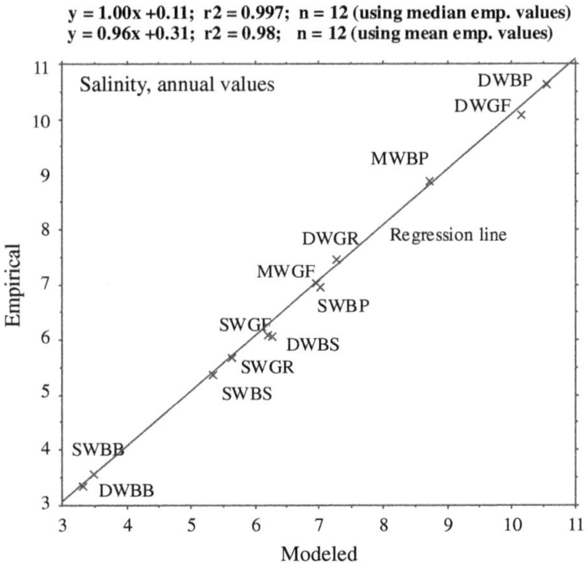

Fig. 2.11 A comparison between empirical and modeled annual values for the salinity using the CoastMab-model. The regression based on annual median empirical data gives an r^2-value of 0.997 and a slope of 1.00 and the regression based on annual mean empirical data gives an r^2-value of 0.98 and a slope of 0.96. The most uncertain empirical value is the data from MWBP, where the CV-value is 0.34 (from Håkanson and Bryhn, 2008b)

Table 2.2). But, of course, the results in Fig. 2.11 are not a result of a blind test, rather a result achieved after calibrations. The results for the mean annual values are also good (slope = 0.96 and r^2 = 0.98). Also the results for the monthly data are good: slope = 0.94 and r^2 = 0.97 (see Håkanson and Bryhn, 2008b).

2.3.3 Comments

To understand how the Baltic Sea system, or any aquatic system, responds to changes in, e.g., loading of toxins, salt or nutrients, it is imperative to have a dynamic process-based perspective quantifying the factors and functions regulating inflow, outflow and internal transport processes and retention rates. This section has demonstrated that this modeling using the theoretical wave base rather than traditional temperature data to define the surface-water, middle-water and deep-water compartments can give excellent correspondence between empirical and modeled data for the salinity for the Baltic Sea system. It is often stressed in contexts of marine eutrophication that it is important to develop practically useful general dynamic mass-balance models based on the ecosystem perspective to be able to give realistic evaluations of how systems will respond to changes in nutrient loading or other remedial actions (see Smith, 2003) and this modeling is meant to do that.

New morphometric data and new hypsographic and volume curves (from Håkanson and Bryhn, 2008b) for the Baltic Sea and the defined sub-basins based on digitized bathymetric data have been used. The basic aim has been to present data on the fluxes of water and the theoretical retention times for water and salt in the defined sub-basins of the Baltic Sea since those values give fundamental information on how the system reacts to changes in, e.g., nutrient loading. The idea with this modeling, and the results presented in this work, is that these water fluxes, water retention rates and the algorithms to quantify vertical mixing and diffusion among the defined layers should be structured in such a manner that the model can be used to quantify also fluxes of nutrients, SPM and toxins. This places certain demands on the structure of this model, which are different from oceanographic models, e.g., in quantifying resuspension, mixing and diffusion and in the requirements regarding the accessibility of the necessary driving variables.

The crucial element(s) of this salt budget for the Baltic Sea, beside the uncertainties in the forcing data (river water fluxes, precipitation, etc.), are the distribution coefficients to calculate the water exchange between the sub-basins, the mixing rates and the diffusion rates. These distributions coefficients and rates have been calibrated based on the comprehensive empirical monthly salinity data available for the system for the given period and the calculated fluxes are the water fluxes needed to obtain the given empirical mean water salinities.

2.4 Phosphorus Dynamics in the Baltic Sea

2.4.1 Background on CoastMab for Phosphorus and Nutrient Fluxes to the Baltic Sea

A central aspect of this work concerns the practical application of this modeling as a tool to find the most appropriate remedial strategy to combat the eutrophication

Table 2.5 Transport of phosphorus to the Baltic Sea in the year 2000 (t; from HELCOM, 2000)

Region	Natural	Diffuse	Point sources	Total load
Bothnian Bay (BB)	2,301	966	187	3,454
Bothnian Sea (BS)	1,477	1,001	290	2,768
Gulf of Finland (GF)	1,191	2,112	2,431	5,734
Gulf of Riga (GR)	202	582	335	1,119
Baltic Proper (BP)	2,394	8,940	4,049	15,383
Archipelago Sea[a]	87	707	100	894

[a]In all following calculations, the Archipelago Sea has been included in the Baltic Proper.

in the Baltic Sea. In that context, it is fundamental to try to identify the anthropogenic contributions to the nutrient loading. HELCOM (2000; see Table 2.5) has presented very useful data regarding the natural, diffuse and point source discharges of phosphorus to the sub-basins discussed in this work. Evidently, the natural nutrient fluxes should not be reduced, only a certain part of the anthropogenic fluxes from point sources and diffuse emissions. It should also be noted that in this work we have added the contribution to the Archipelago Sea to the Baltic Proper (and not to the Bothnian Sea), since the Archipelago Sea is not modeled as a separate unit. As a background to the discussion to find the best possible remedial strategy to mitigate the eutrophication in the Baltic Sea, Table 2.6 shows central aspects of the strategy proposed by HELCOM (2007b), which was also accepted by the Baltic Sea States in November 2007. We will challenge the wisdom of that strategy in Chap. 5.

Table 2.7 gives overall budgets for nitrogen and phosphorus for the Baltic Proper. From, e.g., the National Swedish Environmental Protection Agency (SNV, 1993), HELCOM (2000), and Wulff (2006), one might get the impression that about 30–40 kt of TP on average are transported to the Baltic Sea during a year. Jonsson et al.

Table 2.6 Required nutrient reductions according to HELCOM (2007b)

	Phosphorus (t)	Nitrogen (t)
Denmark	16	17,210
Estonia	220	900
Finland	150	1,200
Germany	240	5,620
Latvia	300	2,560
Lithuania	880	11,750
Poland	8,760	62,400
Russia	2,500	6,970
Sweden	290	20,780
Transboundary pool	1,660	3,780
Sum	15,016	133,170

Table 2.7 An overall budget for nitrogen and phosphorus for the Baltic Proper (t/year). The data from SNV (1993) concern mean values for the period between 1982 and 1989; the data from HELCOM (2000) concerns year 2000. HB is Håkanson and Bryhn (2008b)

	Total-N			Total-P		
	SNV	HELCOM	HB	SNV	HELCOM	HB
A. From countries						
Sweden	44,300	46,636		1,780	1,219	
Baltic states	72,600	145,697		1,890	5,408	
Finland	–	35,981		–	1,874	
Russia	–	90,229		–	5,863	
Poland	109,900	191,521		19,100	12,698	
Germany	20,000	20,602		2,750	512	
Denmark	51,000	27,664		7,860	1,193	
Sum inflow from countries	297,800	558,046	≈ 500,000	33,380	28,767	≈ 30,000
B. From processes and water inflow from adjacent basins						
Precipitation	289,900	192,400	≈ 240,000	3,420	–	1,350
Nitrogen fixation	130,000	–		–	–	
Land uplift			448,000			107,000
Inflow from Kattegat			87,000			10,000
Inflow from Bothnian Sea			218,000			10,000
Total inflow			≈ 1,493,000			≈ 158,400
C. Water outflows to adjacent basins						
To the Bothnian Sea			256,000			16,000
To Kattegat			242,000			17,000
Total outflow			≈ 498,000			≈ 33,000
D. Rest terms						
Burial in sediments			(4.125,400)[a] = 501,600			(158,400 – 33,000) = 125,400
Denitrification			(1,493,000 – 498,000 – 501,600) = 493,400			

[a] the nitrogen concentration is about 4 times higher than the phosphorus concentration in the Baltic Sea sediments.

(1990) and Håkanson and Bryhn (2008b) have shown that there are also nutrient fluxes from land uplift. This has, so far, been a neglected nutrient flux in budget calculations for nutrients in the Baltic Sea. Table 2.7 gives data on this important contributor of nutrients. It is essential to include all major transport processes in order to understand the situation in the Baltic Sea, and especially to know how remedial measures reducing nutrient loading to the system will likely change nutrient concentrations in water and sediments.

Table 2.7 also gives data on the outflows of the nutrients and on denitrification (493 kt of N), which compares fairly well with figures given by Larsson et al. (2001) of between 180 and 430 kt/year. The burial in sediments is then 125 kt of TP and 502 kt of TN, calculated as residual terms in this overall budget. In this chapter, we will present new results on the internal phosphorus fluxes in all Baltic Sea basins.

The importance of the internal fluxes and the transport between basins compared to the anthropogenic nutrient input from land has also been shown by Christiansen et al. (1997) in a study of parts of the Kattegat.

2.4.2 Why Not Model N?

Savchuk et al. (2008) estimated pre-industrial nitrogen loadings and the same research group has presented the results behind the Baltic Sea Action Plan (BSAP, 2007), which requires massive reductions in N-input to the Baltic Sea (see Table 2.6). However, Håkanson and Bryhn (2008b) have, based on data from the Baltic Sea and many other coastal areas, demonstrated why remedial actions should not focus on nitrogen and they gave six main reasons:

1. It is not possible to predict how the Baltic Sea system – or most other aquatic systems – would respond to reductions in N-loading since there are many major uncertainties related to (a) the quantification of atmospheric N_2-fixation by cyanobacteria, (b) wet and dry deposition of nitrogen, (c) the algorithm regulating the particulate fraction for nitrogen and hence also (d) sedimentation of particulate nitrogen and (e) denitrification (see Table 2.8).
2. Lowering the N-concentration the Baltic Sea with salinities from 3 to 17 psu is likely to favor the blooming of harmful cyanobacteria, and such events should be avoided. Occasional very high concentrations of cyanobacteria in the Baltic Sea – and elsewhere – may be quantitatively explained by high total phosphorus (TP) concentrations, high temperatures (higher than 15°C) and/or low TN:TP ratios (lower than 15 by weight).
3. There are no general validated mass-balance models for nitrogen which have been tested for independent coastal systems and been demonstrated to yield good predictive power. Any N-model can be tuned, using different calibration constant sets for different systems, to give perfect descriptive power, but such tuning may obscure the true aspects of how natural systems work just like a deodorant covers a bad smell.
4. The general dynamic P-model, CoastMab, presented by Håkanson and Bryhn (2008b) gave good predictions ($r^2 = 0.98$ for modeled annual mean values regressed against empirical data) in all Baltic Sea basins and vertical layers without basin-specific tuning and without taking N-concentrations into account. These findings fundamentally contradict the popular "vicious circle theory"

Table 2.8 Uncertainties in key nutrient fluxes to the Baltic Proper (from Håkanson and Bryhn, 2008b)

A. Nitrogen	
In to Baltic Proper	1,000·tons TN/year
From land uplift	400–600
From rivers	300–600
From the Bothnian Sea	300–400
From Kattegat	100–150
From wet and dry deposition on the water surface	200–300
From nitrogen fixation by cyanobacteria	100–900
Sum annual input of nitrogen	1,400–2,950
B. Phosphorus	
In to Baltic Proper	1,000·tons TP/year
From land uplift	70–120
From rivers	30–40
From the Bothnian Sea	10–15
From Kattegat	10–15
From wet and dry deposition on the water surface	About 1
Sum annual input of phosphorus	120–190

(Vahtera et al., 2007) which asserts that P-diffusion from deep sediments is driven by low oxygen concentrations created by sedimentation of N-limited diatoms. Instead, the P-cycle in the Baltic Sea appears to be largely or completely independent from N-inputs, similar to the case of Lake 227, most other lakes and the Stockholm Archipelago (Håkanson and Bryhn, 2008d; Schindler et al., 2008).

5. Because plankton cells include both nitrogen and phosphorus (given by the standard composition $C_{106}N_{16}P$), because both nutrients are transported to coastal areas by the same rivers, and because there is in many systems a potential for phosphorus-driven atmospheric N_2 fixation by cyanobacteria, one generally finds a marked co-variation between TP and TN-concentrations in aquatic systems, see Fig. 2.12, which is based on data from 495 systems covering very wide ranges in trophic status, size and form, latitudes and salinity (data sources and more details are given Bryhn et al., 2008). It is interesting to note that only 9 of these 495 systems have TN/TP-ratios lower than the Redfield-ratio (based on weight) of 7.2 and that the coefficient of determination is about 0.6 for both the actual and the logarithmic data. When there is a major difference from the general relationship shown by the regression line in Fig. 2.12, there should be specific causal reasons for this, if one first accounts for the scatter related to the inherent uncertainties in

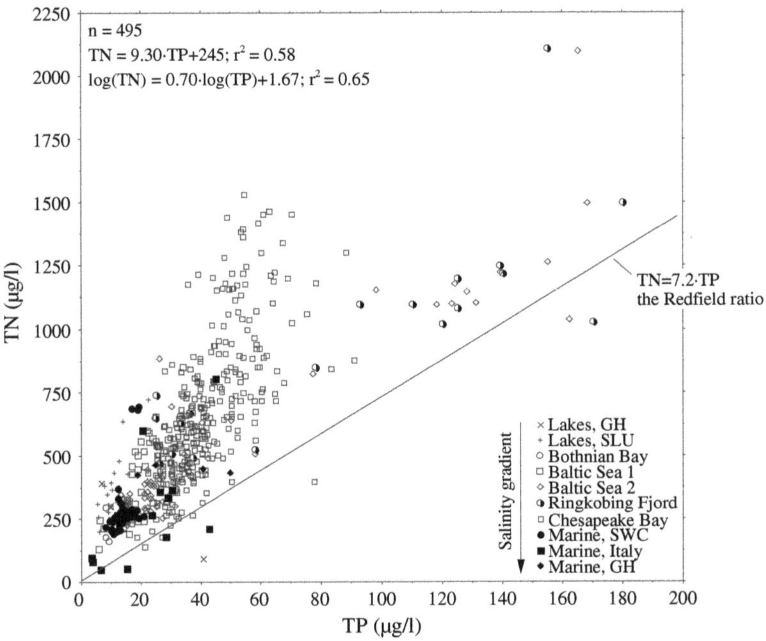

Fig. 2.12 Scatter plot between SW-concentrations of total-P (TP) and total-N (TN) for the growing season from 10 sub-groups constituting a salinity gradient. The figure also gives regressions for the actual data and log-transformed data for the 495 data points (data from Håkanson, 2008)

the data. So, phosphorus rather than nitrogen seems to limit the long-term (growing season period) primary production in the Baltic Sea and in most coastal areas of the world.

6. Primary production (e.g., in g C/m^3/day) cannot be predicted from concentrations (e.g., in mg/m^3) of dissolved nutrients, such as DIN (dissolved inorganic nitrogen), DIP, phosphate, nitrate or ammonia which are frequently below detection and have very high coefficients of variation (CV), but can only be predicted well from total concentrations of nutrients (TN or TP), i.e., from the total pools of the nutrients in the system. The concentrations of dissolved nutrients can be low and approach zero but the system can still maintain a high primary production because primary production is regulated by the regeneration and rapid recycling of dissolved nutrients (Håkanson and Bryhn, 2008b).

The yearly cost of the Baltic Sea Action Plan has been estimated at over 3 billion euros (NEFCO, 2007). According to calculations by the Swedish Department of Agriculture, N-reductions, which Sweden has agreed to undertake in the Baltic Sea Action Plan, cannot be fulfilled unless a large part of the agricultural sector in the country would be permanently shut down, an option that would eliminate tens of thousands of jobs. Sweden, which is presently a net exporter of grain, could become a yearly net importer of millions of tons of grain (Swedish EPA, 2008a), which would be associated with additional environmental pressure and transportation costs. Due to the uncertainties shown in Table 2.8 for the important nitrogen fluxes for the Baltic Sea, no one can predict the environmental outcome of these costly N-reductions in a scientifically relevant manner, while the chances increase that cyanobacteria and N_2 fixation may be favored. The same argument may apply to most/all coastal areas since the uncertainties in the major transport processes would generally be high and this also explains why there are no generally validated mass-balance models available for nitrogen.

Upgrading the phosphorus treatment of municipal sewage in former East Bloc countries could reduce the P-loading and the trophic state of the Baltic Sea to conditions prevailing one century ago. Based on costs for building Water treatment plants in the Baltic States and the St. Petersburg area (20,000 euros/ton P; NEFCO, 2007), the alternative action motivated in Håkanson and Bryhn (2008b; about 10,000 t phosphorus per year) would cost about 0.3–0.4 billion euros/year, or about 10% of the cost of the Baltic Sea Action Plan. There is thus a strong case for P-abatement to the Baltic Sea and other eutrophicated coastal areas and estuaries for which general phosphorus mass-balance models with a unitary set of calibration constants can be used to predict the outcome. We will discuss these issues in greater detail in Chap. 5.

2.4.3 Phosphorus Modeling in CoastMab

The transport processes (sedimentation, resuspension, burial, diffusion, mixing, biouptake, etc.) for phosphorus quantified in the CoastMab-model are general and apply for all substances in all/most aquatic systems (see Fig. 2.13), but there are also substance-specific parts (mainly related to the particulate fraction and the criteria for diffusion from sediments). So, these processes have the same names for all systems and for all substances:

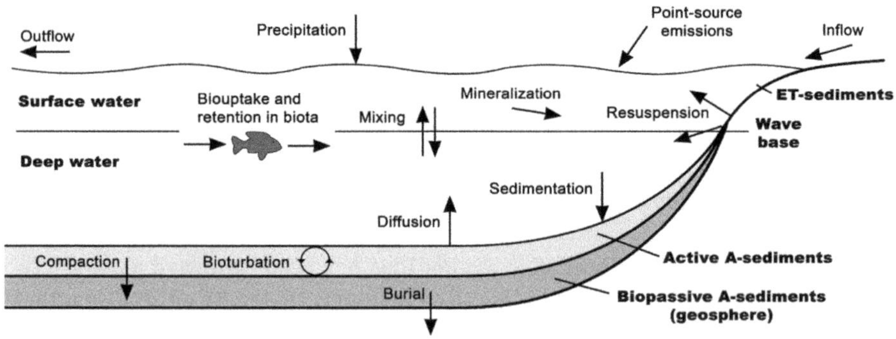

Fig. 2.13 Illustration of general transport processes to, within and from aquatic systems. "ET-sediments" refers to erosion and transportation areas where fine sediments and particulate forms of nutrients are resuspended. "Active A-sediments" are biologically active sediment areas where fine sediments are continuously being deposited (accumulation areas)

- Sedimentation is the flux from water to sediments or to deeper water layers of suspended particles and nutrients attached to such particles.
- Resuspension is the advective flux from sediments back to water, mainly driven by wind/wave action and slope processes.
- Diffusion is the flux from sediments back to water or from water layers with high concentrations of dissolved substances to connected layers with lower concentrations. Diffusion is triggered by concentrations gradients, which would often be influenced by small-scale advective processes; even after long calm periods, there are currents related to the rotation of the earth, the variations of low and high-pressures, temperature variations between day and night, etc.; it should be noted that it is difficult to measure water velocities lower than 1–2 cm/s in natural aquatic systems.
- Mixing (or large-scale advective transport processes) is the transport between, e.g., surface-water layers and deeper water layers related to changes in stratification (variations in temperature and/or salinity).
- Mineralization (and regeneration of nutrients in dissolved forms) is the decomposition of organic particles by bacteria.
- Primary production is creation of living suspended biomass from sunlight and nutrients.
- Biouptake is the uptake of the substance in biota. In the CoastWeb-model, we first calculate biouptake in all types of organisms with short turnover times (phytoplankton, bacterioplankton, benthic algae and herbivorous zooplankton) and from this biouptake in all types of organisms with long turnover times (i.e., fish, zoobenthos, predatory zooplankton, jellyfish and macrophytes) to account for the fact that phosphorus circulating in the system will be retained in these organisms and the retention times for phosphorus in these organisms are calculated from the turnover times of the organisms.
- Burial is the sediment transport of matter from the biosphere to the geosphere often of matter from the technosphere.
- Outflow is the flux out of the system of water and everything dissolved and suspended in the water.

2.4.3.1 New Features in the Phosphorus Modeling

The Dynamic SPM-Model Replaces the SPM-Regression

In the previous version of the CoastMab-model for the Baltic Sea (Håkanson and Bryhn, 2008b), the concentrations of suspended particulate matter, SPM, was calculated from modeled TP-concentrations from the following regression:

$$\log(\mathrm{SPM}) = 1.56 \cdot \log(\mathrm{TP}) - 1.64 \tag{2.5}$$

($r^2 = 0.90$; n = 51 system with salinities < 15 psu; $p < 0.001$)

It should be stressed that this approach works quite well for the conditions in the surface-water (SW) layer (see Håkanson and Bryhn, 2008b), but less well for predictions of SPM in deep-water (DW) layers, where the total concentrations of phosphorus generally are high because the fraction of dissolved phosphorus would be high and sedimentation of particulate phosphorus relatively small. So, the true SPM-value in the DW-layers can be expected to be lower than predicted by this regression and SPM should be better estimated using the dynamic SPM-model, which has been used in this work. This also means that the regression approach would give too high sedimentation values and too high values of phosphorus burial, which actually causes a compensatory effect so that the TP-concentrations in the DW-layer and in accumulation area sediments are predicted quite well (but partly for the wrong reason) with Eq. (2.5).

The first new feature in this modeling concerns the replacement of Eq. (2.5) to predict SPM with a dynamic SPM-model, which will be explained in the Sect. 2.5.

The Effects of Land Uplift

During the testing of the basic CoastMab-model for phosphorus for the Baltic Sea, it became evident that the clay particles added to the system from land uplift, influenced the system in several ways. This was quantified by means of the "clay factor", which was defined in the following way (for example for the Gulf of Finland):

$$Y_{LUGF} = (F_{TPtribGF} + F_{TPprecGF} + F_{TPSWBPGF} + F_{TPLUGE}) / (F_{TPtribGF} + F_{TPprecGF} + F_{TPSWBPGF}) \quad (2.6)$$

Y_{LU} is an important variable in the modeling of the Baltic Sea and it describes the influence of land uplift on several important processes regulating TP-fluxes. It is defined in the same way for all sub-basins. Y_{LU} is basically the ratio between the total TP-inflow when there is land uplift (Tot+LU) divided by the total inflow if there is no land uplift (Tot), i.e., (Tot+LU)/Tot. So, if LU = 0, this ratio is 1.

The basic CoastMab-model predicted well in basins with a small land uplift, and in coastal areas where the total impact from land uplift (i.e., land uplift in mm/year multiplied with the area above the theoretical wave base in km^2) was small compared to other fluxes, e.g., in small estuaries with large tributaries, or in small coastal areas where the TP-inflow from the sea was significantly more important than the TP contribution from land uplift (see Håkanson and Eklund, 2007). In the basins with the highest potential impact from land uplift, i.e., the Bothnian Bay and the Bothnian Sea, the basic model gave poorer predictions (see Håkanson and Bryhn, 2008b) in spite of the fact that it had previously been extensively tested and the algorithms should be general and reliable. The basic model simply did not account for the influence of land uplift in a realistic manner in the Bothnian Bay and the Bothnian Sea.

It was hypothesized that the "clay factor", Y_{LU}, could influence the system in different ways. The phosphorus, clay particles and iron added to the system from land uplift could potentially:

1. Increase SPM-values and the PF-value (particulate fraction) for phosphorus.
2. Increase settling velocities for particulate phosphorus and SPM.

3. Much of the more consolidated materials from land uplift could move downward to deeper areas rather than to the surface-water area, and this could be more pronounced in deeper basins than in shallower basins, i.e., the distribution of the material from land uplift could depend on the mean depth, the form factor and/or the dynamic ratio. Many such alternatives have also been tested (see Håkanson and Bryhn, 2008b).
4. The older material from land uplift would be more consolidated than the recently deposited materials. This would imply the Y_{LU} could influence the burial rate.
5. The materials from land uplift might also influence the SPM-concentration and the Secchi depth differently than material from more normal allochthonous and autochthonous sources.

The phosphorus in the sediments from land uplift would be predominately in particulate form, and most of this phosphorus would not be available to phytoplankton biouptake.

Equation (2.6) was also used to quantify sedimentation in the MW- and DW-layers.

The "clay" factor has been modified in this work (compared to Håkanson and Bryhn, 2008b) to account for the fact that the different sub-basins have different forms (form factors, see Fig. 2.4). This should imply that basins with a high bottom area above the theoretical wave base, i.e., basins with a relatively small form factor (V_d; such as the Baltic Proper with a V_d of 0.40) should be relatively more influenced by nutrients, clay particles and iron added to the system from land uplift than basins with larger V_d-values. That is, in basins with high values of the form factor (such as the Gulf of Riga with a V_d of 1.31) relatively more of the materials from land uplift should be transported downward below the theoretical wave base. This argument is relatively easy to understand by looking at the hypsographic curves in Fig. 2.4. This means that in this modeling, we have used the "clay factor" in the following modified manner:

$$Y_{LU} = (\mathrm{Tot} + \mathrm{LU}) / \mathrm{Tot}) \cdot \left(\mathrm{Vd}_{\mathrm{const}} / V_d\right) \tag{2.7}$$

We have carried out many calibrations/iterations to find a general value for the V_d-constant (Vd_{const}) and in all following simulation, we will use a value of 0.76, which is quite close to a mean/median value for these basins. We have also added a boundary condition, since the "clay" factor should never by smaller than unity, accordingly: if $Y_{LU} < 1$ then 1 else Y_{LU}.

This means that the Y_{LU}-values for the different basins are about 1 for the Gulf of Riga and the Gulf of Finland and hence also that the direct influences of land uplift are small in these two sub-areas with the smallest land uplift (0.625 and 1.6 mm/year, respectively) and with relatively high form factors (1.31 and 1.04, respectively). The "clay" effects should be larger in the areas with the highest land uplift, the Bothnian Bay (8.5 mm/year), the Bothnian Sea (7.25 mm/year) and the Baltic Proper (2.25 mm/year); the form factors for these three systems are 0.84, 0.61 and 0.40, respectively. This gives the "clay" factors 7.5, 4.5 and 4.7, respectively.

So, the impact of land uplift is by far the highest in the Bothnian Bay, which should be expected.

Calculations of Burial

This modification is also related to the change in SPM-modeling and it has to do with the quantification of burial (F_{Bur}), which is the transport of phosphorus from surficial A-sediments to sediment layers deeper than 10 cm. It was previously given by:

$$F_{Bur} = M_{TPADW} \cdot (1.386/Age_{ADW}) \cdot Y_{LU} \tag{2.8}$$

Where 1.386 is the half-life constant, M_{TPADW} is the mass (g) of TP in the given sediment compartment and Y_{LU} is the "clay factor" related to land uplift (as discussed before). The half-life constant (which is generally used in radioecology in contexts related to physical half-lives of radionuclides) has been replaced everywhere in this modeling by 1 (one) so that the burial rate is given by 1/T rather than 1.386/T. The age of TP in the given compartment was calculated from the sedimentation of particulate phosphorus assuming that the newly deposited material had a TP-concentration of 2 mg/g dw. This approach also includes two boundary conditions. If there is very little sedimentation, e.g., 0.01 cm/year, and the calculated age of the 0–10 cm sediment layer is 1,000 years, the TP in the sediments could not be available for diffusive upward transport for such a long time. Previously, we used a boundary age of 20 years. If the age of the A-sediments is shorter than 1 year, we used a second boundary age of 12 months.

In the new approach, we calculate sedimentation automatically by the dynamic SPM-model (and not indirectly from sedimentation of phosphorus), we use the modified "clay" factor for all compartments connected to the theoretical wave base (but not for the DW-compartment in the Gulf of Finland and the Baltic Proper), we have simplified the basic algorithm by omitting the half-life constant (1.386) and we have modified the algorithm for the boundary conditions since this should describe the retention and burial of phosphorus rather that the retention and burial of SPM. This means that burial for phosphorus is given by:

$$F_{Bur} = M_{TPADW} \cdot (1/Age_{TPADW}) \cdot Y_{LU} \tag{2.9}$$

The age (in months) of phosphorus in the sediments, Age_{TPADW}, is given by:

$$Age_{TPADW} = (36 + 5 \cdot (Age_{SPMADW}/36 - 1)) \tag{2.10}$$

Where Age_{SPMADW} is the age (in months) of SPM (calculated from the dynamic model using Eq. 2.11), 36 months is a "normal" retention age for phosphorus on accumulation area sediments (0–10 cm) and 5 is the amplitude value in the equation. Age_{SPMADW} is calculated by the dynamic SPM-model for sedimentation in cm/year (Sed_{SPMDW}):

$$\mathrm{Age}_{\mathrm{SPMADW}} = 12 \cdot 10/\mathrm{Sed}_{\mathrm{SPMADW}} \tag{2.11}$$

This means that if, e.g., the annual sedimentation ($\mathrm{Sed}_{\mathrm{SPMADW}}$) is 0.1 cm/year, a typical value for the Baltic Sea (which we will discuss later in this chapter), $\mathrm{Age}_{\mathrm{SPMADW}} = 1{,}200$ months and $\mathrm{Age}_{\mathrm{TPADW}} = 198$ months; if $\mathrm{Sed}_{\mathrm{SPMADW}} = 1$ cm/year, $\mathrm{Age}_{\mathrm{SPMADW}} = 120$ months and $\mathrm{Age}_{\mathrm{TPADW}} = 48$ months. This approach has been used for all Baltic Sea basins.

Boundary Conditions and Improved Algorithm for the Particulate Fraction

In the previous model, there were no defined boundary conditions for the particulate fraction (PF) for phosphorus in water, and hence no limits for the dissolved fraction of phosphorus (DF = 1 – PF). However, it is evident that all phosphorus in the water phase cannot appear in dissolved form. If this were the case there would be no sedimentation of particulate phosphorus and the TP-concentration in the sediments would approach zero. Since long, it is well known that phosphorus can appear in sediments in many different forms (see Fig. 2.14) and that the TP-concentration in Baltic Sea glacial clays seldom are lower than 0.3–0.4 mg/g dw (see Cato, 1977; Emelyanov, 1988, 2001; Jonsson, 1992). This means that there should be a boundary condition for the PF-value in water and in all following simulations, we have put that at 0.99, i.e., DF should always be ≥ 0.01.

In this modeling, we have used an algorithm for the PF-value for phosphorus which is based on three principles: (1) the PF-value should increase when phosphorous in dissolved form is being taken up by and retained in plankton, (2) the PF-value should increase with increased resuspension of particulate phosphorus and (3) the PF-value should increase with increased SPM-concentrations, which are modeled separately from TP-fluxes by CoastMab for SPM. The improved algorithm

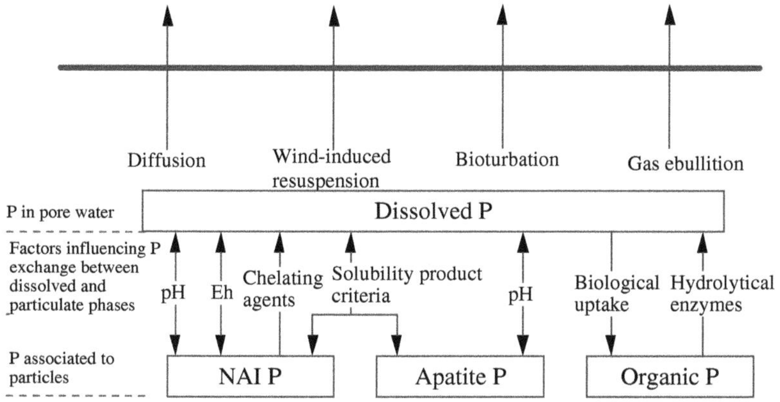

Fig. 2.14 Dominating processes regulating the exchange of phosphorus between water and sediments (NAI P = non apatite inorganic phosphorus; modified from Håkanson and Jansson, 1983)

for the PF-value is described in more detail in section The Particulate Fraction for Phosphorus.

Calculations of Biouptake and Retention in Biota

Since the modeling presented in this work includes a foodweb model, it is possible for the first time in Baltic Sea studies, as far as we are aware, to also account for the biouptake and retention of phosphorus in all the functional groups within the framework of a comprehensive and widely tested mass-balance modeling approach including all major abiotic transport processes. We have done preliminary calculations of the uptake and retention of phosphorus in all functional groups and those calculations indicate that there are only very small changes in predicted phosphorus concentrations in water and sediments between those more complicated calculations and simplified calculations where one would only differentiate between phosphorus uptake and retention in organisms with long turnover times (i.e., fish, zoobenthos, predatory zooplankton, jellyfish and macrophytes; see Table 2.9) and in organisms with short turnover times (i.e., phytoplankton, bacterioplankton, herbivorous zooplankton and benthic algae). So, in the following, we will present results based on the simplified approach, since this will only marginally influence calculated TP-concentrations or the production and foodweb characteristics related to the functional groups. This is one of several simplifications needed to keep the model as small as possible.

To calculate the TP-uptake and retention first in biota with short turnover times, this modeling uses a similar approach as presented by Håkanson and Boulion (2002a). This means that the uptake and retention in biota is:

$$M_{TPBioTPS}(t) = M_{TPBioS}(t - dt) + (F_{TPBioupS} - F_{TPBioretS}) \cdot dt \qquad (2.12)$$

$M_{TPBioS}(t)$ is the mass (amount) of TP in organisms with short turnover times (g). The uptake of TP in these organisms (g/month) is calculated using:

(1) The dimensional moderator for the influence of the light conditions (Y_{DayL}) as given by Håkanson and Bryhn (2008b), see Eq. (2.15).
(2) The biouptake rate is given by the inverse of the mean turnover time for organisms with short turnover time ($1/T_{BioS}$), see Eq. (2.16)
(3) The dissolved fraction of phosphorus in the surface-water layer ($DF_{SW} = 1 - PF_{SW}$; see section The Particulate Fraction for Phosphorus).

This means that the biouptake of phosphorus by these organisms is given by:

$$F_{TPbioupS} = M_{TPSW} \cdot Y_{DayL} \cdot (365/(12 \cdot T_{BioS})) \cdot DF_{SW} \qquad (2.13)$$

The flux of TP out of this compartment is given by:

$$F_{TPbioretS} = M_{TPBioS} \cdot (365/(12 \cdot T_{BioS})) \qquad (2.14)$$

Table 2.9 Metabolic efficiency ratios (MER) for key secondary functional groups (MER = PR/CON, dimensionless)

	CON	PR	RES	FAE	MER	T	CR	Consumes	Consumed by
Zooherb	100	24	36	40	0.24	6.0	0.17	Phytopl., Bacteriopl., Zoopred,	Prey fish, Jellyfish
Zoopred	100	32	48	20	0.32	11.0	0.091	Zooherb	Prey fish, Jellyfish
Zoobenthos	100	15	35	50	0.15	128	0.015	Macrophytes, Benthic algae	Prey fish
Prey fish	100	16	64	20	0.16	300	0.016	Zoobent., Zooherb, Zoopred	Pred. fish
Predatory fish	100	25	55	20	0.25	900	0.0013 –0.02	Prey fish	Pred. fish
Jellyfish	100	22	56	20	0.22	120	0.008	Zoopred, Zooherb	None
Phytoplankton						3.2			
Bacterioplankton						2.8			
Benthic algae						4.0			
Macrophytes						300			

The MER-value is calculated from the mass-balance equation, CON = PR + RES + FAE, where CON = consumption, PR = production, RES = respiration, FAE = unassimilated food (faeces), all dimensionless, and T = turnover time (= BM/PR, days; BM = biomass). The actual consumption rate constant, CR, expresses reduction of prey organism biomass per unit of time. The jellyfish values on PR, RES and FAE are from Schneider (1989) while the corresponding values for the other functional groups mainly are from Winberg (1985) and Håkanson and Boulion (2002a), except for the turnover time for predatory fish, which has been set to twice the value of 450 days used for lakes. This is based on the information that the main type of predatory fish in the Baltic Sea, cod, starts out as a zooplanktivore, then become mainly a benthivore and does not feed only on prey fish even at a size of about 60 cm (see Fig. 2.47, later).

The dimensional moderator based on the number of hours with daylight (HDL; Table 2.3) each month is Y_{DayL}.

$$Y_{DayLBB} = HDL/12 \tag{2.15}$$

The average turnover time of the organisms with short turnover times (T_{BioS} in days) is calculated from the individual turnover times (T_{BA}, T_{PH}, T_{BP} and T_{ZH}) given in Table 2.9 adjusted to the biomasses calculated by the CoastWeb-model. That is:

$$\begin{aligned} T_{BioS} = {} & T_{BA} \cdot (M_{BA}/M_{Stot}) + T_{PH} \cdot (M_{PH}/M_{Stot}) + T_{BP} \cdot (M_{BP}/M_{Stot}) \\ & + T_{ZH} \cdot (M_{ZH}/M_{Stot}) \end{aligned} \tag{2.16}$$

Where M_{BA}, M_{PH}, M_{BP} and M_{ZH} are the calculated monthly biomasses (kg ww) of the given organisms and M_{Stot} is the total biomass of these organisms ($= M_{BA} + M_{PH} + M_{BP} + M_{ZH}$). See Table 2.11 for a compilation of abbreviations used throughout this work.

The biouptake and retention of phosphorus in organisms with long turnover times is calculated from their consumption of the organisms with short turnover times. This is given by:

$$M_{TPBioL}(t) = M_{TPBioL}(t - dt) + (F_{TPBioSL} - F_{TPBioLSW}) \cdot dt \tag{2.17}$$

$M_{TPBioL}(t)$ is the mass (amount) of TP in organisms with long turnover times (g). The uptake of TP in these organisms (g/month) is calculated from:

$$F_{TPBioSL} = M_{TPBioS} \cdot (365/(12 \cdot T_{TPBioL})) \tag{2.18}$$

The flux of TP out of this compartment to the surface-water layer is given by:

$$F_{TPBioLSW} = M_{TPBioL} \cdot (365/(12 \cdot T_{TPBioL})) \tag{2.19}$$

The average turnover time of the organisms with long turnover times (T_{BioL} in days) is calculated in the same manner as for the organisms with short turnover times, i.e.:

$$\begin{aligned} T_{TPBioL} = {} & (T_{MA} \cdot (M_{MA}/M_{Ltot}) + T_{PY} \cdot (M_{PY}/M_{Ltot}) + T_{PD} \cdot (M_{PD}/M_{Ltot}) \\ & + T_{ZB} \cdot (M_{ZB}/M_{Ltot}) + T_{JE} \cdot (M_{JE}/M_{Ltot}) + T_{ZP} \cdot (M_{ZP}/M_{Ltot})) \end{aligned} \tag{2.20}$$

The total biomass of these organisms is $M_{Ltot} = (M_{MA} + M_{PY} + M_{PD} + M_{JE} + M_{ZP})$.

Calculations of TP-Concentrations in Sediments

To use the dynamic SPM-model also means that we can define and predict the TP-concentrations in the accumulation-area sediments (TP_{Ased}) in a more direct and mechanistically correct way from dynamically modeled TP-amounts in A-sediments (M_{TPAsed}) and dynamically modeled SPM-amounts in A-sediments ($M_{SPMAsed}$) as:

$$TP_{Ased} = M_{TPAsed}/M_{SPMAdsed} \tag{2.21}$$

This approach also opens up an important avenue to critically control model predictions; if both M_{TPAsed} and $M_{SPMAsed}$ are correctly modeled, if all TP-concentrations in water are correctly modeled (which will be tested against empirical data) and if all SPM-concentrations in water are correctly modeled (which will be tested against mainly Secchi depth data but also scattered empirical SPM-values), also the TP-concentrations in A-sediments should be correctly modeled, i.e., fall in the critical range between 0.36 and 2.0 mg TP/g dw.

These are the changes carried out relative to the CoastMab-model described by Håkanson and Bryhn (2008b). The new results will be presented briefly in the next section.

2.4.3.2 Empirical Data Versus Modeled Values

Figure 2.15a presents the modeled annual TP-concentrations in water against the corresponding empirical annual data. The results in Fig. 2.15a are about the same as reported before for the slope, which is closer to 1 (0.96 compared to 0.92); the coefficient of determination (r^2) is 0.98 (compared to 0.98).

The modeled mean annual TP-concentrations in A-sediments (0–10 cm) are given below and all these data fall within or very close to the requested range (0.36–2.0 mg TP/g dw; MW = middle water; DW = deep water; BP = Baltic Proper; BS = Bothnian Sea; BB = Bothnian Bay; GF = Gulf of Finland; GR = Gulf of Riga) and are close to what has been reported before (Håkanson and Bryhn, 2008b). Emelyanov (1988) has presented maps on phosphorus in sediments based on many sediment samples from large parts of the Baltic Proper and our modeled values agree very well with those empirical data:

	AMWBP	ADWBP	AMWGF	ADWGF	ADWBS	ADWBB	ADWGR
Now	1.37	0.51	0.50	0.67	0.80	0.35	1.01
Before	1.73	0.61	1.19	1.34	1.00	0.46	0.38

The differences are related to the differences in SPM-modeling. It should be noted that the new TP-concentrations in the accumulation-areas sediments in the Baltic Proper and the Gulf of Finland are lower than calculated before. This means that the new values for the TP-contribution from land uplift to the systems will be lower (107 t/year as compared to 133 t/year for the Baltic Proper)

One way to find out how good the results for the TP-concentrations are, is to carry out an Emp_1 versus Emp_2 test according to Håkanson and Peters (1995), i.e., to split the existing empirical data into two empirical data files and carry out the same type of regression analyses as in Fig. 2.15a with the difference that the empirical data on the y-axis are replaced half the existing data and the modeled data on the x-axis are replaced by the file based on the other half of the empirical data. The data selected for the two empirical files have been done so at random. One cannot generally expect to achieve better results for modeled values than for a parallel set of empirical data.

The results of this test are given in Fig. 2.15b for the annual data. One can note that the slope based on annual data is 1.46, which is clearly higher than the ideal slope of 1, and that the r^2-value is 0.95, which is the same as the value from Fig. 2.15a. It should be stressed that some of the data-pairs used for this Emp_1-to-Emp_2-test are based on thousands of individual samples (e.g., from the

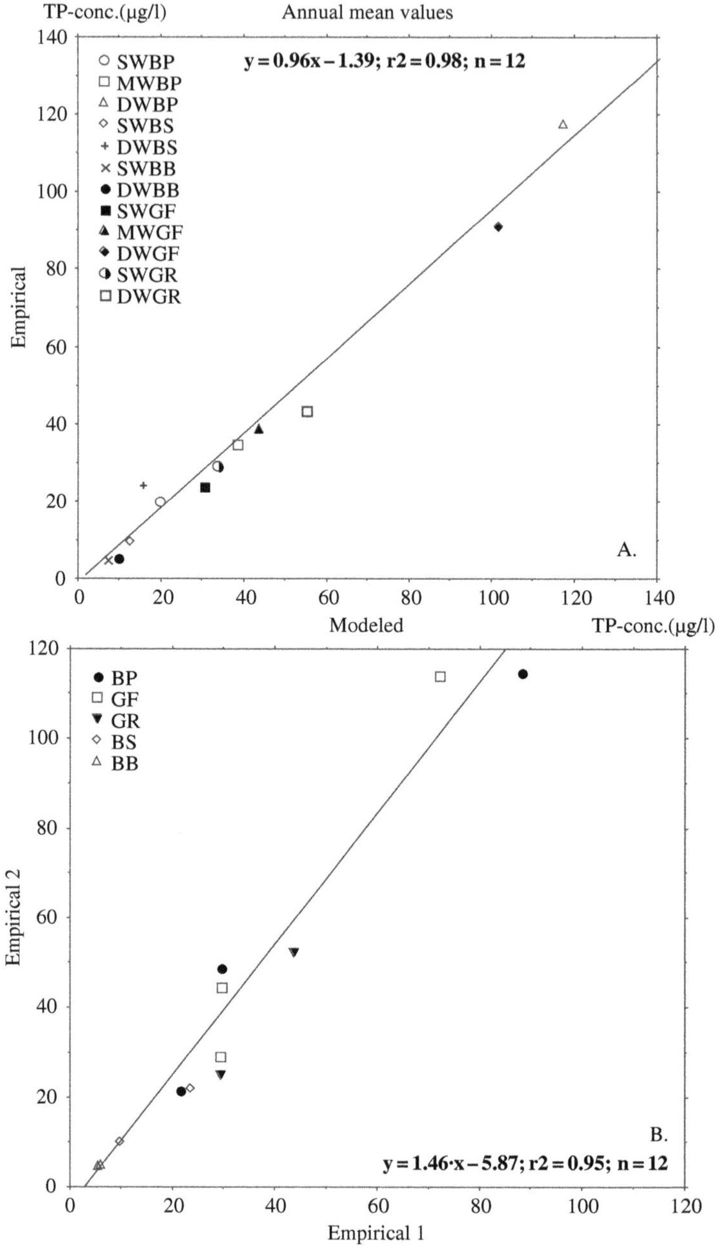

Fig. 2.15 (**a**) Comparison between empirical and modeled annual values on mean TP-concentrations in water. The regression gives an r^2-value of 0.98 and a slope of 0.96. (**b**) A comparison between annual mean TP-concentrations from two parallel datasets from the five sub-basins in the Baltic Sea. The r^2 is 0.95 and the slope 1.46 when the data in the two dataset were randomly distributed to the two datasets

SW-layer in the Baltic Proper), whereas some of the empirical data are based on just 4 empirical data split into two samples (the minimum requirement has been that there should be at least 2 empirical data for each data-pair in the regression in Fig. 2.15b. The scatter around the regression line is largest for the same data, i.e., for the DW-compartments, because these mean empirical values are not so reliable.

It should be stressed that one cannot expect perfect predictions, because:

(1) There are evident uncertainties in the empirical data (see Håkanson and Bryhn, 2008b, for more details),
(2) The monthly inflows of TP from the tributaries are not based on reliable empirical data on water discharge, but on standardized monthly curves calculated from latitude, annual precipitation and annual water discharge.

2.4.3.3 Phosphorus Fluxes

The annual fluxes of phosphorus, as calculated from the modified CoastMab-model, in Baltic Proper are exemplified in Fig. 2.16. More detailed tables and figures on TP-fluxes are given by Håkanson and Bryhn (2008b). These fluxes give information

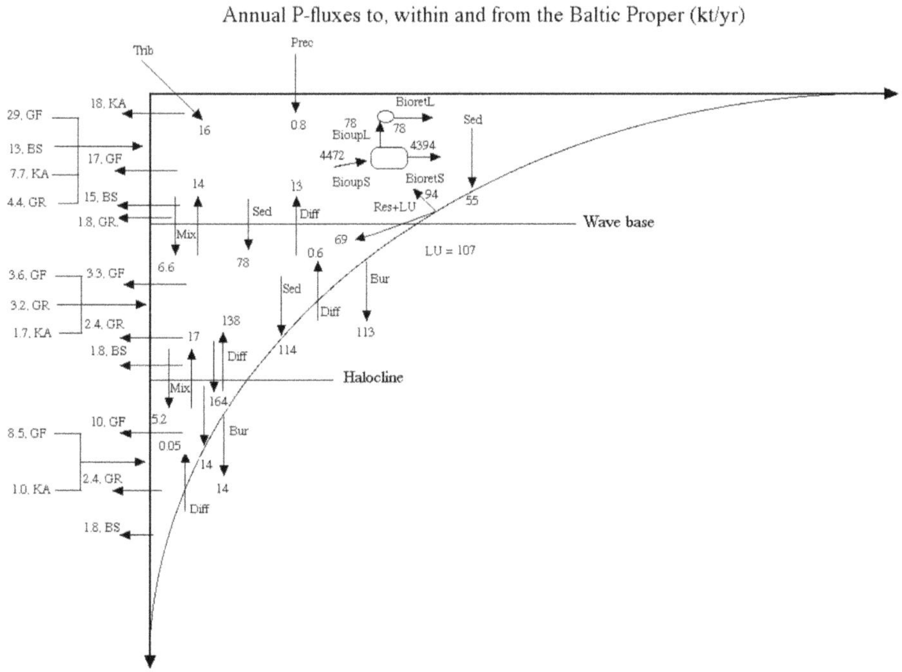

Fig. 2.16 Calculated annual fluxes of phosphorus (in kt/year) to, from and within the Baltic Proper

of fundamental importance related to how the entire Baltic Sea system reacts to changes in phosphorus loading.

It should be noted that the phosphorus fluxes to and from organisms with short turnover times (BioupS and BioretS) are very large in all basins compared to all other fluxes, but the amounts of TP found in biota is small compared to what is found in other compartments. This illustrates the classical difference between "flux and amount". In the ranking of the annual fluxes for the Baltic Proper from Fig. 2.16, it is evident that the most dominating fluxes are the ones to and from biota with short turnover times (about 4,500 kt/year), whereas the average total amount of TP in phytoplankton in BP is just about 30 kt. Looking at the TP-fluxes to the Baltic Proper, land uplift is the most dominating one (107 kt/year), followed by inflow from the Gulf of Finland (29 kt/year), tributary inflow (16 kt/year), inflow from the Bothnian Sea (13 kt/year) and total inflow from Kattegat (10 kt/year). Sedimentation in the SW-layer is also important, 78 kt/year to the MW-layer and 55 kt/year to the ET-sediments. Sedimentation in the DW-layer is relatively small (14 kt/year) since most (about 1%; see Table 2.10) of the phosphorus in the DW-layer is in dissolved form. This also implies that diffusion of phosphorus from the ADW-sediments is small in the Baltic Proper (only 0.05 kt/year). The largest diffusive flux in the water is from the DW-compartment to the MW-compartment (138 kt/year). Diffusion from AMW-sediments is 0.6 kt/year in the Baltic Proper. Burial, i.e., the transport of TP from the sediment biosphere to the sediment geosphere is 114 and 14 kt/year, respectively, from the MW- and DW-zones in the Baltic Proper. These data gives similar order-of-magnitude values as before (Håkanson and Bryhn, 2008b) and the differences have to do with the change in SPM-modeling.

Table 2.10 Modeled mean monthly particulate fractions (PF) for phosphorus in all compartments (SW, MW and DW) and all five sub-basins in the Baltic Sea (BB, BS, BP, GF and GR). Simulation time 481 months

Month	DWBB	DWBP	DWBS	DWGF	DWGR	MWBP	MWGF	SWBB	SWBP	SWBS	SWGF	SWGR
1	0.43	0.81	0.01	0.58	0.07	0.56	0.20	0.59	0.72	0.67	0.43	0.55
2	0.43	0.34	0.01	0.58	0.03	0.54	0.09	0.46	0.73	0.70	0.39	0.58
3	0.34	0.37	0.01	0.59	0.02	0.55	0.05	0.55	0.76	0.72	0.40	0.60
4	0.45	0.50	0.01	0.37	0.09	0.57	0.23	0.66	0.80	0.74	0.56	0.64
5	0.48	0.58	0.01	0.68	0.09	0.47	0.24	0.73	0.78	0.80	0.63	0.70
6	0.42	0.36	0.01	0.06	0.08	0.03	0.23	0.52	0.54	0.57	0.61	0.66
7	0.37	0.17	0.01	0.04	0.10	0.02	0.25	0.59	0.5	0.53	0.62	0.64
8	0.35	0.11	0.01	0.03	0.10	0.01	0.25	0.57	0.47	0.49	0.60	0.63
9	0.01	0.17	0.01	0.04	0	0.02	0.01	0.64	0.45	0.49	0.44	0.46
10	0.02	0.85	0.01	0.06	0	0.03	0.01	0.76	0.46	0.50	0.40	0.44
11	0.39	0.88	0.01	0.67	0.09	0.52	0.23	0.69	0.70	0.76	0.48	0.61
12	0.42	0.84	0.01	0.59	0.08	0.56	0.2	0.62	0.71	0.67	0.41	0.55
MV	0.34	0.50	0.01	0.36	0.06	0.32	0.17	0.62	0.64	0.64	0.50	0.59
M50	0.41	0.44	0.01	0.48	0.08	0.50	0.22	0.61	0.71	0.67	0.46	0.61
SD	0.16	0.29	0.00	0.29	0.04	0.27	0.10	0.09	0.14	0.11	0.10	0.08
CV	0.46	0.58	0.00	0.80	0.62	0.83	0.58	0.14	0.22	0.18	0.20	0.13

The Particulate Fraction for Phosphorus

A very important part of most mass-balance models for chemical substances, including nutrients, is the distribution coefficient. Traditionally (see Santschi and Honeyman, 1991; Erel and Stolper, 1993; Benoit et. al., 1994; Warren and Zimmerman, 1994; Weyhenmeyer, 1996; Gustafsson and Gschwend, 1997), the K_d-concept is used in these contexts; K_d is the ratio between the particulate (C'_{Par} in g/kg dw) and the dissolved (C_{Diss} in g/1) phases, i.e., $K_d = C'_{Par}/C_{Diss}$. K_d is often given in l/kg. This means that the dissolved fraction can be written as:

$$DF = 1/(1 + K_d \cdot SPM \cdot 10^{-6}) \qquad (2.22)$$

Where SPM is the amount of suspended particulate matter in the water in mg/l. It is essential to distinguish between the dissolved and the particulate fractions for all substances. It is especially important to do so for the key nutrients in water management since phytoplankton take up the dissolved fractions and only the particulate fractions can settle out by gravity. This means that there are different transport routes for the two fractions. The SPM-concentration influences the distribution of the nutrients into these two fractions. The settling velocity for the particulate fraction in m/year may be turned into a sedimentation rate (dimension 1/time) by division with the mean depth of the system or a defined part of the system. The sedimentation rate regulates sedimentation, and hence also internal loading in the given system. From Eq. (2.22), it is evident that the SPM-concentration will affect also the particulate fraction of phosphorus.

So, K_d describes particle affinity and represents the chemical equilibrium of numerous processes such as sorption onto particulate matter (see Weber et al., 1991), precipitation and dissolution (Salomons and Förstner, 1984). Depending on the reversibility of these processes, it should be noted that K_d should not be regarded as a constant but rather as a variable. Factors influencing the K_d-equilibrium are, e.g., pH (Balistrieri and Murray, 1983; Tessier et al., 1989; You et al., 1989), salinity (Koelmans and Lijklema, 1992; Turner et al., 1993; Turner, 1996), SPM (Li et al.,. 1984; Hawley et al., 1986; Balls, 1989; You et al., 1989; Yan et al., 1991; Muller et al., 1994; Quémerais et al., 1998), redox conditions (Balistrieri et al., 1992; Pohl and Hennings, 1999), biogenic Si (Boyle and Birks, 1999) and the concentration of dissolved organic matter (DOC) (Watras et al., 1995a, b, 1998; Shafer et al., 1999). Examples of substances for which K_d have been either determined or modeled are trace metals (Balls, 1988; Honeyman and Santschi, 1988; Balls, 1989; Benoit et al., 1994; Benoit, 1995; Watras et al., 1995b; Turner, 1996), organic micropollutants (Turner et al., 1999; Zhou et al., 1999) and radionuclides (Santschi and Honeyman, 1991; Carroll and Harms, 1999). In spite of (or maybe because of) their larger simplicity, statistical models for K_d might yield as good predictions, or even better, than models based on thermodynamics (e.g., Koelmans and Lijklema, 1992; Koelmans and Radovanovic, 1998).

The particulate TP-fraction in the SW-compartment (PF_{SW}) depends on and increases with the biouptake of dissolved phosphorus from the water (i.e., with the mass of phosphorus bound to plankton, M_{TPBioS}); and PF_{SW} increases with increasing resuspension (DC_{resSW}), which supplies particulate phosphorus to the system; and PF increases with increased SPM-concentrations in the water. This is given by:

$$PF_{SW} = Y_{SPMPFSW} \cdot (DC_{ResTPSW} + (M_{TPBio}/(M_{TPSW} + M_{TPBio}))) \qquad (2.23)$$

The particulate fraction for phosphorus in the MW-compartment (PF_{MW}) is given by:

$$\text{If Strat} > 1 \text{ then } PF_{MW} = Y_{SPMPFMW} \cdot DC_{ResTPMW} \text{ else } PF_{MW} = Y_{SPMPFMW} \cdot DC_{ResTPMW} \cdot (\text{Strat}/\text{MWT})^{0.5} \qquad (2.24)$$

The Strat-value is 1 when the system is homothermal. When the temperature difference between the SW-compartment and the MW-compartment is higher than 4°C, i.e., when the system is stratified and there is reduced oxygenation of the MW-layer, PF_{MW} should decrease and this is given by this equation.

PF_{DW} should depend on the theoretical water retention time (and the related oxygenation of the DW-layer) of the DW-compartment (T_{DW}) and this is given by:

$$PF_{DW} = Y_{SPMPFDW} \cdot PF_{MW}/T_{DW} \qquad (2.25)$$

So, if the theoretical water retention time (T_{DW}) in the DW-compartment is quick (e.g., 1 month), PF_{DW} should be close to PF_{MW}. Note that T_{DW} on average varies from 1.8 months in the Gulf of Riga to 151 months in the Bothnian Sea (see Fig. 2.10). So, if T_{DW} is long, oxygenation should be low and much phosphorus should appear in dissolved form (DF = 1 – PF).

The resuspended fraction is calculated in the same manner in the SW-, MW- and DW-compartments. For the SW-compartment, we have (for example for the Bothnian Sea, BS):

$$DC_{ResTPSWBS} = F_{TPETSWBS}/(F_{TPprecBS} + F_{TPETSWBS} + F_{TPSWBPBS} + F_{TPtribBS} + F_{xTPDWSWBS} + F_{DTPDWSWBS}) \qquad (2.26)$$

This is simply the resuspension flux ($F_{TPETSWBS}$) in relation to all other fluxes into this SW-layer. The resuspended fraction in the DW-layer in the Bothnian Sea is then given by:

$$DC_{ResTPDWBS} = (F_{TPETDWBS})/(F_{dTPADWBS} + F_{TPETDWBS} + F_{TPDWBPBS} + F_{TPSWDWBS} + F_{xTPSWDWBS}) \qquad (2.27)$$

The new algorithm quantifying the increase in PF for phosphorus from increased SPM-concentrations is given by:

$$Y_{SPMPF} = (1 + 0.15 \cdot (\text{SPM}/3 - 1)) \qquad (2.28)$$

The norm-value in this dimensionless moderator is 3 (mg/l) since this is the normal value for the Baltic Sea according to Pustelnikov (1977). The amplitude value of 0.15 gives the range for how variations in SPM influence the PF-value. If, e.g., SPM is 5 mg/l, as often in the Gulf of Finland and the Gulf of Riga, Y_{SPMPF} is 1.1 and the PF-value 10% higher than in situations when SPM is 3 mg/l. Note that this dimensionless moderator has been applied in the same manner for all PF-values in all layers and that the corresponding SPM-concentrations have been calculated by the dynamic CoastMab-model.

It is well established from empirical data from many systems that the PF-value for phosphorus in mainly the surface water in aquatic systems is about 0.56 (see Håkanson and Bryhn, 2008b). This means that the algorithm for PF in the SW-layers in the Baltic Sea can be compared to and controlled against this empirical reference value in terms of order of magnitude values. Table 2.10 gives the modeled PF-values on a monthly basis in all twelve compartments in the Baltic Sea under default conditions. From this table, one can note that:

1. The PF-values in the SW-layer generally varies between 0.4 and 0.8 with high average values in the Bothnian Bay (BB; 0.62), the Bothnian Sea (BS; 0.64) and the Baltic Proper Bay (BP; 0.64) and lower values in the Gulf of Riga (GR; 0.59) and the Gulf of Finland (GF; 0.50). There is also a seasonal pattern with higher PF-values in GR and GF during the growing season, and peak values related to the water turnover in spring and fall in BB, BS and BP. The highest value is 0.80 for May in BS (which has a high land uplift); the lowest PF-values in the SW-layer are 0.39 in February in GF.
2. The values in the MW-layer vary much more than in the SW-layer, the CV-value is 0.83 in BP and 0.58 in GF, compared to much lower values in the SW-layer (0.13–0.22).
3. The PF-values in the DW-zone are 0.01 in the Baltic Proper during most of the year. This means that there should be a low sedimentation of particulate phosphorus, and hence fairly low concentrations of TP in sediments. This is also evident if one looks at the empirical data from the DW-zone in the Baltic Proper compiled by Håkanson and Bryhn (2008b). The TP-concentrations in these sediments are about 0.51 mg/g dw, which should be compared to a characteristic TP-concentration in glacial Baltic Sea sediments of 0.36 mg/g dw. So, the diffusion of TP from these sediments should be low because the pool of phosphorus available for diffusion is small and related to the difference between 0.51 and 0.36 (mg/g dw).

Only when the PF-values are high can there be a high sedimentation of particulate phosphorus, and vice versa. The data given in Table 2.10 can in many ways be seen as an interpretational key to the transport processes for phosphorus in the Baltic Sea, and there are several ways to check the validity of the data given in Table 2.10. The first and foremost criteria is to see how well the modeled TP-concentrations using these algorithms for the particulate fraction correspond to the measured data of phosphorus in water and sediments. It should also be stressed that the dissolved forms as defined here from the particulate fraction is not necessarily the same thing as phosphate. There are several different dissolved forms of phosphorus often abbreviated as DP (DIP + DOP), and in Sect. 2.6.2, we will see how well these predicted mean monthly DF-values for the DW-layer in the Baltic Proper correspond to empirical data not on DP but on the ratio between phosphate and total phosphorus.

It should also be noted that the PF-values given in Table 2.10 correspond quite well to the data compiled by Håkanson and Bryhn (2008b).

2.5 Production and Sedimentation of Particles – CoastMab for SPM

The dynamic SPM-model (CoastMab for SPM) has been described by Håkanson (2006). That book also presented results of blind tests, sensitivity and uncertainty analyses and comparisons between empirical and modeled values. The model gave very good results for the tested 17 Baltic Sea coastal areas. The mean error when empirical data on sedimentation (from sediment traps) were compared to modeled values was 0.075, the median error was – 0.05, the standard deviation was 0.48 and the corresponding error/uncertainty for the empirical data was 1.0, as given by the coefficient of variation. This means that the uncertainties in the empirical data set the limit for further improvements of model predictions. The error for the modeled values was defined from the ratio between modeled and empirical data –1, so that the error is zero when modeled values correspond to empirical data.

However, the CoastMab-model for SPM has not been used before for such large areas as the major basins of the Baltic Sea and this section will describe the model.

2.5.1 Basic Structure

The structure of the dynamic model for basins with two water compartments, such as the Gulf of Riga, the Bothnian Bay and the Bothnian Sea, is shown in Fig. 2.17. There are four main compartments: surface water, deep water, areas where processes of fine sediment erosion and transport dominate the bottom dynamic conditions (ET-areas) and the accumulation areas with continuous deposition of fine materials. By definition there is no resuspension and diffusion of SPM from accumulation areas. There are six sources for SPM:

1. Primary production, which causes increasing biomasses for all types of plankton (phytoplankton, bacterioplankton and herbivorous zooplankton) influencing SPM in the water. This is calculated automatically by the CoastWeb-model and those values are incorporated in the CoastMab-model.

The structure of the dynamic SPM model for coastal areas (CoastMab)

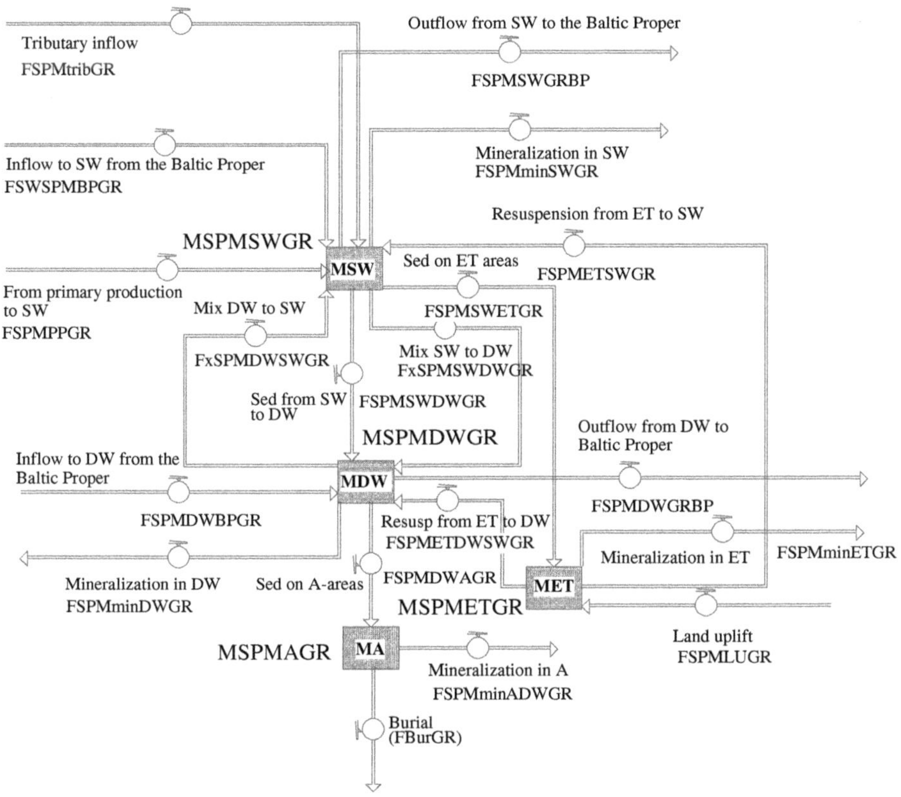

Fig. 2.17 A general outline of the structure of the CoastMab-model for suspended particulate matter (SPM) for basins (such as the Gulf of Riga, GR) with two water compartments (SW and DW)

2. Inflow of SPM to coastal surface water from the outside sea, i.e., the Kattegat to the Baltic Proper and from the Baltic Proper to the sub-basins connected to the Baltic Proper.
3. Inflow of SPM to the deep-water layer (e.g., from Baltic Proper or the Kattegat).
4. Land uplift, which is a special case for the Baltic Sea. The sub-model for land uplift was given by Håkanson and Bryhn (2008b) and has not been not changed in the following calculations.
5. Emissions of SPM from point sources. In Chap. 5, we will exemplify this by a scenario with emissions from fish cage farms. The sub-model for these emissions will be given in Chap. 5. This flux is not shown in Fig. 2.17.
6. Tributary inflow.

The amount of matter deposited on ET-areas may be resuspended by, e.g., wind/wave action or slope processes, so resuspension is an important internal process influencing the SPM-flux in coastal areas. The resuspended matter can be transported either back to the surface water ($F_{SPMETSW}$) or to the deep water ($F_{SPMETDW}$). How much that will go in either direction is regulated by a distribution coefficient calculated from the form (V_d = the form factor) of the coastal area. Other internal processes are mineralization, i.e., the bacterial decomposition of organic SPM-particles in water and sediments. Since the CoastWeb-model calculates the biomass of bacterioplankton, mineralization is calculated in a new way using model-predicted values of bacterioplankton biomass, as will be explained below. The model also accounts for mixing, i.e., the transport from the deep-water layer to the surface-water layer or from surface water to deep water.

All equations are compiled in Table A.1 (in the appendix) using data from the Gulf of Riga; Table 2.11 gives an overview of abbreviations.

2.5.2 Primary Production of SPM

The amount of SPM added to each system each month from primary production in g/month is calculated from the CoastWeb-model. The idea is to account also for production of all types of particles including bacterioplankton and herbivorous zooplankton. So, the added amount of SPM (in the Gulf of Riga in this example; $M_{SPMprodGR}$ in g each month) is given by:

$$M_{SPMprodGR} = (M_{BPGR} + M_{PHGR} + M_{ZHGR}) \cdot 1{,}000 \tag{2.29}$$

The biomasses of bacterioplankton, phytoplankton and herbivorous zooplankton (M_{BPGR}, M_{PHGR} and M_{ZHGR}) are given in kg in the CoastWeb-model. This means that the SPM-concentration (in mg/l) in the SW-compartment (in the Gulf of Riga) is given by:

$$SPM_{SWGR} = (M_{SPMSWGR} + M_{SPMprodGR})/V_{SWGR} \tag{2.30}$$

Where V_{SWGR} is the SW-volume (m^3) and $M_{SPMSWGR}$ (g) is the mass of SPM in the SW-compartment calculated by the CoastMab-model.

2.5.3 Inflow of SPM from the Sea and from Tributaries

The inflow of SPM to the surface water from the Baltic Proper to the Gulf of Riga in this example is calculated from the surface-water flow (Q_{SWBPGR} in m^3/month), which is calculated from the mass-balance for salt and the concentration of SPM in the SW-layer in the Baltic Proper (SPM_{SWBP} mg/l = g/m^3), which is calculated from CoastMab-model for SPM. This means that the SW-inflow of SPM from the Baltic Proper is given by:

$$F_{SPMSWBPGR} = Q_{SWBPGR} \cdot SPM_{SWBP} \tag{2.31}$$

The deep-water inflow of SPM from the Baltic Proper is quantified in the same way:

$$F_{SPMDWBPGR} = Q_{DWBPGR} \cdot SPM_{DWBP} \tag{2.32}$$

Since there are no reliable data on SPM in the Kattegat or in the tributaries to the Baltic Sea available to us, we have estimated SPM-concentrations from existing data on TP-inflow. Generally, in systems highly influenced by anthropogenic nutrient loading, one can assume that the TP-concentration is 2 mg/g dw of SPM (see Håkanson, 2006). In all the following calculations, we have modified this rather crude approach to estimate tributary SPM-concentrations. Instead, we have assumed that the TP-concentration related to SPM should not be a constant for all tributaries but should rather vary with the relationship between the natural TP-load and the total TP-load from tributaries. Such data are given in Table 2.5. From this table, one can see that this ratio is 0.16 for the Baltic Proper (BP), 0.67 for BB, 0.53 for BS, 0.21 for GF and 0.18 for GR. In the dimensional moderator below (Eq. 2.33), we have used 0.15 as a boundary condition for this ratio, which implies that we assume that the TP-concentration is 2 mg/g dw in tributaries where the ratio is 0.15. The amplitude value has been set to 0.1. The moderator, Y_{Trib}, is then written as:

$$Y_{Trib} = (1{,}000/2) \cdot (1 - 0.1 \cdot ((TP_{NatBP}/TP_{TotBP})/0.15 - 1)) \tag{2.33}$$

This means that if TP_{NatBP}/TP_{TotBP} is 0.67 (as in BB), Y_{Trib} is 327 and the SPM-concentration in the tributaries to BB is 327 times the TP-concentration. If TP_{NatBP}/TP_{TotBP} is 0.16 (the value for BP), Y_{Trib} is 497 and more SPM (497 times the TP-concentration) is added from these tributaries. This value has also been applied for the SPM-inflow from Kattegat to the Baltic Proper.

2.5.4 Sedimentation

Sedimentation of SPM depends on:

1. A default settling velocity, v_{Def}, which is set to 72 m/year for planktonic materials, SPM and the carrier particles for the particulate fraction for phosphorus in highly productive systems (Håkanson, 2006; Håkanson and Bryhn, 2008a). The default settling velocity is changed into a rate (1/month) by division with the mean depth of the surface-water layer (D_{SW}) for sedimentation in this layer and by the mean depth of the middle-water layer (D_{MW}) and the deep-water layer (D_{DW}) for sedimentation in these layers.
2. The SPM-concentration will also influence the settling velocity – the greater the aggregation of suspended particles, the bigger the flocs and the faster the settling velocity (Kranck, 1973, 1979; Lick et al., 1992). This is expressed by a dimensionless moderator (Y_{SPM}).

3. The salinity of the water will also influence the settling velocity – the higher the salinity, the greater the aggregation, the bigger the flocs and the faster the settling velocity (Kranck, 1973, 1979). This is expressed by a dimensionless moderator for salinity (Y_{Sal}) operating on the default settling velocity.
4. Burban et al. (1989, 1990) have demonstrated that changes in turbulence are very important for the fall velocity of suspended particles. Generally, there is more turbulence, which keeps the particles suspended, and hence causes lower settling rates, in the surface water than in the calmer deep-water compartment. The turbulence is also generally greater in large and shallow basins (with high dynamic ratios, DR) compared to small and deep basins. When there is ice cover, wind/wave-induced turbulence will also be reduced. In this modeling, three dimensionless moderators (Y_{TDW}, Y_{Tcrit} and Y_{DR}) related to the theoretical water retention time, ice conditions and the dynamic ratio are used to quantify how turbulence is likely to influence the settling velocity in the SW-, MW- and DW-compartments.
5. The settling velocity also depends on the amount of resuspended matter. The resuspended particles have already been deposited and aggregated and they have also generally been influenced by benthic activities, which will create a "gluing effect", and they have a comparatively short distance to fall after being resuspended (see Håkanson and Jansson, 1983). The longer the particles have stayed on the bottom areas, the larger the potential gluing effect and the faster the settling velocity if the particles are resuspended. The resuspended fraction is calculated in the model and the resuspended particles settle out faster. This is expressed by another dimensionless moderator (Y_{Res}).

Sedimentation from the SW-compartment to the ET-areas ($F_{SPMSWET}$) is given by:

$$F_{SPMSWET} = M_{SPMSW} \cdot ((v_{Def} \cdot Y_{SPMSW} \cdot Y_{SalSW} \cdot Y_{Tcrit})/D_{SW}) \cdot ET \cdot ((1 - DC_{ResSW}) + Y_{ResSW} \cdot DC_{ResSW}) \quad (2.34)$$

M_{SPMSW} = The mass of SPM in the SW-compartment (g); calculated automatically in the CoastMab-model using Euler's method.

V_{Def} = The default settling velocity (6 m/month).

Y_{SPMSW} = The dimensionless moderator expressing how SPM-concentrations in the SW-layer influence aggregation and sedimentation of SPM.

Y_{SalSW} = The dimensionless moderator expressing how the salinity in the SW-layer influences aggregation and sedimentation of SPM.

Y_{Tcrit} = The dimensionless moderator expressing the temperature criteria for how the ice would influence the turbulence of the system and hence also sedimentation of SPM.

ET = The fraction of ET-areas (ET = $Area_{ET}$/Area), i.e., a dimensionless measure of the ET-area above the theoretical wave base.

DC_{ResSW} = The distribution coefficient for the resuspended SPM-fraction.

Y_{ResSW} = The dimensionless moderator expressing how much faster the resuspended matter will settle out compared to the primary material (which have not been resuspended).

These expressions will be explained in the following text.

2.5.4.1 SPM Influences on Sedimentation

The higher the concentration of suspended particles in the water (SPM in mg/l), the faster the settling velocity. This is given by the following dimensionless moderator, which is used for all layers:

$$Y_{SPM} = (1 + 0.75 \cdot (SPM/50 - 1)) \quad (2.35)$$

The amplitude value (0.75) is calibrated in such a manner that a change in SPM by a factor of 10, e.g., from 2 mg/l (which is a typical value for relatively low-productive systems) to 20 mg/l (which is typical for highly productive systems), will cause a change in the settling velocity by a factor of 2. The borderline value for the moderator is 50 mg/l, since it is unlikely that marine systems (entire coastal areas on a monthly time scale) will have higher mean monthly SPM-values than that. In this modeling, particulate phosphorus has a default settling velocity of 72 m/year in systems with a SPM-value of 50 mg/l, and in systems with lower SPM-concentrations the fall velocity is lower, as expressed by Eq. (2.35).

In traditional mass-balance models, one would multiply an amount (kg) by a rate (1/month) to get a flux (i.e., amount·rate). In this modeling, one multiplies kg·(1/month)·Y (= amount·rate·moderator), where Y is a dimensionless moderator quantifying how an environmental variable (like SPM) influences the given flux (e.g., sedimentation). Instead of building a large mechanistic sub-model for how environmental factors influence given rates, this technique uses a simple, general algorithm for the moderator. Empirical data can be used for the calibration and test of the moderator. The dimensionless moderator defined by Eq. (2.35) uses a borderline value, i.e., a realistic maximum value of SPM = 50, to define when the moderator, Y_{SPM}, attains the value of 1. For all SPM-values smaller than the borderline value, Y_{SPM} is smaller than unity. One can also build normal-value moderators in such a way that the Y_{SPM} is 1 for the "normal" value and higher or lower than 1 for SPM-values higher and lower than the defined normal value (e.g., SPM = 5; see Håkanson and Peters, 1995). The amplitude value regulates the change in Y_{SPM} when the actual SPM-value differs from the borderline value and/or the normal value.

2.5.4.2 Influences of Salinity on Sedimentation

The salinity influences the aggregation and sedimentation of suspended particulate matter, including particulate phosphorus. The dimensionless moderator for salinity (Y_{Sal}) is given by:

$$\text{If salinity} < 1\,\text{psu then } Y_{Sal} = 1 \text{ else } Y_{Sal} = (1 + 1 \cdot (\text{Sal}/1 - 1)) = 1 \cdot \text{Sal}/1 = \text{Sal} \quad (2.36)$$

The norm-value of the moderator is 1 psu and the amplitude value is 1. This means that if the salinity changes from 5 to 10 psu, the moderator (Y_{Sal}) changes from 5 to 10 and the settling velocity increases by a factor of 2. This moderator attains a value higher than one for brackish and marine systems.

2.5.4.3 Influences of the Potential Turbulence on Sedimentation

The dimensionless moderator for the dynamic ratio (DR; the potential turbulence), Y_{DR}, is given by:

$$\text{If DR} < 0.26 \text{ then } Y_{DR} = 1 \text{ else } Y_{DR} = 0.26/\text{DR} \quad (2.37)$$

Basins with a DR-value of 0.26 (see Håkanson and Jansson, 1983) are likely to have a minimum of ET-areas (15% of the area) and the higher the DR-value, the larger the ET-area relative to the mean depth and the higher the potential turbulence and the lower the settling velocity. For example, for the Gulf of Finland, DR is 4.7 and hence Y_{DR} is 0.055, which is 4.7 times lower than in a system with a DR-value of 0.26. This also means that the settling velocity is 4.7 times lower, if everything else is constant.

The dimensionless moderator expressing how the ice would influence the turbulence of the system, Y_{Tcrit} in Eq. (2.38), is defined by:

If SWT < 0.9°C (the monthly boundary temperature criteria for the ice effects; this value has been changed from 0.8 to 0.9) then:

$$\text{If SWT} < 0.9 \text{ then } Y_{Tcrit} = ((5 - \text{SWT})/(\text{SWT} + 0.5)) \cdot Y_{DR} \text{ else } Y_{Tcrit} = Y_{DR} \quad (2.38)$$

Equation (2.38) gives a successive increase in the settling rate with increasing ice cover and when the monthly SWT-value is zero, the equation gives a factor of 10 higher sedimentation rate due to the reduction in wind-generated wave turbulence.

In situations when there are several consecutive months with ice cover, or with SWT-values lower than 0.9°C, this should influence the settling of SPM and particulate phosphorus substantially, and this is what this algorithm is meant to quantitatively describe. So, the potential turbulence should be related to the wind/wave activity, and large and shallow systems with high dynamic ratios should have more turbulence, which will keep the particles suspended in water for longer periods of time. One can compare sedimentation in a bottle in the laboratory under calm conditions when the bottle rests on a table (as this is described by Stokes' law) compared to sedimentation in the bottle when it is shaken. The analogy with the bottle can be extended one step further: If the bottle contains home-produced wine, one can add the clay mineral bentonite as a clarifier. This corresponds to the "clay effect" related to the clay materials added to the Baltic Sea system from land uplift.

The potential turbulence in the MW- and DW-layers will also influence the settling velocities of SPM in these layers. This is calculated from the theoretical water retention times by (the same approach is used both for MW and DW):

$$\text{If } T_{DW} \text{ or } T_{MW} < 7 \text{ (days) then}$$
$$Y_{TDW} \text{ or } Y_{TMW} = 1 \text{ else } Y_{TDW} = (T_{DWBP}/7)^{0.5} \text{ or } Y_{TMW} = (T_{MWBP}/7)^{0.5} \quad (2.39)$$

This means that if T_{DW} or T_{MW} is 30 days, Y_{TDW} or Y_{TMW} is 2.07 and sedimentation of SPM a factor of 2.07 faster compared to a situation when the theoretical MW or DW water retention time is 7 days; when T_{DW} is 365 days, Y_{TDW} is 7.2 and sedimentation of SPM a factor of 7.2 faster.

2.5.4.4 The Resuspended Fraction and the "Clay" Effects

Resuspended particles generally settle out more rapidly than particles originating directly from autochthonous or allochthonous sources. The resuspended fraction of SPM in the SW-compartment is calculated by means of the distribution coefficient (DC_{ResSW}), which is defined by the ratio between resuspension from ET-areas to surface water relative to all fluxes to the SW-compartment.

The resuspended fraction of TP in the MW-compartment (or in the DW-compartment in sub-basins with only two water compartments) is calculated in the same way.

The resuspended fraction in the DW-compartments in the Baltic Proper and the Gulf of Finland are not calculated, since there is no direct wind-induced resuspension flow to these layers.

The dimensionless moderator expressing how much faster resuspended particles settle compared to primary particles is given by:

$$Y_{Res} = Y_{LU} \cdot ((12/\text{Strat}) + 1)^{0.5} \quad (2.40)$$

Where $((12/\text{Strat}) + 1)^{0.5}$ is dimensionless expression equal to $T_{ET}/1$, where T_{ET} is the mean retention time (the mean age = T_{ET}) of SPM on the ET-areas in months and 1 is a reference age (1 month). When the system is homothermal and Strat is 1, $T_{ET} = Y_{Res}$ is 3.6 months and the resuspended particles settle out 3.6 times faster than the primary materials. When the system is stratified, Strat can approach 0.1 (a highly stratified system) and the T_{ET}-value can approach 12 months, which means that resuspended particles which have stayed that long on the bottom would be more consolidated (including gluing effects from zoobenthos) and would settle out 3.3 (12/3.6) times faster than under homothermal conditions.

The "clay" effects, as calculated by Y_{LU}, are important in the modeling of the Baltic Sea and the approach was discussed in section The Effects of Land Uplift.

2.5.5 Resuspension

By definition, the materials settling on the ET-areas will not stay permanently where they were deposited but will be resuspended by mainly wind/wave activity. If the age of the material (T_{ET}) is set to a very long period, e.g., 10 years, these areas will function as accumulation areas; if, on the other hand, the age is set to 1 week or less, they will act more as erosion areas.

Resuspension of SPM back to surface water from ET-areas, F_{ETSW} (g SPM/ month), is given by:

$$F_{SPMETSW} = M_{SPMET} \cdot R_{Res} \cdot (1 - V_d/3) \tag{2.41}$$

If SWT < 0.9°C (the boundary condition for the ice effects) then:

$$R_{Res} = (SWT + 0.2) \cdot 1/((12/Strat)) \text{ else } R_{Res} = 1/(12/Strat) \tag{2.42}$$

There may be resuspension from current activities and slope processes also under ice, so the monthly resuspension rate, R_{Res}, should not be zero. With this approach, the resuspension is a factor of 5 lower under ice, when SWT = 0, compared to situations when monthly SWT-values are higher than 0.9°C.

Resuspension from ET-areas to MW- or DW-areas below the theoretical wave base areas, F_{ETSWDW}, is given by:

$$F_{SPMETDW} = M_{SPMET} \cdot R_{Res} \cdot (V_d/3) \tag{2.43}$$

M_{SPMET} is the total amount of resuspendable matter on ET-areas (g). V_d is the form factor. Note that $V_d/3$ is used as a distribution coefficient to regulate how much of the resuspended material from ET-areas that will go the surface water or to the DW-compartment. If the basin is U-shaped, V_d is about 3 (i.e., $D_{Max} \approx D_{MV}$) and all resuspended matter from ET-areas will flow to the deeper areas. If, on the other hand, the basin is shallow and V_d small, most resuspended matter will go to the surface-water compartment. R_{Res} is the resuspension rate (1/month) related to the age of the material on the ET-areas.

The ordinary differential equation describing the fluxes of SPM to and from the ET-areas is given below:

$$M_{SPMET}(t) = M_{SPMET}(t - dt) + (F_{SPMLU} + F_{SPMSWET} - F_{SPMETDW} - F_{SPMETSW}) \cdot dt \tag{2.44}$$

The four monthly SPM-fluxes are:

1. F_{SPMLU} = SPM from land uplift (see Eq. 2.45).
2. $F_{SPMSWET}$ = Sedimentation of SPM from SW to ET (see Eq. 2.34).
3. $F_{SPMETDW}$ = Resuspension flux from ET to DW (Eq. 2.43).
4. $F_{SPMETSW}$ = Resuspension from ET to SW (Eq. 2.41).

The SPM-transport from land uplift (F_{SPMLU}) is calculated in the same manner as the TP-transport from land uplift (from Håkanson and Bryhn, 2008b). F_{SPMLU} is estimated using a method based on the hypsographic curve of the basin. If there is no land uplift, and the form and hyposographic curve is constant, the particles deposited above the theoretical wave base will not stay permanently where they were first deposited but these particles will be resuspended in cycles and transported downwards to the areas of continuous sedimentation below the theoretical wave base, where the materials may be resuspended mainly due to slope processes (see Håkanson and Jansson, 1983). The bottom deposits on the E-areas are generally relatively coarse and/or consolidated with a relatively low water content, especially in shallow regions. In the T-areas closer to the theoretical wave base, the sediments are generally less consolidated and finer. Down to the theoretical wave base there is no net deposition of matter and phosphorus if there is no land uplift, but below the theoretical wave base there is net deposition and the sedimentation rate generally varies from zero at the theoretical wave base to maximum values in deep-water areas and in topographically sheltered areas.

F_{SPMLU} (g dw/month) for example for the Baltic Proper is calculated from:

$$\begin{aligned} F_{SPMLUBP} = {} & Y_{TribBP} \cdot 12 \cdot (Area_{ETBP} + Area_{LU}) \cdot 0.001 \cdot LR_{BP} \cdot \\ & ((1 - (75 - 15)/100)) \cdot (d_{BP} + 0.2) \cdot (((1 - Area_{EBP}/Area_{ETBP}) \cdot \\ & TP_{AMWsedBP} + Area_{EBP}/Area_{ETBP} \cdot TP_{clay})) \cdot 1{,}000 \cdot 10^{-6} \end{aligned} \quad (2.45)$$

Y_{TribBP} gives the calculation constant from TP to SPM in the Baltic Proper (see Eq. 2.33). The areas above the theoretical wave base ($Area_{ET}$) are given for all sub-basins in Table 2.1. The fraction of the E-areas above the theoretical wave base ($Area_E/Area_{ET}$) is calculated from the hyposgraphic curve and the corresponding depth given by the ETA-diagram. $Area_{LU}$ in Eq. (2.45) is the newly raised areas above the theoretical wave base (about 4 km^2 in BP) and $Area_{ET}$ is the area above the theoretical wave base where there will be increased erosion/winnowing of the sediments due to land uplift. The water content of the more compacted clays above the wave base is set to be 15% lower than the recently deposited sediments below the wave base (75% in the 0–10 cm sediments below the theoretical wave base in the Baltic Sea) and the bulk density (d in g/cm^3) is set to be 0.2 units higher than in the recently deposited sediments. The bulk density (d) is calculated from Eq. (2.46). The values for the land rise (LR in mm/year) in the different sub-basins are given in Table 2.1. The concentration of TP in the sediments (TP_{AMWBP} in mg/g dw) is calculated automatically in the CoastMab-model; the TP-concentration in the more consolidated clays (TP_{Clay}) in the shallower regions is set to 0.36 mg/g dw (i.e., the reference value for the TP-concentration in clay in the Baltic Sea) as a default value in all the following calculations (see Håkanson and Bryhn, 2008b).

The bulk density of the A-sediments (d in g ww/cm^3) is calculated in the CoastMab-model using a standard formula (from Håkanson and Jansson, 1983) based on the water content (W) and IG (in % ww; abbreviated as IG*). That is:

$$d = 260/(100 + 1.6 \cdot (W + IG^* \cdot ((100 - W)/100))) \quad (2.46)$$

Based on empirical data mainly from Jonsson (1992), the water content in the top decimeter of accumulation-area sediments in the Baltic Sea is (as mentioned before) set to 75% ww as a default value for all basins in all following simulations using the CoastMab-model and the IG-value to 12% dw.

One can rightfully argue that this approach would give too low values on the TP-flux from land uplift during a year with more storms than usual, and lower values in years with less storms than usual. This approach is meant to give normal, typical values for the period 1997–2006 just like the modeled values for the TP-inflow from rivers.

2.5.6 Mixing

To quantify mixing, i.e., the upward and downward advective transport of SPM between the given layers, we use empirical data on water temperatures (from HELCOM) and dynamically modeled salinities. Since this is modeling on a monthly basis, and since these systems circulate due to the fact that the surface water becomes colder than the water in the middle- or the deep-water layers in the winter, this modeling accounts for how such water temperature variations regulate mixing. The greater the difference in mean monthly temperatures between two adjacent layers, the smaller the advective mixing. This is quantified by the following approach exemplified for the upward mixing (in kg/months) between DW and SW in GR (the Gulf of Riga):

$$F_{xSPMDWSWGR} = M_{SPMDWGR} \cdot R_{xSWDWGR} \cdot V_{SWGR}/V_{MWGR} \quad (2.47)$$

The downward mixing transport of SPM from SW to DW in GR is then given by:

$$F_{xSPMSWDWGR} = M_{SPMSWGR} \cdot R_{xSWDWGR} \quad (2.48)$$

$M_{SPMSWGR}$ = The mass of SPM (g) in the SW-layer in GR.
$M_{SPMDWGR}$ = The mass of SPM (g) in the DW-layer in GR.
V_{SWGR}/V_{DWGR} = The ratio between the volume (m^3) of the SW-layer and the DW-layer. This ratio is included in Eq. (2.47) to obtain the same water transport in both directions across the depth of the theoretical wave base (here D_{wb} in GR).
$R_{xSWMWGR}$ = The mixing rate for SW to DW in GR (1/month).

The salinity (Sal) also affects the density of the water and hence also stratification and mixing. The influences of salinity (Sal_{DWGR} and Sal_{SWGR} are the salinities in psu in the SW- and DW-layers) on the mixing rate is given by:

$$\text{If } Sal_{DWGR} > Sal_{SWGR} \text{ then } R_{xSWDWGR} = R_{MixdefGR} \cdot (1/(1 + Sal_{DWGR} - Sal_{SWGR})) \wedge R_{Mixexp} \text{ else} R_{xSWDWGR} = R_{MixdefGR} \quad (2.49)$$

$R_{MixdefGR}$ is the default mixing rate in GR (1/month); $R_{MixdefGR}$ is calculated from the fraction of erosion plus transport areas for fine sediments (ET_{GR}, dimensionless; see Eq. 2.50) – it is assumed that systems with large ET-areas (i.e., systems dominated by resuspension) should be more turbulent with more mixing. $R_{MixdefGR}$ is also influenced by the temperature stratification, which is calculated from the difference in monthly temperatures between two connected compartments. ET_{GR} is defined from:

$$ET_{GR} = (A_{GR} - A_{DwbGR})/A_{GR} \quad (2.50)$$

A_{GR} = The total water area of the system (here GR, in m^2).
A_{DwbGR} = The area below the theoretical wave base (in GR in m^2).
R_{Mixexp} = The mixing rate exponent in Eq. (2.49), which is set to 2 as a general default value (the larger the value of this exponent, the smaller the mixing, and vice versa).

The monthly temperature-dependent stratification is calculated from:

$$\begin{aligned} Strat_{GR} &= \text{if } ABS(SWT_{GR} - DWT_{GR}) < 4^\circ\,C \text{ then} \\ Strat_{GR} &= (1 + R_{Mixconst}/(1/R_{Mixconst} + ABS(SWT_{GR} \\ &\quad - DWT_{GR}))) \text{ else } StratGR = 1/ABS(SWT_{GR} - DWT_{GR}) \end{aligned} \quad (2.51)$$

Where the mixing rate constant, $R_{Mixconst,}$ is set to 1 as a default value. SWT_{GR} is the SW-temperature in GR in °C. This means that the default mixing rate, e.g., for GR (1/month) is given by:

$$R_{MixdefGR} = Strat_{GR} \cdot ET_{GR}/12 \quad (2.52)$$

The value of 1 for the mixing rate constant ($R_{Mixcont}$) has been derived from many calibrations. So, from Eq. (2.51), one can see that if the difference between the SW- and DW-temperatures is, e.g., 6°C, the value for $Strat_{GR}$ is 2/7 = 0.29; if the temperature difference is 3°C, $Strat_{GR}$ is 0.5 and there is more mixing in the system; if the temperature difference is zero, $Strat_{GR}$ is 2 and there is intensive mixing in the system.

2.5.7 Mineralization

Mineralization is the loss of the organic degradable fraction of SPM by bacterial decomposition. The value used for the mineralization rate, R_{Min}, regulates the total

Fig. 2.18 A characteristic vertical distribution of bacterioplankton in a mesotrophic lake (modified from Wetzel, 1983)

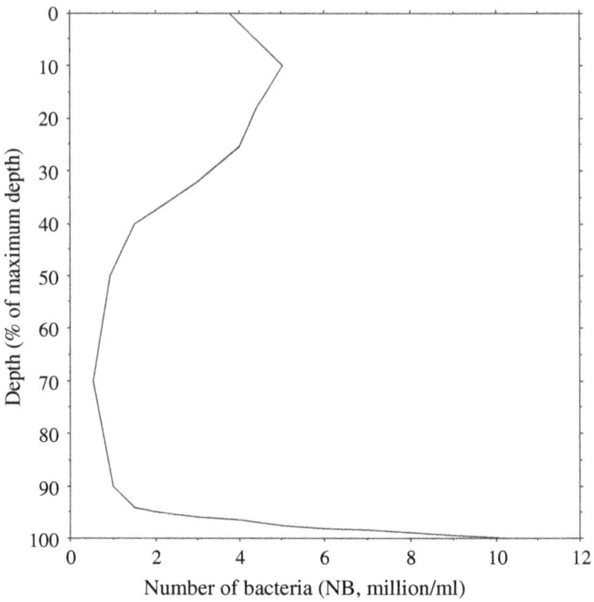

amount of SPM being lost each month in a given compartment. From Fig. 2.18, one can see that bacteria can generally be found in the entire water mass, although the highest bacterial biomasses often appear in the sediments, close to the bottom and near the water surface. This is a "normal" situation. For example, Kuznetsov (1970) has shown that there may be significant differences among aquatic systems, and seasonally within systems, in this pattern.

The bacterial degradation is a function of the temperature (SWT = surface-water temperature in °C and DWT = deep-water temperatures in °C). The loss of SPM from mineralization in surface water is:

$$F_{MinSPMSW} = M_{SPMSW} \cdot R_{Min} \cdot Y_{ET} \cdot (SWT/9)^{1.2} \tag{2.53}$$

Where 9°C is a reference temperature related to duration of the growing season (see Håkanson and Boulion, 2002a); the mineralization rate (R_{Min}) used in all model simulations uses modeled values on the biomass of bacterioplankton (from the CoastWeb-model) and is given in Eq. (2.54). The ratio SWT/9 is a simple dimensionless moderator and the exponent 1.2 stresses the non-linear temperature dependence of bacterial decomposition (see, e.g., Törnblom and Rydin, 1998). Y_{ET} is a dimensionless moderator quantifying in a simple manner a more complicated phenomena related to the fact that resuspended particles are older and more likely to have been mineralized and have a lower organic content than primary particles (see Håkanson, 2006 and Eq. 2.53). The mass (= amount) of SPM in the surface water (M_{SPMSW}) is calculated automatically for each basin by the CoastMab-model.

$$R_{Min} = (M_{BP}/NBM_{BP}) \cdot 0.01 \qquad (2.54)$$

The mineralization constant, 0.01 (per month), has been derived from calibrations. M_{BP} is the actual, modeled biomass of bacterioplankton (kg ww) in the given basin and NBM_{BP} is the corresponding norm-value (the normal biomass of bacterioplankton; this is explained in detail in Chap. 3). The higher the actual bacterioplankton biomass as compared to the norm-value, the higher the potential mineralization.

The organic content is generally highest in the material collected in the surface-water sediment traps (dominated by primary materials) and lowest in the sediment samples taken at greater water depths where the sediments have been decomposed (mineralized) to a larger extent. This demonstrates that bacterial decomposition is important in understanding changes in SPM-values and that resuspended matter should be expected to have a lower organic content than primary materials. Håkanson (2006) has also shown that there is a statistically significant correlation between the variation among systems in ET-areas and the organic content of accumulation area sediments (IG = loss on ignition) – the higher the ET-value, the lower the organic content. That information lies behind the dimensionless moderator Y_{ET}, which is meant to quantify that the mineralization rate should be higher for systems dominated by primary materials and lower for systems dominated by resuspension and SPM which has already been mineralized. In this modeling, ET varies between 0.15 and 0.99, and Y_{ET} is defined by:

$$Y_{ET} = (0.99/ET) \qquad (2.55)$$

This means that $Y_{ET} = 1$ for basins dominated by resuspension (ET = 0.99) and $Y_{ET} = 6.6$ in systems with a minimum of resuspension (ET = 0.15). For such basins, the mineralization rate is also 6.6 times higher.

The mineralization loss from the deep-water compartment is then:

$$F_{MinSPMDW} = M_{SPMDW} \cdot R_{Min} \cdot Y_{ET} \cdot (DWT/9)^{1.2} \qquad (2.56)$$

The mineralization is higher in the sediments, where the biomass of the bacteria should be expected to be many times higher than in water (see Fig. 2.18 and Håkanson and Jansson, 1983). The basic idea in setting the mineralization rate for sediments is to make sure that there is a realistic relationship between sedimentation and burial in all sediment compartments. The difference should be regulated by the substrate decomposition from mineralization and losses of matter released from gas ebullition in the sediments. In systems where much SPM emanate from primary production, the fraction of degradable organic matter should be relatively high and hence also the difference between sedimentation and burial, and vice versa for systems where SPM from land uplift is high. These rather complicated processes, for which there are no available validated models (to the best of our knowledge), have been handled in a simple manner in this modeling. We have tested several alternative approaches and used the following alternative for all sediment compartments in all basins.

So, mineralization in all the sediment compartments is given by:

$$F_{MinSPMsed} = M_{SPMADW} \cdot R_{Minsed}/Y_{LU} \qquad (2.57)$$

Where M_{SPMADW} is the mass of SPM (in this case from the ADW-compartment), R_{Minsed} is given by the default the mineralization constant (i.e., 0.01) times a general sediment constant of 30 (since there are more bacteria in the sediments than in the water; 30 is a calibration constant applied for all sediment compartments). Y_{LU} is the dimensionless moderator for the influences of land uplift in the given system. If the land uplift is high, more older already mineralized sediments would be found in the given sediment compartment and the mineralization would be lower.

2.5.8 SPM Outflow

Outflow and inflow of SPM are treated in similar ways. The outflow of SPM from the surface water in, for example, the Gulf of Riga to the Baltic Proper is calculated from the surface water flow (Q_{SWGRBP} in m^3/month), which is derived from the mass-balance for salt and the concentration of SPM in the SW-layer in the Gulf of Riga (SPM_{SWGR} mg/l = g/m^3), which is calculated here from CoastMab-model for SPM in this basin. This means that the SW-outflow of SPM from the Gulf of Riga is given by:

$$F_{SPMSWGRBP} = Q_{SWGRBP} \cdot SPM_{SWGR} \qquad (2.58)$$

The deep-water outflow of SPM from the Gulf of Riga is quantified in the same way:

$$F_{SPMDWGRBP} = Q_{DWGRBP} \cdot SPM_{DWGR} \qquad (2.59)$$

2.5.9 The Panel of Driving Variables

Table 2.11 gives a compilation of abbreviations used in this modeling and Table 2.12 gives the panel of driving variables for the dynamic SPM-model. These are the coastal-area specific variables needed to run the dynamic SPM-model. No other parts of the model should be changed unless there are good reasons to do so.

2.5.10 Testing of Model Predictions

There are few reliable empirical SPM-data valid for the compartments defined in this modeling, i.e., for mean monthly SPM-concentrations in SW-, MW- and DW-layers in the Baltic Proper, the Bothnian Sea, the Bothnian Bay, the Gulf of Finland

Table 2.11 Abbreviations and dimensions of the most commonly used concepts and variables in this modeling. Note that we have tried to use as simple and self-explanatory abbreviations as possible. Greek letters have been banned

A. Organisms
BA = Benthic algae
BE = Zoobenthos
BP = Bacterioplankton
JE = Jellyfish
MA = Macrophytes
PD = Predatory fish
PH = Phytoplankton
PY = Prey fish
ZH = Zooplankton, herbivores
ZP = Zooplankton, predators

B. Driving variables
Area = Coastal area (m^2)
D_{MV} = Mean depth (m)
D_{Max} = Maximum depth (m)
SWT = Surface-water temperature (°C)
TP = Total phosphorus (μg/l)

C. Foodweb interactions
BM = Biomass (kg ww), e.g., BM_{BP}
CON = Consumption (kg ww/month), e.g., CON_{PHZH} (PH eaten by ZH)
CR = Actual consumption rate (1/month), e.g., CR_{ZPPY} (ZP eaten by PY)
EL = Elimination (kg ww/month), e.g., EL_{ZP}
ER = Erosion (kg ww/month), e.g., ER_{BA}
Fish = Fishing, total (kg ww/month), e.g., $Fish_{PD}$
IPR = Initial production (kg ww/month), e.g., IPR_{ZHZP} (ZH eaten by ZP)
NBM = Normal biomass (kg ww), e.g., NBM_{BE}
NCR = Normal consumption rate (1/month), e.g., NCR_{PY}
MER = Metabolic efficiency ratio (dim. less), e.g., MER_{PYPD} (PY eaten by PD)
NR = Number of first order food choices (dim. less)
PR = Production (kg ww/month), e.g., PR_{PD} (PR defined by the ratio BM/T)
T = Turnover time (month), e.g., T_{PH}

D. Mass-balance (= CoastMab)
A_{Sec} = Area above Secchi depth (m^2)
BL = Biota with long turnover times
BS = Biota with short turnover times
C = Concentration, e.g., of phosphorus C_{TP} (μg/l)
DF = Dissolved fraction of phosphorus (dim. less)
DP = Dissolved phosphorus (μg/l)
DR = Dynamic ratio (dim. less)
ET = Areas of fine sediment erosion & transport (dim. less)
F = Flux (g/month), e.g., F_{ETSW} (from ET to SW)
M = Mass (g), e.g., M_{ET}
PF = Particulate fraction of phosphorus (dim. less)
R = Rate (1/month)
SW = Surface water
MW = Middle water
DW = Deep water
SWT = SW temperature (°C)
MWT = MW temperature (°C)
DWT = DW temperature (°C)
V_d = Volume development (= form factor, dim. less)
Y = Dimensionless moderator

E. Other abbreviations
Chl = Chlorophyll-a concentration (μg/l)
dw = Dry weight
IG = Sediment organic content (= loss on ignition, % dw)
PrimP = Primary phytoplankton production
Prec = Mean annual precipitation (mm/year)
Sal = Salinity (psu)
Sec = Secchi depth (m)
SPM = Suspended particulate matter (mg/l)
W = Sediment water content (% ww)
ww = Wet weight

Table 2.12 Panel of driving variables for the dynamic SPM-model

A. Morphometric parameters
1. Hypsographic curve
B. Chemical variables
2. Data on salinity, TP-concentration and SPM-concentration in the inflowing water to the coastal area
3. Data and tributary inflow of TP and SPM.
C. Other variables
4. Land uplift
5. Tributary water discharge or latitude and annual precipitation and evaporation

and the Gulf of Riga. However, there are scattered empirical SPM-data that can be used for indicative purposes to test the order-of-magnitude values of the model predictions, and in the following, we will test the predicted values using several alternative approaches.

First, it should be stressed that the basic SPM-model has been extensively tested for smaller coastal areas in the Baltic Sea and provided excellent predictions, generally within the uncertainty bands of the empirical data (see Håkanson, 2006).

A default value of 3 mg/l (from Pustelnikov, 1977) has often been used for SPM in the Baltic Sea. Håkanson and Eckhéll (2005) presented a rather comprehensive empirical study on SPM-variations in the Baltic Proper under different wind situations, at different sampling depths and under stratified and non-stratified conditions. The mean and median values were both 2.3 mg/l, with a coefficient of variation (CV) of 0.67, a very high value. Since the primary production is higher, as revealed by the chlorophyll-a concentrations in Fig. 2.19, and the water clarity lower (see Fig. 2.20), in the Gulf of Finland (GF) and the Gulf of Riga (GR), as compared to the Baltic Proper (BP), one should expect higher SPM-concentrations in GF and GR. The opposite should be true in the Bothnian Bay (BB) and the Bothnian Sea (BS).

The modeled SPM-concentrations are given in Table 2.13. One can note that the mean value for the entire surface-water layer in the Baltic Proper is 2.3 mg/l, which corresponds very well with the empirical data given by Håkanson and Eckhéll (2005). The mean SPM-values for all five surface-water compartments are (in mg/l):

BP	BS	BB	GF	GR
2.3	1.2	0.9	3.3	3.3

This pattern among the basins would also agree well with what one might expect from the maps shown in Figs. 2.19 and 2.20.

From relatively crude and uncertain data from HELCOM (1998), one can also note that approximately 7,500 kt SPM per year should be transported into the Baltic

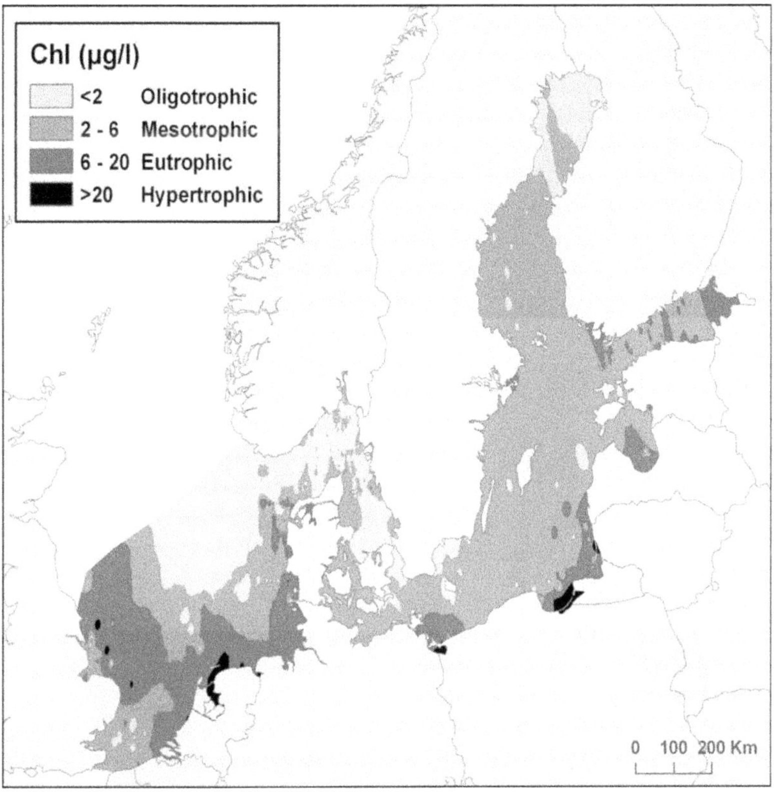

Fig. 2.19 Areal distribution of chlorophyll-a concentrations in the Baltic Sea and parts of the North Sea during the growing season (May–September) in the upper 10 m water column for the period from 1990 to 2005 (from Håkanson and Bryhn, 2008b)

Sea by rivers, most coming from Poland, Russia and Sweden. In the next section, we will give more detailed information on all SPM-fluxes (monthly and annually) to, within and from all the given basins in the Baltic Sea, and from these results, one can note that the order-of-magnitude values for the total SPM-inflow are higher than, but within a factor of 2, from the values from HELCOM (kt SPM per year).

BP	BS	BB	GF	GR	Sum
8,140	1,040	1,130	2,770	550	14,630

It should be stressed that SPM-concentrations in rivers generally vary very much; CV-values are often as high as 2.6. Håkanson (2006) gave a compilation of CV-values for SPM based on data from UK rivers and the median CV was 1.71. This means that 1,124 measurements are needed to calculate the mean SPM-concentration with an error smaller than 10% of the mean. The general sampling formula is given by (from Håkanson, 1984):

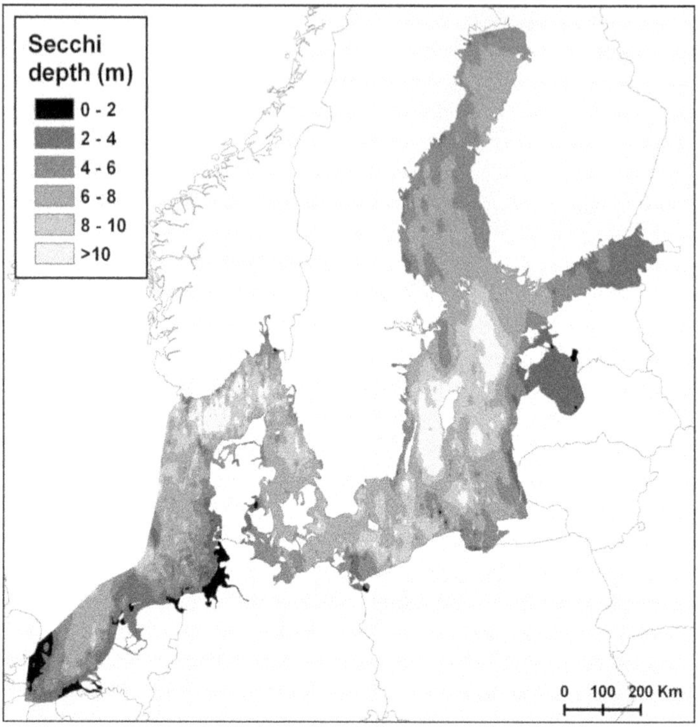

Fig. 2.20 Average annual Secchi depths in the Baltic Sea and parts of the North Sea in the upper 10 m water column for the period from 1990 to 2005 (from Håkanson and Bryhn, 2008b)

Table 2.13 Modeled monthly SPM-concentrations (mg/l) in the Baltic Sea

Month	1	2	3	4	5	6	7	8	9	10	11	12	MV	M50	SD	CV
DWBB	3.0	1.2	0.3	0.2	0.4	2.1	1.1	0.6	0.5	1.6	2.6	2.7	1.4	1.2	1.0	0.74
DWBP	1.9	2.2	2.4	2.6	2.7	2.4	1.9	1.5	1.3	1.1	1.2	1.6	1.9	1.9	0.6	0.30
DWBS	2.3	2.5	2.7	2.8	2.9	2.8	2.4	1.9	1.5	1.2	1.4	2.0	2.2	2.3	0.6	0.27
DWGF	2.0	2.0	1.8	2.0	2.4	2.5	2.5	2.4	2.2	1.9	1.8	1.9	2.1	2.0	0.2	0.12
DWGR	3.6	3.8	3.8	3.9	4.1	4.2	4.1	3.9	3.2	2.5	2.7	3.4	3.6	3.8	0.6	0.15
MWBP	1.1	1.1	1.2	1.3	1.3	1.0	0.8	0.7	0.6	0.5	0.7	0.9	0.9	1.0	0.3	0.29
MWGF	3.4	3.5	3.3	3.1	3.2	3.3	3.4	3.5	3.6	3.4	3.4	3.4	3.4	3.4	0.1	0.04
SWBB	1.6	1.6	1.1	0.7	0.4	0.5	0.7	0.8	0.8	0.8	1.0	1.3	0.9	0.8	0.4	0.39
SWBP	2.1	2.3	2.4	2.5	2.6	2.7	2.5	2.3	2.0	1.8	1.7	1.9	2.2	2.3	0.3	0.15
SWBS	1.2	1.3	1.4	1.3	1.4	1.4	1.2	1.0	0.8	0.7	0.8	1.2	1.1	1.2	0.3	0.23
SWGF	3.3	3.2	2.6	2.4	2.8	3.3	3.6	3.7	3.7	3.6	3.5	3.4	3.3	3.4	0.4	0.14
SWGR	3.0	3.0	2.9	3.0	3.1	3.4	3.4	3.4	3.4	3.4	3.2	3.1	3.2	3.2	0.2	0.06
SWKA	4.6	4.6	4.6	4.6	4.6	4.6	4.6	4.6	4.6	4.6	4.6	4.6	4.6	4.6	0.0	0.00

$$n = (t \cdot CV/L)^2 + 1 \quad (2.60)$$

Where t = Student's t, which specifies the probability level of the estimated mean (usually 95%; strictly, this approach is only valid for variables from normal frequency distributions), and CV = coefficient of variation within a given ecosystem. L is the level of error accepted in the mean value. For example, L = 0.1 implies 10% error so that the measured mean will be expected to lie within 10% of the expected mean with the probability assumed in determining t. Since one often determines the mean value with 95% certainty ($p = 0.05$), the t-value is set to 1.96.

In Baltic archipelago areas, one can generally expect a relatively high sedimentation (Håkanson, 1999; Persson and Jansson, 1988). The existence of the strong Coriolis-driven coastal currents implies that relatively little material is transported to the deep, open parts of the Baltic Sea since these currents move the suspended particulates along and into the coasts. This also means that allochthonous matter from the rivers entering the Baltic Sea, the autochthonous production in the coastal areas and the resuspension in the shallow coastal areas together with the dominating coastal currents create an environment for high sedimentation within coastal areas.

Measurements of sedimentation at individual sites in the open Baltic Proper (see, e.g., Jonsson, 1992 and Mattila et al., 2006) have indicated that sedimentation values are generally around 0.1–0.15 cm/year and have been remarkably steady over several thousand years (Jonsson, 1992). Table 2.14 presents the results from our calculations using the dynamic SPM-model. One can note that our results are very close indeed to the empirical data: the mean value for sedimentation in the deep-water layer is 0.10 cm/year and in the middle-water layer 0.11 cm/year in the Baltic Proper. The mean values for the seven accumulation areas are (cm/year):

DWBP	MWBP	BS	BB	DWGF	MWGF	GR
0.10	0.11	0.12	0.24	0.04	0.04	0.02

So, our results are in good overall agreement with empirical data considering the problems related to the time- and area-comparability of the data. It should be stressed that one must expect major differences in sedimentation (= deposition) of matter within these sub-basins. The rule is that sedimentation should vary from zero at the theoretical wave base, increase with water depth (sediment focusing), and show an areal distribution pattern reflecting the dominating hydrological flow pattern (Håkanson and Jansson, 1983).

The sedimentation values are relatively high in the Bothnian Bay (0.24 cm/year) and the Bothnian Sea (0.12 cm/year) with the highest land uplift and lower in the Gulf of Finland (0.04 and 0.02 cm/year) with a relatively low impact of land uplift. There are also clear seasonal patterns with increasing sedimentation from increased resuspension under homothermal conditions (see Table 2.14). There are also major differences among different months in the different basins, from 0.01 cm/year in the Gulf of Riga to 0.46 cm/year in February in the Bothnian Bay. Note that there may

Table 2.14 Monthly sedimentation rates (in cm/year) for the seven sediment compartments in the Baltic Sea at steady-state (simulation time 481 months)

Month	ADWBP	ADWGF	AMWBP	AMWGF	ABB	ABS	AGR
1	0.09	0.01	0.14	0.02	0.34	0.10	0.02
2	0.10	0.10	0.14	0.08	0.46	0.10	0.02
3	0.11	0.13	0.14	0.11	0.20	0.11	0.04
4	0.12	0.05	0.15	0.05	0.16	0.24	0.02
5	0.14	0.02	0.14	0.02	0.20	0.13	0.01
6	0.17	0.02	0.09	0.02	0.30	0.14	0.01
7	0.13	0.02	0.06	0.02	0.14	0.14	0.01
8	0.11	0.02	0.06	0.02	0.10	0.13	0.01
9	0.09	0.02	0.06	0.02	0.09	0.11	0.01
10	0.08	0.01	0.07	0.02	0.21	0.10	0.01
11	0.06	0.01	0.10	0.02	0.34	0.07	0.01
12	0.07	0.01	0.13	0.02	0.35	0.09	0.02
MV	0.11	0.035	0.11	0.04	0.24	0.12	0.016
M50	0.11	0.020	0.12	0.02	0.21	0.11	0.010
SD	0.03	0.04	0.04	0.03	0.12	0.04	0.01
CV	0.29	1.13	0.35	0.86	0.48	0.35	0.57

be large variations among years related to storms and changes in run-off. These data apply for the default period (1997–2006).

The corresponding settling velocities for SPM are given in Table 2.15. The mean annual values vary from 0.3 to 8 m/month. Generally, the settling velocities are highest in the DW-compartment if land uplift is high (the highest value is 23 m/month in DWBB), and lowest in the SW-compartments if the salinity is low (the lowest value is 0.3 m/month in SWBB).

Yet another way to test the reliability of the predicted SPM-values is to compare sedimentation and burial. We do not have any dynamic model for organic matter, but it is evident that the mineralization in the sediments will create a difference

Table 2.15 Monthly settling velocities (in m/month) for the seven water compartments in the Baltic Sea at steady-state (simulation time 481 months)

Month	1	2	3	4	5	6	7	8	9	10	11	12	MV	M50	SD	CV
DWBB	2.7	23.4	21.6	18.7	15.0	1.9	1.8	2.0	2.1	2.4	2.7	2.7	8.1	2.7	8.8	1.1
DWBP	2.9	2.9	2.9	3.0	3.0	2.8	2.7	2.8	2.8	2.8	2.9	2.9	2.9	2.9	0.1	0.03
DWBS	2.8	2.8	2.8	8.6	3.0	3.2	3.1	3.0	2.9	2.9	2.8	2.8	3.4	2.9	1.6	0.5
DWGF	0.9	6.3	9.1	3.0	0.9	0.9	0.9	0.9	0.9	0.9	0.9	0.9	2.2	0.9	2.7	1.2
DWGR	0.8	0.7	2.5	0.7	0.7	0.7	0.7	0.7	0.7	0.8	0.8	0.8	0.9	0.7	0.5	0.6
MWBP	2.3	2.3	2.3	2.3	2.3	2.1	2.1	2.2	2.2	2.3	2.3	2.3	2.2	2.3	0.1	0.03
MWGF	0.7	4.7	6.8	2.2	0.7	0.7	0.7	0.7	0.7	0.7	0.7	0.7	1.6	0.7	2.0	1.2
SWBB	0.3	3.1	3.0	2.6	2.2	0.3	0.3	0.3	0.3	0.3	0.3	0.3	1.1	0.3	1.2	1.1
SWBP	0.4	0.4	0.4	0.4	0.4	0.4	0.4	0.4	0.4	0.4	0.4	0.4	0.4	0.4	0.01	0.02
SWBS	0.5	0.5	0.5	1.5	0.5	0.5	0.5	0.5	0.5	0.5	0.5	0.5	0.6	0.5	0.3	0.5
SWGF	0.6	4.2	5.9	1.9	0.6	0.6	0.6	0.6	0.6	0.6	0.6	0.6	1.5	0.6	1.8	1.2
SWGR	0.5	0.5	1.6	0.5	0.5	0.5	0.5	0.5	0.5	0.5	0.5	0.5	0.6	0.5	0.3	0.6

Table 2.16 Annual fluxes (kt/year) of SPM related to burial, sedimentation, tributary inflow, bioproduction and land uplift in the Baltic Sea. Note that if one also accounts for the net inflow of SPM to the deep-water layer in the Baltic Proper, the burial percent is 0.9 and not –0.7

	Burial	Sedimentation	Burial (%)	Trib inflow	Prim. prod. (PP)	Land uplift (LU)	Total in	PP (%)
ADWBP	22,236	22,090	–0.7	8,109	73,893	53,351	135,353	54.6
AMWBP	13,784	15,474	10.9					
ADWGF	127	209	39.5	2,708	7,540	3,524	13,772	54.7
AMWGF	367	702	47.7					
BB	8,246	9,446	12.7	830	3,748	12,323	16,901	22.2
BS	13,075	15,099	13.4	861	26,304	22,549	49,715	52.9
GR	85	156	45.8	547	4,323	1,426	6,296	68.7

between sedimentation and burial; the more degradable organic matter in a give sediment compartment, the greater the potential difference between sedimentation and burial. We have addressed that issue in Table 2.16. This table first gives the annual values for burial and sedimentation in the seven sediment compartments and a calculation of the burial transport related to the sedimentation (%Burial). One can note that this percentage is highest in the Gulf of Finland (47.7%), lowest in the deep-water sediments in the Baltic Proper and between 10 and 46% in the other sediment compartments. The table also gives the percentage of SPM produced in the system; this is related to the production of phytoplankton, bacterioplankton and herbivorous zooplankton. One can note that the basin with the highest fraction of organic matter produced within the system is the Gulf of Riga (68.7%) and the basins with the lowest amount of SPM produced within the system is the Bothnian Bay (22.2%), which also has the highest land uplift. Both deep-water compartments in the Baltic Proper and the Gulf of Finland have clearly lower percentage of burial than the corresponding middle-water compartments (– 0.7 and 10.9% in BP and 39.5 and 47.7% in GF) because the organic matter decomposed in the DW-compartments have been subject to bacterial substrate decomposition longer than in the shallower MW-compartments.

These tests should give credibility to the results presented in the following section, which will give fluxes and amounts of SPM.

2.5.11 SPM-Fluxes and Amounts

The monthly fluxes of SPM, as calculated with the CoastMab-model with consideration to the "clay effects", in all sub-basins in the Baltic Sea are complied in Table A.2. These monthly fluxes give information of fundamental importance related to how the Baltic Sea system functions and we have compiled this information into Figs. 2.21, 2.22, 2.23, 2.24 and 2.25, which show the annual SPM-fluxes to, within and from each basin.

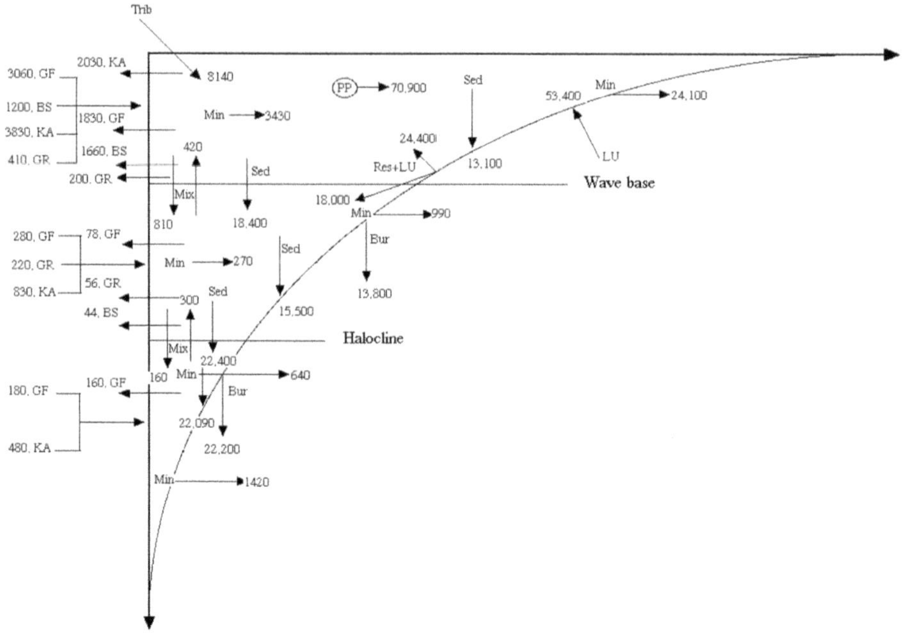

Fig. 2.21 Annual fluxes of SPM (in kt/year) to, from and within the Baltic Proper

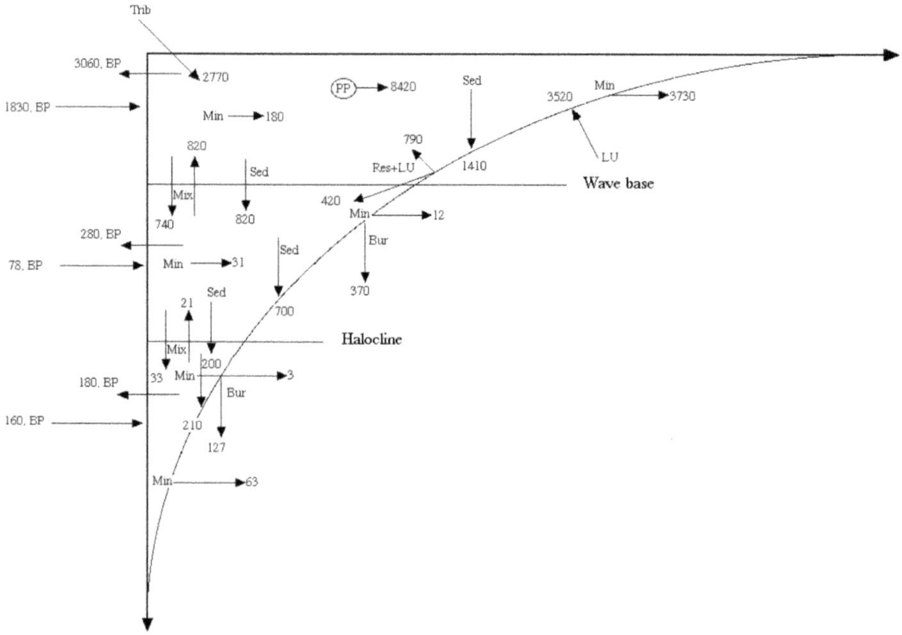

Fig. 2.22 Annual fluxes of SPM (in kt/year) to, from and within the Gulf of Finland

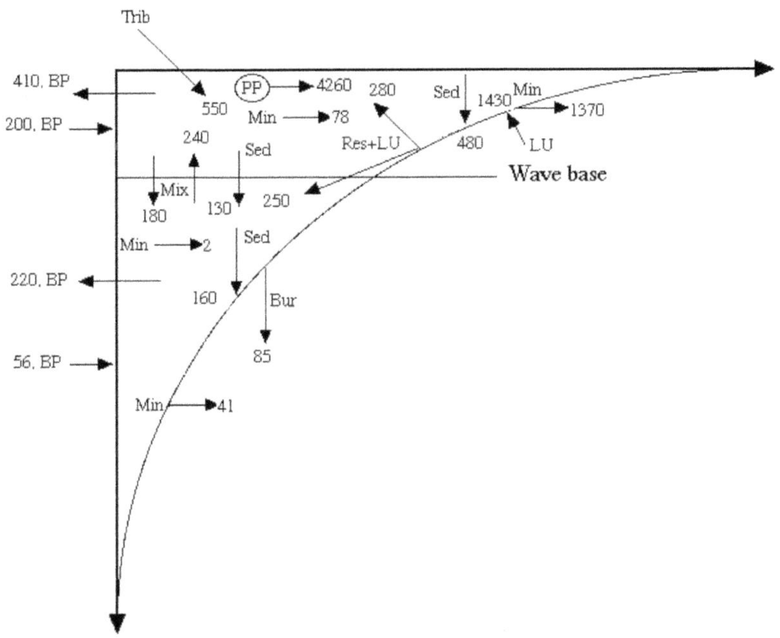

Fig. 2.23 Annual fluxes of SPM (in kt/year) to, from and within the Gulf of Riga

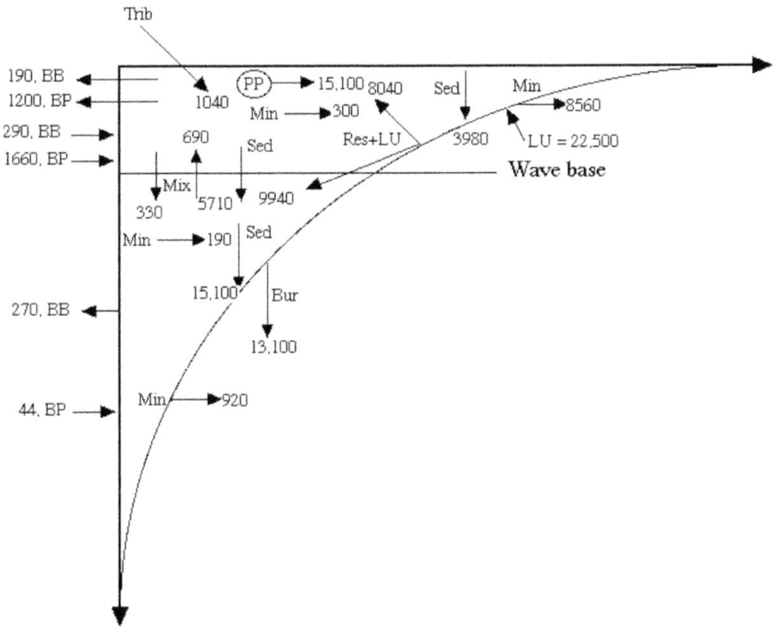

Fig. 2.24 Annual fluxes of SPM (in kt/year) to, from and within the Bothnian Sea

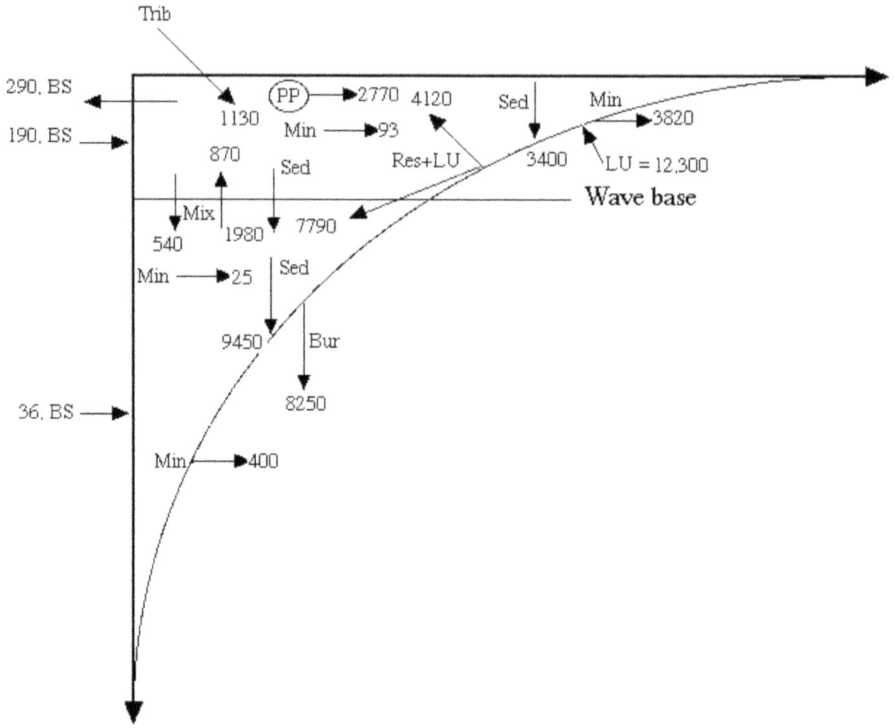

Fig. 2.25 Annual fluxes of SPM (in kt/year) to, from and within the Bothnian Bay

It should be noted that the fluxes to and from plankton (phytoplankton, bacterioplankton and herbivorous zooplankton) are larger in all basins compared to SPM from land uplift, except in the Bothnian Bay with the highest land uplift and the lowest primary production. The amounts of SPM found in plankton are, however, relatively small compared to what is found in other compartments. This is shown in Table 2.22 using monthly data on SPM-amounts for all basins and all compartments. It illustrates the classical difference between "flux and amount". In the ranking of the annual fluxes for the Baltic Proper in Fig. 2.21, it is evident that the most dominating SPM-fluxes are the ones to and from plankton (about 74,000 kt/year), whereas the average total amount of SPM in plankton in BP is just 1,000 kt as compared to over 450,000 kt in accumulation area sediments in the DW-zone. Looking at the abiotic SPM-fluxes to the Baltic Proper, land uplift is the most dominating one (53,400 kt/year), followed by tributary inflow (8,140 kt/year), inflow from the Kattegat (3,830 kt/year) and inflow from the Gulf of Finland (3,060 kt/year). Sedimentation in the SW-layer is also important with 13,050 kt/year. Sedimentation of SPM from the MW to the DW-layer is 22,400 kt/year and from the SW to the MW-layer 18,400 kt/year. The flux related to internal loading (resuspension plus land uplift) is 24,400 kt/year from ET-areas to the SW-layer and 18,000 kt/year to the MW-layer. Burial, i.e., the transport of SPM from the sediment biosphere to the sediment geosphere is 13,800 and 22,200 kt/year, respectively, from the MW- and DW-zones in the Baltic Proper.

It is also interesting to note from Table 2.17 that there are, on average, 447,000 kt of SPM in the accumulation area sediments in the deep-water zone (ADW) in the Baltic Proper and 272,000 kt in middle-water zone in BP. These are the two highest values. The two lowest values are SPM in the small DW-volume in the Gulf of Finland (40 kt) and in plankton in the Bothnian Bay (40 kt).

2.5.12 Comments

Previous knowledge regarding the SPM-concentration, its variation and the factors influencing variations among and within sites was very limited and a value of 3 mg/l

Table 2.17 Compilation of SPM-amounts various months at steady-state (simulation time 481 months) in the Baltic Sea and a ranking based on the mean monthly values. Values in million tons SPM (= 10^{12} g). BioS = biota with short turnover times (plankton)

Month	3	3	4	5	6	7	8	9	10	11	12	MV
ADWBP	449	448	448	448	448	448	448	448	447	447	447	447
MWBP	272	272	272	272	272	272	272	272	272	272	272	272
ABS	233	234	234	234	234	234	234	234	234	234	234	234
ABB	148	149	148	148	148	147	147	146	146	147	147	148
MWGF	127	127	127	127	127	127	127	127	127	127	127	127
AGR	48	48	48	48	48	48	48	48	48	48	48	48
ADWGF	41	41	41	41	41	41	41	41	41	41	41	41
ETBP	30	28	26	26	27	30	33	36	38	39	35	32
SWBP	16	17	18	18	17	16	14	12	11	10	12	14
ETBS	11	10	10	9	10	11	13	14	15	14	12	11
ETBB	6.3	7.3	7.7	7.8	6.8	7.3	8.0	8.7	8.3	7.3	6.5	5.8
DWBP	5.9	6.4	7.0	7.2	6.4	5.1	4.0	3.4	3.1	3.2	4.2	5.2
DWBS	5.3	5.7	5.9	6.2	6.0	5.0	4.1	3.2	2.5	2.9	4.3	4.8
SWBS	3.5	3.7	3.5	3.7	3.5	2.8	2.2	1.7	1.4	1.8	2.9	3.3
MWDP	3.5	3.7	4.0	3.8	3.1	2.4	2.1	1.9	1.7	2.0	2.7	3.2
SWGF	2.6	2.1	2.0	2.3	2.6	2.8	2.9	2.9	2.8	2.7	2.6	2.7
SWBB	1.1	0.3	0.2	0.3	0.9	1.0	0.8	0.7	0.9	1.4	1.7	2.0
DWBB	0.5	0.1	0.1	0.2	0.9	0.5	0.3	0.2	0.7	1.1	1.2	1.3
SWGR	1.1	1.1	1.1	1.1	1.2	1.2	1.2	1.2	1.1	1.1	1.1	1.1
BioSBP	0.8	0.9	1.1	1.9	2.5	2.0	2.9	2.5	1.9	1.3	1.2	1.0
ETGF	1.0	1.3	1.4	1.2	1.0	1.0	0.9	0.9	1.0	1.0	0.9	0.9
MWGF	0.7	0.7	0.6	0.6	0.7	0.7	0.7	0.7	0.7	0.7	0.7	0.7
ETGR	0.4	0.4	0.4	0.4	0.4	0.4	0.4	0.4	0.5	0.5	0.5	0.4
BioSGF	0.0	0.0	0.1	0.2	0.3	0.2	0.2	0.3	0.3	0.2	0.2	0.1
BioSBS	0.1	0.1	0.1	0.2	0.8	0.5	0.6	0.5	0.4	0.2	0.3	0.1
DWGR	0.1	0.1	0.1	0.1	0.1	0.1	0.1	0.1	0.0	0.0	0.1	0.1
BioSBGR	0.0	0.0	0.1	0.1	0.1	0.1	0.1	0.2	0.2	0.1	0.1	0.1
DWGF	0.04	0.04	0.04	0.05	0.05	0.05	0.05	0.04	0.04	0.04	0.04	0.04
BioSBB	0.01	0.00	0.01	0.01	0.09	0.11	0.11	0.11	0.09	0.09	0.10	0.04

(from Pustelnikov, 1977) for the Baltic Sea. The results discussed here represent a step forward in understanding and predicting SPM in the Baltic Sea and also in other similar systems. Evidently, it would have been preferable to have access to an even larger database, but it is very demanding (in terms of costs, manpower, ships, etc.) to collect such data, especially under storms.

It should also be noted that bioturbation, fish movements (Meijer et al., 1990), currents (Lemmin and Imboden, 1987) and slope processes (Håkanson and Jansson, 1983) might all influence the SPM-concentrations and how SPM varies among and within sites as well as boat traffic, trawling and dredging. These factors have, however, not been accounted for in this modeling, which does not concern sites but entire basins.

2.6 Predicting Chlorophyll-a Concentrations

Values of chlorophyll-a concentrations in the surface-water layer drive the CoastWeb-model, which means that it is very important to model chlorophyll as accurately as possible. This section will first describe the approach we have used to model chlorophyll and then present results describing how well modeled values correspond to measured data.

Typical chlorophyll-a concentrations for the Baltic Sea and parts of the North Sea are shown Fig. 2.19. Values lower than 2 μg/l (oligotrophic conditions) are found in the northern parts of the Bothnian Bay and the outer parts of the North Sea, while values higher than 20 μg/l (hypertrophic conditions) are often found in, e.g., the Vistula and Oder lagoons.

Generally, chlorophyll-a concentrations are predicted from light conditions (or water temperatures) and nutrient concentrations (e.g., Dillon and Rigler, 1974; Smith, 1979; Evans et al., 1996; Håkanson and Bryhn, 2008a). Concentrations of chlorophyll-a represent one of the most important bioindicators related to eutrophication. Håkanson and Bryhn (2008b) discuss several approaches to predict chlorophyll in the surface-water layer:

(1) From a regression based on empirical TN-concentrations and light conditions.
(2) From a regression based on modeled or empirical TP-concentrations, light, salinity and boundary conditions related to surface-water temperatures. This is the approach applied in this work and the following calculation will use dynamically modeled TP-concentrations and salinities, light data from Table 2.3 and empirical water temperature data to predict chlorophyll.
(3) From dynamically modeled TP-fluxes and production values for phytoplankton. This approach gave good results but it will not be used here because we will instead model biouptake of phosphorus in organisms with short turnover times (phytoplankton, bacterioplanklton, herbivorous zooplankton and benthic algae) and in organisms with long turn over times (fish, predatory zooplankton, zoobenthos, macrophytes and jellyfish).

From comprehensive empirical data covering very wide domains in trophic state and salinity for the growing season, it has been demonstrated (see Bryhn et al., 2008) that the salinity influences the Chl/TP-ratio (see Fig. 2.8). The model we use basically predicts chlorophyll from TP and salinity as Chl = $Y_{Sal} \cdot TP$, where Y_{Sal} is a dimensionless moderator quantifying how variations in salinity would generally influence variations in chlorophyll.

To obtain seasonal/monthly variations, the following predictions will use (1) a dimensionless moderator (Y_{DayL} = HDL/12) based on the number of hours with daylight each month (from Table 2.3; see Eq. 2.15), (2) modeled values of the dissolved fraction of phosphorus, since this is the only fraction that can be taken up by phytoplankton and since values of the dissolved fraction of phosphorus in the SW-layer (DF_{SW}; dim. less) are automatically calculated by the CoastMab-model for phosphorus and are thus available for predicting chlorophyll, and (3) use a boundary condition related to low water temperatures given by (example for the Bothnian Bay):

$$\text{If } SWT_{BB} > 4°C \text{ then } Y_{SWTBB} = 1 \text{ else } Y_{SWTBB} = (SWT_{BB} + 0.1)/4 \qquad (2.61)$$

This water temperature moderator will not influence modeled chlorophyll values when the surface-water temperature is higher than 4°C, but it will lower predicted chlorophyll values during the winter time, and since there is also primary production under ice, the constant 0.1 is added. This moderator has been used and motivated before (see Håkanson, and Eklund, 2007).

This means that using this approach, Chl (μg/l) is predicted from:

$$Chl_{Mod} = TP_{SW} \cdot DF_{SW} \cdot Y_{DayL} \cdot Y_{Sal} \cdot Y_{SWT} \qquad (2.62)$$

TP_{SW} = TP-concentration in SW-water in μg/l.
Y_{Sal} = Dimensionless moderator salinity influences on chlorophyll calculated from:

Y1 = if Sal < 2.5 psu then (0.20 – 0.1·(Sal/2.5 – 1)) else (0.20+0.02·(Sal/2.5–1))
Y2 = if Sal< 12.5 then Y1 else (0.28 – 0.1·(Sal/12.5 – 1))
Y3 = if Sal > 40 then (0.06 – 0.1·(Sal/40 – 1)) else Y2
Y4 = if Y3 < 0.012 then 0.012 elseY3.

Figure 2.26 compares the modeled values with the confidence bands related to ± one standard deviation of the mean monthly empirical chlorophyll values. Most modeled values for most months are within or close to these uncertainty bands. It should be stressed that the empirical chlorophyll values from the Bothnian Sea and the Bothnian Bay are very uncertain and mainly based on data provided from a few sampling stations given by Hajdu et al. (2007). Since there are no reliable monthly data from BS and BB available to us, the uncertainty bands in Fig. 2.26 are based on the assumption that the uncertainty is 0.8 μg/l, which is a typical value

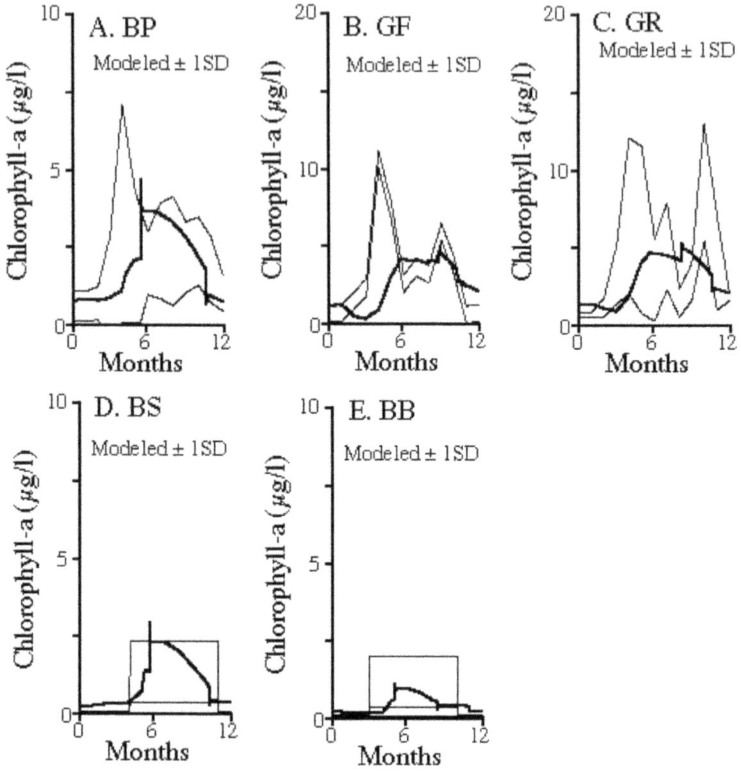

Fig. 2.26 Modeled monthly data on chlorophyll-a concentrations (*thick lines*) against ± 1 standard deviation (*thin lines*) of the mean empirical monthly data for the five sub-basins in the Baltic Sea

for the Baltic Proper. From Fig. 2.26, it is also evident that for individual months the uncertainty in the mean monthly chlorophyll value can be significantly larger than 0.8 μg/l.

From the results in Fig. 2.26, one can conclude that the modeled concentrations of chlorophyll-a are in good general agreement with the empirical data.

2.7 Predicting Water Clarity and Secchi Depth

The Secchi depth is an important variable, not just for the CoastWeb-model, but for aquatic systems, since the water clarity defines the depth of the photic zone, and in all following calculations we will use two Secchi depths to define the entire depth of the photic zone (see Håkanson and Peters, 1995). Figure 2.27 shows that the Secchi depth is generally a function of both allochthonous input, as given by data on water color on the y-axis, and autochthonous production, as given by data on TP-concentrations on the x-axis. These basic principles should apply generally

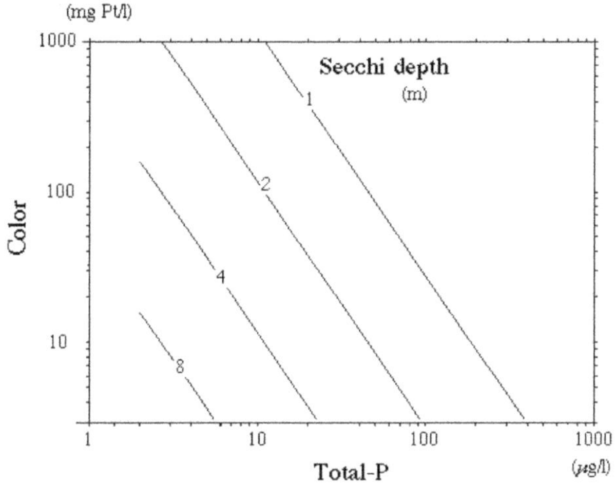

Fig. 2.27 The relationship between lake color, TP-concentrations and Secchi depth (from Håkanson and Boulion, 2002a)

and it means that the Secchi depth in the Baltic Sea basins should also be related to color values, or to data on freshwater inflow (Q), since colored substances are transported to the system from the tributaries, or by the salinities, since these reflect the freshwater inflow in relation to the saltwater inflow to any given basin. We do not have access to any reliable data on water color, but we have access to very reliable data on salinities. So, in predicting Secchi depth, we have used the modeled salinities from the CoastMab-model.

There exists a close relationship between SPM, Secchi depth and salinity (see Håkanson, 2006) – the higher the salinity, the higher the aggregation of suspended particles, the larger the particles and the higher the water clarity. An SPM-concentration of 10 mg/l would imply turbid conditions in a freshwater system, but relatively clearer water in a saline system. This is shown in Fig. 2.28 (the curves in the figure are calculated from Eq. 2.63).

The following approach is used to calculate how the salinity influences the Secchi depth (from Håkanson, 2006). In this approach, it is assumed that the Secchi depth can never be higher than 200 m because even distilled water will scatter light and set a limit to the Secchi depth. The relationship between Secchi depth (Sec in m), SPM_{SW} (mg/l) and salinity (psu) is first given by:

$$Sec_{Ref} = 10 \wedge (-((10 \wedge (0.15 \cdot \log(1 + Sal_{SW}) + 0.3) - 1)) + 0.5) \cdot (\log(SPM_{SW}) + 0.3)/2 + (10 \wedge (0.15 \cdot \log(1 + Sal_{SW}) + 0.3) - 1))) \quad (2.63)$$

The SPM-concentrations in the SW-layer (SPM_{SW} in mg/l) are predicted from the dynamic SPM-model. It should be noted that these value are also influenced by the "clay factor", Y_{LU}. Values on the salinity in the SW-layer, Sal_{SW}, are dynamically modeled using CoastMab for salt.

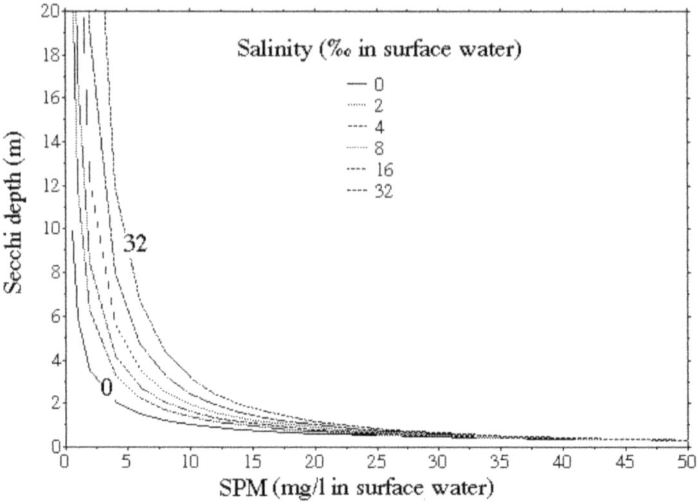

Fig. 2.28 Illustration of the relationship between Secchi depth, SPM in surface water and salinity in surface water using log-data (from Håkanson, 2006)

Next, we add the impact of "color" (or more generally allochthonous influences) as a combination of two indicators of this since we do not have access to any dynamic model to predict water color. One indication of color or allochthonous influences is the fraction of freshwater compared to water from other sources and the other indicator is simply the salinity of the system compared to the salinity in the Kattegat (Sal_{Ka}, which is set to 17.6 psu, as a default value, as explained in Sect. 2.3). This gives, for example for Secchi depth in the Bothnian Bay (BB):

$$Sec_{BB} = Y_{SalSecBB} \cdot Sec_{Ref} \tag{2.64}$$

Sec_{Ref} is given by Eq. (2.63) and Y_{SalSec}, is defined as:

$$Y_{SalSec} = ((Sal_{SWBB}/Sal_{KA}) + ((Q_{SWBSBB})/(Q_{tribBB}+Q_{SWBSBB}+Q_{DWBSBB})))/2 \tag{2.65}$$

The results in Fig. 2.29 compare modeled values with measured data for all five basins. For the Baltic Proper, the predicted Secchi depths are close to the empirical values and to the uncertainty bands given by ± 1 standard deviation for the empirical data; close to the uncertainty bands for the empirical mean value in the Gulf of Finland; within the uncertainty band for the empirical data for the Gulf of Riga; and close to measured values also in the Bothnian Sea and the Bothnian Bay (although the empirical Secchi depths are fairly uncertain in BS and BB).

These results give further empirical support to the general validity and predictive power of the CoastMab-modeling.

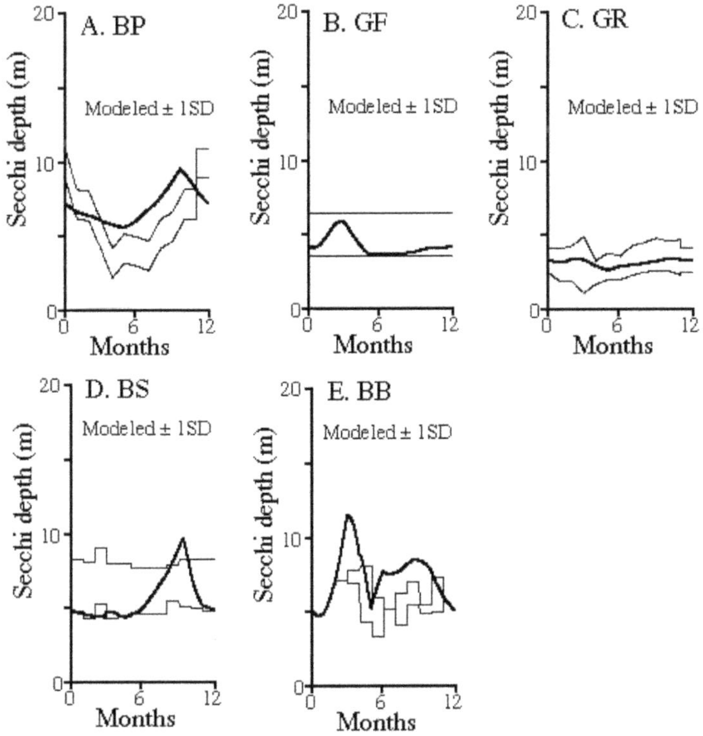

Fig. 2.29 Modeled monthly data on Secchi depth (*thick lines*) against ± 1 standard deviation (*thin lines*) of the mean empirical monthly data for the five sub-basins in the Baltic Sea

2.8 Trends in Nutrient, Chlorophyll and Oxygen Concentrations

2.8.1 Background and Aim

To present a comprehensive picture of the trophic state of the Baltic Sea system, it is essential to take into account the historical perspective of the eutrophication and the fish production potential of this large estuary. One hundred years ago, the nutrient losses from human activity ware substantial in the catchment. Natural fertilizers were used in agriculture, and horses were intensively used for transportation in urban and rural areas. Sewage systems were constructed to prevent outbreaks of cholera and other diseases in the cities but sewage treatment was absent or very ineffective in many areas until after the Second World War (Savchuk et al., 2008). While human activity contributed to large nutrient inputs already one century ago, the effects from this activity did not become apparent in the open sea until the late 1960s when cyanobacterial blooms became a recurring nuisance (Finni et al., 2001). Historical total phosphorus loading estimates compared to present levels have been continuously revised downwards. An early estimate from Larsson et al. (1985)

indicated that the TP-loading had increased by a factor of 8 since before the twentieth century. The development of dynamic nutrient models allowed Schernewski and Neumann (2005) to conclude that the loading had increased over the past century by a factor of 4.5, while Savchuk et al. (2008) calculated a three-fold increase. Håkanson and Bryhn (2008b) presented the first dynamic mass-balance model for phosphorus with a unitary set of calibration constants which also delivered good predictions for all of the five major basins of the Baltic Sea, and estimated that the TP-loading had increased with about 50% during the last 100 years.

In this section, we will present time-series of HELCOM data (ICES, 2009) on TP-, TN- and O_2-concentrations in the Baltic Proper as well as on the PO_4/TP ratio to determine which types of changes have occurred. We will also present subsets of some of these data, from the Eastern Gotland Deep (limited by coordinates 56°10′N, 18°29′E and 57°46′N, 20°59′E), which has been intensively studied by many researchers (e.g., Matthiesen et al., 1998; Kotilainen et al., 2002; Gustafsson and Stigebrandt, 2007; Eilola et al., 2009). The idea is to see how representative this area is for conditions in the whole Baltic Proper. Nutrient data from surface waters will, in addition, be compared with chlorophyll data from Håkanson and Bryhn (2008b), which have been updated with 2007 data from ICES (2009).

2.8.2 Trends in Chlorophyll and Phosphorus Concentrations and the Dissolved Fraction for Phosphorus

Chlorophyll data from SWBP are displayed in Fig. 2.30. The trend is increasing also after the adjustment with data from 2007 ($r^2 = 0.0004$, $p = 0.009$, $n = 14{,}697$). Medians have fluctuated quite strongly, and high values have generally been recorded in the mid-1980s and the early 1990s. It should be noted that data are particularly scarce from years 1974, 1975, 2005 and 2007.

Figure 2.31 shows TP-concentrations in the surface- and middle-water layers in the Baltic Proper (TP_{SWBP} and TP_{MWBP}) in the period 1968–2008. The TP-trend has been increasing in SWBP 1968–2008 ($r^2 = 0.03$, $p < 0.001$, $n = 48{,}978$) despite substantial efforts to decrease the TP-loading. The highest TP-concentrations have generally been measured during recent years (2004–2007) and for the period 1983–1992. Particularly high concentrations were recorded in 2005, when there was also a very accentuated cyanobacterial bloom (Håkanson et al., 2009). Changes in TP-concentrations in MWBP (Fig. 2.31b) have been fairly similar to those in SWBP. The TP-concentrations have also been significantly increasing in this layer ($r^2 = 0.02$, $p < 0.001$, $n = 14{,}830$) and the highest concentrations have been measured during recent years (1998–2007) and during 1983–1991.

The TP-trend in DWBP (Fig. 2.32a) has also been increasing during 1968–2008 ($r^2 = 0.02$, $p < 0.001$, $n = 22{,}693$), and there is no sign of long-term decreases here, just like in MWBP (Fig. 2.31b) and SWBP (Fig. 2.30a). Furthermore, the slight temporary TP-decrease in the 1990s displayed in Fig. 2.32a actually started *before* the major salt-water inflow to the Kattegat in 1993 (Fig. 1.9), and during a long

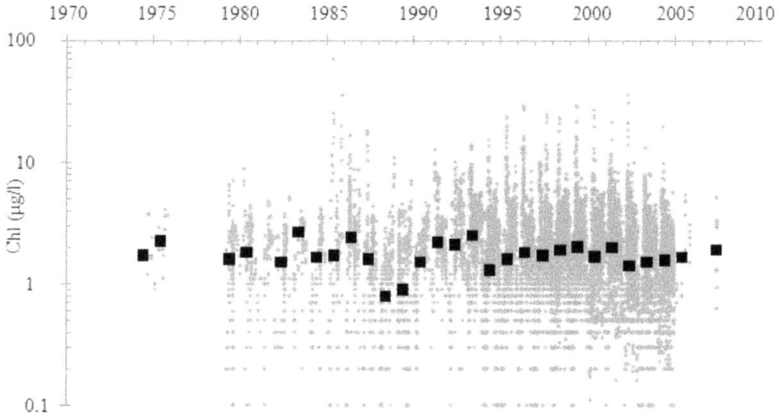

Fig. 2.30 Chlorophyll (Chl) concentrations in 1974–2007 in the surface-water layer (SW) of the Baltic Proper (BP). n = 14,697. *Black squares* denote yearly medians

stagnation period (also found by Gustafsson and Stigebrandt, 2007). The highest concentrations have generally been measured during 1999–2007. However, the TP-trend in the deep-water layer of the Eastern Gotland Deep (Fig. 2.32b) shows a quite different pattern than TP in the deep-water layer in the entire Baltic Proper. The trend 1968–2008 is significantly *negative* ($r^2 = 0.01$, $p < 0.001$, $n = 5{,}496$) and year-to-year variations are also much more pronounced at this restricted area than in the entire Baltic Proper. Thus, the changes in TP-concentrations in the Eastern Gotland Deep must be interpreted with particular care when placed in relation to any large-scale changes in the entire Baltic Proper.

Figure 2.33 shows yearly changes in median TP-concentrations ($\Delta TP/\Delta t$) below the halocline of the Eastern Gotland Deep compared to DWBP. TP data from the Eastern Gotland Deep constituted a subset of the DWBP data, and $\Delta TP/\Delta t$ data from these two areas were significantly correlated ($r^2 = 0.39$). Nevertheless, the regression slope in Fig. 2.23 (0.23) indicates that $\Delta TP/\Delta t$ data from the Eastern Gotland Deep generally exaggerated TP-changes below the halocline in the entire Baltic Proper by a factor of four. It is also worth noting that TP increased in DWBP during some years when it decreased in the sub-halocline Eastern Gotland Deep, and vice versa. When the regression in Fig. 2.33 regarding changes in phosphorus concentrations was instead run for changes in PO_4-concentrations, the slope remained at 0.23 although the r^2 value decreased to 0.38.

Phosphate rather poorly represents the full pool of dissolved phosphorus in water as this is modeled by the CoastMab-model, where the dissolved fraction (DF) should be determined from filtration and the fraction remaining on the filter is defined as the particulate fraction (PF); and hence the dissolved fraction is DF = 1-PF. There are also other fractions of phosphorus included in the dissolved fraction, e.g., DOP (dissolved organic phosphorus). Still, given the fact that there are more than 22,000 empirical data available, it may be informative to study the PO_4/TP-ratio in the

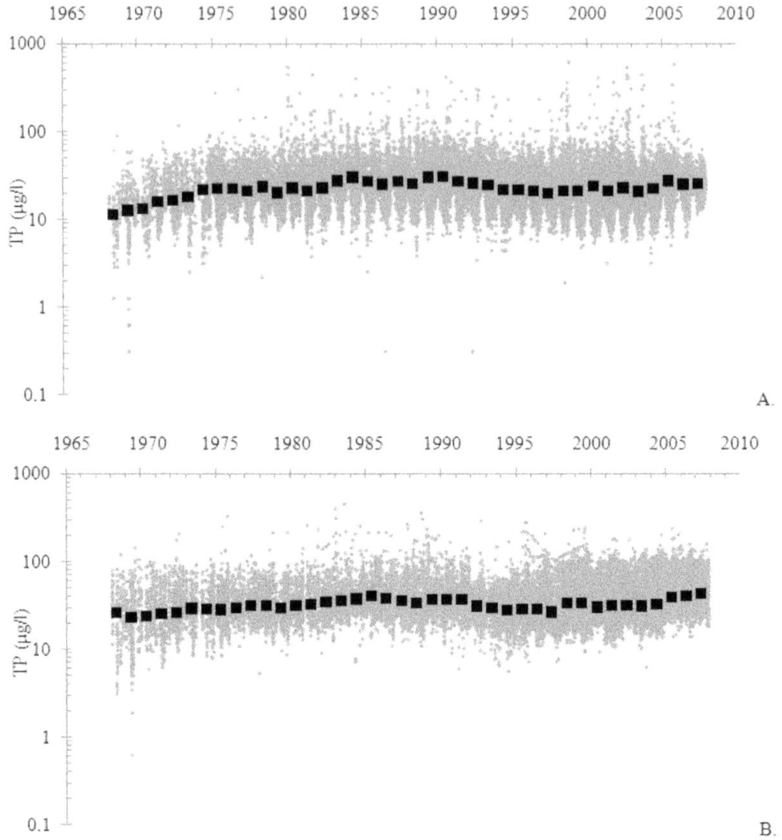

Fig. 2.31 Total phosphorus (TP) concentrations in 1968–2008 in (**a**) the surface-water layer in the Baltic Proper. (TP_{SWBP} in µg/l; n = 104,401) and (**b**) in the middle-water layer in the Baltic Proper (TP_{MWBP} in µg/l; (n = 19,402). *Black squares* denote yearly medians

deep-water layer of the Baltic Proper to determine how this ratio has varied during 1968–2008. This could give a hint about how much phosphorus is available for sedimentation (i.e., about the particulate fraction, PF), and the PO_4/TP-ratio may also provide a baseline below which it is unrealistic to find dissolved fraction (DF) values. Furthermore, increasing PO_4/TP-ratios may indicate that diffusion has increased, while decreasing PO_4/TP may imply that major tributary inflows to the Baltic Sea or sedimentation from SWBP and MWBP has had a great impact on the conditions in the deep-water layer.

Figure 2.34 shows that the annual medians of PO_4/TP in the Baltic Proper 1968–2008 have fluctuated close to 1, i.e., that TP has typically consisted of more than 90% phosphate. Note, however, that in 5.6% of the data (1,316 data) the PO_4/TP-ratio was equal to 1, which would mean that all of the phosphorus consisted of phosphate and that there was no sedimentation of particulate phosphorus. In some additional 782 (3.3%) of the data points, PO_4/TP was greater than one. This

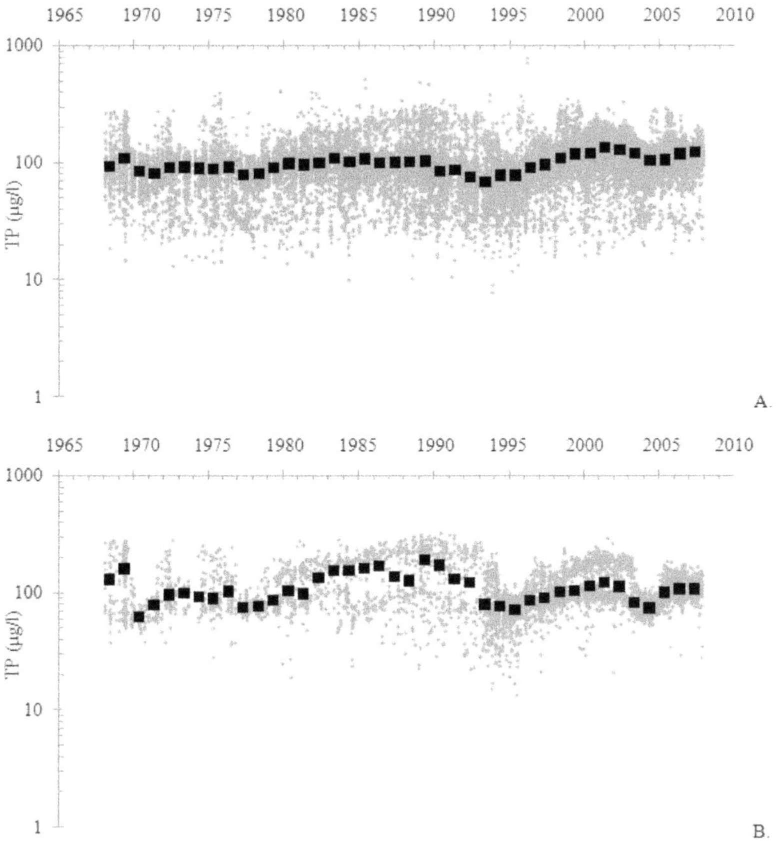

Fig. 2.32 Total phosphorus (TP) concentrations in 1968–2008 in (**a**) the deep-water layer in the Baltic Proper. (TP_{DWBP} in µg/l; n = 23,379) and (**b**) in the deep-water layer in the East Gotland Deep (in µg/l; (n = 5,496). *Black squares* denote yearly medians

would mean that PO_4-concentrations were higher than TP-concentrations, which is clearly impossible and demonstrates that the inherent uncertainties in the phosphate determinations are large. Håkanson and Bryhn (2008b) also showed that a typical coefficient of variation (CV = SD/MV; SD = standard deviation; MV = mean value) for DIP, which is mainly phosphate in the Baltic Sea is 0.54, which means that 113 data would be needed to determine the mean value with an error smaller than 10% of the mean (with a 95% certainty).

The PO_4/TP-trend during 1968–2008 was weakly but significantly positive ($r^2 = 0.005$, $p < 0.001$, $n = 22{,}693$), indicating that the mean PO_4/TP increased marginally from 1968 to 2008. The changes are, however, very small and uncertain when the period 1998–2008 (with little oxygen-rich salt-water inflow from Kattegat) is compared to 1970–1978 (with higher oxygen inputs from frequent major inflows from Kattegat). Note that the higher inflows of oxygenated salt-water from Kattegat have not caused much lower PO_4/TP-ratios in the deep-water layer of the Baltic

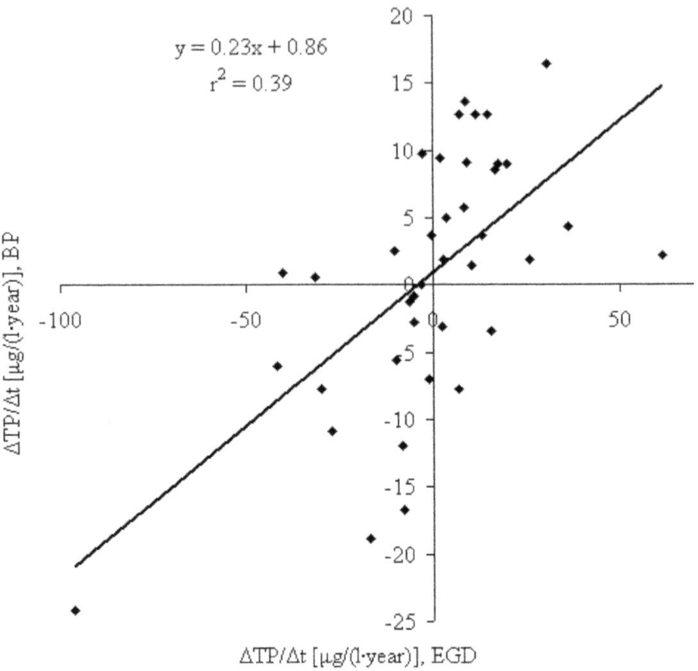

Fig. 2.33 Annual changes in sub-halocline total phosphorus concentrations (ΔTP/Δt) in the Eastern Gotland Deep (EGD) compared to in the Baltic Proper (BP), 1969–2007

Proper, and hence not higher sedimentation and not lower phosphorus diffusion from sediments.

Figure 2.34b describes the conditions for the PO_4/TP-ratio in the deep-water layer in the Eastern Gotland Deep, which is often used as a reference site meant to be representative for the entire Baltic Proper. The median PO_4/TP during 1968–2008 was 0.93, and also many of the uncertainty bars in Fig. 2.34b stretch above the unrealistic value 1. The time trend in all PO_4/TP-data behind Fig. 2.34b is not positive, as in Fig. 2.34a for the entire Baltic Proper, but significantly *negative* ($r^2 = 0.002$, $p < 0.001$, $n = 5{,}496$), indicating that mean PO_4/TP-values have decreased very slightly from 0.89 in 1968 to 0.87 in 2008 despite deteriorating oxygen conditions.

2.8.3 Trends in Oxygen and Nitrogen Concentrations

O_2-concentrations in the deep-water layer in Baltic Proper and at the Eastern Gotland Deep 1960–2008 are displayed in Fig. 2.35. The overall trend is significantly negative in the DWBP while it is significantly positive in the Eastern Gotland Deep, regardless of whether non-transformed or log(x+0.01)-transformed O_2-data are analyzed. Year-to-year variations are also much more pronounced at the Eastern

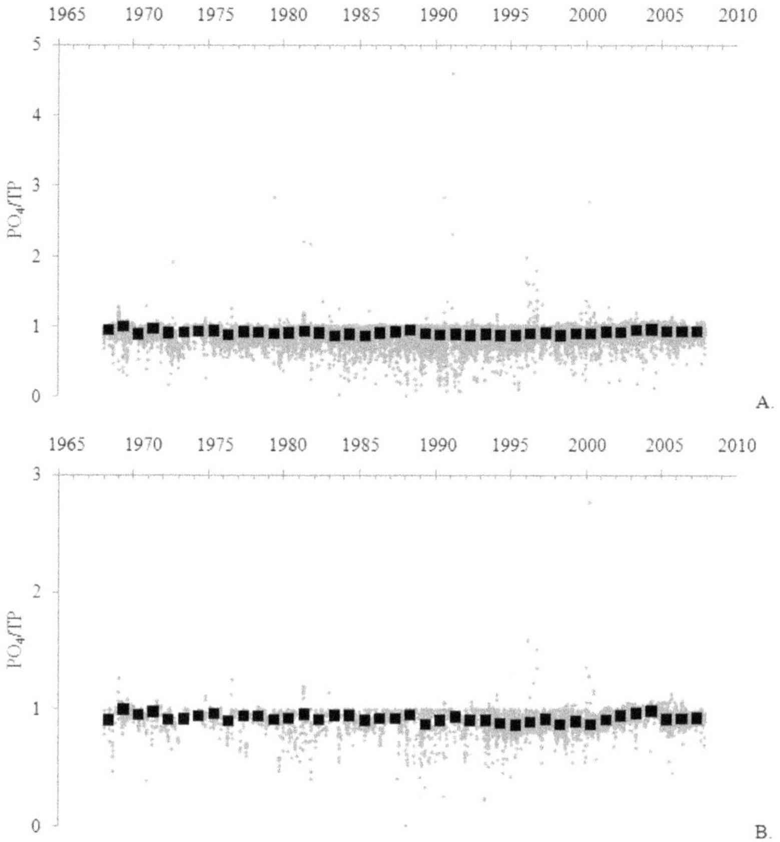

Fig. 2.34 Phosphate (PO_4) to total phosphorus (TP) ratios in the deep-water layers in 1968–2008 in (**a**) the Baltic Proper (n = 23,379) and (**b**) in the East Gotland Deep (n = 5,496). *Black squares* denote yearly medians

Gotland Deep than in the entire Baltic Proper. Thus, neither TP-concentrations nor O_2-concentrations in the Eastern Gotland Deep may be seen as representative for the prevailing conditions in the deep-water layer of the Baltic Proper. Furthermore, there is a short-term increase in O_2-concentrations in the early 1990s in both Figs. 2.34a and 2.35b, which start before the major inflow from Kattegat in 1993 and towards the end a long stagnation period, similar to the temporal trend shift in for TP-concentrations in Fig. 2.31. Since the 1960s, the yearly median O_2-concentrations in DWBP have been above the critical level of 2 ml/l only during a short period in the 1990s, before and after the major inflow in 1993 (Fig. 2.35a). Thus, conditions for zoobenthos in the sediments in DWBP have been very poor. Data from the Eastern Gotland Deep apparently provide an non-representative picture in this respect as well, indicating that the O_2-concentrations in deep-water layer have been above the critical level during more extended periods since 1960 than in DWBP.

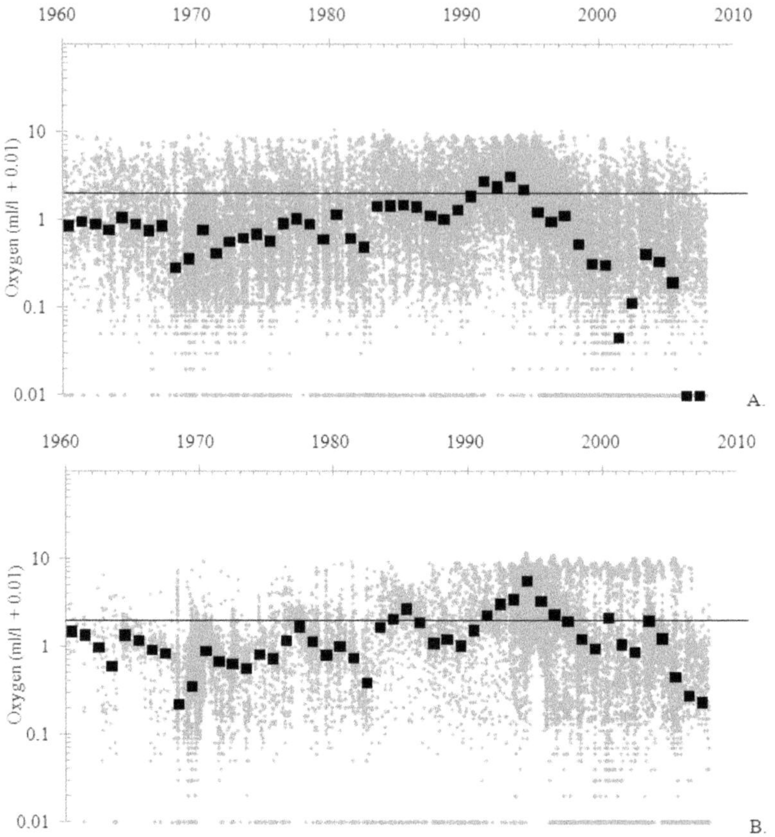

Fig. 2.35 Oxygen concentrations in deep-water layer 1968–2008 in (**a**) The Baltic Proper (n = 54,995) and in (**b**) the East Gotland Deep (n = 24,023). *Black squares* denote yearly medians. The *solid horizontal* line marks 2 ml/l, the critical concentration for zoobenthos in water and sediments

Annual changes in sub-halocline O_2-concentrations of the Eastern Gotland Deep and the entire Baltic Proper are compared in Fig. 2.36. As with changes in TP- and PO_4-concentrations, changes in O_2-concentrations were mutually correlated ($r^2 = 0.24$) in the two water bodies (Fig. 2.36), which is hardly surprising since the Eastern Gotland Deep is a part of the BP. Nevertheless, some years showed positive changes in DWBP while concentrations decreased in the sub-halocline Eastern Gotland Deep, and vice versa. The regression slope (0.24) suggests that changes in O_2-concentrations in deep waters of the Eastern Gotland were normally a factor of four higher in relation to changes in DWBP.

The regression slopes in Figs. 2.33 and 2.36 show that year-to-year variations in TP- and O_2-concentrations of DWBP would be exaggerated by a factor of four if data from the sub-halocline Eastern Gotland Deep would be used to represent conditions in DWBP. During some years, changes in TP- and/or O_2-concentrations of DWBP occurred in the opposite direction compared to changes in the Eastern

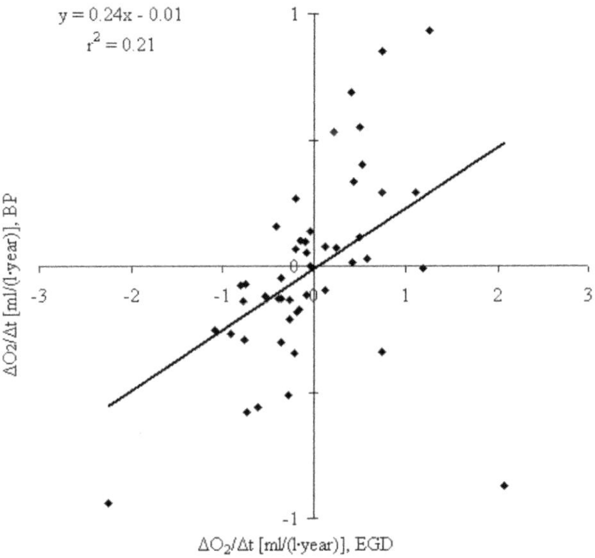

Fig. 2.36 Annual changes in sub-halocline oxygen concentrations ($\Delta O_2/\Delta t$) in the Eastern Gotland Deep (EGD) compared to in the Baltic Proper (BP), 1969–2007

Gotland Deep. This was also the case for PO_4-data. It is therefore rather pointless to try to explain large-scale ecosystem changes in the Baltic Sea with changes in TP-, PO_4- and O_2 concentrations in the sub-halocline Eastern Gotland Deep, such as Gustafsson and Stigebrandt (2007) have done.

Figure 2.37 shows TN-concentrations 1970–2008 in SWBP and it should be noted that data from the first 3 years in the series are relatively scarce. TN has *increased* since 1970 ($r^2 = 0.03$, $p < 0.001$, $n = 45{,}083$). A very small but statistically significant long-term decrease can, however, be established for a more recent period, 1995–2008 ($r^2 = 0.0007$, $p < 0.001$, $n = 27{,}610$). The TN-concentrations

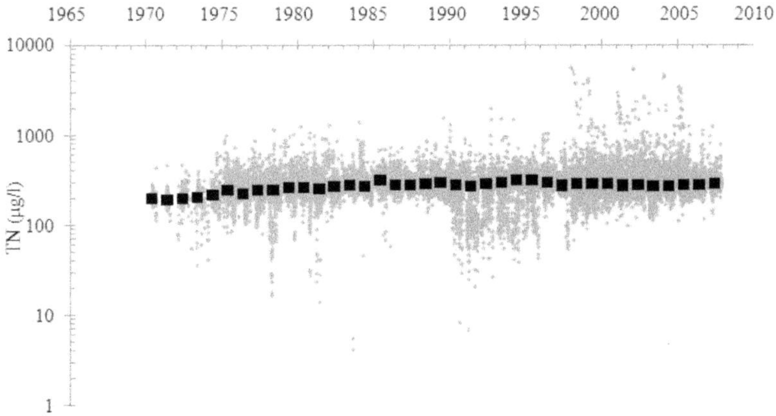

Fig. 2.37 Total nitrogen (TN) concentrations in the surface layer of the Baltic Proper 1970–2008 ($n = 55{,}655$). *Black squares* denote yearly medians

were generally high in 1985 and 1988–1996. It should be stressed that many measurements, which poorly reflect average conditions, have also been recorded after 1998.

2.8.4 *Trends in Total Phosphorus and Freshwater Inflow from the Catchment*

Accurate data on TP-loading is the most crucial determinant for successful eutrophication modeling of aquatic ecosystems (Håkanson, 1999). The freshwater inflow (Q_{Trib}) is another important modeling component, especially in marine environments where ambient salinities, crucial for sedimentation and other processes are a function of the inflowing freshwater and the inflowing seawater (as discussed in Sect. 2.3). Table 2.18 presents the freshwater inflow (Q_{Trib}) and TP-loadings to the five basins of the Baltic Sea investigated in this book. Data on TP-inflow from 1970 to 1993 emanate from Stålnacke et al. (1999) and concerns the loadings from rivers. Point sources along the coastline were reported to contribute with 12.5 kt/year during this period to an area, which included Kattegat and the Belt Sea (Stålnacke et al., 1999. These two areas received about 28% of the TP-loading to the Baltic Sea plus Kattegat and the Belt Sea in the early 1980s (HELCOM, 1987) and 16% of the loading in 1990 (HELCOM, 1993), so in Table 2.18, we have estimated that TP from coastal point sources to the Baltic Sea was $12.5 \cdot (1 - [0.28 + 0.16] / 2) = 9.8$ kt/year on average. The mean annual freshwater inflow in Stålnacke et al. (1999; our estimate from graphs in this source is 449 km^3) corresponds very well with the long time series given in HELCOM (2008a; the mean annual inflow is 452 km^3). TP-loading data from 1994 to 2005 in Table 2.18 have been taken from HELCOM (2007c), and this source provides the total loading to the Baltic Sea plus the Belt Sea and Kattegat. In 1995, the latter two areas received 10.4% of the TP-loading to the area covered by HELCOM (2007c) according to HELCOM (1998) while this figure was 8.4% for the year 2000 (HELCOM, 2003). Thus, Table 2.18 contains TP-loadings from HELCOM (2007c) minus $(10.4\% + 8.4\%)/2 = 9.4\%$.

One can note from Table 2.18 that the TP-loading was very high in the 1980s, at a mean value of 48 kt/year, lower in the 1970s (40 kt/year), lower yet in the 1990s (36 kt/year) and that the lowest loadings were recorded during the first 6 years of the present decade (mean value: 28 kt/year). The freshwater inflow has also varied considerably, and was about 420 km^3/year in the 1970s, 480 km^3/year in the 1980s, 460 km^3/year in the 1990s and 430 km^3/year during 2000–2007. Data from Table 2.18 have been used to calculate the mean, flow-weighted TP-concentration in the inflow from the catchment to the Baltic Sea 1970–2005 and the result is displayed in Fig. 2.38. It is also clear from this graph that the most accentuated TP-pollution to the Baltic Sea occurred in the 1980s. A trend analysis showed that the mean TP-concentration in the inflow has decreased significantly during 1970–2005 ($r^2 = 0.64$, $p < 0.001$). The 1970–2005 trend was also significant for the TP-load ($r^2 = 0.32$, $p < 0.001$) while the 1970–2007 trend in freshwater inflow was insignificant ($p = 0.32$).

Table 2.18 Total phosphorus (TP) loading and freshwater inflow (Q_{Trib}) to the Baltic Sea

Year	TP-loading (kt)	Q_{Trib} (km^3)
1970	43	421
1971	38	398
1972	36	379
1973	34	365
1974	45	467
1975	41	453
1976	34	340
1977	45	461
1978	43	441
1979	45	455
1980	50	448
1981	53	545
1982	45	477
1983	42	463
1984	42	464
1985	48	483
1986	50	463
1987	50	517
1988	54	498
1989	49	478
1990	46	469
1991	43	443
1992	40	464
1993	40	456
1994	33	447
1995	36	482
1996	26	367
1997	31	421
1998	36	553
1999	33	470
2000	33	515
2001	30	490
2002	30	428
2003	21	339
2004	26	431
2005	26	438
2006	N.A.	358
2007	N.A.	432

N.A., Not available.
TP-loading data for 1970–1993 from Stålnacke et al. (1999), TP-loading data for 1994–2005 from HELCOM (2007c), Q_{Trib}-data from HELCOM (2008).

The percentage of the TP-loading to the Baltic Sea going to each of the sub-basins is given in Table 2.19. The largest part of the loading went directly into the Baltic Proper (BP). The second most heavily loaded basin was the Gulf of Finland (GF), followed by the Bothnian Bay (BB) and the Bothnian Sea (BS) while the Gulf of Riga received the smallest part of the total inputs among the sub-basins.

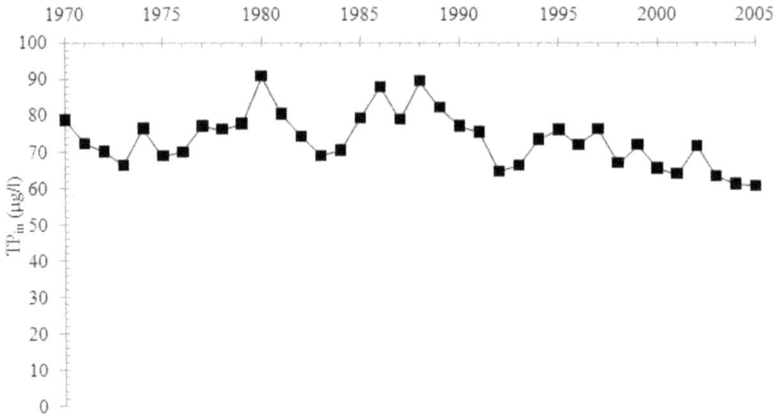

Fig. 2.38 Mean flow-weighted TP-concentrations in the inflow (TP_{in}) from the catchment to the Baltic Sea, 1970–2005 (data from Table 2.18)

There was no evident time-trend in any of the distributions in Table 2.19, and it is possible that a substantial part of the variation in this table is due to measurement uncertainties regarding TP-concentrations and water fluxes.

Table 2.19 The distribution (in %) among sub-basins of waterborne phosphorus inputs to the Baltic Sea

Sub-basin	Early 1980s[a]	1990[b]	Mean value	1995[c]	2000[d]	Mean value
BB	11.1	6.1	8.6	8.5	10.9	9.7
BS	8.5	5.9	7.2	6.7	9.6	8.1
BP	62.9	48.5	55.7	54.2	60.7	57.4
GF	14.3	30.7	22.5	24.2	14.2	19.2
GR	3.2	8.8	6.0	6.4	4.6	5.5

Data sources: HELCOM ([a]1987, [b]1993, [c]1998, [d]2005).

2.8.5 Comments

The findings in Sect. 2.8 raise the following important points:

1. Chlorophyll concentrations in SWBP have decreased slightly during the last three decades and TP-concentrations have also decreased during the same period.
2. TP-concentrations in SWBP, MWBP and in DWBP have, however, been increasing quite continuously 1968–2008 despite intensive efforts to decrease the TP-loading to the Baltic Sea (see Sect. 2.8.5). Further on in this book, some possible reasons for this will be quantitatively explored: (a) whether TP-diffusion from hypoxic sediments below the halocline has offset an expected decrease in

TP-concentrations, (b) whether diminished saltwater inflow from the Kattegat has had an impact on TP-concentrations and (c) whether increased temperatures have decreased the ice-cover and increased the erosion and resuspension of P-rich materials from sediments above the theoretical wave base.

3. The Eastern Gotland Deep does not provide representative data for the Baltic Proper as a whole since the significant time trends in TP, O_2 and PO_4/TP in the deep-water layers often show different signs for the two different volumes. Year-to-year variations are much stronger in the Eastern Gotland Deep than in the Baltic Proper, so it is rather pointless to correlate temporal slopes in TP and O_2 in the Eastern Gotland Deep with large-scale ecosystem changes, such as Gustafsson and Stigebrandt (2007) have done. This was evident even though as much as 24% of the TP and PO_4 data and 44% of the O_2 data in DWBP were measured in the Eastern Gotland Deep.
4. Drastic changes in the PO_4/TP-ratio have neither occurred in the Eastern Gotland Deep nor in the rest of the DWBP. Phosphate diffusion may thus have been quite stable during the last four decades.
5. TP-concentrations in DWBP in the early 1990s unexpectedly preceded the major salt-water inflow from Kattegat in 1993, which occurred after a long stagnation period. Thus, the major salt-water inflows should be one out of many factors, which may alter environmental conditions in the Baltic Proper and connected basins.
6. Phosphate has constituted about 91% of the total phosphorus pool in DWBP 1968–2008. This means that less than 9% of TP in these waters could be found in settling particles, since there are other forms of dissolved phosphorus, e.g., DOP.

Claims have frequently been made in Baltic Sea studies that the conditions at individual stations and/or smaller areas are representative for the conditions in the entire Baltic Proper, or even worse, the Baltic Sea. From the maps given in Fig. 2.19 for the areal distributions of chlorophyll-a concentrations in the upper 10 m in the Baltic Sea, or in Fig. 2.20 showing a similar map for the Secchi depths, or from similar maps for TP- and TN-concentrations shown by Håkanson and Bryhn (2008b), it is evident that there are large areal variations among individual sites and smaller coastal areas in these nutrient concentrations and bioindicators. From the figures showing vertical variations, e.g., Fig. 2.2 and many similar figures given in Håkanson and Bryhn (2008b), it is clear that there are also major vertical variations in all these variables among sites in the Baltic Proper or even higher variations among sites in the Baltic Sea. In this section, we have shown that there are also major differences in the temporal trends if we compare the data from the entire Baltic Proper with data from smaller areas or individual sites in the Baltic Proper, which are often claimed to represent the conditions in the Baltic Proper. For the future, it would be good scientific practice to give adequate information in these contexts so that claims for changes in the Baltic Proper are based on data from as many sites as possible covering the entire area as evenly as possible and results related to sites and smaller areas would be referred to in an adequate manner. It seems that this could have prevented many alarmistic and erroneous results, e.g., related to the "flip", i.e., a sudden change in TP-concentrations in the Baltic Proper claimed in Swedish media and in several reports (see, e.g., Swedish Environmental Advisory Council, 2005; Wulff, 2006) or the "vicious circle" theory supported by many Baltic Sea researchers and administrators (Vahtera et al., 2007), which has been shown (see Håkanson and Bryhn, 2008b) to be largely incorrect. Such alarmistic reports generally find a fertile ground in the mass-media and there must be literally thousands of such articles written during the last decades in the Baltic States on the "extremely severe eutrophication situation in the Baltic Sea". They are all incorrect, which can be seen from the thousands of empirical data presented in this section. During the last three decades there is, for example, a relatively stable level of TN-concentrations in the SW-layer in the Baltic Proper. The chlorophyll-a concentrations in the surface-water of the Baltic Proper show a decreasing trend (see also Håkanson and Bryhn, 2008b). As a matter of fact, the eutrophication in the Baltic Proper has slightly improved for 30 years, but given the massive amount of money spent to improve the conditions by nitrogen reductions, one can raise a question about the efficiency of those measures. So, how should a better remedial strategy look? We will discuss that in Chap. 5 and also discuss possible reasons why TP-concentrations have not decreased thus far.

2.8.6 Trends in Ice Cover, Water Temperatures and Comments Related to Sediment Resuspension

2.8.6.1 Background

During winter, the Baltic Sea is partially and some basins (such as the Bothnian Bay) completely covered by ice, which puts a particular stress on the foodweb since a considerable part of the key functional organisms thrives in surface waters. Furthermore, the ice affects the nutrient supply to the water column since ice reduces wind/wave-induced erosion and resuspension of sediments. This is illustrated in Fig. 2.39 for the conditions in the Bothnian Sea. Figure 2.39a shows conditions without ice and the geographical distribution of ET- and A-sediments, i.e., of the areas above and below the theoretical wave base. The theoretical wave base and the relative distribution of E-, T- and A-areas are determined by the effective fetch, which is the mean length of the free water surface over which the wind blows. Thus, when the effective fetch is long, large bottom areas can have their sediments eroded and resuspended by waves (Håkanson and Jansson, 1983). Figure 2.39b illustrates ice conditions during a relatively mild winter. The Bothnian Sea first freezes along the coastline, i.e., above the shallowest areas, and the ice then gradually extends towards the deeper, central parts of the system. Ice-growth is usually, but not always, more accentuated in the western part than in the eastern part. The least likely part to freeze is the central and southern part of this sub-basin (SMHI and FIMR, 1982; Seinä and Palusuo, 1996). In the central part of the basin, there is then only a small relative extent of ET-areas, which is illustrated in Fig. 2.39. So, a partial ice cover implies that open water areas – and thereby the effective fetch – have decreased compared

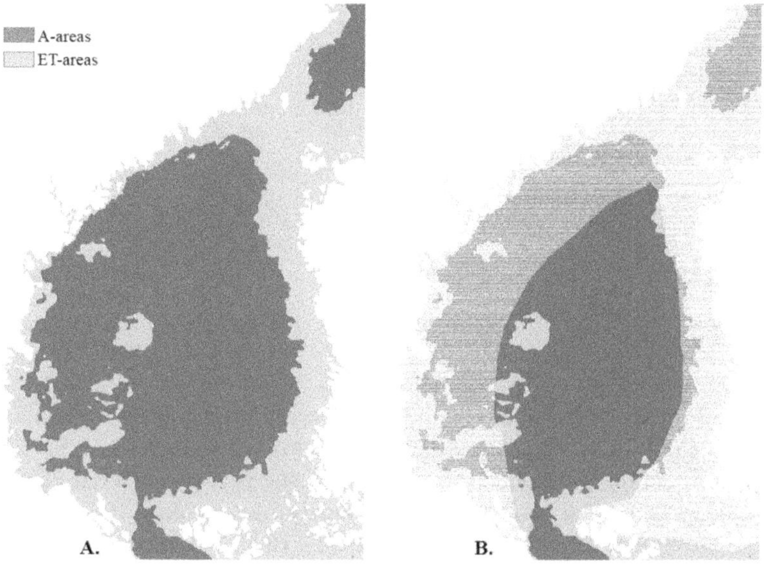

Fig. 2.39 Illustration of ice cover and bottom dynamic conditions (E, T, A) in the Bothnian Sea. (**a**). Ice free conditions (summer). (**b**) Maximum ice extent (*shaded areas*) and open waters during a very mild winter. A-areas = areas of fine sediment accumulation, ET-areas = areas of fine sediment erosion and transportation

to ice-free conditions. In the case illustrated in Fig. 2.39b, waves cannot reach as deep as in Fig. 2.39a, so even some of the ET-areas in Fig. 2.39b would be inaccessible to erosion and resuspension by waves. Less wave action and calmer, less turbulent conditions also imply that the sedimentation rate of suspended particles will increase. In addition, land uplift (see Fig. 2.5) is larger in the colder, northern part of the Baltic Sea than in the warmer, southern part, and land uplift supplies large quantities of particles and nutrients to the system every year (Jonsson et al., 1990; Håkanson and Bryhn, 2008b). Thus, year-to-year variations in water temperatures and ice formation should influence sedimentation, resuspension and concentrations of TP (and TN) in the Baltic Sea. This will be discussed more in Chap. 5, but a basis for those discussions will be given in this section. A possible warmer climate in the future would also affect the ice formation as well as TP-concentrations, and in Chap. 5 we will run scenarios to illustrate this and how such changes would likely influence production and biomasses of key functional groups of the Baltic Sea ecosystem.

2.8.6.2 Ice Conditions

The maximum extent of the ice-cover in the Baltic Sea is usually observed in February or March (SMHI and FIMR, 1982). The extent and monthly development of the ice during a normal winter is illustrated in Fig. 2.40. This type of mountain-like shape of the curve is a common feature during mild and severe winters as well

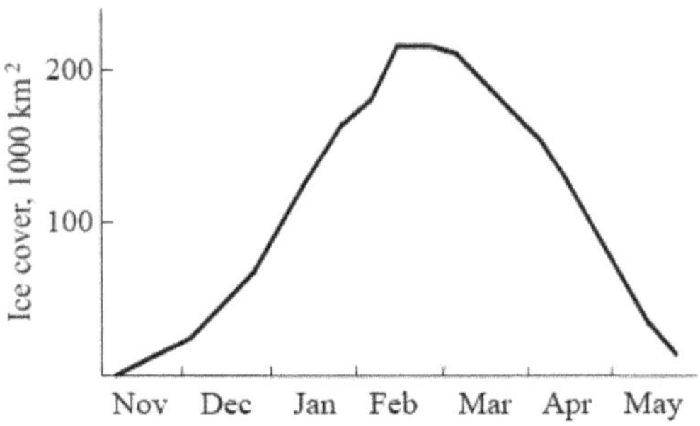

Fig. 2.40 The monthly evolution of an average ice season in the Baltic Sea (data from Seinä and Palusuo, 1996; Meier et al., 2004)

(SMHI and FIMR, 1982; Meier et al., 2004). During an average winter (maximum ice extent: 217,000 km^2; Seinä and Palusuo, 1996), all of the sub-basins are covered by ice except for the Baltic Proper, which is only partially covered. Figure 2.41 describes the year-to-year variations in maximum ice extent since 1960. It is evident that the winters since 1990 have had significantly ($p < 0.05$) smaller maximum ice extents compared to preceding decades.

The Bothnian Bay freezes completely even during "mild" winters and is normally frozen for more than 4 months. Half of the bay is usually covered during more than 5 months and the northern part for more than 6 months (SMHI and FIMR, 1982). The first freezing starts between early October and late November. A complete ice

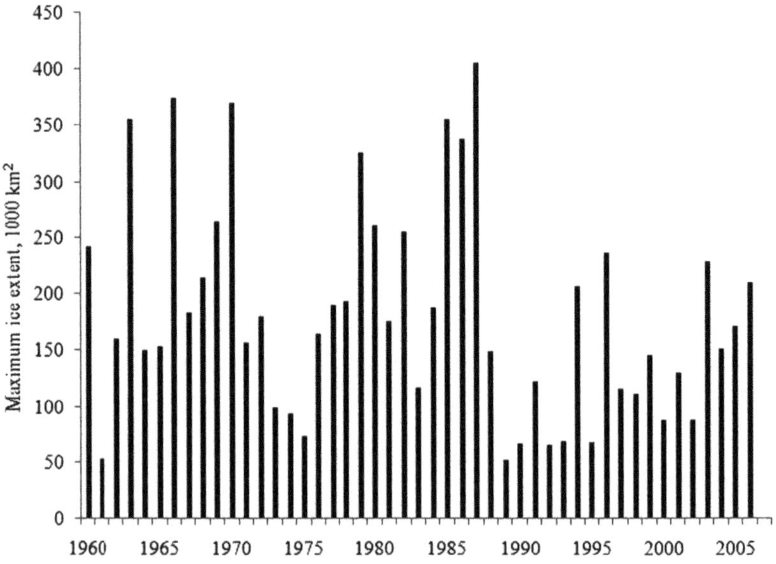

Fig. 2.41 The maximum ice extent 1960–2006. Data until 1995 from Seinä and Palusuo (1996) and for subsequent years from SMHI (2007)

coverage is reached between January and March (Seinä and Palusuo, 1996). Ice-break in the south normally occurs in early May (and ranges in time from mid-March to late May) and in the north it occurs in late May but may delay well into June (SMHI and FIMR, 1982).

The sub-basin with the second earliest ice formation during winter is the Gulf of Finland. It is usually completely ice-covered for 2 months every winter, half of it for another month and the eastern part of it for more than 4 months. During extremely mild winters, such as the ones in 1930, 1961 and 1989, only the eastern part of the bay and some other coastal stretches are covered. Freezing normally starts in late November and completes in early February, and the ice-break typically occurs in the western part in April and in early May in the north-eastern part.

The Gulf of Riga is also completely covered by ice during average winters and local variations are limited in this relatively small bay. A typical ice-season lasts for 2.5–3.5 months, from late December or January until early or late April. The Bothnian Sea is larger and the spatial variations greater than in the Gulf of Riga. The ice-cover is normally present from 2 to 5 months, from mid-December until early May in the northern part, and from late February until mid-April in the southern part.

The Baltic Proper is the largest basin and also shows great spatial variability with respect to the ice-cover. During a normal winter, the Archipelago Sea is covered for 2 to 3 months from late December or January until April. The area above 39.4°N, the western coastline down to Listerlandet in south-eastern Sweden, and the eastern coastline down to Gdansk (except for waters outside the Kaliningrad Peninsula) may then be ice-covered for 1 month or more, usually from January or early February until March or April (SMHI and FIMR, 1982). The least likely areas to freeze are located in the south-eastern part, between Sweden and Poland (Seinä and Palusuo, 1996).

2.8.6.3 Temperatures

Temperature data from surface waters have been taken from ICES (2009) and there are particularly good data available for the Baltic Proper. Monthly mean values showed higher correlations ($r^2 > 0.80$) between sub-basins than monthly medians, so the mean values were considered more reliable and representative for monthly conditions in each sub-basin. Figure 2.42 shows monthly temperatures in SWBP (the surface-water layer in the Baltic Proper) 1960–2007. Both summer and winter temperatures appear to have increased during the period after 1990 compared to the period before, and a t-test showed that the difference in monthly means was significant ($p < 0.05$) analogously with the trend for maximum ice extent (Fig. 2.41).

Mean temperatures of several months were strongly correlated with the log-value of the maximum ice extent (which had a log-normal distribution) from 1960 to 2006. The strongest co-variation regarding maximum ice extent was found with mean temperatures in March ($r^2 = 0.81$), followed by February ($r^2 = 0.73$), May ($r^2 = 0.70$), April ($r^2 = 0.60$), June ($r^2 = 0.52$), January (0.49), July ($r^2 = 0.36$), October ($r^2 = 0.35$), September ($r^2 = 0.35$), August ($r^2 = 0.30$), and September from the previous

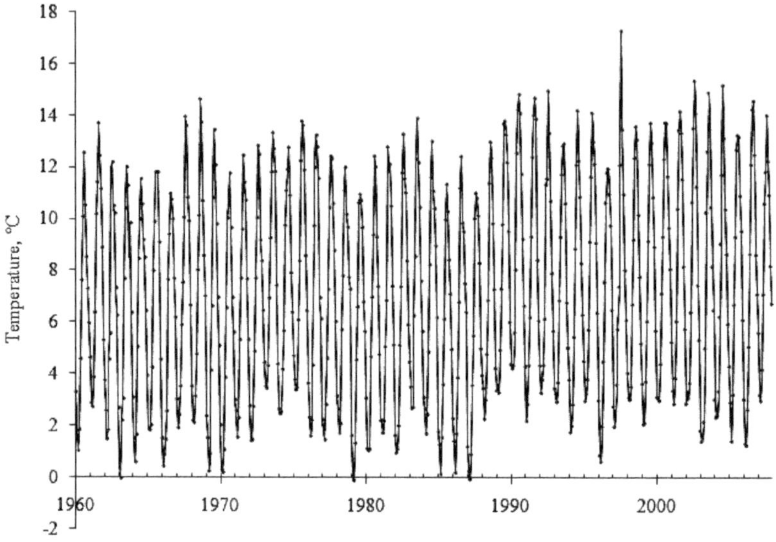

Fig. 2.42 Mean monthly temperatures in the surface-water of the Bothnian Proper (SWBP) 1960–2007

year ($r^2 = 0.13$). A forward stepwise multi-variate regression only allowed March as a determining factor, which suggested a stable, yearly recurring pattern in temperature difference between the different months. The strong correlation with March and February has a straightforward explanation; these are the months during which the maximum ice cover is usually recorded (SMHI and FIMR, 1982). Omstedt and Chen (2001) noted high correlations between air temperatures and the maximum ice extent during various periods from 1720 and onwards, which should clear suspicions about a high degree of autocorrelation between these variables which were non-stationary between 1960 and 2006. Omstedt and Chen (2001) also found average air temperatures between December and February to be useful for predicting the maximum ice extent with similar or stronger correlations in relation to the highest ones mentioned in this section, which may be a reflection of air temperature data being more abundant than water temperature data.

Patterns in monthly surface-water temperatures (SWT) in the five sub-basins are displayed in Fig. 2.43. All sub-basins generally have minimum temperatures in March and maximum temperatures in August. The Bothnian Bay (Fig. 2.43a) stood out from the other sub-basins in the sense that January–April temperatures were comparatively stable, and that the difference between June and July temperatures was comparatively pronounced, which may be an effect of the enduring ice cover in the area. The Bothnian Bay and the Bothnian Sea (Fig. 2.43a, b) had rather similar minimum and maximum temperatures, which were lower than in the Baltic Proper (Fig. 2.43c). The Gulf of Finland (Fig. 2.43d) and the Gulf of Riga (Fig. 2.43e) had similar summer temperatures as in the Baltic Proper, but lower winter temperatures.

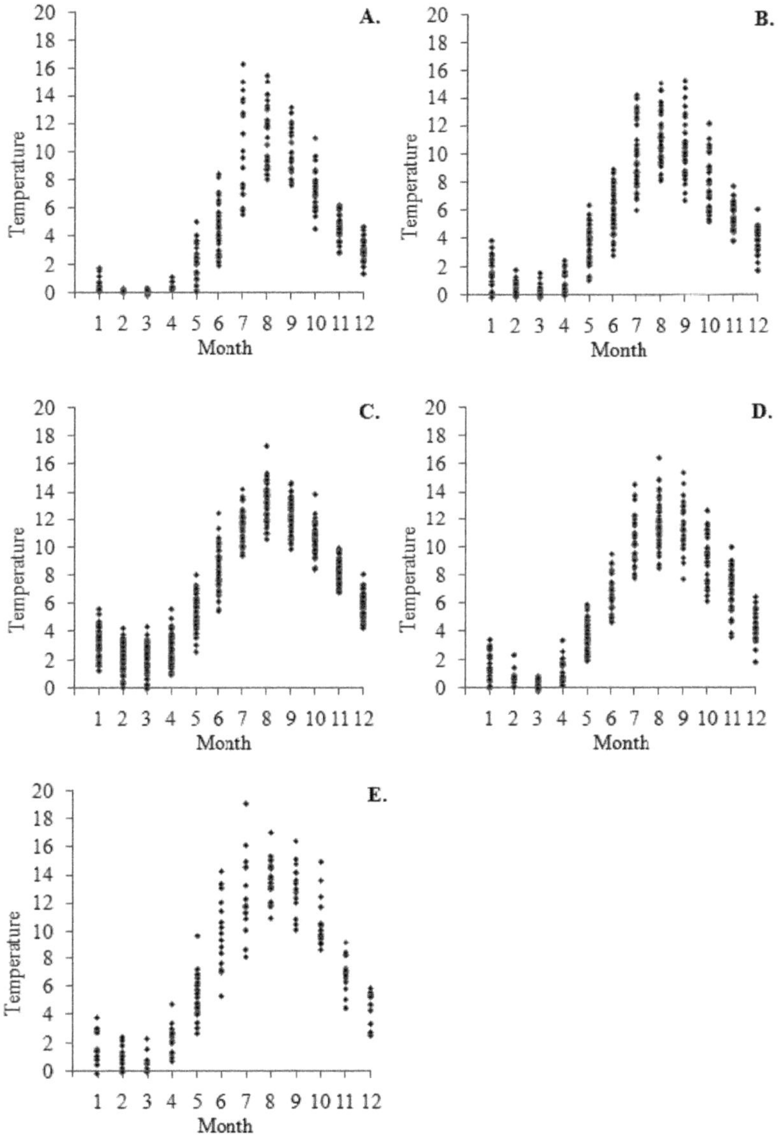

Fig. 2.43 Seasonal surface-water temperature patterns (in °C) in five sub-basins of the Baltic Sea, 1960–2007. (**a**) SWBB. (**b**) SWBS. (**c**) SWBP. (**d**) SWGF. (**e**) SWGR

Because of data scarcity in four of the sub-basins, four non-linear temperature models with SWT_{BP} as an x-variable were constructed to fill gaps in the time-series (Table 2.20). SWT_{BB} had to be constructed with month-specific coefficients to capture the special temperature pattern in the Bothnian Bay, whereas models for the Bothnian Sea, the Gulf of Finland and the Gulf of Riga were more straightforward. It should be noted that the lower boundary conditions for the three latter models reflect the general temperature difference between these basins and the Baltic Proper with respect to the lower temperature range; this difference is greater for the Bothnian Sea than for the Gulf of Finland, and greater yet than for the Gulf of Riga, which can also be extracted from Fig. 2.43. Model equations and test results against empirical data are given in Table 2.20. The modeled values in this table yielded r^2-values from 0.88 to 0.90 when regressed against empirical data, and the regression slopes were very close to 1, which should indicate that these models are fairly reliable.

2.8.6.4 Concluding Comments

Results in this section show that surface-water temperatures in the Baltic Proper have increased in a significant way since 1990 compared to the period 1960–1989 and that the maximum ice extent in the Baltic Sea has become significantly smaller during this period. Decreases in ice extent may have a profound impact on the sedimentation, erosion and resuspension of particulate phosphorus and other substances added to the Baltic Sea system every year from land uplift.

From the trend analyses given in this section, one could argue that with these data it would be possible to carry out a more holistic trend analysis using the CoastWeb-model. There are, however, several reasons why it is not so meaningful to do that. First, one could never hope to obtain good predictions of how the monthly or yearly TP-concentrations have changes in the five sub-basins (and 12 layers) in the Baltic Sea for the period between 1960 and 2008 in spite of the fact that we have presented time series of data from this period on empirical TP-concentrations, salt-water intrusions, temperature data and TP-loading. This information is simply not sufficient because:

- One should also expect major temporal changes in wind frequencies within sites in one basin and among sites in the five basins. This is exemplified in Fig. 2.44

Table 2.20 Temperature models for surface water layers in four sub-basins of the Baltic Sea. Test results from linear regressions between modeled data and empirical data (Mod-Emp) are displayed in the last two columns

Sub-basin	Model	r^2, Mod-Emp	Slope, Mod-Emp
Bothnian Bay	$SWT_{SWBB} = SWT_{SWBP} / Y_{SWTSWBB}$ where $Y_{SWTSWBB}$ has the values 5.0, 34, 110, 6.3, 2.5, 1.9, 1.1, 1.2, 1.2, 1.5, 1.8, and 2.0 for months 1–12.	0.89	0.99
Bothnian Sea	$SWT_{SWBS} =$ if ($SWT_{SWBP} > 2.3$) then ($SWT_{WBP} - 2.3$) else 0	0.90	1.00
Gulf of Finland	$SWT_{SWGF} =$ if ($SWT_{SWBP} > 2$) then ($1.05 \cdot SWT_{SWBP} - 2.1$) else 0	0.89	0.99
Gulf of Riga	$SWT_{SWGR} =$ if ($SWT_{SWBP} > 1.77$) then ($1.18 \cdot SWT_{SWBP} - 2.09$) else 0	0.88	1.00

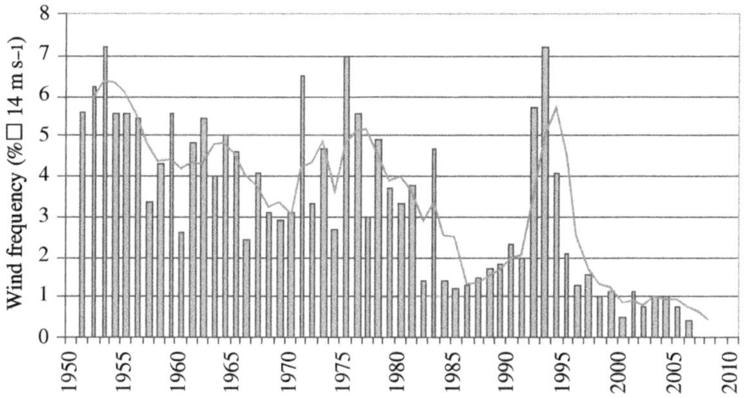

Fig. 2.44 Frequency of wind speed ≥ 14 m/s at Gotska Sandön. The *line* indicates running 3-year average (from Karlsson et al., 2009; based on unpublished data from SMHI)

using data from one site in the Baltic Proper (Gotska Sandön). Temporal and areal variations in the wind climate as large as those shown in Fig. 2.44 would influence resuspension, internal loading, mixing and stratification and hence also the measured and modeled TP-concentrations and we do not have access to adequate wind data for the five basins for the given period to account for such variations. This means that for periods with high winds, one should except higher TP-concentrations in water than the model predicts and corresponding changes in Secchi depth, chlorophyll-a concentrations and primary and secondary production.

- There are also major uncertainties in the annual data for the TP-loading for the tributaries to the given basins; even if the total annual loading would be correct for certain years (which can not be known only assumed) to a given basin, the TP-loading may not be reliable for other basins and there would still remain uncertainties for other years. Since the total annual TP-loading varies from about 21 kt TP/year for 2002 to 54 kt TP/year for 1988 (see Table 2.18) and since uncertainties of 5–10 kt/year would have marked effects on the measured and modeled TP-concentrations (see Håkanson and Bryhn, 2008a), one should expect major differences when modeled TP-concentrations are compared to empirical data.
- The data on the annual TP-loading given in Table 2.18 are very uncertain for the loading from many small and patchy diffuse sources and point sources. Maybe the total actual TP-loading should by a factor of 1.25–1.5 higher than the values given in Table 2.18, at least for some of the basins some of the years, but probably not for all of the basins all of the years. This means that also the uncertainty related to many small and more diffuse TP-fluxes may be up to 10 kt TP/year some of the years.

We have illustrated the role of such uncertainties in Fig. 2.45 where we have added 10 kt TP/year to the default values for the TP-loading shown in Table 2.18

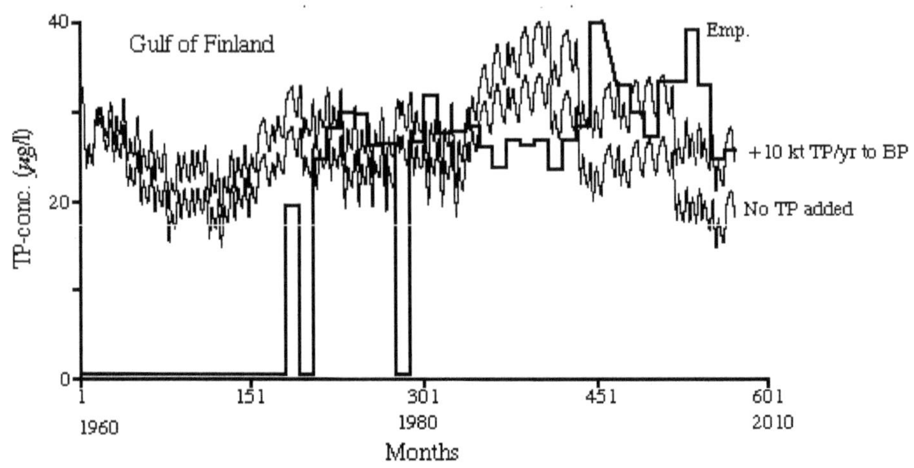

Fig. 2.45 A simulation to indicate how the TP-concentrations in the SW-layer in the Gulf of Finland would change if 10 kt TP/year is added every year to the default tributary inflow to the Baltic Proper using the data for the period 1960 and 2008 from Table 2.18

and adjusted those total TP-fluxes to the Baltic Proper (from the data in Table 2.5) and calculated the effects for the TP-concentrations in the surface-water layer, not in the Baltic Proper but in the Gulf of Finland. One can see that the empirical TP-data for the SW-layer in the Gulf of Finland for certain periods fall close to the two curves (with 10 kt TP/year added and not added) and for other periods the two curves fall outside the line given by the mean empirical data. The distance between the two curves indicate what this uncertainty for the TP-inflow to the Baltic Proper means for the conditions in the Gulf of Finland. Hence, the discrepancies between the empirical data and the modeled values may then be related to factors not accounted for in the modeling, such as variations in wind frequencies and temporal variations in TP-fluxes to the given basin and to other basins.

So, to carry out a meaningful trend analysis would require information which is, to the best of our knowledge, missing today on at least these three important factors regulating the temporal variation in measured TP-concentrations in the Baltic Sea basins.

In coming sections of this book, we will, on the other hand:

- Test the hypothesis that the decreasing ice extent has influenced the TP-concentrations during recent decades.
- Use results from such tests as a basis for temperature scenario, i.e., to predict likely effects on the trophic state and the biomasses of the various functional groups from changes in temperatures.

2.9 Background on Flora and Fauna of the Baltic Sea

It should be stressed that we have scrutinized the literature for Baltic Sea data on biomasses for the functional groups that may be used as empirical reference values for the modeled values from the CoastWeb-model, but the results are largely discouraging. The target biomasses concern the two functional groups of fish (prey fish and predatory fish) and for these two we have, however, access to more reliable empirical data than for the other functional groups. For the primary producers (phytoplankton, benthic algae and macrophytes), bacterioplankton, zooplankton and zoobenthos, there are only scattered data from a few investigated sites and such data are often, as shown earlier in this chapter for oxygen and phosphorus, poorly time- and area-compatible with the modeled values, which represent mean, characteristic values for the five main coastal areas in the Baltic Sea for the period from 1997 to 2006. For example, we have searched for and not found any reliable, time- and area-compatible data for:

- Biomasses of macrophytes in the five main basins.
- Biomasses of coastal fish in the five main basins.
- Relationships between predatory and non-predatory zooplankton.
- Reliable diet data for many seasons of the year for fish (such as is those available for the North Sea at http://www.cefas.co.uk).

The combination of marine and freshwater flora and fauna in the Baltic Sea is unique. For many groups, the community composition is mainly determined by the salinity, which results in a transition from a dominance of marine species in the south and in the open-water areas to a dominance of freshwater species in the north and in many of the more sheltered coastal areas.

There are large differences between enclosed coastal areas and the open sea. The shallow and sheltered bays along the coast are rather similar to inland lakes, with extensive beds of reed and other macrophytes of freshwater origin (Münsterhjelm, 1987). The fauna too is often of freshwater character. Typical freshwater species of zoobenthos are abundant, and many species of freshwater fish may also be found in the coastal areas of the Baltic Sea. The bioproduction is generally very high in these areas, and importantly, these areas warm up quickly in the spring, which allows many warm-water species to spawn. In these shallow areas, all functional groups of primary producers (i.e., phytoplankton, benthic algae and macrophytes) are present.

Primary production, biomasses of animals and the number of species decrease rapidly to the north in the Baltic Sea (see Kautsky and Kautsky, 2000, and references therein). This is usually attributed to, on the one hand the decrease in salinity, and on the other hand, to differences in growing season (lower temperatures at higher latitudes). The different trophic states of the basins also have clear effects.

2.9.1 The Fish Community

The Baltic Sea fish community consists of some 100 species (ICES, 2008a, b, c). The history of the Baltic Sea and the migration patterns of fish have been described elsewhere (e.g., Ojaveer and Lehtonen, 2001). In this section, we will focus on the dominant species. The composition of the fish fauna in the Baltic Sea is highly dependent on salinity. The saline parts are dominated by salt-water species, such as cod, herring and sprat, whereas the less saline parts are dominated by freshwater species, such as roach and perch (e.g., Lappalainen et al., 2000). Cod, herring and sprat form the bulk of the fish community, with the other fish generally contributing with a small fraction of the total biomass. In one study, the other fish constituted about 5% of the total fish biomass (Enin et al., 2004). Smelt (*Osmerus eperlanus*; see Table 2.21 for a compilation of fish names in English and Latin) and nine-spined stickleback (*Pungitius pungitius*) were the most abundant non-clupeid species in a study performed in the Gulf of Finland. At levels of maximum of 4 and 2% of the total catch, they were evidently not very abundant (Peltonen et al., 2004). All of the commercially most important species are of marine origin, namely

Table 2.21 A list of the most common fish species in the Baltic Sea and the main areas where the live (if nothing is specified, the species are likely to be found in the entire Baltic Sea)

Name (English)	Latin name	Main area
A. Open Baltic Sea		
Cod	*Gadus morhua*	
Herring	*Clupea harengus*	
Sprat	*Sprattus sprattus*	
B. Coastal areas		
Salmon	*Salmo salar*	
Whitefish	*Coregonus lavaraetus*	*GR, BS, BB*
Vendace	*Coregonus albula*	BB
Perch	*Perca fluviatus*	*BS, BP*
Roach	*Rutilus rutilus*	
Pike	*Esox lucius*	
Bream	*Abramis brama*	
Silver bream	*Blicca bjoerkna*	*BP, GF, GR*
Rudd	*Scardinius erythrophthalmus*	*BP, GF, GR*
Tench	*Tinca tinca*	*BP*
European smelt	*Osmerus eperlanus*	GF, GR
Zander	*Sander lucioperca*	GR
Sticklebacks	*Gasterosteus aculaetus* and *Pungitius pungitius*	*GF, BS, BB*
Bleak	*Alburnus alburnus*	BS, BB.
Flounder	*Platichthys flesus*	*BP, GF, GR*
Turbot	*Psetta maxima*	
European plaice	*Pleuronectes platessa*	
Sand goby	*Pomatoschistus minutus*	
Common goby	*Pomatoschistus microps*	
Round goby	*Neogobius melanostomus*	

cod (*Gadus morhua*), herring (*Clupea harengus*), sprat (*Sprattus sprattus*), flounder (*Platichthys flesus*), turbot (*Psetta maxima*) and salmon (*Salmo salar*). Some of the freshwater fish are high-valued species and, as such, commercially important locally, e.g., vendace (*Coregonus albula*), eel (*Anguilla anguilla*) and zander (*Sander lucioperca*). The coastal fish production depends on suitable spawning habitats, generally shallow bays that provide warm water, shelter and food for the fry (Sandström et al., 2005; Snickars et al., 2009; Sundblad et al., 2009). Recruitment failure of coastal fish, mainly perch (*Perca fluviatilis*) and pike (*Esox lucius*), has been observed in recent years in the more open areas along the Swedish Baltic coast (Sandström and Karås, 2002; Nilsson et al., 2004). The sheltered bays seem more unaffected, making them even more important. The reason(s) for these failures are not yet clarified.

The foodwebs of large marine systems such as the Baltic Sea are not static and it is therefore difficult to isolate typical conditions. Fishing and hunting of large vertebrates have changed the marine foodwebs fundamentally (Jackson et al., 2001). Historically, it is likely that marine mammals regulated the fish abundance (MacKenzie et al., 2002). The marine mammals that inhabit the Baltic Sea today are grey (*Halichoerus grypus*), ringed (*Phoca hispida*) and harbor seals (*Phoca vitulina*), as well as a small, little studied, population of harbor porpoise (*Phocoena phocoena*). Hundred years ago, there were much more seals and porpoise in the Baltic Sea than there are today (Elmgren, 1989b; Hårding and Härkönen, 1999). More recently, there has been a shift from a fish community that used to be dominated by cod to a community that is more dominated by clupeids. This shift appears to be due to a limited inflow of saline and oxygen-rich water from the North Sea, warmer temperatures and a consistently high fishing rate. This has resulted in lower cod recruitment (Jarre-Teichmann et al., 2000), while at the same time the conditions for sprat recruitment have been favorable (Köster and Möllmann, 2000b; Köster et al., 2003a, b; MacKenzie and Köster, 2004). Still, the Baltic Sea foodweb normally seems to be regulated by a bottom-up controlling mechanism (e.g., Rönkkönen et al., 2004).

The composition of the fish communities varies both between and within basins. Generally, the fish community composition varies considerably within the larger coastal areas (Snickars et al., 2009). The open parts of Baltic Proper are dominated by cod, herring and sprat, whereas the coastal fish community is dominated by freshwater species, such as perch, roach, pike, bream, silver bream, rudd and tench (Fiskeriverket, 2009). Herring is the most abundant species in the Gulf of Riga, but there are also many freshwater fish species such as smelt, whitefish, perch and zander (Ojaveer, 2003). The low salinities in the coastal semi-enclosed basins limit the occurrence of marine species. Cod may, however invade the Gulf of Riga in large shoals (Reila, 1979, cited in Ojaveer, 2003). Herring and sprat are the most important species in the open sea of the southernmost parts of the Bothnian Sea, whereas vendace and herring are important in the Bothnian Bay. Sticklebacks, perch, pike, roach and bleak are all common in the coastal areas of the Bothnian Sea and the Bothnian Bay. Salmon, whitefish, herring and vendace are important species for the coastal fisheries (ICES, 2008b; Fiskeriverket, 2009). Percids (perch, zander and ruffe) and cyprinids (roach, bream, white bream, ide, bleak, vimba and ziege) dominated the coastal fish fauna of the Gulf of Finland with more than 95% of the catches (Lappalainen et al., 2000). Generally, the fish community composition varies considerably within the larger coastal areas (Snickars et al., 2009).

The total fish biomass in the Baltic Sea has likely increased due to eutrophication and decreased mammal predation (Thurow, 1997). However, estimating the biomass of the total fish population in the Baltic Sea is difficult, as data for the non-commercial species are rare, and even where and when samplings have been carried out, the methods of sampling may miss some potentially important species (Andersson et al., 2007). It is likely that small bottom-dwelling species are under-estimated and the pelagic species over-estimated. In addition, available data generally represent the catch rather than the actual biomass. Still, the catches of the less important fish species are small compared to the main ones, and for the commercially important species, such as herring, sprat and cod, comparatively reliable local/regional estimations are, however, available. Around 50,000 t have been estimated (Lindquist, 2001) for the less important fish species, to be compared with the about 3,000,000 t of herring, sprat and cod. The dominant "non-ICES" species according to Lindquist's study were pike, zander, perch, whitefish, vendace, European smelt and garpike. Vendace is relatively important in the northern part of the Bothnian Bay. The biomass is estimated to be around 3,500 t, some 2 t are landed annually (Aho et al., 2008). The estimated total biomass of three-spined stickleback was about 25,000 t in the pelagic areas of the Bothnian Sea and the Bothnian Bay (Jurvelius et al., 1996). Other species that are fished commercially are flounder (about 10,000 t landed in 2007) and turbot (40 t), European plaice (170 t caught by Swedish fishermen) and whitefish (560 t in the Baltic Proper and the Bothnian Bay; Aho et al., 2008).

As stressed, biomass estimations for the non-commercial species are rare, and even where samplings have taken place the methods of sampling may miss some potentially important species (Andersson et al. 2007).

Another interesting fish group is the gobies, such as sand goby and common goby, but data are very scarce on many small bottom-dwelling species (such as gobies). There are indications that they could – at least locally – be important. Gobies have been found to consume some 15–30% of the total benthic production locally (Ehrenberg et al., 2005). However, reliable biomass estimates are lacking. The invading round goby might become important in the future and pose a threat to commercially important species through competition (Karlson et al., 2007). Round goby is an exotic species for the Baltic Sea that is spreading at a fast rate. They compete with the native benthic feeders and may become important in the future (e.g., Karlson et al., 2007).

About 8,000–12,000 t of freshwater fish are caught annually in the Baltic Sea, (Lindquist, 2001). As far as we understand, there are no reliable estimates of the coastal prey and predatory fish biomasses in the Baltic Sea. The catch per unit effort (CPUE) in Gulf of Finland ranged from 5.5 in the outer parts and 3 in the inner parts (Lappalainen et al., 2000). Data from Sweden show similar catches with CPUE ranging from 2 to 6 (Fiskeriverket, 2009). We converted CPUE to biomass/ha by multiplying CPUE by 17, as suggested by Heibo and Karås (2005), yielding estimated fish densities between 30 and 100 kg/ha. The mean CPUE in the Swedish coastal fishings was 2.7, based on yearly means from 9 different stations along the coast, yielding 46 kg fish/ha. Most of the coastal production is within the photic zone, so multiplying the area of the photic zone (Al-Hamdani and Reker, 2007) with the estimated density of fish yields an estimated biomass of coastal fish (see Table 2.22). Comparing these biomasses to those of the dominating species, i.e., cod, sprat and herring, leads us to believe that no single coastal species can be considered as important among the "top-three".

The bulk of the migration in the Baltic Sea is from the spawning areas to the feeding areas and to over-wintering areas. Most of the migration takes place within the large main sub-basins. However, cod can migrate at very high densities to the north and the east in search for food. Salmon and trout also migrate between the main sub-basins (Aro, 1989).

Reliable data on the ratio between predatory and non-predatory fish are not known to us for the five main coastal sub-basins. However, between 5 and 22% of the fish caught in the Gulf of Finland were predatory according to Lappalainen et al. (2000). This is similar to the situation in the open sea (see Fig. 2.46). Data from the Swedish coast is at present (Fall 2009) being analyzed by the Swedish Board of Fisheries, so hopefully the situation will become clearer.

2.9.1.1 The Important Species and Functional Groups

Predatory Fish

Adult cod is the main predator on herring and sprat and there is also cannibalism by large predatory cod on small cod (Köster et al., 2003a, b). The amount of juvenile cod in the Baltic Sea is determined by abiotic conditions in the spawning areas, such as the oxygen content and the salinity (Köster et al., 2005; Heikinheimo, 2008), and by biological processes, such as competition for food (Hinrichsen et al., 2002), predation on eggs by clupeids (Köster and Möllmann, 2000a) and, as mentioned, cannibalism by larger cod (Uzars and Plikshs, 2000).

Table 2.22 The photic zone is given as the percentage of the areas of the five main sub-basins where light reaches the bottom. Coastal fish is the estimated biomass of coastal fish species in the sub-basins. The estimate is based on the area of the photic zone and typical biomasses of fish caught during gill-net test fishing

Basin	Photic zone (%)	Coastal fish (kt)
Bothnian Bay (BB)	18	30.5
Bothnian Sea (BS)	14	51.0
Gulf of Finland (GF)	10	14.3
Gulf of Riga (GR)	14	10.6
Baltic Proper (BP)	7	68.8
Total	–	175

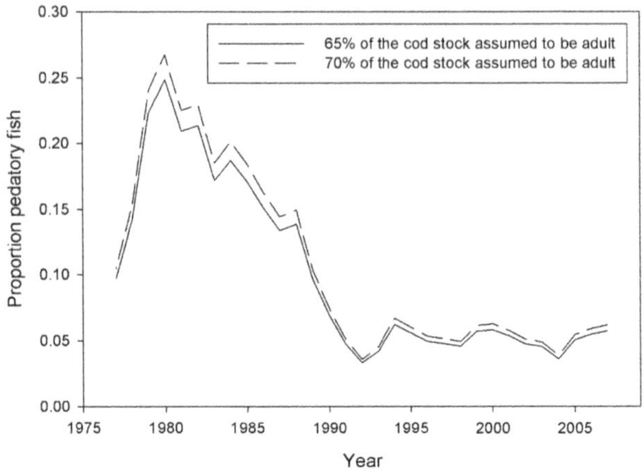

Fig. 2.46 Our estimation of the proportion of predatory fish in the Baltic Sea in the years between 1975 and 2008 assuming that 65 and 75% of the cod stock is assumed to be predatory

So, the reproduction success of cod depends on the salinity and the oxygen concentrations for the fertilization and survival of eggs and larvae. Thus, the recruitment of cod in the Baltic Sea depends primarily on the "reproductive volume", which is the volume of water in the deeper basins that has high enough oxygen levels and salinity (Westin and Nissling, 1991; Sparholt, 1996; MacKenzie et al., 2000; Köster et al., 2003b). In periods with few intrusions of saline oxygen-rich water from the Kattegat, the salinity and the oxygen conditions have only promoted successful spawning, egg fertilization and egg development in the Bornholm Deep. In the past, these favorable conditions for the cod reproduction were also found in the Gdansk and Gotland Deeps. It has also been argued that the increased primary production during the last 100 years has caused oxygen depletion in the system, which has lowered the reproductive success for cod in the Baltic Proper (Hansson and Rudstam, 1990).

Homing is not very clear in Baltic Sea cod (see Aro, 2000); homing might help them to again exploit their old spawning grounds when the conditions become favorable. Besides being necessary for spawning success, a high oxygen level also leads to an increased production of zoobenthos, which is the preferred food for young cod (see Fig. 2.47). In addition, adult cod avoids low oxygen levels (Waller and Boettger, 2001) and they eat less under oxygen stress and therefore grow at a lower rate (Chabot et al., 2001).

Predatory cod generally eat what is available. At times, this might include zoobenthos (Ojaveer, 2003). Herring, sprat and small cod represent, however, the common diet for predatory fish in the Bothnian Sea (Köster and Möllman, 2000a; Köster et al., 2003a). In periods of high abundance of cod, there is a large competition for food and thus the growth is limited by the resources (Uzars et al., 2001).

The distribution of the predatory cod varies both seasonally and spatially because of fluctuating environmental conditions, which affect their spawning, feeding and migrations (Aro, 2000). Cod spawning takes place during the summer. However, the timing of the spawning may vary between years by several months (Wieland et al., 2000). This makes cod migration difficult to predict, but it is likely that most of the migrations are within basins, such as the Baltic Proper (e.g., Sparholt et al., 1991). Larvae are spread by water currents (Hinrichsen et al., 2001).

Prey Fish

Herring and sprat can feed either by capturing particular food objects or by filter-feeding more indiscriminately (Blaxter and Hunter, 1982). Herring in the Baltic Sea have, however, been shown to feed selectively on larger prey, in one case reproducing female zooplankton (Flinkman et al., 1992). Both sprat and herring feed on zooplankton (see Fig. 2.47) – sprat is mainly a zooplanktivore independent of size. Herring may turn to other food sources such as zoobenthos, especially during autumn and winter (Fig. 2.47; see also Arrhenius and Hansson, 1993; Möllmann et al., 2004). Herring and sprat eat more intensively during summer and spring than during the autumn and winter (Möllmann et al., 2004). Sprat and herring seem to affect each other through competition (Casini et al., 2006). In the CoastWeb-modeling, however, we consider the functional group of prey fish and not individual species.

There are distinct stocks of herring in the Bothnian Sea, the Bothnian Bay and two different stocks in the Baltic Proper. The stock in the south-western part of the Baltic Proper is shared with spring spawning herring in the Kattegat and the Skagerrak (Aho et al., 2008). Both spring and autumn spawning herring are found in the Baltic Sea, though autumn spawners seem to be become more rare (Ojaveer, 2003).

Herring shows marked preferences for certain zooplankton (e.g., Viitasalo et al., 2001). Newly hatched herring eat copepod larva and eggs of planktonic organisms. In the autumn, when they are around 6 cm, they begin to also eat mysids and amphipods. 1+ herring (i.e., herring older than 1 year but younger than 2 years) eat mainly copepods (*Pseudocalanus*, *Limnocalanus*, *Eurytemora*, *Temora*, *Centropages*, *Acartia*), but cladocerans also play a central role as feed. Mysids, amphipods and other invertebrates are also consumed (Ojaveer, 2003; Möllman and Köster, 2002; Casini et al., 2004). Herring may also eat cod eggs (Köster and Möllman, 2000b), but do not feed at temperatures lower than 2°C (Ojaveer, 2003).

Larval sprat feed on diatoms, flagellates, eggs and larva of copepods, e.g., *Temora*. Large larvae feed exclusively on zooplankton. When they grow larger than about 30 mm, sprat eats mainly *Termora longicornis* at all times of the year, *Pseudocalanus* in the winter, and *Eurytemora* and *Acartia* in the summer and autumn months. Cladocerans (*Bosmina, Evadne* and *Podon*) occur in the diet during the summer (Möllman and Köster, 1999, 2002; Viitasalo et al., 2001; Ojaveer, 2003; Casini et al., 2004). Sprat occasionally also feeds on cod eggs (Köster and Möllman, 2000a).

Young cod move actively to shallower feeding grounds (Ojaveer, 2003) and the small cod starts feeding at a length of about 4–5 mm. At that size, they eat mainly eggs and young stages of copepoda (see references in Ojaveer, 2003). 0+ cod mainly eats mysids. Other important food organisms are *Pontoporeia*, *Saduria,* and other invertebrates. When the cod grows larger, the proportion of fish increases in the diet

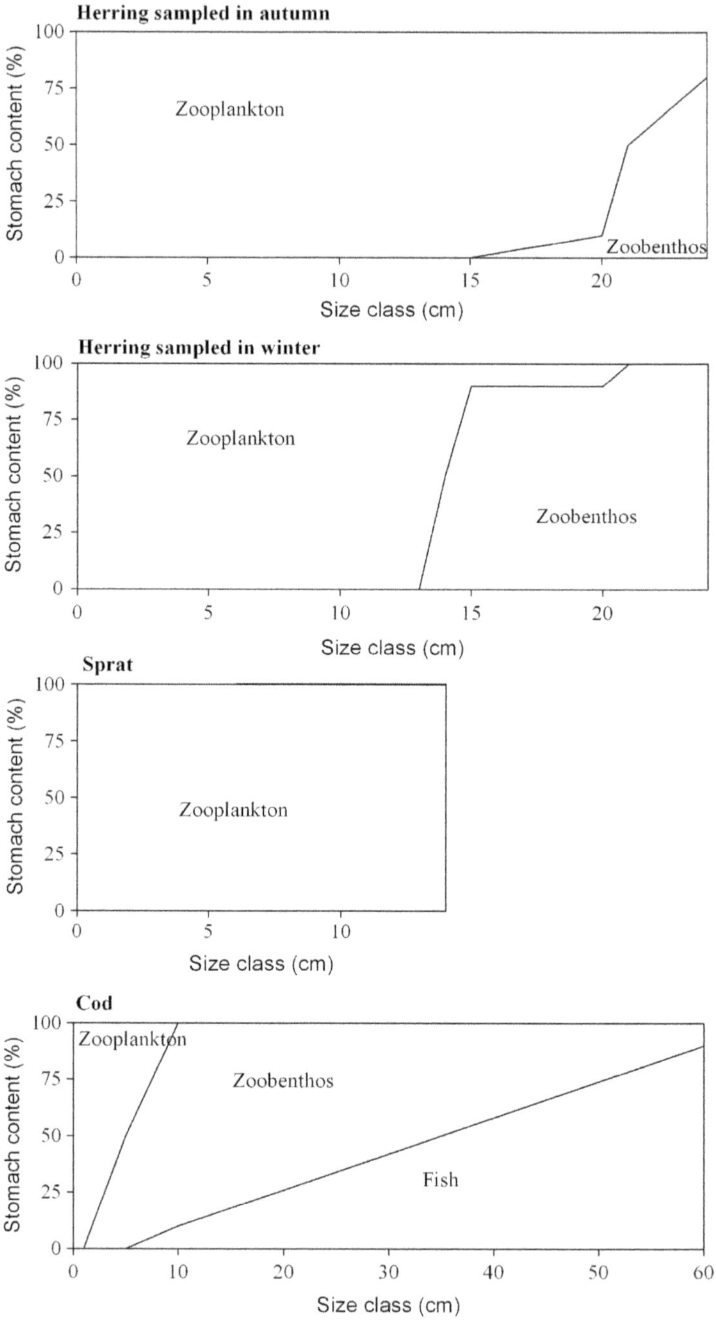

Fig. 2.47 A compilation of information on the relationship between size class (cm) and stomach content for (*upper*) herring sampled in the fall, herring sampled in the winter, sprat, and (*lower figure*) cod using data from Elmgren (1984), Arrhenius and Hansson (1993), Hüssy et al. (1997), Uzars et al. (2000), Uzars and Plikshs (2000), Ojaveer (2003), Casini et al. (2004) and Möllmann et al. (2004)

(see Fig. 2.47; Hüssy et al., 1997; Ojaveer, 2003). The successive change in food choices for cod, from pelagic prey (zooplankton) to benthic prey, are illustrated in Fig. 2.27.

Primary Producers

Many species of macrophytes are found in the Baltic Sea. Along the Swedish coast no less than 14 species of charophytes have been found. The occurrence of macrophytes is affected by the salinity, the Secchi depth (water clarity) and the morphomertry of the coastal area (Blindow, 2000; Håkanson and Boulion, 2002a). They tend to abound in the inner most sheltered parts of archipelago areas. Such areas can have a wide variety of vegetation (Appelgren and Mattila, 2005). These so-called flads are very productive and important for the reproduction of warm-water fish species (Karås, 1999; Sandström et al., 2005). There are, however, no reliable biomass estimates for macrophytes in the five main basins of the Baltic Sea (Harvey et al., 2003; HELCOM, 2008c).

Bladderwrack used to be the dominant species of benthic algae in the Baltic Sea, but has in many places been replaced by filamentous algae. These are opportunistic and fast growing, which should be to their advantage in the Baltic Sea under present conditions (Pihl et al., 1999). Filamentous algae are important habitats for benthic fauna (Salovius and Kraufvelin, 2004), but grazers prefer the filamentous algae to the bladderwrack, as the latter has chemical defences (Goecker and Kåll, 2003).

Phytoplankton species composition is determined by, e.g., nutrient availability and salinity. Therefore gradual changes occur from the southwest to the northeast in the Baltic Sea. Primary production varies greatly both within and among years (HELCOM, 2002; Wasmund and Uhlig, 2003; HELCOM, 2008b). Phytoplankton biomass, mostly diatoms, increases rapidly in the spring due to increasing light and available nutrients. Typically, this spring bloom starts in March in the western Baltic Proper. This bloom is delayed as one moves northwards and tend to occur in May–June in the Bothnian Sea and the Bothnian Bay. In late summer (July to August), another bloom dominates in the Baltic Sea, especially in the Baltic Proper, namely that of blue-green algae (cyanobacteria). These nitrogen-fixing species also tend to occur in summer blooms in the Bothnian Sea, the Gulf of Riga and the Gulf of Finland. It should be noted that few of the negative (toxic) effects of cyanobacteria that have been observed in laboratory conditions have been observed in the Baltic Sea (Karjalainen et al., 2007).

The Reducer, Bacterioplankton

In our literature survey, no reliable data on bacterioplankton biomasses or dominating species in the five main sub-basins have been found.

Zoobenthos

The Baltic Sea is, compared top many other coastal systems, comparatively poor in benthic species. The species richness of zoobenthos decreases from over 1,600 species in the Skagerrak to only 500 in the western Baltic Sea, 80 in the southern and less than 20 species in the northern parts (Bonsdorff, 2006).

The combined effects of low salinity, low concentrations of dissolved oxygen and relatively high sedimentation of organic matter have been shown to be important for benthic fauna community composition. The richness decreases and the composition changes with decreasing salinity (Laine, 2003). A low richness and low biomasses zoobenthos are also found in areas with low concentrations of dissolved oxygen, such as the deep basins (below the halocline) and also in some areas affected by effluents from industries (Karlson et al., 2002). Recolonization of these areas can be observed after major inflow events of saline oxygen rich water in the deep basins and also after improved water treatment in the coastal areas (e.g., HELCOM, 2008c; Karlsson et al., 2009).

The benthic community on hard bottom substrates is dominated by the blue mussel (*Mytilus edulis),* while deposit feeders and burrowing forms, such as *Macoma balthica,* dominate on soft bottom areas (Voipio, 1981). The blue mussels cover suitable substrates from the water surface to more than 30 m depth in some places. Even at the outer boundaries of their salinity tolerance (4.5 psu), dense beds may be found, albeit with much smaller size ranges (Westerbom et al., 2002). Few fish species appear to eat mussels in the Baltic Sea, and the mussels do not seem to have many competitors. It is the mainly non-burrowing forms of the four most common species in the Baltic Sea that could be important for fish food. Predation pressure on the mussel-beds has been considered to be low or almost non-existent. Yet, in one study the diet of roach consisted of molluscs to 95% and was dominated by blue mussel (Lappalainen et al., 2004). Mussels can either feed selectively or non-selectively, depending on abundance and quality of food (Arifin and Bendell-Young, 2001).

The crustacean *Idotea balthica* is an important grazer on the extensive stands of *Fucus* (Kautsky and Kautsky, 2000). Besides *Idotea, Fucus vesiculosus* host the richest benthic communities in the Baltic Sea, with snails, mussels, different crustaceans and insect larvae (Kautsky and Kautsky, 2000). The isopod *Saduria entomon*, the amphipods *Monoporeia affinis* and *Pontoporeia spp.,* and the bivalve *Macoma balthica* form the bulk of the benthic communities of the Bothnian Bay and the Bothnian Sea (Kautsky and Kautsky, 2000; Laine, 2003). Mysids, e.g., *Mysis mixta* and *Neomysis integer* are common in many parts of the Baltic Sea (see Viherluoto, 2001). They feed on many zooplankton taxa, phytoplankton as well as detritus (Viherluoto, 2001; Viherluoto and Viitasalo, 2001).

Different countries around the Baltic Sea use different measures and methods to assess zoobenthos both for coastal areas and the open sea (HELCOM, 2008c). Therefore, it is difficult to make an assessment of the biomass of the benthic fauna in the Baltic Sea.

Zooplankton

The species composition of both phytoplankton and zooplankton is mainly regulated by the salinity gradient. Marine species dominate in the southwest, whereas freshwater species are more common in the brackish waters of the northeast. Besides salinity, the temperature also affects the abundance of the most dominant zooplankton groups (e.g., Möllmann et al., 2000). An increase in the number of species of freshwater origin caused by increased river runoff in comparatively shallow northern areas of the Baltic Sea has been observed (Viitasalo et al., 1995; Vuorinen et al., 1998).

The abundance of small zooplankton species seem to be controlled to a large extent by climatic variability during the winters whereas the larger zooplankton seems to be controlled by the grazing of herring and sprat (Dippner et al., 2001; Kornilovs et al., 2001). Baltic herring and sprat form a major part of the prey fish populations (see Fig. 1.10) and together with the mysids are the dominating zooplanktivores. However, a predatory zooplankton species invaded the Baltic Sea in the early nineties and can now be found in the Baltic Sea (Gorokhova et al., 2000). Other predatory zooplankton species have also invaded the Baltic Sea (e.g., Radionova and Panov, 2006). These species feed on herbivorous zooplankton and play a similar functional role as the pelagic mysids in the foodweb (Lehtiniemi and Lindén, 2006) – thus increasing the competition for zooplankton as food for the prey fish. This has sparked an interest in predatory zooplankton, but information on long-term trends and biomasses of predatory zooplankton are scarce. Time-series on the biomasses of zooplankton are only available for a few restricted areas (ICES, 2007).

2.10 Conclusions

Traditional hydrodynamic or oceanographic models to calculate water fluxes to, within and out of coastal areas generally use water temperature data (the thermocline) or the salinity (the halocline) to differentiate between different water layers. This chapter has motivated another approach, the theoretical wave base as calculated from process-based sedimentological criteria, to differentiate between the surface-water layer and lower vertical layers and this approach gives one characteristic value for each basin. Morphometric data for the Baltic Sea and the defined sub-basins, and hypsographic and volume curves based on digitized bathymetric data, have been used in the CoastMab-modeling. The basic aim of this chapter has been to present empirical data on the conditions in the Baltic Sea.

This chapter has used extensive databases from the Baltic Sea on total phosphorus (TP), the PO_4/TP-ratio, total nitrogen (TN), oxygen concentrations, tributary water dischage, TP-loading, ice cover, chlorophyll-a concentrations and water temperatures. One can conclude that:

1. There was an increasing TP-trend in SWBP 1968–1975, but after that changes have been less conspicuous and after 1990, the TP-trend has actually been slightly decreasing. Highest TP-concentrations were generally measured during 1990 and the preceding decade.

2. The slight temporary TP-decrease in the 1990s started before the major salt-water inflow to the Kattegat in 1993 and during a long stagnation period. The TP-trend in the deep-water layer of the Eastern Gotland Deep shows quite a different pattern than TP in the deep-water layer in the entire Baltic Proper. The trend 1968–2008 is significantly negative and year-to-year variations are also much more pronounced at this site than in the entire Baltic Proper. Thus, the changes in TP-concentrations in the Eastern Gotland Deep must be interpreted with particular care when placed in relation to any large-scale changes in the entire Baltic Proper.
3. Phosphate does not represent the full pool of dissolved phosphorus in water as this is modeled by the CoastMab-model, where the dissolved fraction (DF) should be determined from filtration and the fraction remaining on the filter is defined as the particulate fraction (PF), and hence the dissolved fraction is DF = 1-PF. There are also other fractions of phosphorus included in the dissolved fraction, e.g., DOP (dissolved organic phosphorus). Still, given the fact that there are more than 23,000 empirical data available, it is informative to study how the PO_4/TP-ratio in the deep-water layer of the Baltic Proper has varied during 1968–2008. It has been shown that the annual medians of PO_4/TP in the Baltic Proper 1968–2008 have fluctuated close to 1, i.e., that TP has typically consisted of more than 90% phosphate. The higher inflows of oxygenated salt-water from Kattegat have not caused much lower PO_4/TP-ratios in the deep-water layer of the Baltic Proper, and hence not markedly higher sedimentation or lower phosphorus diffusion from the DW-sediments. Note, however, that in 3.3% of the data points, PO_4/TP was greater than one. This is impossible and it indicates the limited reliability of the phosphate data.
4. TP in MWBP and in DWBP have been increasing quite continuously 1968–2008 despite the slight oligotrophication in SWBP 1990–2008. This contradicts the "vicious circle theory" (Vahtera et al., 2007) about the long-term regulation of primary production by both nitrogen and phosphorus. This theory relies on the assumption that the TP-flux from surface waters to deep waters and vice versa is intensive on an annual scale.
5. The overall trend in the oxygen concentration is significantly negative in the DWBP while it is significantly positive in the Eastern Gotland Deep. Year-to-year variations for oxygen are also much more pronounced at the Eastern Gotland Deep than in the entire Baltic Proper. Thus, neither TP-concentrations nor O_2-concentrations in the Eastern Gotland Deep may be seen as representative for the prevailing conditions in the deep-water layer of the Baltic Proper.
6. The TN-concentrations in the SW-layer have slightly increased since 1975 in the Baltic Proper contrary to decreasing TP- and chlorophyll-concentrations. The small general decline in TN-concentrations since the mid-1990s may be a combined effect of nitrogen abatement and a dampened long-term intensity of P-driven fixation of atmospheric nitrogen.
7. The TP-loading was very high in the 1980s, at a mean value of 48 kt/year, lower in the 1970s (40 kt/year), lower yet in the 1990s (36 kt/year) and the lowest

loadings were recorded during the first 6 years of the present decade (mean value: 28 kt/year). The freshwater inflow has also varied considerably, and was about 420 km^3/year in the 1970s, 480 km^3/year in the 1980s, 460 km^3/year in the 1990s and 430 km^3/year during 2000–2007.
8. This chapter has shown that surface-water temperatures in the Baltic Proper have increased in a significant way since 1990 compared to the period 1960–1989 and that the maximum ice extent in the Baltic Sea has become significantly smaller during this period. A decrease in ice extent may have a profound impact on the turbulence in the system, on sedimentation, erosion and resuspension of particulate phosphorus. So, a possible warmer climate in the future would also affect the ice formation as well as TP-concentrations, and in Chap. 5 we will run scenarios to illustrate this and how such changes would likely influence production and biomasses of key functional groups of the Baltic Sea ecosystem.

An aim of the first part of this chapter was to present data on the fluxes of water in the defined sub-basins of the Baltic Sea since those values give fundamental information on how the system reacts to changes in, e.g., nutrient loading. This places certain demands on the structure of this model, which is different from oceanographic models, e.g., in quantifying resuspension, mixing and diffusion and in the requirements regarding the accessibility of the necessary driving variables.

In this chapter, we have also presented empirically-based models which have been added to the process-based dynamic CoastMab-model (for salt, phosphorus and SPM). These are the sub-models for Secchi depth and chlorophyll-a concentrations. When tested against empirical data for the Baltic Proper, there was reasonably good overall correspondence between predicted values for Secchi depth and chlorphyll-a concentrations and the dynamic SPM-model predicts sedimentation, SPM-concentrations and burial in accordance with existing, but rather scattered, data.

The published Baltic Sea literature on biomasses for the key functional groups has been studied. The idea with the literature review was to try to find reliable empirical reference values to be compared with the modeled values from the CoastWeb-model. In this chapter, we have presented results from the literature review and the results are largely discouraging since we have found but few reliable data for the model tests. The target biomasses concern the two functional groups of fish (prey fish and predatory fish) and for these two targets, we have, however, access to the most reliable empirical data. For the primary producers (phytoplankton, benthic algae and macrophytes), bacterioplankton, zooplankton and zoobenthos, there are only scattered data from a few investigated sites and such data may be poorly time- and area-compatible with the modeled default values, which represent mean values for the five main coastal areas in the Baltic Sea in the period from 1997 to 2006.

From the literature data for the Baltic Sea presented mainly in Fig. 1.10, Figs. 2.46 and 2.47, the following conclusions can be drawn for the testing of the model in subsequent chapters:

- In the years around 1980, the total biomass of cod in the Baltic Sea was about 1,000 kt ww. Out of this 65–70% should be adult cod (see Fig. 2.46); adult cod also eats zoobenthos with decreasing amounts from about 5 cm length to more than 65 cm length and there are corresponding increases in prey fish as the main food for the larger cod; cod with a length of about 50 cm would consume about 65% prey fish.
- For the default period, i.e., for the years around 2,000, the total biomass of all types of cod in the Baltic Sea was about 200 kt (see Fig. 1.10).
- Predatory fish in the Baltic Sea would mainly comprise (about 85–90%) of predatory cod and the rest would are other species, such as salmon, pike, big perch, etc.
- This means that in the years around 1980, one would expect that the total biomass of predatory fish in the Baltic Sea would be about: $1{,}000 \cdot (0.675) \cdot (0.65) \cdot (1/0.875) = 501$ or 500 kt ww and that the uncertainty in this figure may well be $\pm$ 100 kt ww. For the years around 2,000, we have $200 \cdot (0.675) \cdot (0.65) \cdot (1/0.875)$ or about 100 kt ww.
- The proportion of predatory fish in the years around 2,000 was about 6–7% of the total fish biomass.
- The total biomass of prey fish has fluctuated very much indeed during the last 4 decades. Typically, the annual biomasses for the dominating species of prey fish vary around 2,500 and 5,000 kt ww.

All these empirically-based data related to Baltic Sea fishery will be used in subsequent tests of the model predictions. As a background to those discussions, it should also be stressed that the fishing was very intensive in the years around 2,000.

Chapter 3
The CoastWeb-Model – Structures and Set-Up

The aquatic ecosystem:

> *It is a world of hill and valley, plain and grove, just like ours, and, on the whole, its inhabitants are not unlike us. They are perhaps a little more candid in their manner of devouring each other. They do not, as far as I know, talk about the laws of supply and demand, or promulgate doctrines about the survival of the fittest. Not that I blame us – far from it. If I am to be devoured, I would rather have it done tactfully by a minister with golden words in his rotund mouth.*
>
> H.T. Sheringham (1912).

3.1 Introduction and Aim

The aim of this chapter is to present the basic structure of the CoastWeb-model. Most of the structures and equations for the coastal version of the model are the same as for the LakeWeb-model (see Håkanson and Boulion, 2002a), but there are also differences related to the transformation of a lake model to a model for the entire Baltic Sea. It is certainly important which organisms and processes are accounted for in models of this kind, and it is equally important to explain and motivate how this is done. To keep the text short and to the point, we will present a series of overview figures. We will also discuss how the CoastWeb-model differs from other seemingly similar models.

Chapter 2 discussed fundamental abiotic transport pathways for substances in water and sediments (inflow, outflow and internal transport processes for salt, phosphorus and SPM) using the CoastMab-model. CoastMab is an integral part of the CoastWeb-model.

3.1.1 Key Biotic Parts and Concepts

Figure 3.1 gives a compilation of key biological concepts. Phytoplankton, benthic algae and macrophytes are primary producers because they transform inorganic carbon to organic carbon by means of solar energy or by using energy derived

L. Håkanson et al., *The Fish Production Potential of the Baltic Sea*, Environmental Science and Engineering, DOI 10.1007/978-3-642-11562-2_3,

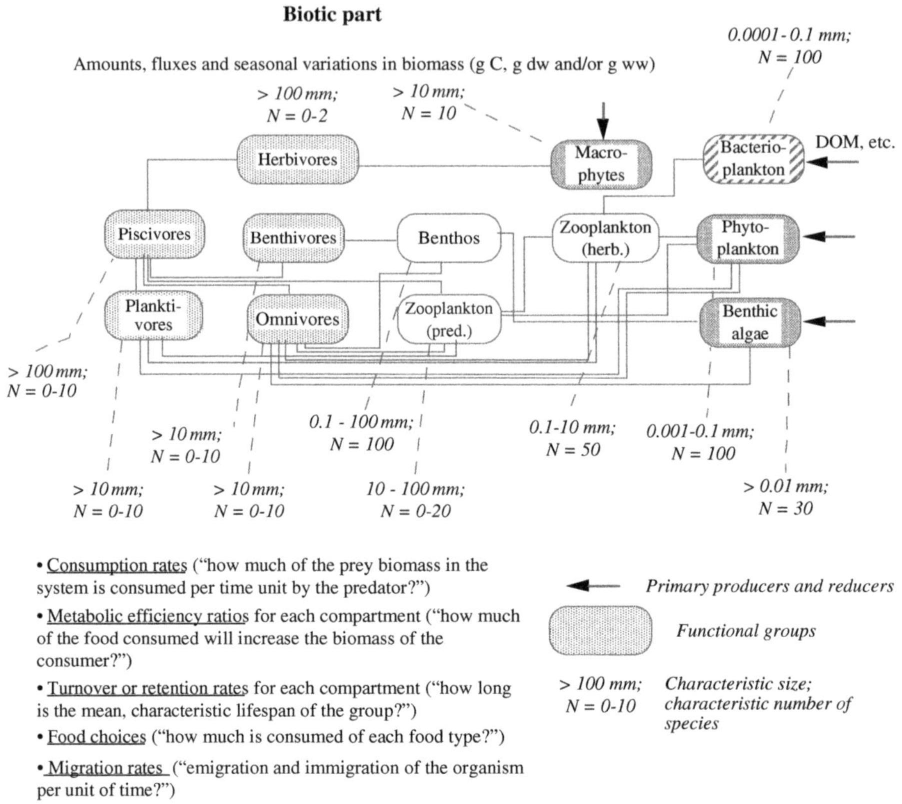

Fig. 3.1 Key functional groups and central concepts related to feeding (modified from Håkanson and Boulion, 2002a)

from oxidation of reduced substances. Bacterioplankton (a reducer) generally uses organic matter as a source of energy (see Fig. 3.2). Figure 3.1 also gives characteristic number of species in aquatic ecosystems (such as lakes, lagoons and smaller coastal areas) and values illustrating the normal size (or size range) of the various groups. Generally, there are few species of the big animals/plants, and vice versa. Figure 2.7 gave the relationship between salinity and number of species. The CoastWeb-model is constructed to account for key functional and structural groups and it does not deal with specific species. Functional groups of fish are piscivores, here called predatory fish, eating other fish, here called prey fish; the latter group includes, omnivores (eating "everything"), planktivores (eating plankton), benthivores (eating zoobenthos) and herbivores (eating plants). Note that the herbivores are not included in this version of the CoastWeb-model because they constitute only a small fraction of the prey fish in the Baltic Sea (see Chap. 2). The CoastWeb-model is simplified in such a manner that it basically only separates piscivores and all other types of fish (the prey fish). However, the model is constructed so that it is possible to define food habits and feeding behavior for prey fish in a flexible manner.

There are four very important concepts in the CoastWeb-model related to eating and food choices (see Fig. 3.1):

(1) The consumption rates (CR), which mathematically describe "how much of the prey biomass that is being consumed per time unit by the predator"; if, e.g., predatory zooplankton reduce 10% of the biomass of its prey, herbivorous zooplankton, per time unit (1 month in the following modeling), the consumption rate is 0.1 (per month).
(2) The metabolic efficiency ratios (MER, also sometimes called the "efficiency of growth") are used to quantify for each functional group "how much of the food consumed will increase the biomass of the consumer"; if, e.g., MER = 0.2, it means that 20% of the food consumed per time unit (= month) actually increases the biomass of the consumer.
(3) The turnover times (T) (T = BM/PR; BM = biomass in kg ww, wet weight; PR = production in kg ww/month) for each functional group. The turnover time describes "how long the median, characteristic lifespan of the organisms in the functional group is".
(4) Food choices – "how much is consumed of each food type"? The menu for each functional group will be discussed for all functional groups included in CoastWeb.

Figure 3.2 gives an overview of bacterioplankton (for further details, see, e.g., Kuznetsov, 1970). For more detailed modeling of bacterioplankton than in CoastWeb, it may be important to differentiate between the source of the carbon supply (autotrophs utilize CO_2, heterotrophs also need organic carbon), the energy supply (chemotrophs use energy derived from the oxidation of reduced substances, and phototrophs take energy from solar radiation), and the electron acceptors (i.e., whether oxic or anoxic conditions prevail).

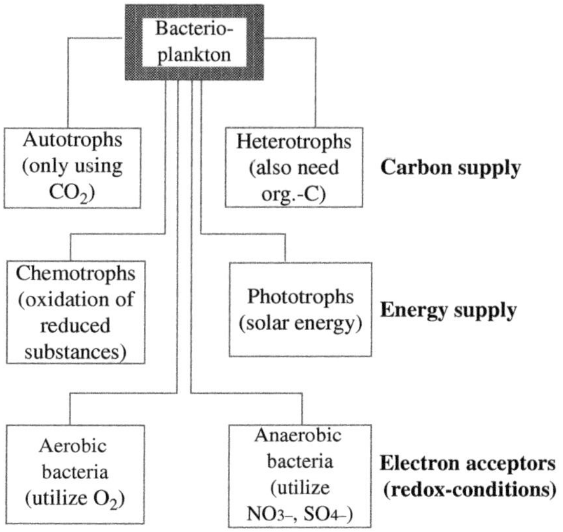

Fig. 3.2 Functional criteria related to bacteria in aquatic systems (see also Wetzel, 1983). Note that there are many organisms with combined energy supplies (i.e., they function as autotrophs plus phototrophs plus aerobic bacteria)

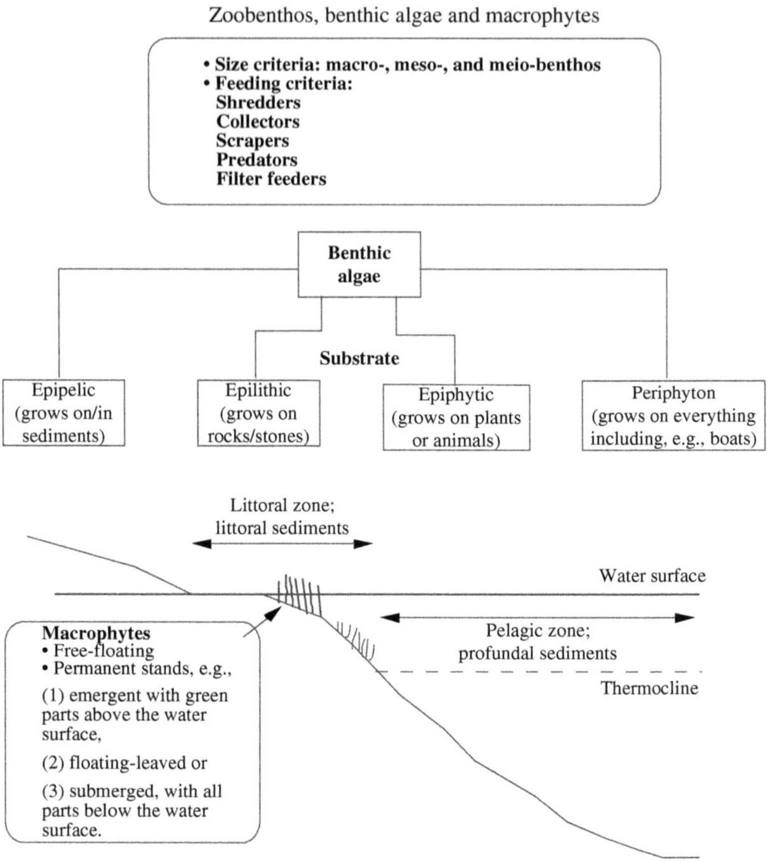

Fig. 3.3 Compilation of concepts related to zoobenthos, benthic algae and macrophytes (see also Vollenweider, 1968, 1976; Cummings, 1973; Brinkhurst, 1974) (modified from Håkanson and Boulion, 2002a)

Figure 3.3 gives a compilation of animals and plants living on or in close contact with the sediments. The three main groups are:

1. Macrophytes (2 main categories, free-floating and with permanent stand, the latter includes 3 groups, emergent, with floating leaves and submerged).
2. Zoobenthos (several groups with different feeding behavior, such as shredders, collectors, scrapers, predators and filter feeders).
3. Benthic algae (different groups related to different substrates, e.g., epipelic, epilitic, epiphytic and periphyton).

Note that Fig. 3.3 also defines the littoral zone (the shore zone) and the pelagic zone (the open water area). Sediments from the pelagic zone are often referred to as profundal.

Zoobenthos represent a staple food for many types of prey fish (the benthivores) and they play a very important role in contexts related to the fish production potential of aquatic systems such as the Baltic Sea. The sediments are their habitat and zoobenthos generally live down to about 5–15 cm sediment depth (see Fig. 3.4). The larger animals (macro- and meiofauna) die if the oxygen concentration in the sediments becomes lower than about 2 mg/l. This will stop the bioturbation (= the biological mixing of the sediments) and laminated (layered and unmixed) sediments may appear. The continuous sedimentation will cause the sediment layer to grow upward so that the bioturbation limit, i.e., the limit between the upper biological layer and the lower biopassive (or geological) layer moves upward (see Håkanson and Jansson, 1983 and Jonsson, 1992 for further information on sediments).

Seston is another term for suspended particulate matter (SPM), and seston may be differentiated into various categories, such as particulate organic matter (POM), particulate inorganic matter (PIM). In many systems about 50% of POM is detritus, i.e., dead plankton, remains of macrophytes, faeces, etc.

In this modeling, we will also include jellyfish. The jellyfish sub-model will be presented in a subsequent section. In Chap. 5, we will discuss a scenario using the jellyfish sub-model and the idea is to study how a possible invasion of jellyfish might influence the fish production of the Baltic Sea system.

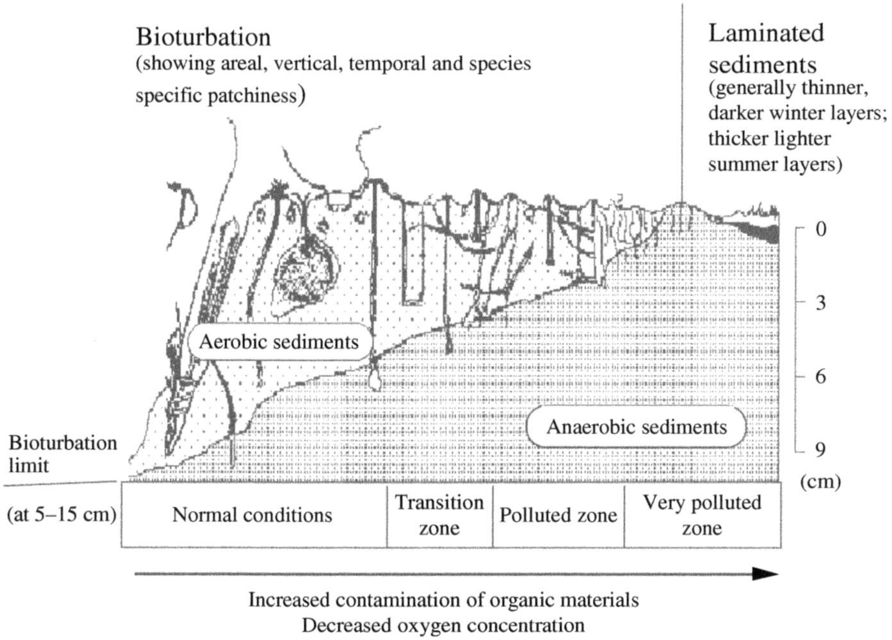

Fig. 3.4 If the oxygen concentration reaches a threshold value of 2 mg/l, zoobenthos die, bioturbation ceases and laminated sediments appear (figure modified from Pearson and Rosenberg, 1976)

3.2 A Brief Comparison with Other Foodweb Models

First, it should be stressed that different models for fish production in aquatic systems generally focus on different targets and use different scales. Fundamentally different approaches concern, e.g., physiological models, models for individual species and models based on functional groups, approaches using different time scales (hours to years), different spatial scales (data from individual sites or mean values from entire ecosystems), different driving variables (e.g., online climatological data or map parameters) and approaches using statistical methods or models based on ordinary differential equations (compartment or box models) or partial differential equations (2-dimensional or 3-dimensional distributed models), see Peters (1991), Monte (1996) and Mace (2001) for discussions on ecosystem modeling. To make a thorough model comparison is beyond the scope of this work.

In a strict sense, there is no such thing as a general (= generic) ecosystem model, which works equally well for all ecosystems and at all scales because all models need to be tested against reliable, independent empirical data and the data used in such tests must of necessity belong to a restricted domain. The ultimate obstacle in achieving predictive power and general validity for a model is to find the most appropriate simplifications, and/or omit small and irrelevant processes related to the given target variables to be predicted (Monte, 1995, 1996; Monte et al., 1997; Peters, 1991; Håkanson and Peters, 1995).

It should be stressed that the point we want to make with the model comparison given below between CoastWeb and Ecopath/Ecosim (see, e.g., Christensen et al., 2000; Walters et al., 1997, 2000; Sandberg et al., 2000; Harvey et al., 2003) is that these two modeling approaches are different, not competing, but complementary. Both of these modeling approaches can calculate production (in kg ww per time unit) and biomasses (in kg ww), but CoastWeb has a separate sub-model to calculate inflow, outflow and internal processes (sedimentation, resuspension, diffusion, etc.) of salt/water fluxes, phosphorus and SPM so that the phosphorus concentration can be related to nutrient sources (such as point sources and river inflow) and used to calculate chlorophyll concentrations, which in turn regulate primary production. Ecopath/Ecosim has no general, validated mass-balance model like CoastMab creating a quantitative link to pollution sources. On the other hand, Ecopath/Ecosim is designed to handle more detailed foodweb interactions than CoastWeb. The point we would like to make is that these models, like all models, have benefits and limitations.

There are many differences between Ecopath/Ecosim and CoastWeb. The models have different structures, driving variables and users:

User knowledge. To use Ecopath/Ecosim for aquatic systems requires detailed knowledge of the studied organisms. CoastWeb is an ecosystem model, which may be used without expert knowledge on ecosystem processes and without comprehensive and detailed data on food choices, constants, production and consumption.

Basic structures. Ecopath/Ecosim is a tool kit and does not include any general well-tested foodwebs, but rather the building blocs to construct such foodwebs. So, Ecopath/Ecosim is a platform for construction, parameterization and analyses of mass-balances of trophic interactions. It can also be used for terrestrial ecosystems.

CoastWeb is, on the other hand, an open foodweb model with a mass-balance model for phosphorus. It can only be used for aquatic systems but it is structured in such a manner that it may be used for most lakes, rivers and coastal areas. The processes and functional groups in the model are general.

Mode of operation. Ecopath can calculate biomass accumulation (as a difference between given start and end values). So, it is not steady-state, but it is not dynamic either. When ecosystems have undergone massive changes, two or more models may be needed. To do modeling over time, one needs Ecosim. To do spatially distributed modeling, one needs Ecospace. When seasonal changes are important different models may have to be constructed for each month, season, or for extreme situations ("summer" vs "winter"). However, Ecosim has a "forcing routine", which allows for manually forced changes in the conditions of the consumer groups. CoastWeb is created to study the dynamics of foodweb interactions and therefore also to find those phenomena one needs to know in Ecopath before one constructs the model. Instead of having to construct different models for different seasons, one gets the dynamic response from the system with CoastWeb.

System demands. In Ecopath, the ecosystem should be defined so that the interactions within the system add up to a larger flow than the interactions between the given system and the adjacent system(s). In practice, this means that the import to and export from a system should not exceed the sum of the transfer between the sub-systems. If necessary, one or more groups originally left outside the system may have to be included in order to achieve this. In CoastWeb, there is no restriction on the size of different flows within or between systems. If a coastal area is to be modeled, and the export/import largely exceeds internal nutrient flows, there is nothing in the model that hinders this from being modeled.

Units. Ecopath can be run with either energy-related units or nutrient related units. But if Ecosim and Ecospace are used, energy is the only choice. In CoastWeb, the transport of phosphorus (in g/week) is modelled in the mass-balance model for phosphorus to and from the different compartments and in the foodweb part of the model, one calculates flows of matter in kg ww per week to and from the different functional groups.

Required input. As a rule, three of the four basic input parameters, biomass, production/biomass ratio, consumption/biomass ratio and ecotrophic efficiency, must be at hand to run the Ecopath model. This must be done for all Ecopath groups (e.g., benthic fish, large zooplankton, benthos, phytoplankton, detritus, etc). If some data are missing, the program might be able to estimate (via a mathematical iteration process) the missing information. If not, the user will be warned by a message and the program will halt. An alternative input is also available. Diet composition must be entered for all consumers. One also needs a lot of boundary conditions (migration). There are also routines for providing fishery information and economical information concerning fisheries. To run LakeWeb, one needs data on area, mean depth, maximum depth, latitude, altitude, characteristic values of water colour and pH and characteristic concentration of total phosphorus in the tributary(ies). The CoastWeb-model also requires data on salinity.

Uncertainty analyses. Ecopath handles uncertainties via a re-sampling routine (Ecoranger). One can enter ranges and mean values for all basic parameters (or use default values) and after selecting frequency distributions (uniform, triangular or normal), the program runs Monte Carlo simulations. CoastWeb can be tested in the same manner, but it is not restricted to those distributions since it can also use other transformations.

Sensitivity analyses. Ecopath has a simple sensitivity routine that varies all basic input parameters from −50 to +50% and checks what this means at each step for each of the input parameters on all of the "missing" basic parameters for each group in the system. In CoastWeb, sensitivity analyses can be made via Monte Carlo simulations where all parameters are varied, one at a time, according to their own estimated or known variance and range.

Foodweb outputs. Ecopath outputs the total throughput (in t/km^2/year) for many variables. It also gives all flows by trophic levels, the trophic impacts, etc. Ecosim uses time series data for biomasses, available from single species stock assessments. Ecopath/Ecosim is thus built more on traditional stock assessments. It has been used for exploring the effects of changes in fishing efforts. CoastWeb outputs all flows (kg ww/week) and biomasses (kg ww) summed or instantaneously for any arbitrary period of the simulation period.

Simulation periods. Ecosim uses 10 years as default simulation period; the maximum is 100 years. One can save the end state and continue with a new, thus enabling long-term simulations. By default, there are 100 time steps per year. This can be changed, but it is not recommended. CoastWeb has no maximum simulation period.

Integration methods. Ecosim uses Adams-Basforth or 4th Runge-Kutta. LakeWeb and CoastWeb use Euler or Runge-Kutta

Spatial distribution. With the addition of Ecospace, one can add spatial distribution. One can then assign habitats to different cells in a user defined grid map of the modeled ecosystem. With probabilities rated for movement, one can then follow migration. CoastWeb is a compartment model and has currently no spatially distribution, but this can be added. Those effects would have to be handled in the same way, with probabilities of migration. This can then be governed by environmental changes, food availability, etc.

In summary: Ecopath/Ecosim is constructed as a tool for scientists; it requires lots of data on the system and attempts to fit the data into a foodweb. Ecosim, like Ecopath, only describes feeding actions. No linkage is made with the surrounding environmental factors. If detailed data are available, one can get very detailed output on foodweb interactions. CoastWeb is a tool for scientists and managers seeking knowledge of an ecosystem in relation to its physical environment and how remedial actions may influence the system, e.g., the production and biomass of fish.

3.3 General Outline of CoastWeb

Figure 3.5 gives a general and more comprehensive outline of the CoastWeb-model, and Fig. 3.6 gives an overview of the interrelationships among the ten functional groups. All equations are compiled in Table A.3. Figure 3.5 also shows that there are several abiotic sub-models in the CoastWeb-model, one for the depth of the

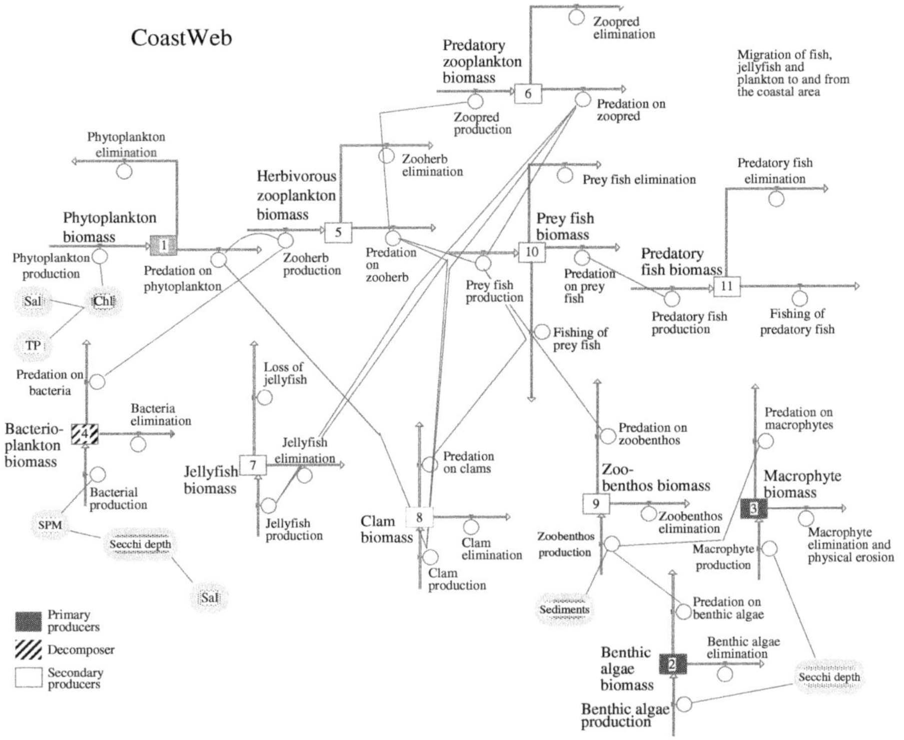

Fig. 3.5 An outline of CoastWeb, a model to quantify all important foodweb interactions for functional groups including biotic/abiotic feedbacks in a general manner. TP, total phosphorus; Chl, chlorophyll; SPM, suspended particulate matter; Sal, salinity

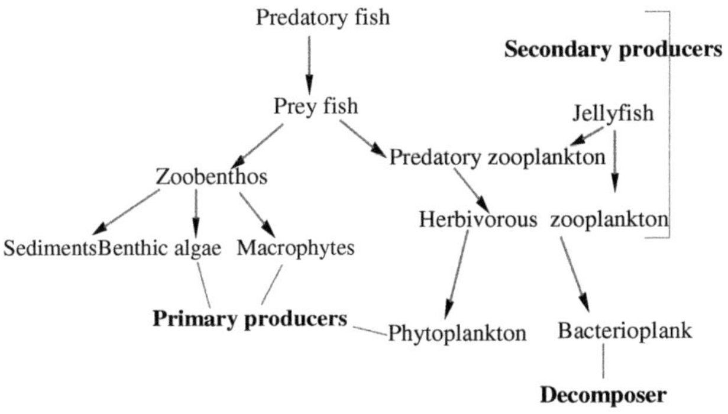

Fig. 3.6 Illustration of the ten functional groups of organisms (phytoplankton, bacterioplankton, benthic algae, macrophytes, herbivorous zooplankton, predatory zooplankton, jellyfish, zoobenthos, prey fish and predatory fish) included in the CoastWeb-model

photic zone (i.e., the sub-model for the Secchi depth, see Chap. 2), CoastMab for phosphorus, CoastMab for salinity and water fluxes, CoastMab for SPM, and the sub-model to predict chlorophyll-a concentrations, which actually drives the CoastWeb-model. To run the model, there is a need also for other driving variables, such as three morphometric parameters (coastal area, mean depth and maximum depth) and water temperature data. If empirical data are not available on water temperatures, such data can be predicted by the temperature sub-model given in Håkanson (2006). The depth of the photic zone (Secchi depth) is certainly one of the most important variables in aquatic ecosystem contexts. It is one of the key variables regulating primary production, and hence also secondary production. Errors in Secchi depth will cause errors in all subsequent calculations.

Table 3.1 stresses the very important role that total phosphorus (TP) concentrations play in aquatic sciences, water management and in this modeling. This table is given here because the empirical models in Table 3.1 play a paramount role in the CoastWeb-model and this will be explained in the coming text.

We have also complemented the regression models given in Table 3.1 by several new algorithms to transfer the lake model to a model for marine areas (see Table 3.2).

The next section will present the basic set-up of each primary production unit in the CoastWeb-model.

3.3.1 Basic Mathematical Structure of Each Unit

There are evident similarities in the set-up used for the different primary producers (as well as for the reducer bacterioplankton), and also clear similarities in the model structure for the secondary producers.

3.3.2 Primary Units

Figure 3.7 illustrates the principle set-up of each primary compartment (e.g., phytoplankton) in the CoastWeb-model. The following processes and factors are accounted for. The basic equation is given by:

$$BM_{PU}(t) = BM_{PU}(t - dt) + (IPR_{PU} - CON_{PUSU} - EL_{PU}) \cdot dt \qquad (3.1)$$

BM_{PU} is the biomass (BM) of the primary unit (PU; in kg ww). The initial BM_{PU}-value (BM_0) is set equal to the normal (= norm) biomass (NBM_{PU} in kg ww), as calculated from the empirical equation for the given unit from Table 3.2. That is:

$$BM_0 = NBM_{PU} \qquad (3.2)$$

Since this is a primary unit, the production is directly related to abiotic limiting factors, such as the concentration of phosphorus, light/temperature, etc. Other

Table 3.1 Many biological variables whose determination normally require extensive and expensive field and laboratory work may be estimated or predicted from one key abiotic state variable, the TP-concentration (in $\mu g/l = mg/m^3$). Some variables may be predicted with great precision (higher r^2), others with much less

y-value	Equation	Range	r^2	n	Units
Chlorophyll (summer mean)	$=0.28 \cdot TP^{0.96}$	2.5–100	0.77	77	mg ww/m^3
Chlorophyll (summer max.)	$=0.64 \cdot TP^{1.05}$	2.5–100	0.81	50	mg ww/m^3
Max. prim. prod. (TP>10)	$=20 \cdot TP-71$	7–200	0.95	38	mg C/m^3·d
Max. prim. prod. (TP<10)	$=0.85 \cdot TP^{1.4}$				mg C/m^3·d
Mean prim. prod. (TP>10)	$=10 \cdot TP-79$	7–200	0.94	38	m C/m^3·d
Mean prim. prod. (TP<10)	$=0.85 \cdot TP^{1.4}$				mg C/m^3·d
Phytoplankton	$=30 \cdot TP^{1.4}$	3–80	0.88	27	mg ww/m^3
Bacterioplankton	$=0.90 \cdot TP^{0.66}$	3–100	0.83	12	mill./ml
Bacterioplankton	$=10^{(0.973 \cdot (0.27 \cdot \log(Chl)+0.19)-0.438)}$				mg ww/m^{-3}
Zooplankton, herbivores	$=0.77 \cdot 38 \cdot TP^{0.64}$	3–80	0.86	12	mg ww/m^3
Zooplankton, predators	$=0.23 \cdot 38 \cdot TP^{0.64}$(the distr. coeff. is 0.77)				mg ww/m^3
Zoobenthos	$=810 \cdot TP^{0.71}$	3–100	0.48	38	mg ww/m^2
Fish	$=590 \cdot TP^{0.71}$	10–550	0.75	18	mg ww/m^2
Fish yield	$=7.1 \cdot TP$	8–550	0.87	21	mg ww/m^2·year
Fish yield	$=0.0023 \cdot PrimP^{0.9}$	170–14,000	0.64	66	mg ww/m^2·year
Prey fish	=DC·fish biomass (default DC=0.73)				mg ww/m^2
Predatory fish	$=(1-DC) \cdot$fish biomass				mg ww/m^2
Macrophyte cover	$= 0.50 \cdot (Sec_{MV}/D_{MV})$			229	%
Macrophytes	$=1.37 \cdot \log(Mac_{cov})+3.58$				g ww/m^2·year
Zooplankton, herbivores	$=0.15 \cdot (PrimP \cdot 1{,}000)^{0.86}$	13–15,000	0.61	42	g ww/m^2·year
Zooplankton, predators	$=0.076 \cdot (PrimP \cdot 1{,}000)^{0.84}$	2–3,000	0.43	42	g ww/m^2·year

From Håkanson and Peters (1995) and Håkanson and Boulion (2002a).
PrimP, primary production (in g ww/m^2·year); Mac_{cov}, Macrophyte cover (in % area); Sec_{MV}, mean annual Secchi depth (in m); D_{MV}, mean depth (in m); DC, distribution coefficient (dimensionless); n, number of lakes used in the regression.

Table 3.2 Equations to predict norm-values, i.e., reference values based on empirical regressions between the listed target variables and TP-concentrations (from Table 3.1) and modifications of these regressions for coastal areas

y-value	Equation	Units
Phytoplankton (IPR_{PH})	$=(1/Y_{SalSec})\cdot(10^{-6})\cdot((2.13\cdot Chl^{0.25}+0.25)^4)\cdot(1/0.45)\cdot(1/0.2)\cdot 30.42\cdot Vol_{2Sec})$	kg ww/month
Phytoplankton (NBM_{PH})	$=Y_{Chl}\cdot(10^{-6})\cdot Vol_{2Sec}\cdot(30\cdot TP_{SW}{}^{1.4})$	kg ww
Phytoplankton (NBM_{PHKA})	$=Y_{ChlKA}\cdot 30\cdot 30^{1.4}$	mg ww/m^3
Benthic algae (NPR_{BA})	$=0.63\cdot(A_{Sec}/A)\cdot PR_{PH}$	kg ww/month
Benthic algae (NBM_{BA})	$=NPR_{BA}\cdot T_{BA}$	kg ww
Bacterioplankton (NBM_{BP})	$=Y_{SPM}\cdot 0.001\cdot Vol_{SW}\cdot 10^{(0.973\cdot(0.27\cdot\log(Chl)+0.19)-0.438)}$	kg ww
Bacterioplankton (NBM_{BPKA})	$=Y_{SPMKA}\cdot 10^{(0.973\cdot(0.27\cdot\log(Chl)+0.19)-0.438)}$	mg ww/m^3
Zooplankton, herbivores (NBM_{ZH})	$=Y_{Chl}\cdot 0.8\cdot 10^{-6}\cdot Vol_{SW}\cdot(38\cdot TP_{SW}{}^{0.64})$	kg ww
Zooplankton, herbivores (NBM_{ZHKA})	$=Y_{ChlKA}\cdot 0.8\cdot(38\cdot 30^{0.64})$	mg ww/m^3
Zooplankton, predators (NBM_{ZP})	$=Y_{Chl}\cdot(1-0.8)\cdot 10^{-6}\cdot Vol_{SW}\cdot(38\cdot TP_{SW}{}^{0.64})$	kg ww
Zooplankton, predators (NBM_{ZPKA})	$=Y_{ChlKA}\cdot(1-0.8)\cdot(38\cdot 30^{0.64})$	mg ww/m^3
Zoobenthos (NBM_{ZB})	$=Y_{ChlZB}\cdot(10^{-6})\cdot 810\cdot(TP_{SW}{}^{0.71})\cdot Area$	kg ww
Jellyfish (NBM_{JE})	$=NBM_{ZP}\cdot 10\cdot Y_{salJE}$	kg ww
Jellyfish (NBM_{KA})	$=NBM_{ZPKA}\cdot 10$	mg ww/m^3
Fish (NBM_{Fish})	$=Y_{Chl}\cdot(10^{-6})\cdot(Area\cdot 590\cdot TP_{SW}{}^{0.71})$	kg ww
Prey fish (NBM_{PY})	$=DC_{PYPD}\cdot SMTH(NBM_{fish}, T_{PY}, NBM_{fish})$	kg ww
Prey fish (NBM_{PYKA})	$=DC_{PYPD}\cdot Y_{ChlKA}\cdot 10^{-6}\cdot(590\cdot 30^{0.71})\cdot Area_{KA}$	kg ww
Predatory fish (NBM_{PD})	$=(1-DC_{PYPD})\cdot SMTH(NBM_{Fish}, T_{PD}, NBM_{Fish})$	kg ww
Predatory fish (NBM_{PDKA})	$=(1-DC_{PYPD})\cdot Y_{ChlKA}\cdot 10^{-6}\cdot(590\cdot 30^{0.71})\cdot Area_{KA}$	kg ww
Macrophyte cover (MA_{Cov})	$=(Area_{2Sec}/Area)\cdot 0.01$	Fraction

Table 3.2 (continued)

y-value	Equation	Units
Macrophytes (NPR_{MA})	$= 0.001 \cdot Area \cdot (1/12) \cdot 10^{(2.472+1.028 \cdot \log(MAcov)-0.516 \cdot 90/(90-Lat))}$	kg ww /month
Macrophytes (NBM_{MA})	$= NPR_{MA}/T_{MA}$	kg ww
$PR_{PH} = BM_{PH}/T_{PH}$	$Y_{Chl} = Chl_{Coast}/Chl_{Lake}$	$Y_{Chlsea} = Chl_{Sea}/Chl_{Coast}$
$Y_{ChlZB} = 1/Yarea_{2Sec}$	$Y_{SalSW} = Sal_{SW}/1$	$Y_{SalSecSW} = Sal_{SW}/12$
Y_{SalJE} = if $Sal_{SW} < 10$ psu then 0 else $Sal_{SW}/10$		$Y_{SPM} = SPM_{SWcoast}/SPM_{SWlake}$
$DC_{PYPD} = (TP_{SW}/(TP_{SW}+22))^{0.4}) > 0.99$ then 0.99; if < 0.9 then 0.9		

Sec, Secchi depth (in m); TP, TP-concentration (μg/l); Chl, concentration of chlorophyll-a (μg/l); Vol_{2Sec}, the coastal volume above two Secchi depths (= the volume of the photic zone, m^3); $Area_{Sec}$, the coastal area above the Secchi depth (m^2); Y, dimensionless moderator; IPR, initial production (kg ww/month); PR, production (kg ww/month); BM, biomass (kg ww); T, turnover time (months); N, normal; e.g., NBM, normal biomass (kg ww); PH, phytoplankton; BA, benthic algae; BP, bacterioplankton; ZH, herbivorous zooplankton; ZP, predatory zooplankton; ZB, zoobenthios; PY, prey fish; PD, predatory fish; JE, jellyfish; SMTH, smoothing function; ww, wet weight; dw, dry weight; SW, surface water; KA, Kattegat.

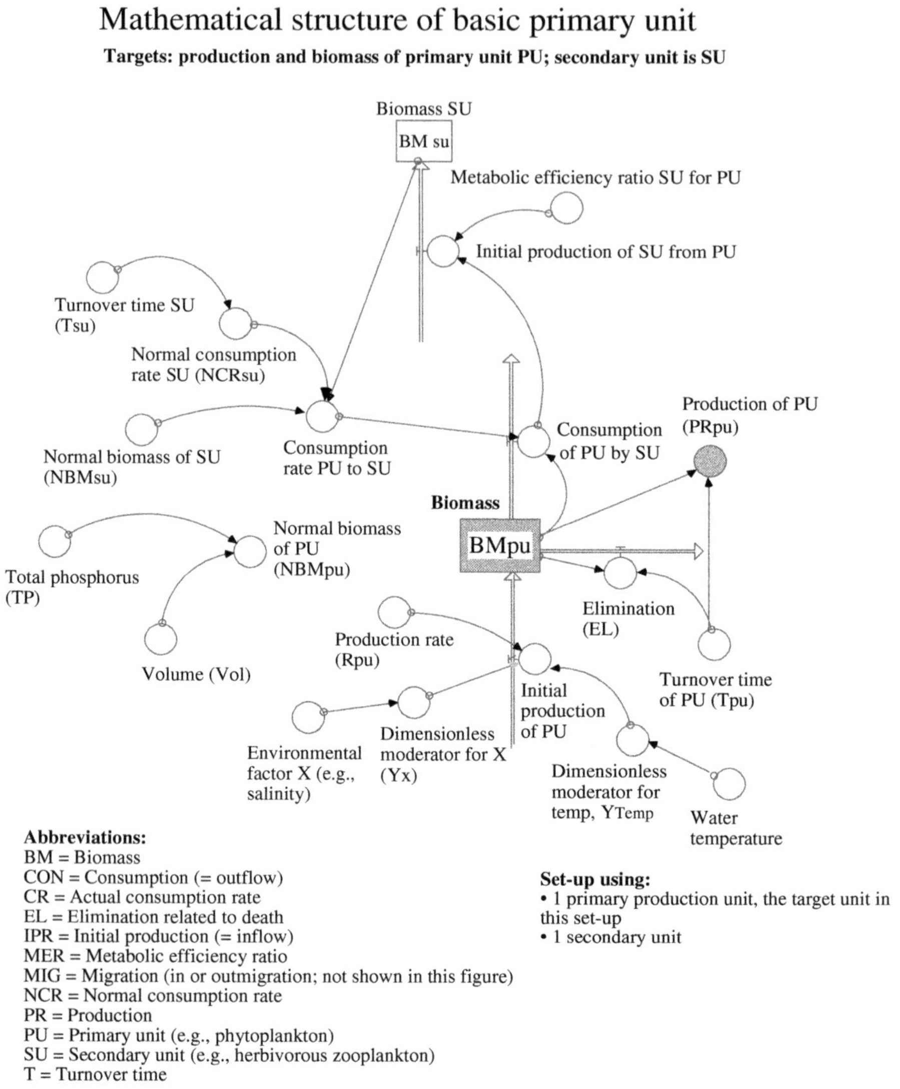

Fig. 3.7 Generalized set-up of the mathematical structure of each primary unit in the CoastWeb-model. The figure also gives abbreviations used throughout this book

environmental factors, such as salinity (see Chap. 2), can also influence the primary production. This is accounted for in the CoastWeb-model in various ways, generally by dimensionless moderator techniques (see Håkanson and Peters, 1995, for more information about dimensionless moderators). That is:

$$\mathrm{IPR_{PU}} = \mathrm{PrimP_{PU}} \cdot \mathrm{Y_X} \cdot \mathrm{Y_{Light}} \tag{3.3}$$

IPR_{PU} = The initial production of the primary unit; generally in kg ww per month.

$PrimP_{PU}$ = The primary production; generally in kg ww per month.

Y_X and Y_{Light} = Dimensionless moderators expressing how changes in environmental conditions, like water chemistry (X), and monthly hours with daylight (light), salinity or temperature influence primary production. Y_X has the following general definition:

$$Y_X = (1 + amp \cdot (X_{Act}/X_{Norm} - 1)) \tag{3.4}$$

Where the amplitude value (amp) quantifies how changes in actual values (X_{Act}) relative to a normal (= reference = norm) value (= X_{Norm}) will influence the production; if $X_{Act} = X_{Norm}$, then $Y_X = 1$.

A simple dimensionless moderator may have the form:

$$Y_{Light} = HDL/Refvalue \tag{3.5}$$

HDL is the monthly numbers of hours with daylight (see Table 2.3) and Refvalue is a reference value (in this modeling, we use 12 h as a reference value, see Eq. 2.15).

The loss of biomass from the compartment is given by three processes, (1) elimination (EL), which is related to the characteristic turnover time of all organisms in the functional group (T_{PU} in months), (2) consumption (= CON, i.e., the predation) by a secondary unit (SU; e.g., zooplankton) and (3) migration (see Sect. 3.4.2). Elimination (EL_{PU} in kg ww per month) is generally given by:

$$EL_{PU} = BM_{PU} \cdot 1/T_{PU} \tag{3.6}$$

LakeWeb used the halflife constant, 1.386 (−ln(0.5)/0.5 = (0.693/0.5; see Håkanson and Peters, 1995) in this algorithm, but the halflife constant has been omitted in CoastWeb and replaced by 1, as motivated in Chap. 2. T_{PU} is the mean, characteristic turnover time (= lifespan) of the organisms in the given compartment. Table 2.9 gave a compilation of turnover times used for all ten groups of organisms included in the CoastWeb-model. These turnover times are important for CoastWeb predictions. The turnover time (T) of a given group of organisms is defined in the traditional way as T = BM/PR, where BM = the biomass of the organism in kg ww and PR = the mass production in kg ww/month. This also means that the requested *production* is often given by the ratio BM/T rather than by the *initial production* (IPR; see Eq. 3.3)

Table 3.3 gives a compilation of useful rules to make calculations between kJ, kcal, g C, g dw and g ww having been used throughout this book for the groups of organisms included in the CoastWeb-model.

An important feature of the model concerns the relationships between the ten groups of organisms and the mathematical techniques to calculate predation and feedbacks. The predation by a secondary unit on the biomass of a primary unit is given by:

Table 3.3 The following calculation constants have been used to transform values for different species given in kcal, g ww and g dw (1 kJ = 4.19 kcal ~ 0.42 g C)

For phytoplankton, bacterioplankton, benthic algae, zoobenthos and fish
1 kcal ~ 0.2 g dw ~ 1 g ww
For zooplankton
1 kcal ~ 0.2 g dw ~ 2 g ww
For macrophytes
1 kcal ~ 0.2 g dw ~ 1.32 g ww

From Håkanson and Boulion (2002a).

$$CON_{PUSU} = BM_{PU} \cdot CR_{SU} \tag{3.7}$$

CON_{PUSU} is the consumption of biomass per time unit (kg ww/month) out of the primary unit from the animals constituting the secondary unit feeding on the primary unit. The *actual* consumption rate, CR_{PU} (1/month) is defined by:

$$CR_{SU} = (NCR_{SU} + NCR_{SU} \cdot (BM_{SU}/NBM_{SU} - 1)) \tag{3.8}$$

NCR_{SU} is the *normal* consumption rate of the given secondary unit (1/month). NBM_{SU} is the normal (= norm = reference) biomass of the secondary unit (1/month); calculated from empirical equations given in Table 3.2.

Basically, the consumption rate for any given organism is related to three factors:

1. The ratio between the actual biomass (BM) and the normal biomass (NBM) of the predator. The higher this ratio, the higher the predation pressure on the given prey.
2. The number of first order food choices (NR; see Fig. 3.8 for illustration of a food-choice panel). The structure of the CoastWeb-model involves several simplifications and there are always either one or two first order food choices. This means that NR is 1 or 2.
3. The inverse of the turnover time (T_{SU}) of the predator, which gives an indication of the predation pressure on the prey. Animals with quick turnover times create a greater predation pressure on their prey than animals with long turnover times (and a higher value of the actual consumption rate, CR_{SU}). So, big animals, like predatory fish, will only eat a relatively small fraction of the total available biomass of its prey per time unit. Small herbivorous zooplankton, on the other hand, are likely to consume a larger percentage of their prey (phytoplankton and bacterioplankton) per unit of time. That is, for animals in the secondary unit, we have:

$$NCR_{SU} = NR_{SU}/T_{SU} \tag{3.9}$$

A consumption rate of 0.2 means that 20% of the biomass of the primary unit is being consumed by the animals making up the secondary unit per time period (here 1 month).

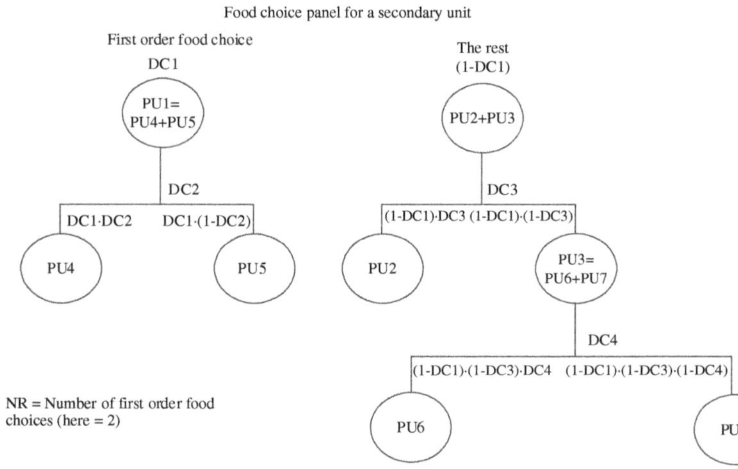

Fig. 3.8 Schematic outline of a food choice panel for a secondary unit with two first order food choices, four second order and two third order food choices

Figure 3.9 illustrates the relationship between NCR_{SU} and T_{SU}, as given by Eq. (3.9). In the CoastWeb-model, we have used the NCR-values determined in this way for all groups of animals except for fish (see also Table 2.9). For prey fish (including planktivores, omnivores and benthivores), CR has been set to $(0.15 \cdot 0.091 + 0.85 \cdot 0.0022) = 0.016$, where 0.091 and 0.0022 are the values for predatory zooplankton, ZP, and predatory fish, PD, respectively. For predatory fish, $CR = Y_{Fish} \cdot 1/T_{PD}$, where Y_{Fish} is a dimensionless moderator related to the trophic state of the system and the amount of food available for predatory fish; CR becomes lower in more productive systems. That is:

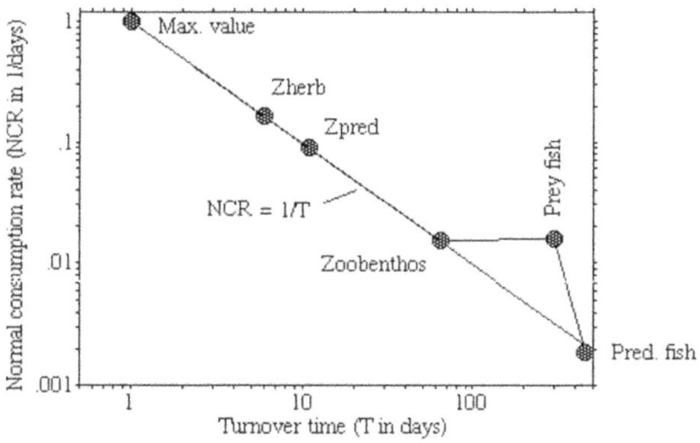

Fig. 3.9 The relationship between the normal consumption rate (NCR) and the turnover time (T). NCR is set to 1/T for all groups with only one food choice. If there are two first order food choices, NCR = NR/T. NR = the number of first order food choices (which is 1 or 2 and never 3 in the CoastWeb-model). The normal consumption rate for predatory fish also depends on the trophic level of the system

$$\text{If } TP_{SW} < 30\mu g/l, \text{ then } Y_{Fish1} = (1 - 2.5 \cdot (NBM_{PYTP}/NBM_{PYref} - 1)) \text{ else } Y_{Fish1} = (1 - 0.4 \cdot (NBM_{PYTP}/NBM_{PYref} - 1)) \text{ and if } Y_{Fish1} < 0.2 \text{ then } Y_{Fish} = 0.2 \text{ else } Y_{Fish} = Y_{Fish1} \quad (3.10)$$

This means that for a system with a TP-concentration of 30 μg/l, Y_{Fish} is 1 and the normal consumption rate is given by the basic approach. For less productive systems, Y_{Fish} will increase; for a system with a TP-concentration of 10 μg/l, Y_{Fish} is 2.67, and the normal consumption rate so much higher. For more productive systems, Y_{Fish} will decrease but the value is never permitted to go below 0.2. NBM_{PYTP} is the calculated normal biomass of prey fish related to the calculated TP-concentration (μg/l), NBM_{PYref} is the biomass of prey fish related to a reference TP-concentration of 30 μg/l (so $NBM_{PYTP} = 10^{-6} \cdot (\text{Area} \cdot 590 \cdot TP_{SW}{}^{0.71})$) and $NBM_{PYref} = (10^{-6} \cdot (\text{Area} \cdot 590 \cdot 30^{0.71}))$; 2.5 is the amplitude value of the moderator for lakes with $TP_{SW} < 30$ μg/l and 0.4 is the amplitude value for more productive systems, and 0.2 a boundary condition in Eq. (3.10).

The characteristic MER-values given in Table 2.9 are used as default values in the CoastWeb-model for the different functional groups or organisms. The MER-value for zoobenthos eating sediments is set to 25% of the MER-value for zoobenthos feeding on benthic algae. The values used for production (PR), respiration (RES), faeces production (FAE) and MER in Table 2.9 are mainly based on data from Winberg (1985).

Prey fish is the most complex group of all in the model. Prey fish feed on three other groups, zoobenthos, herbivorous zooplankton and predatory zooplankton. Neither prey fish nor predatory fish are "permitted" to feed on their own group. Cannibalism exists in natural aquatic systems among fish (see Menshutkin, 1971), and in the CoastWeb-model we calculate net production and net biomasses of all the groups of organisms. This means that predatory cod included among the predatory fish may eat small cod included in the prey fish group, but predatory cod would not eat predatory cod. Using this set-up for prey fish, the calibrations have indicated that unrealistically low production values and biomasses would be obtained for prey fish if one sets the consumption rate for prey fish to 0.0033 ($= 1/T_{PY} = 1/300$; using the dimension per day). A more realistic value should lie between the NCR-values used for predatory zooplankton (ZP) and predatory fish (PD), and we have, as explained, used a value of 0.016. This means that the NCR-value for prey fish is 5 times higher than if the NCR-value would have been determined from $1/T_{PY} = 1/300 = 0.0033$. The main reason for this is related to the structuring of the prey fish compartment, i.e., that all types of prey fish are compiled into one functional group.

Note that Eq. (3.8) quantifies changes in the actual consumption rate of the primary unit related to changes in the biomass of the consumer: the more animals in the secondary unit (the higher BM_{SU}), the higher the actual consumption rate, CR_{SU}. If the actual biomass of the secondary unit is equal to the normal biomass of the secondary unit, then $BM_{SU}/NBM_{SU} = 1$, and $CR_{SU} = NCR_{SU}$. If the actual biomass of the secondary unit, BM_{SU}, is twice the normal biomass of the secondary unit, then $CR_{SU} = 2 \cdot NCR_{SU}$. So, Eq. (3.8) gives a linear increase in consumption with increases in biomass of the secondary unit. If one would like to have non-linear dependences, then Eq. (3.8) may, e.g., be written as:

$$CR_{SU} = (NCR_{SU}+NCR_{SU}\cdot((BM_{SU}/NBM_{SU})^{exp} - 1)) \quad (3.11)$$

Where the exponent, exp, may be given values larger or smaller than 1. For simplicity, we have set the exp-value to 1 in all the following simulations.

To calculate the normal biomass of each primary unit, NBM_{PU}, the equations given in Table 3.2 are used. This means that, e.g., for phytoplankton, $Y_{Chl}\cdot NBM_{PH}$ (values in kg ww), we have:

$$\begin{array}{l} \text{If } TP_{SW} \leq 80\,(\mu\,g/l), \text{ then } NBM_{PH} = Y_{Chl}\cdot 10^6\cdot 30\cdot TP_{SW}1.4\cdot Vol_{SW} \\ \text{If } TP_{SW} > 80, \text{ then } NBM_{PH} = Y_{Chl}\cdot 10^6\cdot 30\cdot TP_{SW}{}^{(1.4-0.1\cdot(TP/80-1))}\cdot Vol^{SW} \end{array} \quad (3.12)$$

Vol_{SW} is the SW-volume in m^3 and TP_{SW} is the mean TP-concentration in $\mu g/l$ ($= mg/m^3$). Y_{Chl} is the dimensionless moderator used to transform a lake regression to a regression for marine systems. It is difined later (see Eq. 3.19).

3.3.3 Secondary Production Units

The set-up for each secondary unit in the CoastWeb-model (there are two zooplankton units, jellyfish, zoobenthos and two fish units) will be discussed in this section and Fig. 3.10 gives an outline of the approach. Note that for simplicity, migration or other fluxes in and out of a given system has not been included in Fig. 3.10. The basic equation is given by:

$$BM_{SU}(t) = BM_{SU}(t-dt) + (IPR_{SUPU1}+IPR_{SUPU2}-CON_{SUPU1}-CON_{SUPU2}-EL_{SU})\cdot dt \quad (3.13)$$

BM_{SU} is the biomass (BM) of the secondary unit (SU; in kg ww). The initial BM_{SU}-value (BM_0) is generally set equal to the normal biomass (NBM_{SU} in kg ww), as calculated from equations in Table 3.2.

In the set-up shown in Fig. 3.10, the initial secondary production (IPR) is related to two fluxes, from primary units 1 and 2 (PU_1 and PU_2). For each flux, IPR is given by:

$$IPR_{SUPU1} = DC_1\cdot CON_{PU1SU}\cdot MER_{PU1SU}\cdot Y_{Temp} \quad (3.14)$$

DC_1 is the first order distribution coefficient (in this set-up there is only one distribution coefficient; NR = 1).

If, however, there are more than two food choices (NR = the number of first order food choices; see Fig. 3.8), the model uses a simple general system to assign weights on such food choices and adjust the consumption rates for the number of food choices. In the CoastWeb-model, there are 3 food choices for zoobenthos (benthic algae, macrophytes and sediments) and also prey fish has a menu of 3 food

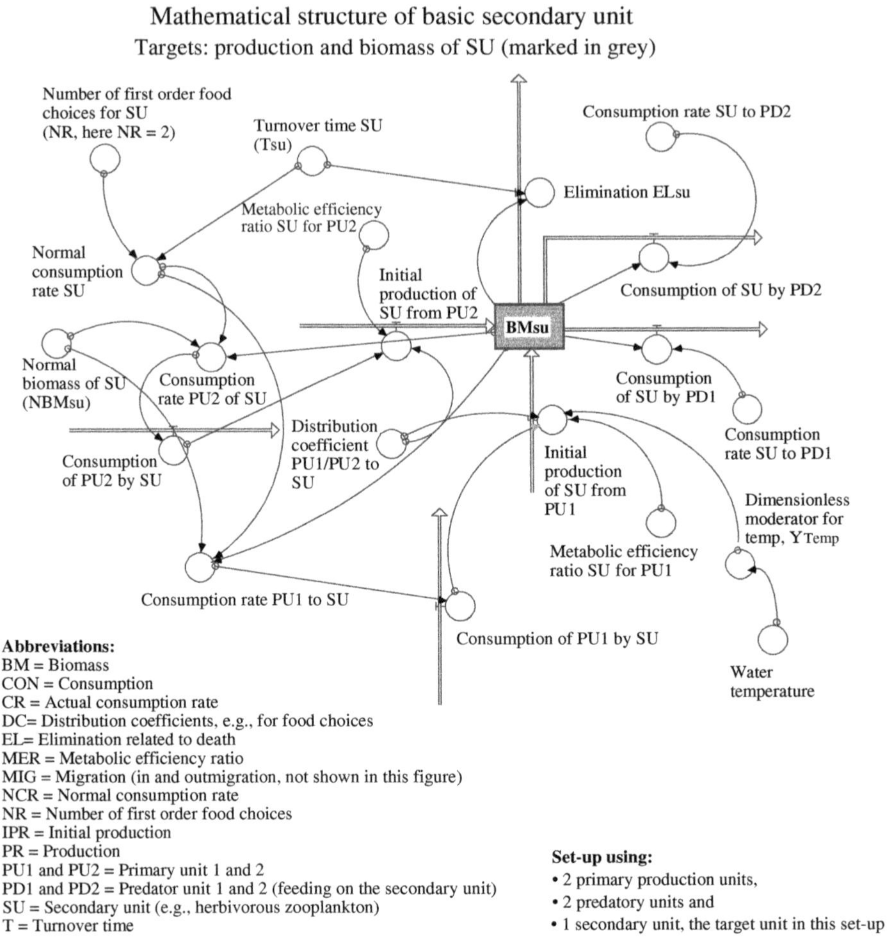

Fig. 3.10 Basic set-up of each secondary unit in the CoastWeb-model. The figure also gives general abbreviations

alternatives (predatory zooplankton, herbivorous zooplankton and zoobenthos). In the schematical model given in Fig. 3.10, the secondary unit has two first order food choices, so NR = 2. So, if there are more than two food choices, they are first differentiated by a distribution coefficient (DC_1) into two first order food choices and then by a second distribution coefficient (DC_2) into second order food choices, etc. (see Fig. 3.8 for illustration).

Y_{Temp} is a simple dimensionless moderator $(SWT/9)^{exp}$ accounting for temperature dependences; 9°C is a standard reference temperature related to the length of the growing season (see Håkanson and Boulion, 2002a) and the exponent varies from 0.25 for zoobenthos, predatory zooplankton and fish, 0.5 for herbivorous zooplankton and jellyfish and 1 for more temperature dependent organisms such as bacterioplankton macrophytes and benthic algae.

CON_{PU1SU} is the consumption (outflow or loss) of biomass from the compartment PU_1 from grazing by the animals in compartment SU (kg ww/month). The grazing rate (= the actual consumption rate, CR_{SUPU1}) is defined in the same manner as was given for the primary unit. That is:

(1) The normal consumption rate is related to the number of first order food choices (NR), and the inverse of the turnover time of the consumer:

$$NCR_{SU} = NR_{SU}/T_{SU} \qquad (3.15)$$

(2) The normal biomass of the secondary unit (NBM_{SU}) is given by an empirical model (see Table 3.2).
(3) The actual biomass of the secondary unit (BM_{SU}) is this is calculated by the CoastWeb-model. That is, Eq. (3.8) may be rewritten for each unit as:

$$CR_{PU1SU} = (NCR_{SU} + NCR_{SU} \cdot (BM_{SU}/NBM_{SU} - 1)) \qquad (3.16)$$

Where CR_{PU1SU} is the actual consumption rate for the secondary unit, SU, feeding on the first primary unit, PU_1, and NCR_{SU} the normal consumption rate for the secondary unit.

The loss of biomass from primary unit 1 from predation by the secondary unit, CON_{PU1SU} is then calculated using Eq. (3.7) as:

$$CON_{PU1SU} = BM_{PU1} \cdot CR_{PU1SU} \qquad (3.17)$$

The metabolic efficiency ratio (dimensionless), MER_{PU1SU}, in Eq. (3.14) gives the fraction of the food that actually increases the biomass of the secondary unit (the consumer). Table 2.9 gave a compilation of characteristic MER-values used for all groups of organisms in the CoastWeb-model. From this table, one can note that the MER-values for plants and animals vary between 0.15 and 0.32. A typical value is 0.2. This means that 20% of the food consumed increases the biomass of the consumer. The MER-vaue for zoobenthos eating sediments is, however, lower (25% of the MER-value for zoobenthos feeding on benthic algae).

The initial production of the secondary unit from consumption of the second primary unit (IPR_{SUPU2}) is handled in the same manner using the same consumption rate but (1-DC_1) instead of DC_1. That is, Eq. (3.16) is valid also for CR_{PU2SU}, so that $CR_{PU1SU} = CR_{PU2SU}$. Then, this initial production is given by:

$$IPR_{SUPU2} = (1 - DC_1) \cdot CON_{PU2SU} \cdot MER_{PU2SU} \cdot (SWT/9)^{exp} \qquad (3.18)$$

Note that this approach involves several simplifications. One can also use specific consumption rates reflecting species-specific feeding preferences in the CoastWeb-model (like switches; see Håkanson and Peters, 1995).

The loss from the compartment PU_2, i.e., CON_{PU2SU}, is calculated in the same manner as the loss from compartment PU_1, CON_{PU1SU} (see Eq. 3.17).

The two transport routes from the secondary unit to the two predatory units feeding on the animals in the secondary unit (PD_1 and PD_2) are handled in the same manner. That is, each flow is given as a function of, (1) the normal consumption rates (here NCR_{SUPD1} or NCR_{SUPD2}), (2) the normal biomasses of the two predatory units (NBM_{PD1} and NBM_{PD2}), as these are given by empirical models (see Table 3.2), and the actual biomasses of the two predatory units (BM_{PD1} and BM_{PD2}), as these are calculated by the CoastWeb-model. This is schematically shown in Fig. 3.3.

Elimination (EL_{SU} = the loss of biomass in kg ww/month from the secondary compartment) is given in the same manner as already described for the primary unit (Eq. 3.16). That is: $EL_{SU} = BM_{SU} \cdot 1/T_{SU}$. Table 2.9 also gave a compilation of turnover times used for all nine groups or organisms included in CoastWeb-model.

The CoastWeb-model also accounts for the fact that fish and other organisms may migrate between habitats for feeding and spawning (see, e.g., Harden-Jones, 1968; Northcote, 1978; McDowall, 1988; Wotton, 1990; Brittain and Brabrand, 2001). The algorithms handling in- and outmigration are new in the CoastWeb-model compared to the LakeWeb-model and will be described and motivated in a following section.

Macrophytes generally appear with relatively high biomasses in lakes and smaller and sheltered coastal areas and they can play several important roles in aquatic ecosystems, e.g., to bind nutrients (see Håkanson and Bryhn, 2008c), as a substrate for zoobenthos (this is handled in CoastWeb in the sub-model for zoobenthos), and by providing shelter for fish and thereby increasing fish production and biomasses. It is important to note that the latter is not related to fish feeding on macrophytes (although single species, like carp, do), but indirectly, by providing more protected living conditions for the small fish. This process is included in the LakeWeb-model but not in this version of the CoastWeb-model for entire, large basins in the Baltic Sea.

3.3.4 Target Variables in CoastWeb

The target variables in this modeling are monthly production values and biomasses for prey and predatory fish, but to be able to predict and understand how the production and biomasses for fish vary seasonally within systems and among different systems, one must also provide as accurate predictions as possible of all defined functional groups of organisms. As a background to the following results and simulations, Fig. 3.11 gives results on production values for these organisms as calculated in the traditional manner from the ratio between the biomass (BM) and the turnover time (T) for the Baltic Proper. The ranking according to annual production values (kg ww/month) is:

Bacterioplankton > Phytoplankton > Herbivorous zooplankton > Benthic algae > Predatory zooplankton > Zoobenthos > Prey fish > Predatory fish > Macrophytes.

There are characteristic seasonal patterns among these organisms: production values (PR = BM/T) for predatory fish and zoobenthos vary relatively little during the year, whereas zooplankton, phytoplankton and benthic algae vary the most, and

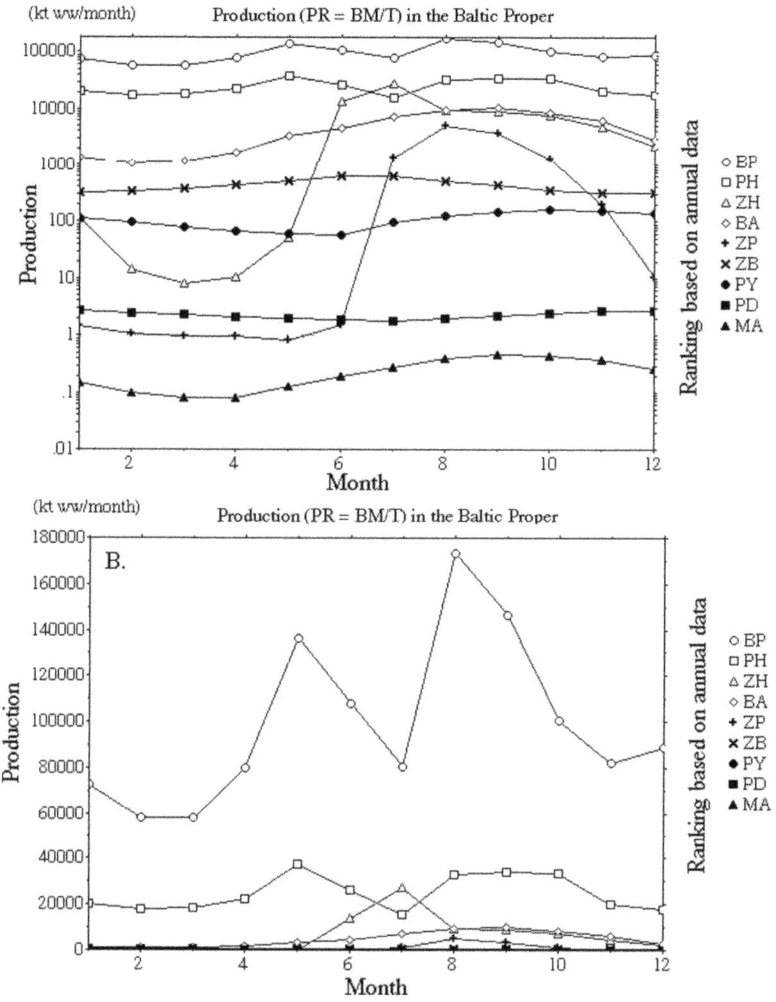

Fig. 3.11 Results using the CoastWeb-model to calculate production values (PR = BM/T) for the Baltic Proper for the nine functional groups of organisms in this system (there are no jellyfish in the default set-up). The figure also gives a ranking based on the annual values

the other groups in-between. These results are based on the calculated biomasses given in Fig. 3.12 for the Baltic Proper for the default period.

It is important to distinguish between production calculated in this manner (as PR = BM/T) and the initial production (IPR) needed to maintain the actual biomasses. At each time step there is elimination, migration and predation. Predation increases the biomass of the predator and reduces the biomass of the prey. This is accounted for in calculating PR, whereas the initial production (IPR) describes the actual initial production. The difference between the initial production and the production (PR = BM/T) are shown in Fig. 3.13 for phytoplankton (a),

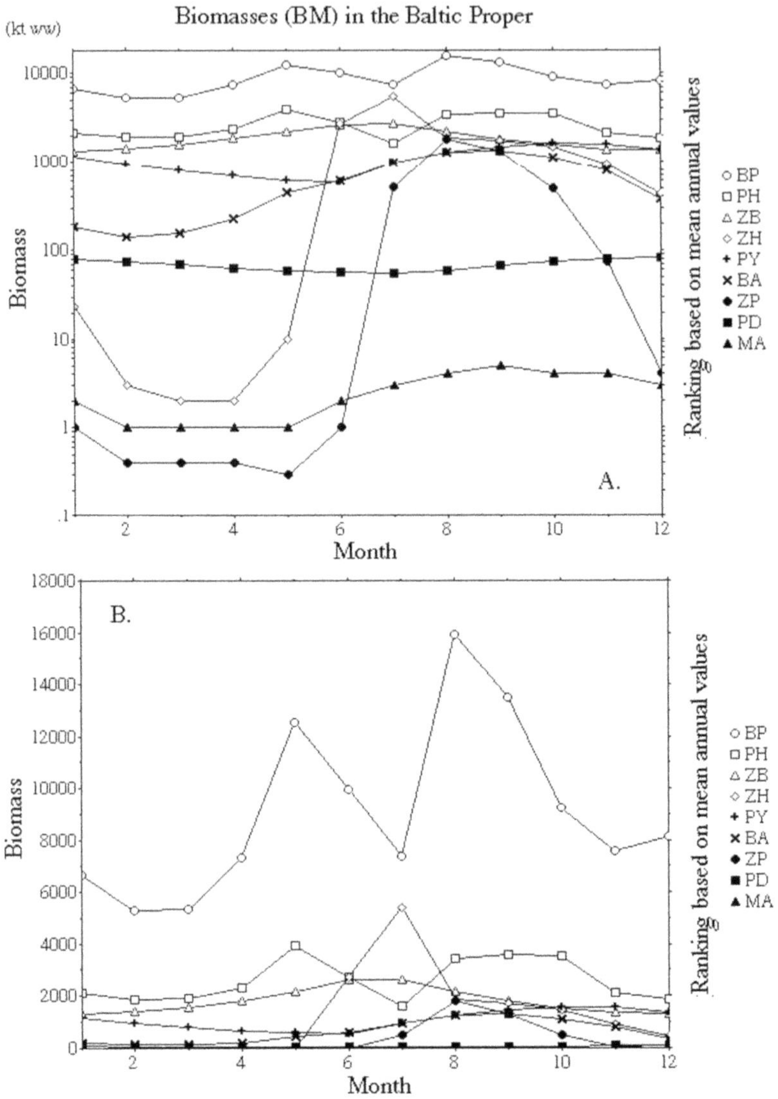

Fig. 3.12 Results (**a** gives logarithmic data, **b** actual values) using the CoastWeb-model to calculate biomasses for the Baltic Proper for the nine functional groups of organisms in this system (there are no jellyfish in the default set-up). The figure also gives a ranking based on the mean annual values

bacterioplankton (b), benthic algae (c), macrophytes (d), herbivorous zooplankton (e), predatory zooplankton (f), zoobenthos (g) and total fish production (h) for the Baltic Proper. From this figure, one can note that there is generally a more marked seasonal pattern for IPR than for PR and that the IPR-values are generally and logically higher than the PR-values.

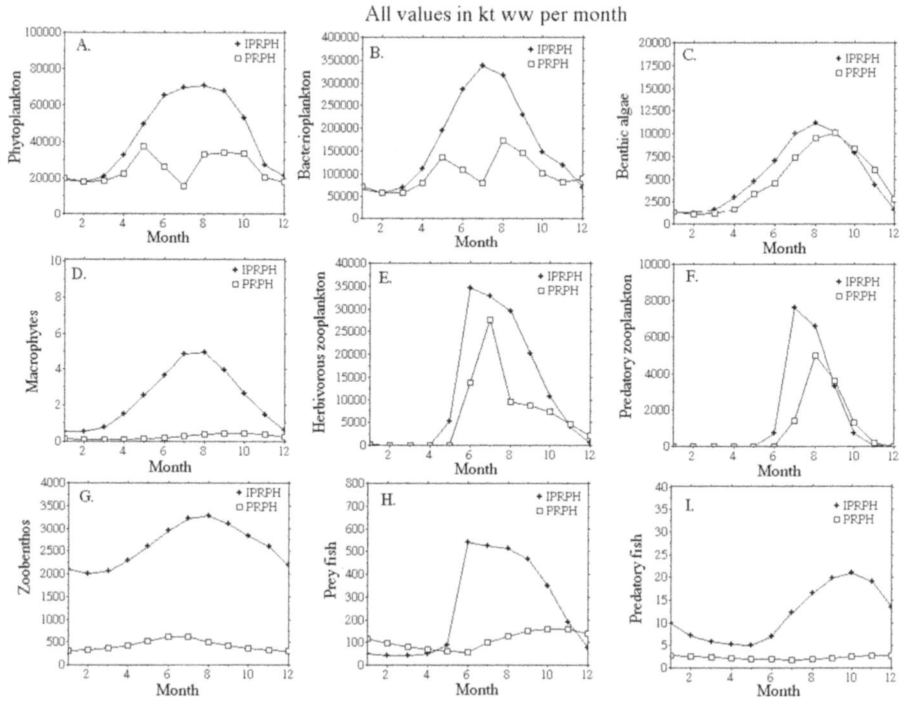

Fig. 3.13 Illustration of the difference between initial production values (IPR in kt ww/month) and production values (PR in kt ww/month), as calculated from the ratio between the biomass (BM) and the turnover time (T), for (**a**) phytoplankton, (**b**) bacterioplankton, (**c**) benthic algae, (**d**) macrophytes, (**e**) herbivorous zooplankton, (**f**) predatory zooplankton, (**g**) zoobenthos, and (**h**) fish (prey fish plus predatory fish) using the CoastWeb-model for the Baltic Proper

3.3.5 Panel of Driving Variables

Same as CoastMab, see Table 2.12.

3.4 Modified Features

This section presents modifications that have been made to adapt LakeWeb to CoastWeb. The role of the salinity and the relationship between primary production, salinity and the amount of SPM were discussed in Chap. 2. In this chapter, we will discuss:

1. The approach to estimate normal biomasses of the functional groups (Sect. 3.4.1).
2. In- and outmigration (immigration and emigration) of organisms between the modeled coastal area and the outside sea (Sect. 3.4.2).

3. Jellyfish as a new functional group (3.4.3).
4. Fishing (3.4.4).
5. Minor modifications (3.4.5).

3.4.1 Predictions of Normal Biomasses of the Functional Groups

Both LakeWeb and CoastWeb use "normal" (= reference = norms) biomasses of all functional groups as a key concept needed for calculations of consumption and migration. For a more detailed description of the normal biomasses and their role in the model, see Håkanson and Boulion (2002a). It should be stressed that in CoastWeb, the TP-concentration affects the foodweb in many ways; primary production is one example. TP also affects the predictions of normal biomasses. The way in which this is done in CoastWeb is highlighted in Table 3.2. All normal biomasses and production values for the functional groups are basically calculated from empirical regressions, most of which include TP-concentrations. Such general regressions as in Table 3.1 for lakes are not, to the best of our knowledge, available for coastal areas. One exception is the regression between primary production and fish yield, which also includes data from marine systems (see Håkanson and Boulion, 2002a, and Chap. 1). This is a major problem, not just to obtain good predictive power using CoastWeb, but also more generally in coastal ecology to know what the "normal" conditions are, given a set of standard abiotic regulating variables, such as TP-concentration, color and Secchi depth for lakes. For coastal areas, the salinity is of paramount importance, not just to the number of species and the prevailing fish stocks, but also to the Secchi depth and the relationship between chlorophyll and nutrient concentrations (see Chap. 2). Since normal biomasses are essential to this foodweb model, the approach here is to modify the normal biomasses used in the LakeWeb-model (see Tables 3.1 and 3.2).

As mentioned in Chap. 2, extensive empirical data have shown that the salinity influences the relationship between Chl and TP (see Fig. 2.8). If the primary production becomes lower at higher salinities, so will the secondary production of zooplankton and fish. Hence, the first step in the approach to estimate coastal norms is to calculate a correction factor (Y_{Chl}) that describes the difference between the concentration of chlorophyll in a lake (Chl_{Lake}) and the concentration of chlorophyll in a given coastal system (Chl_{Coast}) at the same TP-concentration, i.e.:

$$Y_{Chl} = Chl_{Coast}/Chl_{Lake} = TP_{SW}\cdot Y_{Sal}/TP_{SW}\cdot Y_{Sal0} = Y_{Sal}/0.28 \qquad (3.19)$$

One can see from Fig. 2.8 that the salinity moderator for freshwater systems, Y_{Sal0}, is 0.28. In this work, Chl_{Coast} is calculated from the approach discussed in Sect. 2.6, i.e., from dynamically modeled TP-concentrations, salinities, dissolved fractions of phosphorus, monthly hours with daylight and empirical surface-water temperatures. Hence, Y_{Chl} is 1 for lakes and less than 1 for coastal areas that have lower Chl-values than lakes at the same TP. To get the norms for the different functional groups, the corresponding norms in the LakeWeb-model (Table 3.1) are multiplied with Y_{Chl}.

The normal biomass for, e.g., herbivorous zooplankton (NBM_{ZH}) in a given coastal system is calculated as:

$$NBM_{ZH} = Y_{Chl} \cdot DC_{ZHZP} \cdot 10^{-6} \cdot Vol_{SW} \cdot 38 \cdot TP_{SW}{}^{0.64} \tag{3.20}$$

In the LakeWeb-model DC_{ZHZP} is set to 0.77 as a default value (see Håkanson and Boulion, 2002a), but in the following modeling for the Baltic Sea we have, for simplicity, used a default value of 0.8. Vol_{SW} is the coastal SW-volume (m^3) and TP_{SW} is the TP-concentration in the SW-layer (μg/l).

The normal biomass of predatory zooplankton is given as:

$$NBM_{ZP} = Y_{Chl} \cdot (1 - DC_{ZHZP}) \cdot 10^{-6} \cdot Vol_{SW} \cdot 38 \cdot TP_{SW}{}^{0.64} \tag{3.21}$$

The normal biomass for fish (prey plus predatory fish) is in a similar way given by:

$$NBM_{Fish} = Y_{Chl} \cdot 10^{-6} \cdot (Area \cdot 590 \cdot TP_{SW}{}^{0.71}) \tag{3.22}$$

For prey fish, we have:

$$NBM_{PY} = DC_{PYPD} \cdot SMTH(NBM_{fish}, T_{PY}, NBM_{Fish}) \tag{3.23}$$

And for predatory fish:

$$NBM_{PD} = (1 - DC_{PYPD}) \cdot SMTH(NBM_{Fish}, T_{PD}, NBM_{Fish}) \tag{3.24}$$

Where T_{PY} and T_{PD} are the turnover times for prey and predatory fish. The smoothing function (SMTH) is used to adjust the temporal variability to the turnover times of prey fish (T_{PY}) and predatory fish (T_{PD}).

The distribution coefficient regulating the fraction of prey fish in the system is given by Eq. (3.25). The more eutrophic the system is, the higher the fraction of prey fish. That is, clear waters should be expected to have a higher percentage of predatory fish. In the model, we have also used boundary conditions in this algorithm, so that DC_{PYPD} is never permitted to be < 0.9 and never > 0.99 (which means that there should always be at least 10% predatory fish).

$$DC_{PYPD} = (TP_{SW}/(TP_{SW}+22))^{0.4} \tag{3.25}$$

In LakeWeb, the normal biomass of zoobenthos (NBM_{ZB}) is calculated from $810 \cdot (TP_{SW}{}^{0.71})$, see Table 3.1. In CoastWeb for the Baltic Sea, we have:

$$NBM_{ZB} = Y_{ChlZB} \cdot 10^{-6} \cdot Area \cdot 810 \cdot (TP_{SW}{}^{0.71}) \tag{3.26}$$

Area is the entire coastal area (m^2), TP_{SW} is the dynamically modeled monthly TP-concentration in the SW-layer in the given basin. The dimensionless moderator for zoobenthos living on/in the sediments, Y_{ChlZB}, is different from the related dimensionless moderator for organisms in water, Y_{Chl}. This moderator should reflect the difference between freshwater systems and marine systems related to the area of the photic zone (above the depth given by 2·Secchi depth) where benthic algae and macophytes can be found. So, Y_{ChlZB} is simply defined by the ratio between the area above two Secchi depths in the given marine system to the same area in a lake, $Area_{2Seccoast}/Area_{2Seclake}$.

The normal biomass of phytoplankton was given in Eq. (3.12).

The normal biomass of benthic algae (NBM_{BA} in kg ww) is calculated from the normal production of benthic algae (NPR_{BA} kg ww/month) and the turnover time of benthic algae (T_{BA}) by:

$$NBM_{BA} = NPR_{BA} \cdot T_{BA} \tag{3.27}$$

$$NPR_{BA} = 0.63 \cdot (A_{2Sec}/A) \cdot PR_{PH} \tag{3.28}$$

The littoral area (A_{2sec} in m^2) is the part of the bottom shallower than two Secchi depths; the constant 0.63 and the connection to the production of phytoplankton will be explained in more detail in Sect. 4.2.3.1. Note that the algorithm to predict the Secchi depth includes effects of how the salinity influences the water clarity – a higher salinity means an increased production and biomass of benthic algae (if all else is constant).

The normal biomass of bacterioplankton (NBM_{BP}) is estimated from a regression based on chlorophyll (Chl, see Table 3.2), and then modified by a moderator for suspended particulate matter (Y_{SPM}) – the higher the amount of degradable organic suspended matter, the higher the normal biomass of bacterioplankton. For the Baltic Proper (BP), we have:

$$NBM_{BP} = Y_{SPMBP} \cdot 0.001 \cdot Vol_{SW} \cdot 10^{(0.973 \cdot (0.27 \cdot \log(Chl) + 0.19) - 0.438)} \tag{3.29}$$

Where Y_{SPMBP} is given by:

$$Y_{SPMBP} = SPM_{SWcoast}/SPM_{SWlake} \tag{3.30}$$

So, if there is a difference in the SPM-values between a given coastal area and a similar lake, this would influence the normal biomass of bacterioplankton in the coastal system. The SPM-concentration in the surface-water compartment in the coastal area ($SPM_{SWcoast}$) and in a corresponding lake (SPM_{SWlake}) are calculated from the dynamic SPM-model for the Baltic Sea; and the regression between TP and SPM given in Eq. (2.5) is used for lakes.

The same approach is also used to estimate normal biomasses of the functional groups in the sea or basins outside the given coastal area (BM_{sea}; see Table 3.2), which are used to calculate inmigration or inflow (see next section).

The method to calculate the normal biomass for jellyfish will be discussed in Sect. 3.4.3.

3.4.2 Migration In and Out of Coastal Areas

LakeWeb accounts for in- and outmigration of fish to and from the studied lake. Smaller coastal areas in the Baltic Sea have a much more dynamic exchange of water than lakes; a typical theoretical water surface-water retention time for Baltic coastal areas is 5 days and for lakes it is about 1 year (Håkanson, 2000). This affects the in- and outmigration of fish and other organisms to/from smaller coastal areas very much. For large, entire basins in the Baltic Sea, in- and outmigration should be of less importance than in smaller coastal areas, but still, of importance. Evidently, migration of fish is a very complicated issue (Levinton, 2001), and it has to be quantified in the CoastWeb-model for the Baltic Sea, since the aim is to obtain realistic predictions of fish biomasses in the Baltic Sea basins and how the system responds to changes in fishing pressure. Not only prey and predatory fish move around or are being moved around by water currents, but also jellyfish, predatory and herbivorous zooplankton, bacterioplankton and phytoplankton. The new algorithms for in- and outmigration are used for all these functional groups. These new algorithms are based on the following principles:

1. The migration rate (R_{Mig}, per month) in LakeWeb is related to the surface water retention rate ($R_{Mig} = 1/T_{SW}$ in months). This is meant to account for the physical possibilities for organisms to migrate: if there is no inflow or outflow of water, no organisms can migrate or be transported in- and out of the system. Plankton (predatory zooplankton, herbivorous zooplankton, phytoplankton and bacterioplankton) and jellyfish are not likely to migrate in search for food, but are mainly transported by water currents, as given by the surface water exchange rate ($1/T_{SW}$). Jellyfish generally drift rather passively in the free water. Vertical movements are achieved through contraction of the bell (Moen and Svensen, 2004). The deep-water exchange is generally smaller than the surface-water exchange and the focus here is on the water exchange for the productive surface-water layer.
2. Fish can migrate in and out of coastal areas for a number of reasons: as a part of their life cycle, to spawn, mate, etc. (Levinton, 2001) or more seasonally in search for food. This behavior is different for different species. Knowledge of the dominating species in the studied region should be used to define an optimal migration behavior for the functional groups prey and predatory fish for that specific area. For example, in the Baltic Sea, herring is likely to migrate into the smaller coastal areas in great numbers during the spawning season in the spring (see, e.g., Axenrot and Hansson, 2004), which implies that the predatory fish feeding on herring also will migrate into the smaller coastal areas to eat, and that the fish biomass within the coastal areas will increase significantly under such conditions. To be able to account for regional migration patterns, not between smaller coastal areas and the outside sea but between the defined sub-basins used in this work, CoastWeb uses a dimensionless moderator (Y_{Season}). This moderator is multiplied with the default migration rate (R_{Mig}). As in LakeWeb, it is also assumed that the in- or outmigration of fish for coastal areas depend on the relationship between the actual biomass of fish in the coastal area and the normal

biomass (the BM/NBM-ratio). The migration may also sometimes be temperature dependent. Fish grow faster in species-specific temperature ranges and thus need to eat more (Larsson and Berglund, 2005), i.e., some fish might migrate less in cold and very warm water. This influence of temperature should also be accounted for in the Y_{Season}-moderator. An algorithm was added to take into account that the latitude (Lat in °N) probably influences the seasonal migration patterns for fish, with a more pronounced seasonal pattern at high latitudes than at low latitudes with smaller seasonal temperature variations. This has been done in the following manner:

If Lat > 63°N then AV = 1 else AV = (63 – Lat); AV is an averaging function used in the smoothing function (SMTH) below:

$$Y_{SeasonB} = SMTH(Y_{SaeasonA}, AV, 1) \tag{3.31}$$

Where $Y_{SeasonA}$ is the seasonal dimensionless moderator used for all coastal areas at latitudes ≥ 63°N (for which AV = 1; this is an assumed boundary latitude for the given migration algorithm); 1 is the initial value. At lower latitudes, the smoothing function will smooth this curve (Fig. 3.13a). The general moderator for migration is then given by:

$$\begin{aligned}&\text{If } (Y_{SeasonA} - Y_{SeasonB}) \geq 0 \text{ then } Y_{Season} = ((Y_{SeasonA} + Y_{SeasonB})/2) \cdot (Lat/63)\\&\text{else } Y_{Season} = ((Y_{SeasonA} + Y_{SeasonB})/2) \cdot (63/Lat)\end{aligned} \tag{3.32}$$

Y_{Season} is the default dimensionless moderator. It is shown in Fig. 3.14 a for four different latitudes (≥ 63, 56, 45 and 35°N). With this setup, the moderator attains high values in the spring in high latitude coastal areas reflecting regions with strong migrations related to the spawning of the dominating fish species during the spring. This is a suggestion of a general approach that can be used if no information is available on local or regional migratory patterns of fish. If such information is available, it should preferably be used.

3. It would require a very extensive sea-model (compatible to CoastWeb, driven by readily accessible driving variables and generally applicable for any sea) to predict the amount of fish or plankton available for inmigration outside any given coastal area, e.g., the Kattegat outside the Baltic Proper. Since such a model is not available, and since an estimate is needed of how much fish and plankton that is available for immigration, CoastWeb estimates the potentially available fish biomass for inmigration using an estimated normal fish biomass in the sea outside the given coastal area which is calculated from chlorophyll concentrations in the sea outside the coastal area. The inmigration of predatory fish is then calculated in the following manner:

$$\begin{aligned}&\text{If } BM_{PD}/NBM_{PD} > 1 \text{ then } F_{InmigPD} = 0.5 \cdot R_{MigPD} \cdot BM_{PDsea} \text{ else } F_{InmigPD}\\&\qquad = R_{MigPD} \cdot BM_{PDsea}\end{aligned} \tag{3.33}$$

Where R_{MigPD} is the migration rate for predatory fish given by:

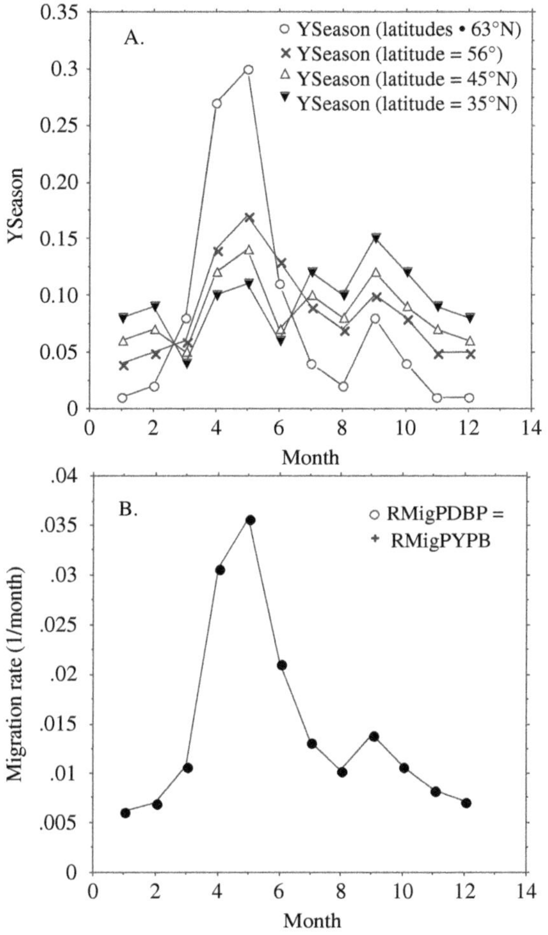

Fig. 3.14 (**a**) Illustration of the seasonal moderator for in- and outmigration, Y_{Season}, for coastal areas at four different latitudes ($\geq$ 63, 56, 45 and 35°N) and (**b**) The default migration rates for prey and predatory fish (R_{MigPY} and R_{MigPD}, respectively) using data for the Baltic Proper area (latitude 58°N); in this modeling these two migration rates are identical

$$R_{MigPD} = 5 \cdot Y_{Season} \cdot (1/T_{SW}) \tag{3.34}$$

If the theoretical surface water retention time, T_{SW}, is 29 months, which is typical for Baltic Proper, the annual migration into the basin represents about 12% of the total biomass of predatory fish or about 5% of the total initial production (see Table 3.5). Since there are no general migration rates to/from these basins available in the literature (to the best of our knowledge), the value of the migration rate constant (5 in Eq. 3.34) is our best estimate based on calibrations and tests against the norms and empirical data (see coming section).

For the inmigration of predatory fish (kg ww/month) from the Kattegat to the Baltic Proper (which is the only "outside sea" to the Baltic Sea), we have estimated the normal biomass of predatory fish in the Kattegat as:

$$NBM_{PDKA} = (1 - DC_{PYPDBP}) \cdot Y_{Chl} \cdot 10^{-6} \cdot (590 \cdot 30^{0.71}) \cdot Area_{KA} \quad (3.35)$$

Where $(590 \cdot 30^{0.71})$ gives the total fish biomass (mg ww/m^2; see Table 3.1); 30 μg/l is the empirical mean TP-concentration for the Kattegat (see Håkanson and Bryhn, 2008b); multiplication with the area of the Kattegat ($22{,}000 \cdot 10^6$ m^2) gives total fish biomass in mg ww; multiplication with 10^{-6} gives the value in kg ww, multiplication with $(1-DC_{PYPDBP})\cdot$ gives the predatory fish biomass; and multiplication with Y_{Chl} gives the adjustment from lakes to marine areas. So, the total biomass of predatory fish in the Bothnian Sea, the Gulf of Finland, the Gulf of Riga and the Kattegat available for inmigration to the Baltic Proper is:

$$BM_{SeaBP} = BM_{PDBS} + BM_{PDGF} + BM_{PDGR} + NBM_{PDKA} \quad (3.36)$$

The inmigration is then:

$$MIG_{InseaBP} = BM_{SeaBP} \cdot (1/T_{SWBP}) \quad (3.37)$$

The outmigration of prey fish from the Kattegat, the Gulf of Finland, the Gulf of Riga and the Bothnian Sea is calculated in the same manner.

For bacterioplankton the total biomasses available for inmigration (= inflow), and the total inmigration are given by (e.g., to the Baltic Proper):

$$BM_{BPsea} = BM_{BPBP} \cdot Y_{SPMseaBP} + BM_{BPBS} + BM_{BPGR} + BM_{BPGF} \quad (3.38)$$

$$\begin{aligned} MIG_{InBPBP} = {} & (1/T_{SWBP}) \cdot (BM_{BPBS} + BM_{BPGR} + BM_{BPGF}) \\ & + Q_{SWKABP} \cdot Y_{SPMKA} \cdot (0.90 \cdot 30^{0.66}) \cdot 10^{-6} \end{aligned} \quad (3.39)$$

Where Y_{SPMKA} is SPM_{sea}/SPM_{coast} and $SPM_{sea} = SPM_{KA}$ is calculated from ($SPM_{sea} = 10^{(1.56 \cdot \log(TPsea) - 1.64)}$; see Eq. 2.5). The TP-concentration in the sea outside the given basins is an obligatory driving variable ($TP_{KA} = 30$ in μg/l is used in these calculations); the normal biomasses for the Gulf of Finland, the Gulf of Riga and the Bothnian Sea are calculated automatically in the CoastWeb-model, as already discussed.

4. Outmigration of predatory fish is calculated in a similar way, e.g., of predatory fish from the Baltic Proper:

$$\begin{aligned} MIG_{OutPDBP} = {} & \text{if } BM_{PDBP}/NBM_{PDBP} > 1 \text{ then } MIG_{OutPDBP} \\ & = R_{MigPDBP} \cdot BM_{PDBP} \text{ else } MIG_{OutPDBP} = 0.5 \cdot R_{MigPDBP} \cdot BM_{PDBP} \end{aligned} \quad (3.40)$$

The same approach is used for prey fish (see Eq. 3.35).

The outmigration (MIG_{Out}, emigration = outflow in kg ww/month) of plankton is given by (e.g., of bacterioplankton from the Baltic Proper):

$$MI_{GOutBPBP} = BM_{BPBP} \cdot (1/T_{SWBP}) \quad (3.41)$$

Figure 3.14b illustrates the migration rate (R_{Mig}) for predatory and prey fish for the Baltic Proper (latitude 58°N; see Table 2.1). A R_{Mig}-value of 0.01 means that 1% of the fish biomass may move either in or out of the coastal area per month.

Table 3.4 gives a compilation of all equations quantifying in- and outmigration. It shows that the basic set-up to calculate migration is applied to all migrating/moving functional groups.

Table 3.5 exemplifies all calculated monthly fluxes of predatory fish, i.e., in- and outmigration, initial production (IPR), production (BM/T), fishing and elimination,

Table 3.4 Compilation of equations used to quantify in- and outmigration of predatory fish, prey fish, jellyfish, predatory zooplankton, herbivorous zooplankton, phytoplankton and bacterioplankton. There is no in- and outmigration for benthic algae or zoobenthos (or macrophytes) in the model. Equations for normal biomasses are given in Table 3.2

$Y_{Chlsea} = Chl_{Sea}/Chl_{Coast}$

Predatory fish (PD); inmigration

If $BM_{PD}/NBM_{PD} < 1$ then $MIG_{InPD} = R_{MigPD} \cdot BM_{PDsea}$ else $MIG_{InPD} = 0.5 \cdot R_{MigPD} \cdot BM_{PDsea}$

Migration rate: $R_{MigPD} = (5 \cdot Y_{Season}/T_{SW})$ (this is the theoretical surface water retention rate)

Dimensionless seasonal moderator for migration: Y_{Season}

If $(Y_{SeasonA} - Y_{SeasonB}) \geq 0$ then $Y_{Season} = ((Y_{SeasonA}+Y_{SeasonB})/2) \cdot (Lat/63)$ else $Y_{Season} = ((Y_{SeasonA}+Y_{SeasonB})/2) \cdot (63/Lat$, where $Y_{SeasonB} = SMTH(Y_{SaeasonA}, AV, 1)$

Predatory fish (PD); outmigration

If $BM_{PD}/NBM_{PD} < 1$ then $MIG_{OutPD} = 0.5 \cdot R_{MigPD} \cdot BM_{PD}$ else $MIG_{OutPD} = R_{MigPD} \cdot BM_{PD}$

Prey fish (PY) inmigration

If $BM_{PDY}/NBM_{PY} < 1$ then $MIG_{InPY} = R_{MigPY} \cdot BM_{PYsea}$ else $MIG_{InPY} = 0.5 \cdot R_{MigPY} \cdot BM_{PYsea}$

Migration rate: $R_{MigPY} = (Y_{Season} \cdot 1/T_{SW})$

Prey fish (PY); outmigration

If $BM_{PY}/NBM_{PY} < 1$ then $MIG_{OutPY} = 0.5 \cdot R_{MigPY} \cdot BM_{PY}$ else $MIG_{OutPY} = R_{MigPY} \cdot BM_{PY}$

Jellyfish (JE)

Migration rate: $R_{MigJE} = 1/T_{SW}$

Inmigration: $MIG_{InJE} = R_{MigJE} \cdot BM_{JEsea}$

Outmigration: $MIG_{OutJE} = R_{MigJE} \cdot BM_{JE}$

Predatory zooplankton (ZP)

Migration rate: $R_{MigZP} = 1/T_{SW}$

Inmigration: $MIG_{InZP} = R_{MigZP} \cdot NBM_{ZPsea}$

Outmigration: $MIG_{OutZP} = R_{MigZP} \cdot BM_{ZP}$

Herbivorous zooplankton (ZH)

Migration rate: $R_{MigZH} = 1/T_{SW}$

Inmigration: $MIG_{InZH} = R_{MigZH} \cdot BM_{ZH}$

Outmigration: $MIG_{OutZH} = R_{MigZH} \cdot BM_{ZH}$

Phytoplankton (PH)

Migration rate: $R_{MigPH} = 1/T_{SW}$

Inmigration: $MIG_{InPH} = R_{MigPH} \cdot BM_{PHsea}$

Outmigration: $MIG_{OutPH} = R_{MigPH} \cdot BM_{PH}$

Bacterioplankton (BP)

Migration rate: $R_{MigBP} = 1/T_{SW}$

Inmigration: $MIG_{InBP} = R_{MigBP} \cdot BM_{BPsea}$

Outmigration: $MIG_{OutBP} = R_{MigBP} \cdot BM_{BP}$

Table 3.5 Calculated monthly and annual biomasses and production values of predatory fish and fluxes (initial production, elimination, fishing, in- and outmigration) of predatory fish in the Baltic Proper under default conditions

Month	Biomass BM_{PD} (kt ww)	IPR_{PD} (kt ww/m)	$Elim_{PD}$ (kt ww/m)	$Fishing_{PD}$ (kt ww/m)	MIG_{InPD} (kt ww/m)	MIG_{OutPD} (kt ww/m)	Production BM_{PD}/T_{PD} (kt ww/m)
Jan	79	10	2.6	13	0.3	0.2	2.7
Feb	74	7.3	2.4	11	0.3	0.3	2.5
Mar	68	5.8	2.2	9.1	0.8	0.7	2.3
Apr	62	5.2	2.0	7.8	1.4	1.0	2.1
May	58	5.1	1.9	6.8	1.3	0.8	2.0
Jun	55	7.0	1.8	6.3	0.8	0.5	1.9
Jul	54	12	1.9	6.7	0.5	0.3	1.8
Aug	58	17	2.1	8.0	0.6	0.4	2.0
Sep	65	20	2.3	10	0.6	0.4	2.2
Oct	73	21	2.6	12	0.5	0.4	2.5
Nov	79	19	2.7	14	0.4	0.3	2.7
Dec	81	14	2.7	14	0.3	0.3	2.8
Annual	67	143	27	118	7.8	5.5	27

for the Baltic Proper. Tables 3.5 and 3.6 for prey fish are included here to show the magnitude of the different fluxes, as given by the model. Evidently, uncertainties in major fluxes are much more decisive for the predictions than uncertainties in minor fluxes. So, it is important to identify the major fluxes and use algorithms that quantify these fluxes as correctly as possible. From Table 3.5, one can note:

- The biomass of predatory fish varies between 54 and 81 kt ww during the year in the Baltic Proper for the default period (1997–2006).
- The initial production in the system is relatively high during summer and fall with a yearly total of 143 kt; the production values are lower, about 27 kt/year under default conditions considering migration and fishing and the fact that the biomass is relatively small (67 kt on average) and the turnover time for predatory fish relatively long (900 days).
- In- and outmigration are relatively low, 7.8 and 5.5 kt/year, respectively. There is a small net inmigration of 2.3 kt/year predatory fish to the Baltic Proper from the Kattegat, the Bothnian Sea, the Gulf of Finland and the Gulf of Riga, according to these calculations (for the default period).
- The total annual fishing of predatory fish in Baltic Proper under default conditions is 118 kt/year, which is significantly higher than the production (27 kt ww/year) but lower than the initial production (143 kt/year).
- The elimination of predatory fish (death, etc.) is 27 kt/year.

It should be stressed that it is sometimes argued in these contexts that: "to achieve a sustainable fishery, the permitted fishing quota must never be higher than the fish production". If we would follow that line of argument for the Baltic Proper (using data for the default period) and permit an annual fishing of 120 kt, Fig. 3.15 shows that the predatory fish biomass would collapse in the Baltic Proper; that also the prey

Table 3.6 Calculated monthly and annual biomasses and production values of prey fish and fluxes (initial production, elimination, fishing, in- and outmigration) of prey fish in the Baltic Proper under default conditions

	Biomass BM_{PY} (kg ww)	IPR_{ZBPY} (kg ww/m)	IPR_{ZHPY} (kg ww/m)	IPR_{ZPPY} (kg ww/m)	From PY to PD (kg ww/m)	$Elim_{PY}$ (kg ww/m)	$Fishing_{PY}$ (kg ww/m)	MIG_{InPY} (kg ww/m)	MIG_{OutPY} (kg ww/m)	Production (BM_{PY}/T_{PY}) (kg ww/m)
Jan	1,142	52	0.9	0.01	52	106	85	2	7	116
Feb	946	44	0.1	0.01	40	88	66	2	8	96
Mar	791	43	0.1	0.01	31	74	51	9	8	80
Apr	678	49	0.2	0.005	25	65	41	16	10	69
May	602	89	29	0.01	21	59	35	13	8	61
Jun	581	541	443	1.5	26	74	42	8	8	59
Jul	978	526	330	56	43	116	68	4	13	99
Aug	1,269	511	298	70	58	140	90	5	17	129
Sep	1,480	466	295	42	74	157	113	5	19	150
Oct	1,588	350	233	11	84	161	129	4	15	161
Nov	1,554	190	103	0.98	82	149	127	3	12	158
Dec	1,378	77	16	0.04	68	128	109	2	8	140
Annual	1,082	2,940	1,747	181	605	1,316	956	73	134	1,317

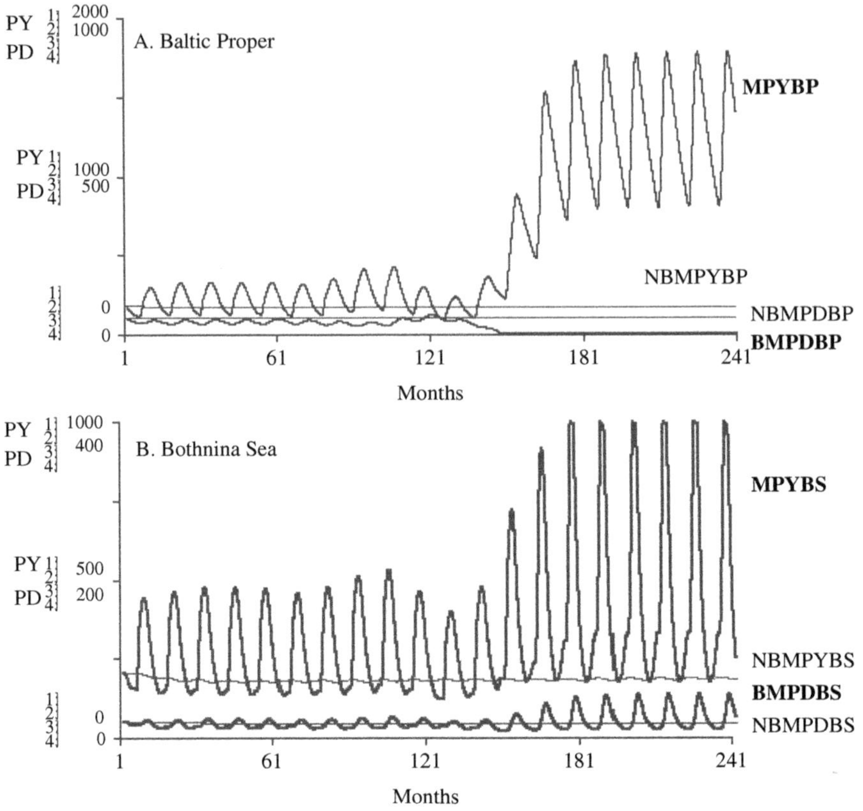

Fig. 3.15 A simulation of the consequences for predatory fish (PD; the two *upper curves*) and prey fish (PY; the two *lower crves*) in the Baltic Proper (BP) and the Bothnian Sea (BS) if the fishing of predatory fish in the Baltic Proper under default conditions from month 61 is set to 120 kt ww/year, which is lower than the initial production of (143 kt/year), higher than the production (27 kt/year) and slightly higher than the present total fishing (118 kt/ww). BM is biomass and NBM is normal biomass. From this figure, one can also note the good correspondence between dynamically modeled biomasses and the modeled norm-values in the initial period (before month 61) and the fishing for predatory fish in the Baltic Proper would collapse

and predatory fishing in the connected basins would be influenced (see Fig. 3.15b using data for the Bothnian Sea); given a lower predatory pressure from the predatory fish, the biomass of the prey fish would increase in the Baltic Proper; the higher biomass of prey fish in the Bothnian Sea would cause a higher biomass of predatory fish in the system. So, and there are connected consequences for the entire trophic cascade in all sub-basins in the Baltic Sea. These results are given here to stress that this modeling can provide data of fundamental importance in settling optimal fish quota and we will give a scenario in Chap. 5 to discuss in greater detail the reasons behind the results in Fig. 3.15. Here, we will just stress that the total permitted fish quota is only one component of the total fishing. There is also illegal fishing, recreational fishing and seals, birds and animals consume fish. The permitted fish quota must include considerations to all types of fishing and the production potential of the entire system. The fish quota should be adjusted to the environmental conditions regulating the fish production potential of the system (i.e., changes in nutrient concentrations, temperature, salinity, etc.) and include a margin for uncertainties in data and modeled values. These aspects will be discussed in Chap. 5.

From Fig. 3.15, one can also note the good correspondence between dynamically modeled biomasses and the modeled norm-values in the initial period (before month 61). In Chap. 4, we will give many such comparisons for basically all functional groups in all sub-basins of the Baltic Sea. This is one important manner in which the modeled values can be critically tested. Ideally, there should be a close (order-of-magnitude) correspondence between the empirically-based norm-values and the values predicted by the CoastWeb-model.

From Table 3.6, which gives the corresponding values for the default period for the Baltic Proper for prey fish feeding on zooplankton and zoobenthos, one can note:

- The biomass of prey fish in this coastal area is about a factor 6 (67/1,081 = 6.2) times higher than for predatory fish and this is what the empirical data shows for the default period around the year 2000 in Fig. 2.46. As should be expected, the biomass for prey fish varies more seasonally than the biomass for predatory fish; the values for prey fish vary between 580 and 1,600 kt ww during the year in the Baltic Proper for the default period.
- The initial production is about 3,000 kt/year and the prey fish production about 1,300 kt/year.
- Inmigration is 70 kt/year and outmigration 130 kt/year, which means a net outmigration of prey fish from the Baltic Proper.
- Zoobenthos is the most important food for prey fish in the Baltic Proper. The initial prey fish production from zoobenthos consumption is 2,900 kt/year, compared to 1,750 kt/year from consumption of herbivorous zooplankton and 180 kt/year from consumption of predatory zooplankton. This indicates that zoobenthos is very important for the total fish production in the Baltic Proper and in the entire Baltic Sea.
- The total annual fishing of prey fish under default conditions is 1,320 kt/year.
- The loss of prey fish (death, etc.) is the same, 1,320 kt/year.

These results can also be used to highlight another scenario in Chap. 5 dealing with the setting and finding of optimal fish quota and optimal fishing rates.

Table 3.7 gives initial results concerning predatory fish biomass, initial production of predatory fish, elimination, total fishing, in- and out-migration and production (PR = BM/T) at different fishing rates, 0.5 (which is the default value in the CoastWeb-model, but also note that the fishing rate is 2 times higher for predatory fish than for prey fish), 0.75, 1, 2, 0.25 and 0.1 applied for all basins in the Baltic Sea for the years around 2000. From this table one can note:

- The maximum total fishing of predatory fish would be 144 kt ww/year in the Baltic Proper if the fishing rate is 0.1; only 48 kt/year if the fishing rate is set as

Table 3.7 Calculated annual biomasses (mean values in kt ww) and production values of predatory fish and fluxes (initial production, elimination, fishing, in- and outmigration) of predatory fish in the Baltic Proper for different fishing rates

Fishing rate constant (1/m)	Mean biomass BM_{PD} (kt ww)	IPR_{PD} (kt ww/year)	$Elim_{PD}$ (kt ww/year)	$Fishing_{PD}$ (kt ww/year)	MIG_{InPD} (kt ww/year)	MIG_{OutPD} (kt ww/year)	Production BM_{PD}/T_{PD} (kt ww/year)
0.5 (default)	67	143	27	118	7.7	5.5	27
0.75	49	113	20	95	6.1	4.0	20
1	39	93	16	79	5.2	3.2	16
2	21	54	9	48	3.8	1.7	9
0.25	104	193	42	141	5.8	16	42
0.1	166	231	67	144	7.7	28	67

higher as 2; the total fishing is 118 kt ww/year under default conditions when the fishing rate is 0.5. It should be noted that these results mean that the effort to catch twice as much predatory fish in the Baltic Proper is much higher if the fishing rate is set too high because there is then less fish in the system to be caught. This may seem like a paradox – the higher the fishing rate, the less fish will be caught for the same effort.

- So, if the fishing rate is very low (0.1), the total biomass of predatory fish available for fishing is 166 kt ww, which is a factor of 8 higher than if the fishing rate is set to 2.0.
- Finding the optimal fishing rate, or the optimal fish quota, means a balance not just between the available biomasses of predatory fish and the fishing effort, but also between the conditions regulating the fish production of the system. The strength of the CoastWeb-model is that it can be used as a tool to do that.

These preliminary results for the default period will be further elaborated in Chap. 5. It should be evident that the CoastWeb-model can be used for many important purposes, but it should also be stressed that there are some issues that one can not address using the CoastWeb-model, e.g., (1) how fishing and migration influence individual species, such as the dominating predatory fish in the Baltic Sea, adult cod, or the dominating prey fish species, such as sprat, herring and young cod or (2) or how local fish stocks within the system may differ significantly from the average functional groups modeled using the CoastWeb-model, which do not provide spatial resolution.

These simulations also demonstrate that zoobenthos is an important food source for prey fish in the Baltic Sea and that threats to the production of zoobenthos (e.g., low oxygen conditions in water and sediments) would be serious to the fish production in the system. In- and outmigration of fish and zooplankton are relatively small transport processes. There are generally no, or only small amounts of, jellyfish in the Baltic Proper, but in areas with higher salinities, jellyfish could abound. The next section focuses on the modeling of jellyfish.

Also note that the "sea" is the Baltic Proper for the Gulf of Riga and the Gulf of Finland; it is the Bothnian Sea for the Bothnian Bay, and the Kattegat, the Gulf of Finland and the Gulf of Riga for the Baltic Proper.

In- and outmigration are not calculated for benthic algae, macrophytes and zoobenthos.

3.4.3 Jellyfish

In Chap. 4, we will give a scenario on the possible consequences for Baltic Sea fishery related to an invasion of jellyfish. The aim of this section is to present the sub-model for jellyfish, which is a new feature in the CoastWeb-model.

Jellyfish (JE) can appear in great numbers and are able to consume substantial amounts of mainly zooplankton, which could reduce the fish production of the system (Schneider and Behrends, 1994; Brodeur et al., 2002; Purcell, 2003). This makes them a potentially important part of the coastal foodweb and they have been included as a secondary production unit in CoastWeb. The default threshold salinity value for invasions of jellyfish in this model has been set to 10; Lucas (2001) gave a value of 14 for a common jellyfish (*Aurelia aurita*). Jellyfish is a predatory zooplankton and could belong to the predatory zooplankton group in the model. However, the medusae-stage of jellyfish, i.e., what we generally mean by jellyfish, is so different from other predatory zooplankton concerning size, abundance, etc. that jellyfish have been assigned an own functional group in CoastWeb. Recent studies have shown that certain types of jellyfish, e.g., the ctenophore *Mnemiopsis leidyi*, occur (see Javidpour et al., 2006; Gorokhova et al., 2009) at salinities down to about 4 psu in the Baltic Sea. This gives further motivates for the jellyfish scenario in this book and when more information becomes available the jellyfish model presented in this section may have to be modified.

Figure 3.16 illustrates the jellyfish sub-model and Table 3.8 gives all equations. This sub-model is built in the same way as all sub-models for functional groups in CoastWeb.

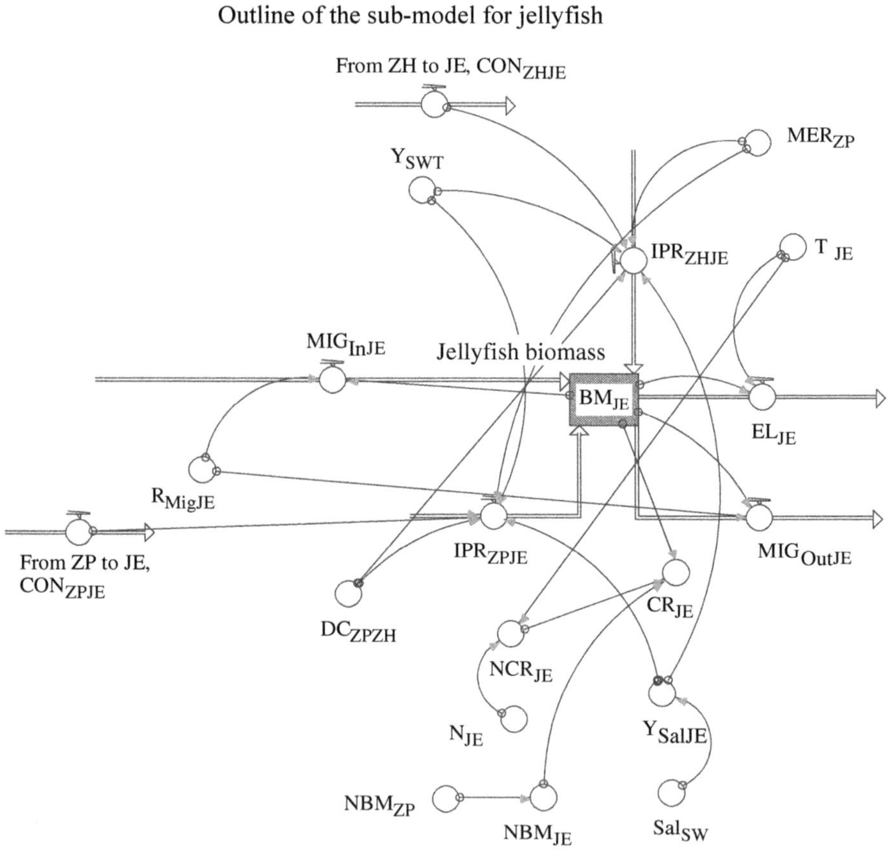

Fig. 3.16 An outline of the new sub-model for jellyfish

Table 3.8 Basic differential equation for production and biomass of jellyfish (JE)

$BM_{JE}(t) = BM_{JE}(t - dt) + (IPR_{ZHJE} + IPR_{ZPJE} + MIG_{InJE} - EL_{JE} - MIG_{OutJE})\cdot dt$
$IPR_{ZHJE} = Y_{SalJE}\cdot(1 - DC_{ZPZH})\cdot CON_{ZHJE}\cdot MER_{ZP}\cdot(Y_{Temp})^{0.5}$ (initial production of JE from eating ZH, kg ww/month)
$IPR_{ZPJE} = Y_{SalJE}\cdot DC_{ZPZH}\cdot CON_{ZPJE}\cdot MER_{ZP}\cdot(Y_{Temp})^{0.5}$ (initial production of JE from eating ZP, kg ww/month)
$MIG_{InJE} = (1/T_{SW})\cdot NBM_{JEsea}$ (inmigration of JE, kg ww/month)
$EL_{JE} = BM_{JE}\cdot 1/T_{JE}$ (elimination of JE, kg ww/month)
$MIG_{OutJE} = (1/T_{SW})\cdot BM_{JE}$ (outmigration of JE, kg ww/month)
$CON_{ZHJE} = BM_{JE}\cdot CR_{JE}$ (flux from ZH to JE, kg ww/month)
$CON_{ZPJE} = BM_{JE}\cdot CR_{JE}$ (flux from ZP to JE, kg ww/month)
$Y_{Chl} = Chl_{Coast}/Chl_{Lake}$ (dim. less correction factor for biomasses in coasts/lakes related to chlorophyll)
$CR_{JE} = (NCR_{JE}+NCR_{JE}\cdot(BM_{JE}/NBM_{JE}-1))$ (consumption rate for JE; if $NBM_{JE} = 0$, then $CR_{JE} = 0$; 1/month)
$DC_{ZPZH} = 0.5$ (distribution coefficient for JE eating ZP or ZH; dim. less)
$MER_{ZP} = 0.32$ (metabolic efficiency ratio for JE eating ZP or ZH; dim. less)
$N_{JE} = 2$ (number of first order food choices; dim. less)
$NBM_{JE} = 10\cdot NBM_{ZP} = 10\cdot Y_{Chlsea}\cdot(1 - DC_{ZHZP})\cdot 10^{-6}\cdot(V_{SW}\cdot 38\cdot TP^{0.64})$ (normal biomass of JE, kg ww)
$R_{ProdJE} = 8.5$ (initial production rate for JE, 1/month)
$NCR_{JE} = N_{JE}/T_{JE}$ (normal consumption rate for JE, kg ww/month)
Sal_{SW} = Surface-water salinity (psu)
$T_{JE} = 120/30.42$ (turnover time for JE in months)
Y_{SalJE} = if $Sal_{SW} < 10$ psu then 0 else 1 (assumed threshold salinity for JE production)
$Y_{Temp} = SWT/9$ (dimensionless moderator for SW-temperature influences on bioproduction)

Jellyfish mainly eat zooplankton (Larson, 1987; Mills, 1995; Hansson, 2006) and hence there is only one food choice for jellyfish: between predatory zooplankton and herbivorous zooplankton. So, the number of first order food choices for jellyfish is $NR_{JE} = 2$, separated by DC_{ZPZH}, and there are no second order alternatives. Jellyfish are also known to consume ichthyoplankton (fish eggs and larvae; Cowan et al., 1996; Suchman and Brodeur, 2005). For simplicity, this is not considered in this modeling where the basic idea is to see if an invasion of jellyfish would likely threat the fish production of the system.

The total consumption of predatory zooplankton by jellyfish is given by $F_{ZPJE} = BM_{ZP}\cdot CR_{JE}$. BM_{ZP} is the available biomass of predatory zooplankton and CR_{JE} is the actual consumption rate (i.e., jellyfish eating its prey). CR_{JE} is given by:

$$CR_{JE} = (NCR_{JE}+NCR_{JE}\cdot(BM_{JE}/NBM_{JE}-1)) \qquad (3.42)$$

NCR_{JE} is the normal consumption rate, BM_{JE} is the actual biomass of jellyfish and NBM_{JE} is the normal biomass. This means that the model quantifies changes in the actual consumption of the prey unit related to changes in the biomass of the consumer: more animals in the secondary unit (a higher BM_{JE} compared to NBM_{JE})

means a higher actual consumption rate constant, CR_{JE}. If the actual biomass of the secondary unit is equal to the normal biomass of the secondary unit, then $BM_{JE}/NBM_{JE} = 1$, and $CR_{JE} = NCR_{JE}$. If the actual biomass of the secondary unit, BM_{JE}, is twice the normal biomass, then $CR_{JE} = 2 \cdot NCR_{JE}$. So, the model gives a linear increase in consumption with increase in biomass of the secondary unit. We will also, as mentioned, use a threshold value for jellyfish and assume that there are only small amounts of jellyfish in the system if the salinity is lower than 10 psu, but that invasions of jellyfish would be possible if the salinity would be higher than the threshold value. However, this will be changed in the jellyfish scenario in Chap. 5.

$$NBM_{JE} = 10 \cdot NBM_{ZB} \cdot Y_{SalJE} \tag{3.43}$$

The normal biomass of jellyfish (NBM_{JE}) is set to be 10 times higher than the normal biomass of predatory zooplankton (NBM_{ZP}) and related to the salinity of the system (Y_{SalJE}).

By using a boundary condition, NBM_{JE} can never be less than 0. The normal consumption rate, NCR_{JE}, is $NCR_{JE} = 2/T_{JE}$. T_{JE} is the turnover time of jellyfish (i.e., of the medusae stage). According to Lucas (2001), medusae of the common jellyfish, *Aurelia aurita*, live for 4–8 months in most environments. In the model, 120 days is used as a default value of turnover time for jellyfish. The initial production (IPR) of jellyfish (JE) from eating predatory zooplankton (ZP) is given by:

$$IPR_{JE} = R_{ProdJE} \cdot Y_{SalJE} \cdot DC_{ZPZH} \cdot CON_{ZPJE} \cdot MER_{ZP} \cdot (SWT/9)^{0.5} \tag{3.44}$$

The distribution coefficient, DC_{ZPZH}, gives the fraction of predatory zooplankton versus herbivorous zooplankton consumed by jellyfish (the default value is 0.5). The MER-value is the amount of the total consumption (CON_{ZPJE}) that will increase the biomass of the jellyfish. The jellyfish digestion of its prey is temperature dependent (Martinussen and Båmstedt, 1999), which also implies that the consumption is temperature dependent. This is accounted for in the CoastWeb-model by Y_{Temp} ($Y_{Temp} = (SWT/9)^{exp}$; SWT = the surface-water temperature in °C; the exponent is set to 0.5 for jellyfish). Y_{SalJE} is a salinity moderator, which works in the following way: If the surface-water salinity (Sal_{SW}) is lower than 10 psu, $Y_{SalJE} = 0$ else $Y_{SalJE} = 1$. The initial production rate for JE (1/month, R_{ProdJE}) has been determined after calibrations against the normal biomass of jellyfish and we use a default value of 8.5 for this rate.

Jellyfish may be eaten by fish and turtles (Legović, 1987) and they can also be consumed by other jellyfish (Martinussen and Båmstedt, 1999). However, this is not accounted for in this set-up of the model. Jellyfish are removed from the coastal system by three processes: elimination, related to the turnover time of jellyfish, and outmigration (= outflow).

In- and outmigration of jellyfish are calculated in the same manner as for plankton from:

$$MIG_{InJE} = (1/T_{SW}) \cdot NBM_{JEsea} \cdot Y_{SalJE} \tag{3.45}$$

$$MIG_{OutJE} = (1/T_{SW}) \cdot BM_{JE} \tag{3.46}$$

The actual biomass of jellyfish in the coastal area, BM_{JE}, is calculated automatically in the model.

Elimination of jellyfish, i.e., the loss of biomass in kg ww/month (EL_{JE}) is given as:

$$EL_{JE} = BM_{JE} \cdot 1/T_{JE} \tag{3.47}$$

The turnover time for jellyfish, T_{JE}, is, as stated set to 120 days.

3.4.4 Fishing

Fishing here means all types of fishing from professional fishermen and leisure-time anglers, birds, seals, etc. The fish production is connected to the primary production by several intermediate trophic links (and this is a focal point in the CoastWeb-modeling), see Fig. 1.6. In quantifying fishing of predatory and prey fish, CoastWeb is based on the following principles:

There is a general fishing rate, which applies to both prey and predatory fish and it describes the monthly total loss of fish from the given system as (illustrating the set-up with examples for predatory fish in the Bothnian Bay, BB):

$$R_{FishBB} = (BM_{PDBB}/NBM_{PDBB}) \cdot (Y_{ArearefBB}) \cdot R_{Fishconst}/12 \tag{3.48}$$

This means that the fishing rate increases in systems with much fish, i.e., if the actual biomass of predatory fish (BM_{PDBB}) is higher than the normal biomass of predatory fish (NBM_{PDBB}), the fishing should increase because more fishermen should be inclined to fish. The fishing rate const, $R_{Fishconst}$, is set to 0.5 (per month) as a default value for prey fish and twice that value for predatory fish, which are generally more attractive for fishermen and may be caught by nets used for the purpose of catching larger fish. The default value for the fishing rate constant has been determined after many calibration rounds from the basic idea that there is a maximum fishing pressure and so that all modeled biomasses of all functional groups in all basins in the Baltic Sea should reflect a high but sustainable fishing. The higher fishing rate for the predatory fish compared to prey fish is also used in the basic LakeWeb-modeling. The general algorithm for the fishing pressure is also assumed to depend on the size/area of the system, so that it is larger in small basins than in large basins. This is given by:

$$Y_{ArearefBB} = (10^{12}/Area_{BB})^{0.5} \tag{3.49}$$

An area of 10^{12} m^2 is used as a reference area in this dimensionless moderator for the fishing pressure (the largest sub-basin in the Baltic Sea is the Baltic Proper with an area of $0.211 \cdot 10^{12}$ m^2). This means that $Y_{Arearef}$ varies from 7.7 for the smallest sub-basin in the Baltic Sea (GR) to 2.2 for the largest basin (BP); and hence, the default fishing rate is 7.7/2.2 = 3.5 times higher in the Gulf of Riga as compared to the Baltic Proper. Evidently, this would vary. This is a default set-up of the modeling, which has been used in the calibration of the model.

This mean that total fishing (kg ww/month) of predatory fish, e.g., in the Bothnian Bay is given by:

$$FISH_{PDBB} = 2 \cdot M_{PDBB} \cdot R_{FishBB} \quad (3.50)$$

For the prey fish it is, $FISH_{PYBB} = M_{PYBB} \cdot R_{FishBB}$. The fishing is quantified in the same manner in all basins.

It should be stressed that in testing the CoastWeb-model, the algorithm to calculate the default fishing rate has focused on the data illustrating the actual documented fishing in mainly the Baltic Proper (see Fig. 3.17). The model should be able to describe those changes in biomasses and fishing as closely as possible and this will be discussed in greater detail in Chap. 4.

3.4.5 Minor Modifications

There are a few minor modifications in transforming LakeWeb into CoastWeb. These changes will be briefly presented in this section:

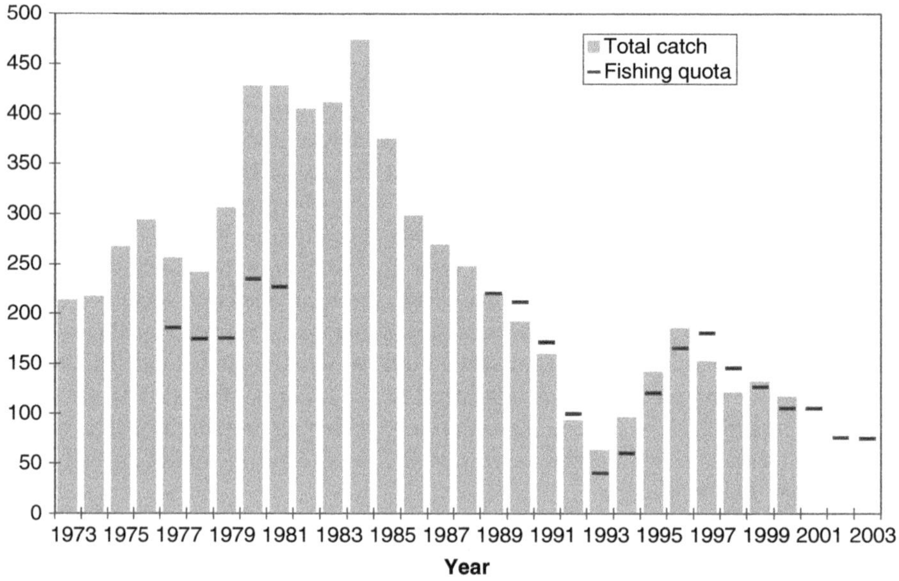

Fig. 3.17 Cod fishing in the Baltic Sea and the fishing quota (TAC) according to the International Baltic Sea Fisheries Commission (IBSFC, 2003)

1. The volumes of the deep-water zone, V_{DW}, or the middle-water layer, V_{MW} in the Baltic Proper and the Gulf of Finland, are not calculated from a general morphometric algorithm but from the actual hypsographic and volume curves for the five sub-basins.
2. The moderator, Y_{Eh1}, for expressing low oxygen stress on zoobenthos related to sedimentation of oxygen-consuming materials on accumulation areas beneath the theoretical wave base (Sed_A) is now simply given by two boundary conditions:

$$\begin{aligned}&\text{If } Sed_A \text{ in cm/year} > 0.75 \text{ then } Y_{Eh1} = 0 \text{ else } Y_{Eh1} = 1 \text{ and}\\&\text{if } Sed_A \text{ in cm/year} < 0.075 \text{ then } Y_{Eh} = (1 - 1\cdot(Sed_A/0.075 - 1))\\&\text{else } Y_{Eh} = 1\end{aligned} \tag{3.51}$$

So, if the mean sedimentation is higher than 0.75 cm/year, which is not the case in the sub-basins of the Baltic Sea (see Table 2.14), Y_{Eh1} is 0 and the initial production of zoobenthos also zero. If the mean sedimentation is low (< 0.075 cm/year), the oxygen consumption should be low, which would favor the survival of the zoobenthos. If, e.g., $Sed_A = 0.01$, Y_{Eh} is 1.87 and the initial production of zoobenthos from eating sediments 1.87 higher than if Sed_A is between 0.075 and 0.75.
3. The sediment pool of bioavailable food for detrivorous zoobenthos is now given by (for the Bothnian Bay):

$$\begin{aligned}Sedpool_{BB} &= ((12-5)/12)\cdot(IG/100)\cdot(1/(1-W/100))\\&\quad\cdot(M_{SPMABB}+M_{SPMETBB})\end{aligned} \tag{3.52}$$

The default organic content (= loss on ignition, IG) for the sediment layer 0–10 cm in all Baltic Sea basins is set to 12% dw (dw = dry weight; based on data from mainly Jonsson, 1992; see Chap. 2); the IG-concentration in glacial clays is generally about 5% dw (see Jonsson, 1992), which means that about 60% of the organic content may be lost by bacterial decomposition, hence the constant ((12–5)/12); (1/(1–W/100)) transforms IG-values from dw to ww; the mass of SPM (M_{SPMA}) on the accumulation areas below the theoretical wave base, the A_{DW}-areas in the Bothnian Bay, the Bothnian Sea and the Gulf of Riga, and the A_{MW}-areas and A_{DW}-areas in the Gulf of Finland and the Baltic Proper together with the SPM on the ET-areas, M_{SPMET}, will constitute the potential sediment pool of food for the zoobenthos.

Below the halocline, at depths larger than 75 m in the Baltic Proper, the biomass of zoobenthos should be relatively small. Generally, the density of zoobenthos decreases exponentially with water depth and the highest value are to be found at water depths above the Secchi depth (see Möller et al., 1985; Håkanson and Rosenberg, 1985; Jonsson, 1992). The oxygen stress is always pronounced below the halocline in the Baltic Proper (see Fig. 2.29) where also laminated sediments abound (see Fig. 3.18a). Figure 3.18b gives the trend for the changes in the extension of laminated sediments in a smaller coastal area, the

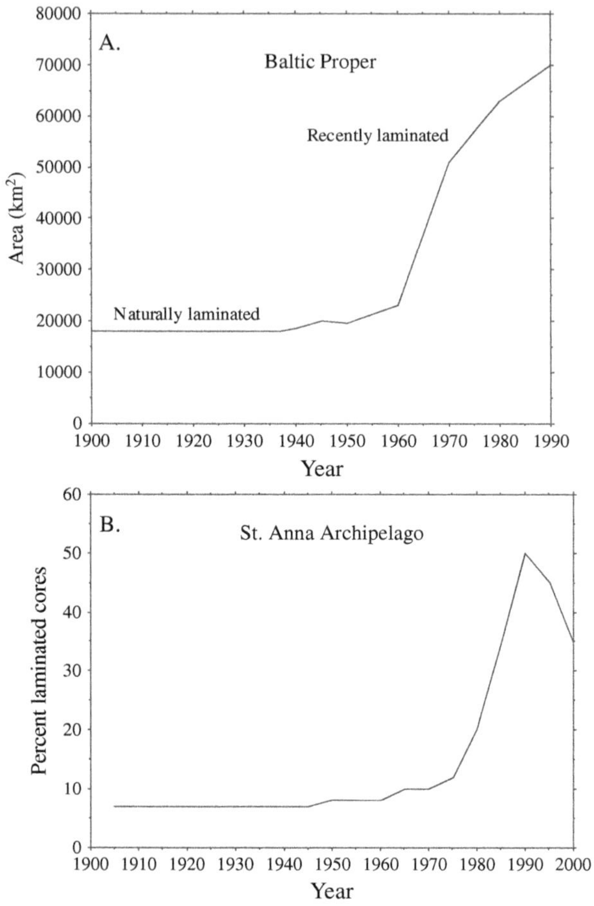

Fig. 3.18 (**a**) Changes in areas with laminated sediments in the Baltic Proper from 1900 to 1990 (data from Jonsson, 1992). (**b**) Changes in areas with laminated sediments in St. Anna archipelago, Sweden (data from Persson, 1999)

St. Anna Archipelago on the Swedish east coast. It should be stressed that the production potential and "biological value" of such coastal areas are very great because all primary producers (phytoplankton, benthic algae and macrophytes) can be found in smaller coastal areas, but not below the halocline at 75 m in the Baltic Proper. It is also interesting to note that this smaller coastal area is recovering. This is also the case for the entire Baltic Proper (see Håkanson and Bryhn, 2008b and Chap. 2).

Under aerobic (= oxic) conditions zoobenthos may create a biological mixing of the sediments down to about 15 cm sediment depth (the bioturbation limit). If the deposition of organic materials increases and hence also the oxygen consumption from bacterial degradation of organic materials, the decreasing oxygen concentration may reach a threshold value of 2 mg/l, when zoobenthos die, bioturbation ceases and laminated sediments appear (see Fig. 3.4).

4. The distribution coefficients regulating the prey fish consumption of either predatory zooplankton or herbivorous zooplankton (DC_{ZPZH}) been set to 0.2 in all Baltic Sea basins to stress that the main zooplankton consumption by prey fish concerns herbivorous zooplankton.
5. The distribution coefficient regulating the prey fish consumption of either zooplankton or zoobenthos has not been changed; the value 0.75 is used both in the LakeWeb-model and the CoastWeb-model. The main reason for this is that, based on data on the fish species constituting the prey fish group in the Baltic Sea (see Chap. 2), zooplankton is an important food for several prey fish species, such as sprat and herring, but small cod, as well as many other species, mainly eat zoobenthos.
6. Influences of macrophyte cover and macro algae on fish production. The higher the macrophyte cover in lakes and smaller coastal areas, the higher the potential fish biomass in the given basin (if everything else is constant), since the macrophytes provide a protected environment for small fish (Sogard and Able, 1991). In this modeling for the large entire basins, we have omitted the dimensionless moderator in the LakeWeb-model, which increases the fish production with increasing macrophyte cover.

3.5 Concluding Comments

This chapter has presented and motivated the building blocks of the CoastMab-model for the Baltic Sea, and also modifications compared to the basic LakeWeb-model. In the next chapter, we will focus specifically on all the functional groups and we will use results from the literature survey presented in Chap. 2 to test the model including the distributions coefficient in the food choice panel. We will also give the detailed modeling set-up for all the functional groups and present simulations to illustrate how the model behaves.

Preliminary simulations indicate that zoobenthos is an important food for prey fish in the Baltic Sea and that threats to the production of zoobenthos (e.g., low oxygen conditions) would be serious to the fish production in the system. In- and outmigration of fish and zooplankton are relatively small transport processes.

In this chapter, we have shown that the predatory fish biomass would likely collapse within 4 years in the Baltic Proper if the permitted fishing of predatory fish would be equal to the total initial production of predatory fish, that also the predatory fishing in the connected Baltic Sea basins would be seriously reduced, and given a much lower predatory pressure from the eliminated predatory fish in the Baltic Proper, the biomass of the prey fish would increase very much. These results have been given here to stress that this modeling can provide data of fundamental importance in settling optimal fish quota and we will give a scenario in Chap. 5 to discuss in greater detail the reasons behind the results in this chapter.

Chapter 4
Modeling of the Different Functional Groups

The aim of this chapter can be stated in one short and simple sentence: It is to present all details related to the understanding and modeling of all functional groups of organisms in the CoastWeb-model. It is, however, difficult to make this presentation short and simple, but we will try.

4.1 Introduction, Aim and Set-Up

The same set-up and structure of the presentations will be given for all functional groups of organisms in each of the following sections. We will:

(1) First give short introductory information on the role of the given functional group.
(2) Present the modeling of each functional group and discuss details, which were not included in the general overview of the CoastWeb-modeling in Chap. 3.
(3) Present comparisons between dynamically modeled production values and biomasses, values from the empirically-based norms (see Table 3.2) and, when available, empirical data. We will also compare modeled values in the different sub-basins to see whether there are logical and realistic differences among the sub-basins.
(4) And also present results from three scenarios to pave the way for Chap. 5, which will discuss the use of the CoastWeb-model in more specific scenarios.

Chapter 2 gave background information on the most dominating species within each functional group present in the Baltic Sea to get an overview of the organisms making up the main biomass of each group; Chap. 2 also presented results of literature surveys aimed at finding as adequate data as possible on all distribution coefficients in the food choice panel for each functional group; and that chapter also discussed ecosystem differences between the five sub-basins in the Baltic Sea, and also between the smaller coastal areas and entire, large basins, as a basis for the structuring of the CoastWeb-modeling.

L. Håkanson et al., *The Fish Production Potential of the Baltic Sea*, Environmental Science and Engineering, DOI 10.1007/978-3-642-11562-2_4,

Scenario 1. In this chapter, we first present results when the TP-loading to the system from tributaries has been reduced in three steps. The first 10-year period gives the default conditions and this will demonstrate that it may take a few (<5) years to reach steady state (from initial conditions) and then for the next years, one can compare the correspondence between modeled values and norm-values. Then, for the next 10 years, the TP-loading to the Baltic Proper has been reduced by 5,000 t/year (from the present default conditions, see Table 2.5) and we will give the dynamic response of the system to such a large and sudden reduction. Evidently, we do not suggest that there will ever be any such large and sudden TP-reduction (month 121). In reality, it would take a long time to implement measures to achieve such massive reductions in phosphorus loading. Our aim has been to demonstrate how the system would likely respond to changes in TP-loading. In month 241 (i.e., after 20 years), we will double the reductions of all tributary TP-fluxes to the Baltic Proper and after 30 years, we will triple the reductions and eliminate 15,000 t TP/year, which corresponds to the amount suggested by HELCOM for the entire Baltic Sea (not just for the Baltic Proper, as in this scenario; see Table 2.6).

Scenario 2. In this scenario, we will change the salt-water inflow from the Kattegat to the Baltic Sea in three steps. The first 10-year period will give the default conditions (just like the previous TP-scenario), during the second 10-year period we will increase the water inflow from the Kattegat to the Baltic Proper by 10% (month 121), then by 20% (month 241) and for the last 10-year period by 40% (month 361). The idea is to study how the system would likely react to such drastic changes in the inflow of saline water. It is well known (see Fig. 1.9) that there may be very large and sudden inflows of saline water from the Kattegat, and this scenario is meant to illustrate how changes in salinity (not oxygen; this will be shown in Chap. 5) would likely influence the Baltic Sea ecosystem, as this is handled by the CoastMab and CoastWeb modeling.

Scenario 3 is a temperature change scenario. Note that we would not like to call this a "climate change" scenario because possible future changes in climatic conditions may influence the Baltic Sea system in many different ways beside causing increases in water temperature, e.g., the wind frequencies and the wind climate may change, the water level may rise, the hydrological regime regulating tributary water discharge may change, etc. (BACC, 2008). This scenario should rather be regarded as a third sensitivity test where we alter one abiotic variable, water temperature, in a systematic and structured manner while we keep all else constant. The first 10-year period will give the default conditions (just like the previous two scenarios), then we will increase the surface-water temperatures (SWT) by 0.5°C every second year and by 0.1°C the other years for the next 10-year period; then for the next 10-year period (starting month 241), we will change SWT by 1 and 0.2°C (month 241); and for the last 10-year period by 2 and 0.4°C (month 361). The middle-water temperatures in basins with three vertical compartments and the deep-water temperatures in basins with two compartments will be changed by 0.75 times the added increase in SWT; the deep-water compartments in the Baltic Proper and the Gulf of Finland (with three vertical compartments) will be increased by 0.5 times the added value in the SW-layer. The idea is to study how the entire system would likely react to such temperature changes, which would significantly reduce the impact of ice in the coldest basins, the Bothnian Bay (BB), the Bothnian Sea (BS) and the Gulf of Finland (GF). One should note that two of these basins (BB and BS) also have very high land uplift and thus high inputs of eroded materials. We assume in this scenario that years with higher water temperatures are followed by years with colder temperatures since it is probably less realistic to assume that the water temperature would be equally high all years. Note, however, that this scenario gives temperature increases in a rather standardized manner. Figure 4.1 gives a compilation of TP-fluxes to, within and from the Bothnian Sea and the Bothnian Bay using the CoastWeb-model. From this figure, one can note that:

- The total annual TP-inflow to the Bothnian Bay (with the highest land uplift of 8.5 mm/year; the value for BS is 7.25 mm/year; see Table 2.1) is 3,460 t/year from tributaries and 2,320 (2,060 + 260) t/year from BS, i.e., 5,780 t/year, while the TP-amount added from land uplift is 39,800 t/year, or a factor of 7 higher than the water inflows. If the ice cover vanishes due to water temperature increases, this will imply that the wind-induced turbulence as well as the water mixing

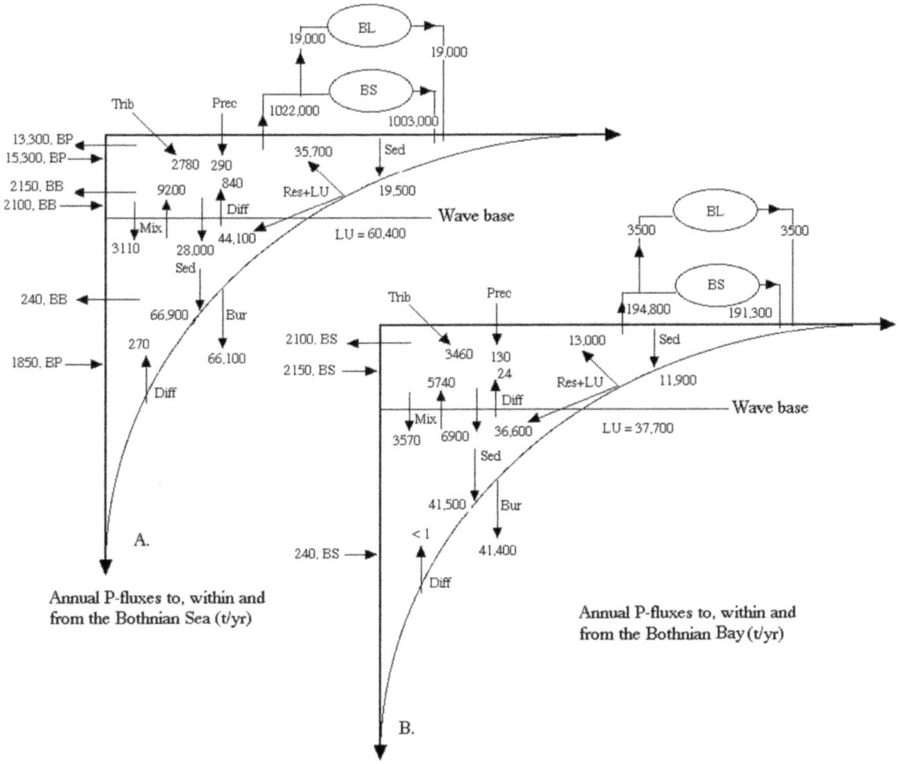

Fig. 4.1 (**a**) Calculated annual fluxes of phosphorus (in t/year) to, from and within the Bothnian Sea. (**b**) Calculated annual fluxes of phosphorus (in t/year) to, from and within the Bothnian Bay

will change sedimentation and resuspension considerably. The ice cover reduces wind-induced turbulence and without ice the suspended particulate matter and particulate phosphorus will settle out much slower (as given by Eq. 2.34). If there are more months with surface-water temperatures close to 4°C, or rather more months when the difference between SWT and DWT or MWT is lower than 4°C, there will be more resuspension (as given by Eq. 2.47). So, if the ice would be totally lost all months in the Bothnian Bay, one should expect significant increases in TP-concentrations in the Bothnian Bay just from the changes in sedimentation and resuspension in this bay. But one should also expect significant TP-increases in this bay because of the losses of ice cover in the Bothnian Sea, the Gulf of Finland and the Gulf of Riga. These bays generally have ice cover as well (as discussed in Chap. 2), which will disappear in this scenario. We will model the influence on ice, sedimentation and resuspension in more details in a temperature scenario in Chap. 5.

This chapter first discusses the primary producers (phytoplankton, benthic algae and macrophytes), then the reducer (bacterioplankton) and finally the secondary producers (zooplankton, zoobenthos and fish; the jellyfish model was presented in Chap. 3). It should be stressed again that many of the structures and equations in the CoastWeb-model are the same as in the LakeWeb-model (see Håkanson and Boulion, 2002a).

4.2 Primary Producers

4.2.1 Phytoplankton – Background

Phytoplankton plays several fundamental roles in aquatic ecosystems (Overbeck, 1972; Fursenko and Kuzmitskaja, 1975; Aizaki et al., 1981; Bird and Kalff, 1984; Currie, 1990; Conan et al., 1999). There are, however, few general quantitative models capturing the most important factors and processes regulating phytoplankton production accounting for such fundamental but complex properties as the depth of the photic zone and seasonal variations. And few such models have, as far as we know, yielded good predictive power over a wide domain of systems from just a few readily accessible driving variables.

The empirically-based norm for phytoplankton biomass is calculated from modeled TP-concentrations in the surface-water layer, the volume of the photic zone (Vol_{2Sec}) and a dimensionless moderator, Y_{Chl}, which is related to the influence of salinity on the relationship between chlorophyll and TP-concentration, as given by the equation in Table 3.2. This norm is, like all regressions and all models, only applicable in a certain defined domain, where it has a certain predictive power, as indicated in Table 3.1 by the ranges in the TP-values and the r^2-values.

The model for depth of the photic zone (2·Secchi depths, see Chap. 2) is an important, integral part of this sub-model.

It should be stressed that the empirical norms based on regressions capture characteristic patterns in many systems. However, they are basically static and may not provide realistic seasonal patterns. In some cases, however, CoastWeb includes also feedbacks in the norms to obtain more realistic reference values for the modeling. This is generally done by using the Secchi depth, which is influenced by dynamically modeled values of SPM and salinities using the CoastMab-model. Also note that the following model tests in this chapter target on comparisons between values for the growing season, since most of the empirical regressions used for the norms are based on data from this period. This means that divergences between modeled values and data from the empirical norms for other seasons of the year are less critical. The CoastWeb-model is, however, meant to provide the best possible estimates of phytoplankton biomasses and production values for all seasons of the year. It is, however, evident that the conditions during the growing season are generally more important for the structure and function of the aquatic foodweb than the conditions during the rest of the year.

The CoastWeb-model is meant to give more information than the empirical norms. Divergences between the modeled values and the data given by the empirical norms or by empirical data from the Baltic Sea should be logical and supported by solid ecosystem theory.

4.2.2 Modeling of Phytoplankton

The following differential equation gives the changes in the biomass of phytoplankton (using the Bothnian Bay, BB, as an example):

$$\begin{aligned} BM_{PHBB}(t) = {} & BM_{PHBB}(t-dt) + (IPR_{PHBB} + MIG_{InPHBB} - EL_{PHBB} \\ & - CON_{PHZHBB} - MIG_{OutPHBB}) \cdot dt \end{aligned} \tag{4.1}$$

BM_{PHBB} = Phytoplankton biomass (kg ww).
IPR_{PHBB} = Initial phytoplankton production (kg ww/month).
MIG_{InPHBB} = Inflow of phytoplankton (from the Bothnian Sea, kg ww/month).
EL_{PHBB} = Elimination or turnover of phytoplankton (kg ww/month).
CON_{PHZHBB} = Consumption of phytoplankton by herbivorous zooplankton (kg ww/month).
$MIG_{OutPHBB}$ = Outflow of phytoplankton (from the Bothnian Bay, kg ww/month).

Elimination products and dead phytoplankton are consumed by bacterioplankton and this is accounted for by the flux to bacterioplankton from suspended particulate matter and is not included in this differential equation. The initial phytoplankton biomass is set equal to the norm-value (NBM_{PH} in kg ww), as calculated by the approach given in Table 3.2. The basic regression in Table 3.1 is, however, only valid for TP-concentrations in the range from 3 to 80 μg/l. At higher TP-values, there is a likely increase in algal shading, and CoastWeb modifies the basic equation to account for this, see Eq. (3.12). Similar relationships have been described by, e.g., Smith (1979), Straskraba (1980) and Chow-Fraser and Trew (1994), but this is of limited interest in the Baltic Sea where the TP-concentrations are generally much lower than 80 μg/l.

The initial phytoplankton production (in BB) is given by $IPR_{PHBB} = PrimP_{BB}$, where PrimP is the primary phytoplankton production (kg ww/month) calculated from:

$$
\begin{aligned}
&\text{If } Sec_{BB} > 1\,\text{m then}\\
&PrimP_{BB} = (10^{-6}\cdot((2.13\cdot Chl_{ModBB}{}^{0.25}+0.25)^4)\cdot(1/0.45)\cdot(1/0.2)\cdot 30.42\cdot\\
&\qquad(1{,}500 - V_{2SecBB})\cdot 10^9)\ \text{else}\\
&PrimP_{BB} = (10^{-6}\cdot((2.13\cdot Chl_{ModBB}{}^{0.25}+0.25)^4)\cdot(1/0.45)\cdot(1/0.2)\cdot 30.42\cdot\\
&\qquad Area_{BB}\cdot(2\cdot Sec_{BB})^2
\end{aligned}
\tag{4.2}
$$

The value is first calculated from modeled chlorophyll data (see Chap. 2) and given in mg C/m^3·d; then transferred to kg ww. Calculation from g C to g dw is given by 1/0.45; calculation from g dw to g ww by 1/0.2 (using a water content of 80% for phytoplankton; see Table 3.3); 30.42 is the average number of days per month. The modeled chlorophyll-values account for changes in TP-concentrations, dissolved phosphorus, light and temperature influences, which was also discussed in Chap. 2.

Migration (kg ww/month), or rather inflow of phytoplankton from the outside sea by water currents (in this example from the Bothnian Sea, BS) is calculated from:

$$MIG_{InPHBB} = R_{MigPHBB}\cdot BM_{PHBS} \tag{4.3}$$

Where $R_{MigPHBB}$ is the migration rate for phytoplankton (= 1/T_{SWBB}, T_{SWBB} is the water turnover time for the SW-water in BB in months) and BM_{PHBS} is the actual biomass of phytoplankton in BS (modeled using CoastWeb). The outflow of phytoplankton is calculated in the same manner using the modeled biomass of phytoplankton in the given system (here BM_{PHBB}).

The elimination or loss of phytoplankton (kg ww/month) is given by:

$$EL_{PHBB} = BM_{PHBB}\cdot 1/T_{PH} \tag{4.4}$$

Where BM_{PHBB} is the dynamically modeled biomass (kg ww) of phytoplankton (here in BB). The turnover time, T_{PH}, for phytoplankton is set to 3.2 days (see Table 2.9).

Phytoplankton consumption by herbivorous zooplankton (CON_{PHZH}, kg ww/month) is given by:

$$CON_{PHZHBB} = BM_{PHBB} \cdot CR_{ZHBB} \tag{4.5}$$

Where CR_{ZHBB} the actual consumption rate (1/month) quantifying the loss of phytoplankton from predation by herbivorous zooplankton in the Bothnian Bay. CR_{ZHBB} is related to the turnover time of herbivorous zooplankton (T_{ZH}; 6 days; see Table 2.9). That is:

$$CR_{ZHBB} = (NCR_{ZHBB} + NCR_{ZHBB} \cdot (BM_{ZHBB}/NBM_{ZHBB} - 1)) \tag{4.6}$$

NCR_{PHZH} = The normal consumption rate (phytoplankton eaten by herbivorous zooplankton), as given by $2/T_{ZH}$ (1/months); the number of first order food choices is 2.
NBM_{ZH} = The normal biomass of herbivorous zooplankton (kg ww; see Table 3.2).

So, the predation pressure on phytoplankton from herbivorous zooplankton is a function of the actual biomass of herbivorous zooplankton (BM_{ZH} or BM_{ZHBB} in the Bothnian Bay) relative to the normal biomass of herbivorous zooplankton (NBM_{ZH}), and the actual consumption rate (CR_{ZH}), and these rates are related to the inverse of the turnover time of herbivorous zooplankton (T_{ZH}).

Phytoplankton production (PR_{PH} in kg ww/month) is defined by the ratio $PR_{PH} = BM_{PH}/T_{PH}$.

With this, the dynamic model for phytoplankton has been described and we will focus on its function.

4.2.3 Testing Modeling Results

The aims of this section are:

1. To test the dynamic model using the general empirical norms (see Table 3.1) as references. So, we will study how the CoastWeb-model predicts monthly values of phytoplankton in all sub-basins of the Baltic Sea compared to the empirical norms, and in this section we will present results on biomasses.
2. To illustrate how the model behaves in situations when gradients are created in a systematic manner for pertinent driving variables in contexts of Baltic Sea management. That is:

   - TP-concentrations will be changed in four 10-year steps by reducing the TP-inflow to the Baltic Proper (default inflow minus 5,000 t TP/year, minus 10,000 t TP/year and minus 15,000 t TP/year). Note that this scenario is of special interest in contexts related to Baltic Sea management plans since man can actually implement – costly (see Chap. 5) – remedial measures and significantly influence the nutrient fluxes from land to water.

- We will also change the water inflow from the Kattegat in four steps (default conditions, a 10% increase, a 20% increase and a 40% increase) and study how this would influence the system, and first phytoplankton. The fluxes of saline water from the Kattegat are mainly driven by large-scale climatological factors (such as variations in high- and low-pressures over the North Atlantic and the Baltic Sea, see BACC, 2008) and this cannot be influenced by actions in the Baltic Sea countries in the same way as nutrient reductions. But it is important to have access to an ecosystem model, which can produce realistic predictions how such changes in salt-water inflow may influence the structure and function of the Baltic Sea ecosystem, including the fish production potential and the criteria for setting fish quota.
- Water temperatures will, as already explained, also be increased in four steps by 0.5 and 0.1°C, 1 and 0.2°C and 2 and 0.4°C in the surface-water layer and by less in the middle-water and deep-water compartments (0.75 and 0.5 of the increases in the surface-water layer, respectively). Man may be able to change the global climate (see IPCC homepage) and hence also Baltic Sea water temperatures, but to create such actions to increase or reduce water temperatures goes much beyond the actions that can actually be carried out to reduce the eutrophication in the Baltic Sea and can only be evaluated in a more long-term perspective (several decades to centuries).

All these changes will be calculated for all functional groups of organisms and the basic idea is (1) to compare dynamically modeled values to the norm-values and (2) to study the response of the entire system to these changes in abiotic driving variables. The target predictions concern the changes in prey and predatory fish but to be able to understand those changes we will present all underlying changes in the system.

The results of all these tests will be summarized in multi-diagram figures.

4.2.3.1 Reductions in TP-Loading

Figure 4.2 first (A) shows how the model predicts TP-concentrations in the SW-layer, (B) annually smoothed chlorophyll-a concentrations, (C) annually smoothed SPM-concentrations and (D) annually smoothed Secchi depths in all five sub-basins of the Baltic Sea under default conditions (for the period 1997–2006). The reason why the monthly values for chlorophyll, SPM and Secchi depth have been smoothed is simply that the seasonal variations for these variables are generally high (see Fig. 4.3 which gives the monthly variations in phytoplankton biomass as an example; this variability is closely related to variations in chlorophyll-a concentrations). The variation is, in fact, so high that it becomes very difficult to observe the results when the curves for the five systems are shown in one figure. The smoothed curves give a much clearer picture of the changes related to these scenarios covering a 40-year period.

The model gives values close to steady-state conditions after an initial period of a few years, and for the last years during the first 10-year period, the model gives the

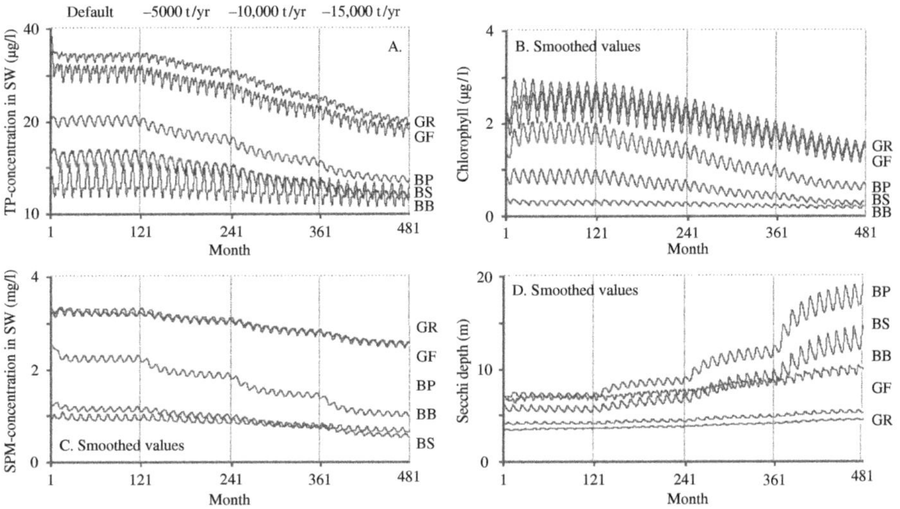

Fig. 4.2 Modeled TP-concentrations (**a**), modeled smoothed values of chlorophyll (**b**), SPM (**c**) and Secchi depth (**d**). The default conditions are shown for the first period (months 1–61), for the second period the annual TP-inflow via tributaries to the Baltic Proper has been reduced by 5,000 t/year, in the third period by 10,000 t/year and in the fourth period (months 361–481) by 15,000 t/year. Note that the values for chlorophyll, SPM and Secchi depth have been smoothed (an annual smoothing) to achieve clarity

references values, which reflect default conditions. From Fig. 4.2a, one can note the logical change in trophic state from the coldest and most oligotrophic sub-basin, the Bothnian Bay, to the most productive sub-basin, the Gulf of Riga, when the tributary TP-inflow to the Baltic Proper (and only to the Baltic Proper) has been reduced. The figure shows that when 5,000, 10,000 and 15,000 t TP/year are eliminated from the tributary TP-inflow to the Baltic Proper, one should expect significant changes in the entire system:

- The changes are logically greatest in the Baltic Proper, but there are clear reductions in the TP-concentrations also in the Bothnian Bay, and corresponding changes in chlorophyll, SPM-concentrations and Secchi depths in this and all other sub-basins.
- The Secchi depth in the Gulf of Finland will, however, not reach the value of 7 m, which may be regarded as a target for Baltic Sea management (see Håkanson and Bryhn, 2008b). Table 4.1 gives a statistical compilation of empirical data on Secchi depths from the Gulf of Finland (mean values, medians, standard deviations, coefficients of variation and number of data) for three interesting periods, 1900–1920, 1920–1980 and 1980–1991. One can note the high CV-values (about 0.35) and that the mean Secchi depth has decreased from 7.1 to 4.8 m during the last 100 years. This is a significant change influencing not just the conditions in the Gulf of Finland but the entire Baltic Sea system. Changes in Secchi depth

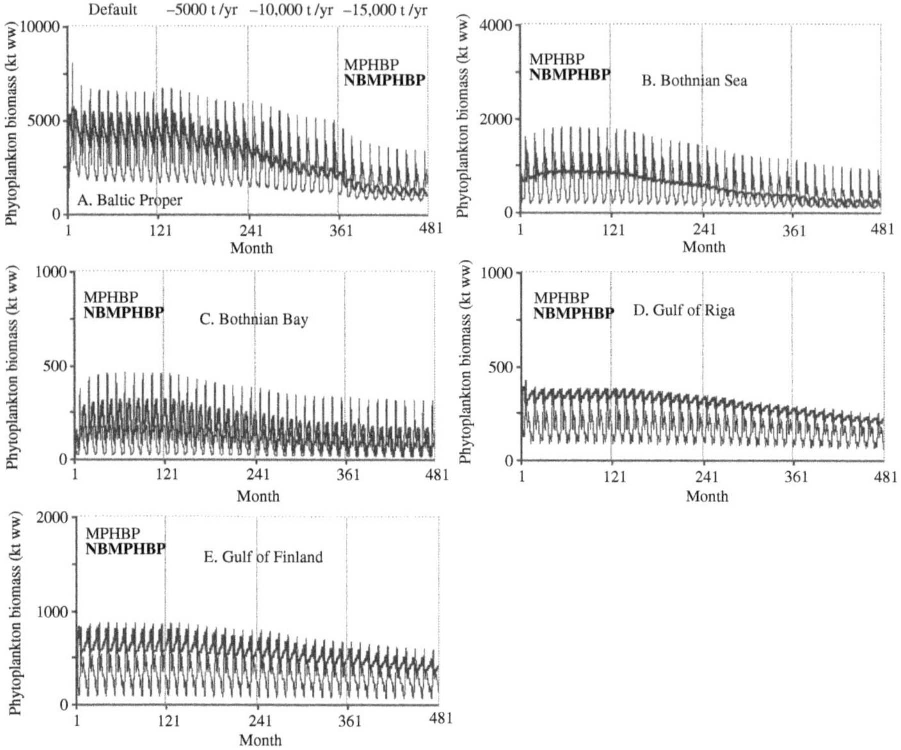

Fig. 4.3 A comparison between dynamically modeled biomasses of phytoplankton and values calculated from the empirically-based norms for the 5 sub-basins in the Baltic Sea in a situation where the default conditions are shown for the first period (months 1–61), for the second period the annual TP-inflow via tributaries to the Baltic Proper has been reduced by 5,000 t/year, in the third period by 10,000 t/year and in the fourth period (months 361–481) by 15,000 t/year

influence the depth of the photic zone and the benthic production, which is highly dependent of water clarity. If there are major changes in primary phytoplankton and benthic algae production and biomasses, one should also expect changes in

Table 4.1 Statistics of Secchi depth measurements from different periods from the Gulf of Finland

	1900–1920	1920–1980	1980–1991
Mean (MV)	7.1	6.3	4.8
Median (M50)	7.0	6.1	4.5
Standard deviation (SD)	2.45	2.2	1.6
Coefficient of variation (CV)	0.35	0.35	0.33
Number of data (n)	123	556	60

secondary production (of zooplankton, zoobenthos and fish). An important question for Baltic Sea management is thus to try to find realistic and cost-effective measures so that the Secchi depth in the Gulf of Finland can again reach 7 m. We will discuss that in more detail in Chap. 5, but here we can conclude that that goal is not reached in this particular scenario, not even if 15,000 t TP/year are reduced to the Baltic Proper.

- The TP-concentration in the Baltic Proper would, however, likely reach a level far lower than in historic times. So, to reduce the tributary TP-inflow by 15,000 t/year to the Baltic Proper can not be regarded as an "optimal" remedial strategy. If 15,000 t TP/year are to be reduced, as suggested by HELCOM (see Table 2.6), this initial scenario indicates that the reductions should be more evenly distributed among the tributaries to the various sub-basins, and especially target on conditions in the catchments of the most eutrophicated basins. The Secchi depth in the Baltic Proper would approach 18 m in this scenario, as compared to about 6 m today, according to Fig. 4.2.

Figure 4.3 gives the monthly variations in phytoplankton biomass in all five sub-basins. The correspondence between dynamically modeled values and norm-values is good for the entire 40-year period. This may be difficult to see from Fig. 4.3 and we show this more clearly in Fig. 4.4, which gives the correspondence between the two curves for the default conditions. Evidently, reductions in TP-loading will cause reductions in primary phytoplankton production and biomasses. This is easy to say but hard to predict and quantify in a realistic manner. One can also note from Fig. 4.3 that the reductions in phytoplankton biomass in this scenario are logically greatest in the Baltic Proper, but marked also in the Bothnian Bay since the Baltic Sea is a system of inter-connected basins. We will demonstrate the corresponding changes for all functional groups included in the CoastWeb-model in the following sections in this chapter, so the changes in the variables given in Fig. 4.2 provide relevant framework for all the following changes related to this oligotrophication scenario.

4.2.3.2 Increasing Salt-Water Inflow

Figure 4.5 gives the results from the "salinity scenario" where the salt-water inflow from the Kattegat to the Baltic Sea has been increased in three steps, as shown in Fig. 4.6a, which gives the changes in surface-water salinities in all five sub-basins for a 40-year period. These changes will cause several changes to the structure and function of the Baltic Sea ecosystem and Fig. 4.5 gives information on TP-concentrations in the SW-layer (B), annually smoothed chlorophyll-a concentrations (C), annually smoothed SPM-concentrations (D) and annually smoothed Secchi depths (E) in all five sub-basins.

- Increases in salinity will cause accentuated aggregation of suspended particles, increase sedimentation of all types of suspended particles (including particulate phosphorus) and reduce SPM-concentrations in water (Fig. 4.5d), and hence

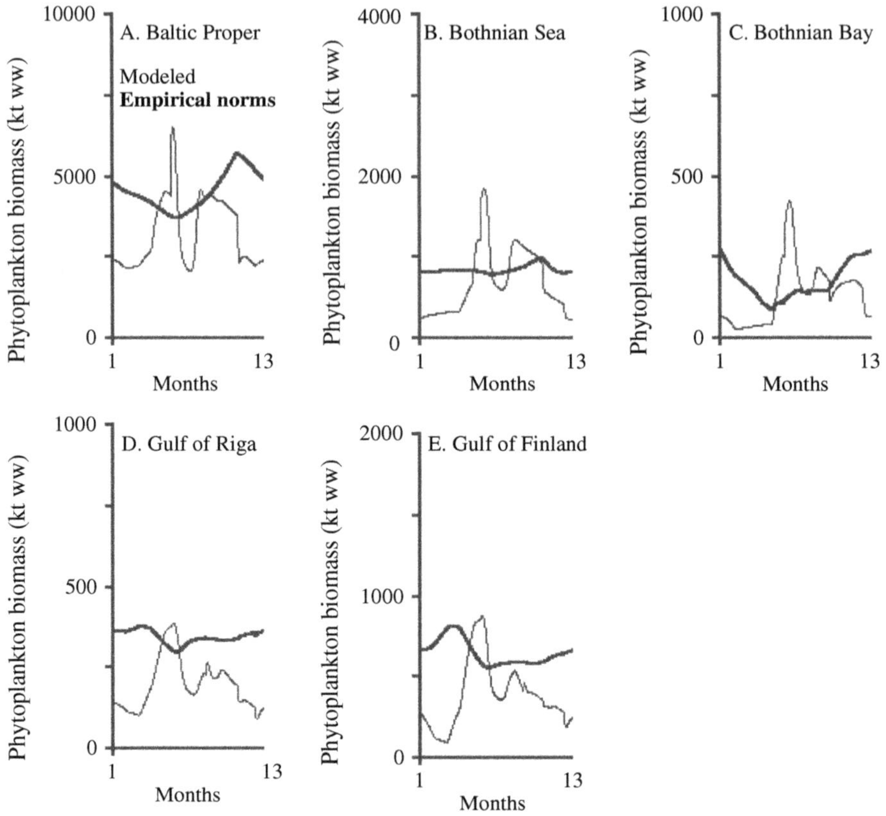

Fig. 4.4 A comparison between modeled empirically-based norm-values (*bolded* reference values for mainly the growing seasons) and dynamically modeled values of phytoplankton biomass at steady-state (simulation time 241 months) under default conditions in the five basins of the Baltic Sea

also increase Secchi depths (Fig. 4.5e). It will also reduce TP-concentrations in water (Fig. 4.5b) and thus also reduce primary production (see Fig. 4.6) and chlorophyll-a concentrations reflecting primary phytoplankton production (Fig. 4.5c).

- Given these changes in salinities, the Secchi depth in the Bothnian Bay, the Bothnian Sea and the Baltic Proper would increase markedly, but the Secchi depth in the Gulf of Finland would not reach the target value of 7 m.

The correspondence between model-predicted values for phytoplankton biomass and the norm-values (bolded in Fig. 4.6) is good in all five sub-basins also in this scenario.

To conclude: Increasing the salinity would create an oligotrophication of the Baltic Sea system.

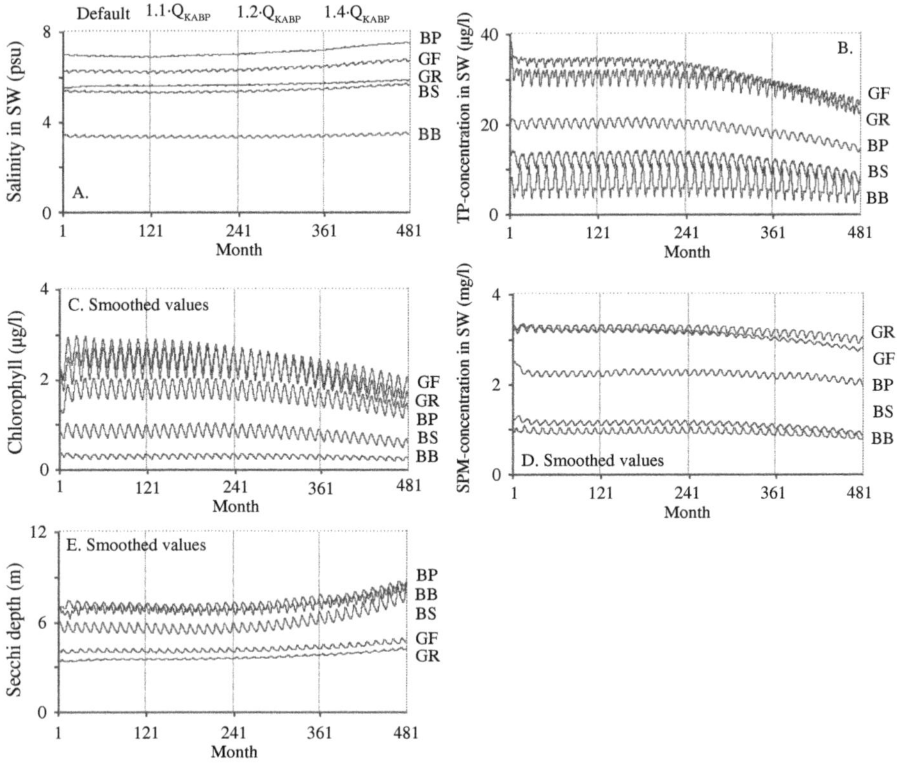

Fig. 4.5 Modeled surface-water salinities (**a**), TP-concentrations (**b**), modeled smoothed values of chlorophyll (**c**), SPM (**d**) and Secchi depth (**e**). The default conditions are shown for the first period (months 1–61), for the second period the monthly saltwater inflow from the Kattegat has been increased by a factor of 1.1, in the third period 1.2 and in the fourth period (months 361–481) by 1.4. Note that the values for chlorophyll, SPM and Secchi depth have been smoothed (an annual smoothing) to achieve clarity

4.2.3.3 Temperature Scenario

Figure 4.7 gives the results from the "temperature scenario" where the water temperatures have been increased in a structured and systematic manner for a 40-year period, as exemplified in Fig. 4.7a for the coldest sub-basin, the Bothnian Bay. In this figure, we have specifically marked the critical surface-water temperature for the ice cover of 0.9°C.

Note that the water temperatures affect many processes, (1) there is less turbulence, and higher sedimentation in situations with ice cover, and vice versa when the ice disappears, (2) the ice cover also affects resuspension and internal loading, and there is less resuspension under ice cover, (3) water temperatures affect stratification and mixing, (4) the temperatures also affects many biotic processes related to the production of functional groups and (5) temperature influences the bacterial decomposition of suspended organic particles. All these processes are accounted for in this

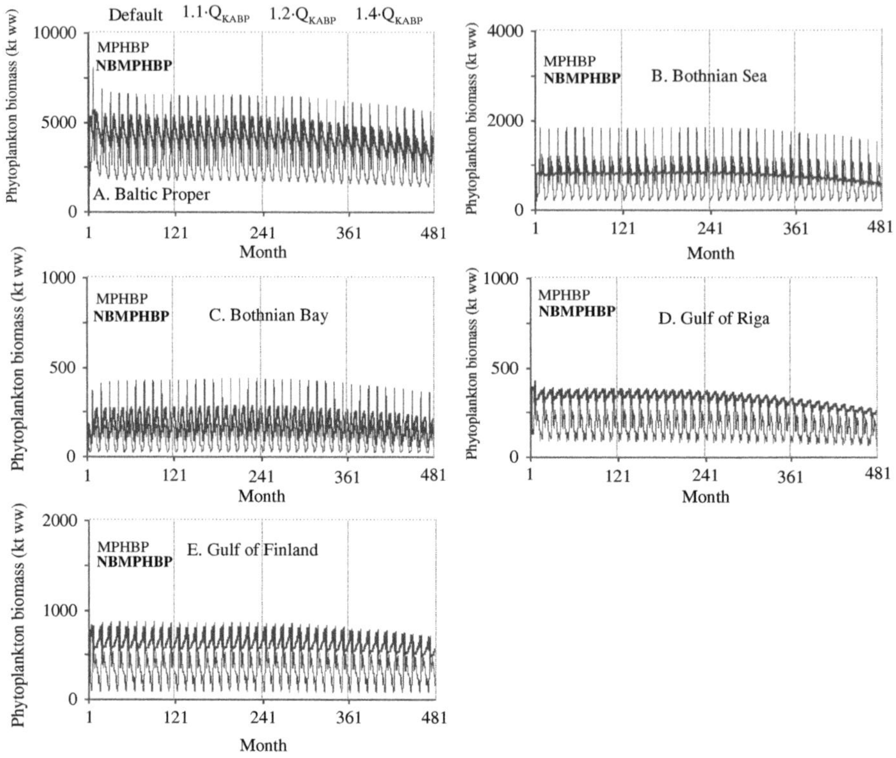

Fig. 4.6 A comparison (a sensitivity test) between dynamically modeled biomasses of phytoplankton and values calculated from the empirically-based norms for the 5 sub-basins in the Baltic Sea in a situation where the default conditions are shown for the first period (months 1–61), for the second period the monthly saltwater inflow from the Kattegat has been increased by a factor of 1.1, in the third period 1.2 and in the fourth period (months 361–481) by 1.4

modeling. So, one should expect that changes in water temperature would cause several changes to the structure and function of the system and Fig. 4.7 gives information on changes in TP-concentrations in the SW-layer (B), annually smoothed chlorophyll-a concentrations (C), annually smoothed SPM-concentrations (D) and annually smoothed Secchi depths (E) in the five sub-basins.

- The given changes in water temperatures would cause more dramatic changes to the Baltic Sea system than given in the two previous scenarios. The TP-concentrations (Fig. 4.7b) would likely increase so that all the sub-basins in the Baltic Sea would be clearly more eutrophic and also more similar in terms of TP-concentrations. This would affect primary production and the chlorophyll-a concentrations would be a factor of 1.5 higher in the Baltic Proper in this 40-year scenario, which does not reflect steady-state conditions (from about 2 μg/l to about 3 μg/l). This means that the system would change from

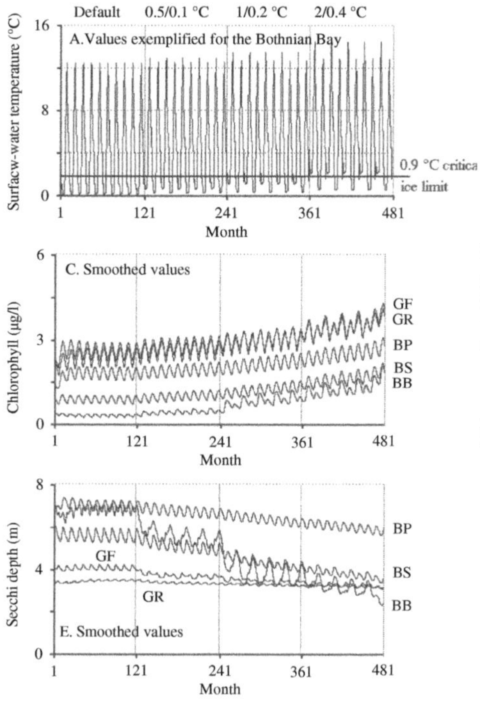

Fig. 4.7 Surface-water temperatures in the Bothnian Bay (**a**), modeled TP-concentrations (**b**), modeled smoothed values of chlorophyll (**c**), SPM (**d**) and Secchi depth (**e**). The default conditions are shown for the first period (months 1–61), for the second period the monthly surface-water temperatures have been increased by 0.5 and 0.1°C every second year, for the third period by 1 and 0.2°C every second year and in the fourth period by 2 and 0.4°C every second year. Note that the values for chlorophyll, SPM and Secchi depth have been smoothed (an annual smoothing) to achieve clarity. Middle- and deep-water temperatures have been increased by 0.75 and 0.5 of the increases given for the surface-water temperatures

mesotrophic/oligotrophic to clearly eutrophic conditions, see Table 4.2, which gives criteria for different trophic categories at different salinities. The SPM-concentrations (Fig. 4.7d) would also increase markedly, especially in the Bothnian Bay, i.e., the coldest sub-basin with the highest land uplift. This also means that the Secchi depths would be significantly reduced (Fig. 4.7e).

The corresponding changes in primary phytoplankton biomasses are given in Fig. 4.8. The correspondence between model-predicted values for phytoplankton biomass and the norm-values (bolded in Fig. 4.8) is good in all five sub-basins also in this case although the norm-values, which mainly depend on changes in TP-concentration increase more than the modeled values in the Bothnian Bay and the Bothnian Sea. The dynamically modeled values would better account for the fact that the phytoplankton biomass would not increase as much as the TP-concentrations since the dynamically modeled value also accounts for predation on

Table 4.2 Characteristic features in (A) freshwater dominated systems, (B) brackish systems and (C) marine coastal systems of different trophic levels (see also OECD, 1982; Håkanson and Jansson, 1983; Wallin et al., 1992; Håkanson and Boulion, 2002a; Håkanson et al., 2007). All data represent characteristic (median) values for the growing season for the surface-water layer

Trophic level	Secchi (m)[a]	Chl-a (μg/l)	Total-N (μg/l)	Total-P (μg/l)	Cyanobacteria (μg ww/l)[b]
A. Freshwater dominated systems, salinity <5 psu					
Oligotrophic	>5	<2	<60	<8	<2.2
Mesotrophic	3–5	2–6	60–180	8–25	2.2–250
Eutrophic	1–3	6–20	180–430	25–60	250–1,400
Hypertrophic	<1	>20	>430	>60	>1,400
B. Brackish systems, salinity 5–20 psu					
Oligotrophic	>8	<2	<70	<10	<9.5
Mesotrophic	4.5–8	2–6	70–220	10–30	9.5–380
Eutrophic	1.5–4.5	6–20	220–650	30–90	380–2,500
Hypertrophic	<1.5	>20	>650	>90	>2,500
C. Marine systems, salinity >20 psu					
Oligotrophic	>11	<2	<110	<15	<55
Mesotrophic	6–11	2–6	110–290	15–40	55–680
Eutrophic	2–6	6–20	290–940	40–130	680–4,040
Hypertrophic	<2	>20	>940	>130	>4,040

Relationships between chlorophyll, TP, TN and salinity calculated from Håkanson (2006).
[a] Secchi depth calculated from Håkanson (2006).
[b] Concentration of cyanobacteria (CB) calculated using the model from Håkanson et al. (2007) when TP/TP is set to 15, surface-water temperature to 17.5°C and the salinity to 2.5, 12.5 and 36, respectively for freshwater, brackish and marine systems.

phytoplankton from herbivorous zooplankton, which increase relatively much in this gradient (see coming section).

To conclude: Increasing the water temperatures may further strongly increase the eutrophication of the Baltic Sea system. This will be discussed this in more detail in Chap. 5.

4.2.4 Benthic Algae

This section will introduce the sub-model for benthic algae in CoastWeb. Figure 4.9 gives an overview of the model and Table A.3 gives a compilation of all equations, which will be explained in this section. Figure 3.3 gave an overview of different types of benthic algae.

Westlake (1980) distinguishes between three groups of periphyton communities. One is attached to different substrates and forms dense belts including young, old and dead cells. The second group consists of numerous filaments or lumps of gelatinous material. The third group includes communities whose members are

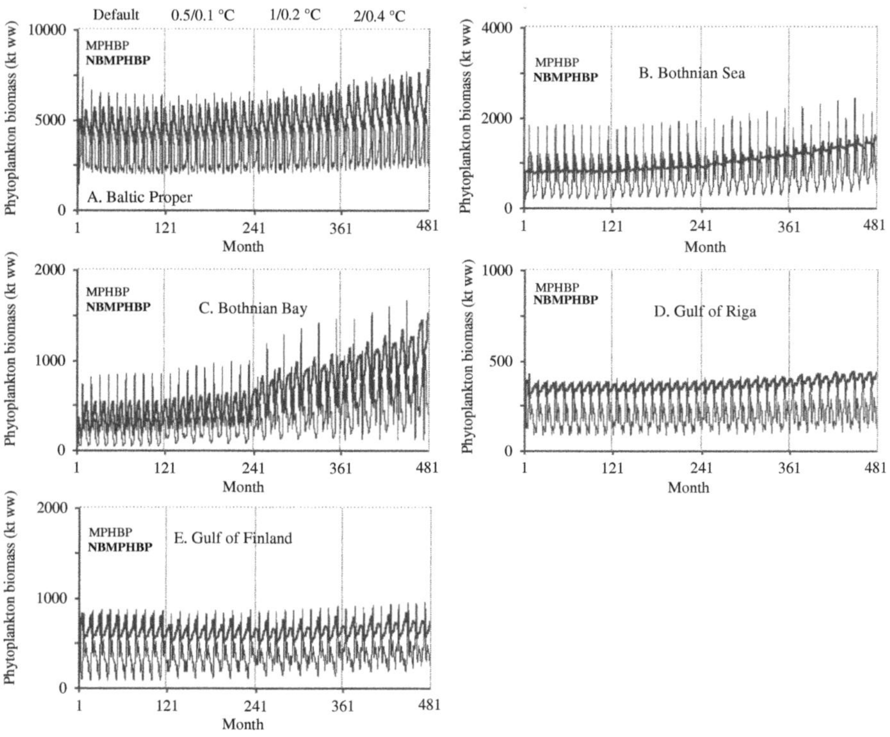

Fig. 4.8 A comparison (a sensitivity test) between dynamically modeled biomasses of phytoplankton and values calculated from the empirically-based norms for the 5 sub-basins in the Baltic Sea in a situation where the default conditions are shown for the first period (months 1–61), for the second period the monthly surface-water temperatures have been increased by 0.5 and 0.1°C every second year, for the third period by 1 and 0.2°C every second year and in the fourth period by 2 and 0.4°C every second year. Middle- and deep-water temperatures have been increased by 0.75 and 0.5 of the increases given for the surface-water temperatures

not strongly attached to the substratum. They are not aggregated and move freely over the substratum, generally the bottom sediments. Microphytobenthos have often been included in this third group. Today, however, microphytobenthos are generally considered as a separate group of autotrophic organisms, and not included among periphyton. Such classification problems partially explain the differences in biomasses presented by different authors for various periphyton communities.

Obviously, the horizontal and vertical distributions of benthic algae strongly depend on the distribution of the illuminated substrates. A map showing the distribution of where more than 1% of the light reaches the bottom of the Baltic Sea is given in Fig. 4.10. In comparison with macrophytes and phytoplankton, benthic algae are confined to a relatively thin surficial sediment layer, within which the concentration of cells can be very high (Wetzel, 1983). Organisms on underwater substrates form heterogeneous and complex associations and colonize almost all types of substratum

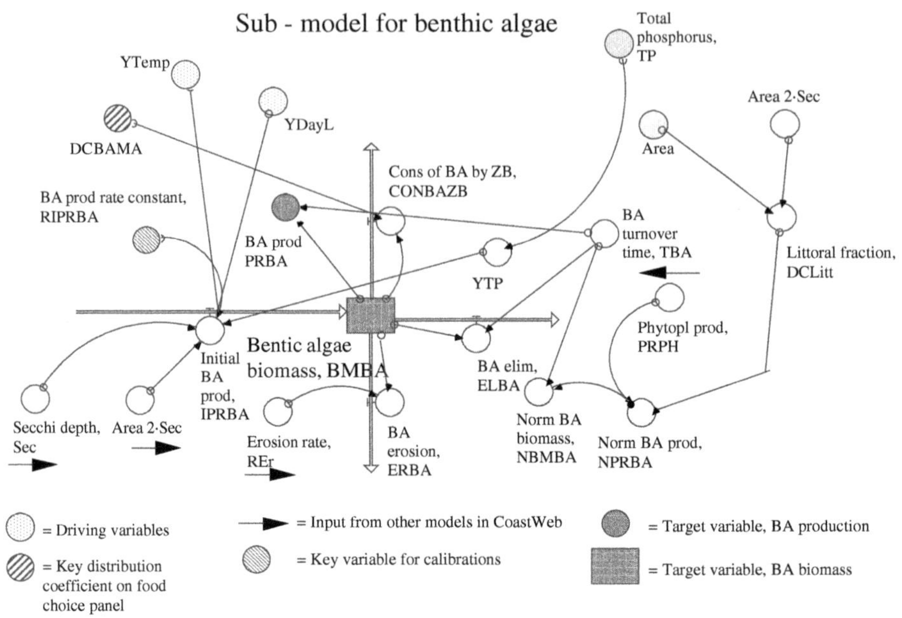

Fig. 4.9 Illustration of the sub-model for benthic algae

in the littoral zone (see, e.g., Vollenweider, 1969; Lalonde and Downing, 1991). The terminology used for the various algae living on different substrates varies, it sometimes seems, with the number of researchers (Vollenweider, 1969; Wetzel, 1964, 1983). Generally, the term "periphyton" is applied for all forms of plants (with the exception of macrophytes) growing on submerged materials. The underwater substratum can be, e.g., sediments, stones, boats, constructions and living organisms.

Filamentous algae are important habitats for benthic fauna (Salovius and Kraufvelin, 2004).

4.2.4.1 Modeling of Benthic Algae

The following ordinary differential equation gives the fluxes (kg ww/month) to and from the compartment "Benthic algae" (BA), which includes all types of bottom-living algae, except macrophytes, which are treated separately in the CoastWeb-model. The two target variables are biomass and production of benthic algae.

$$BM_{BA}(t) = BM_{BA}(t - dt) + (IPR_{BA} - CON_{BAZB} - EL_{BA} - ER_{BA}) \cdot dt \qquad (4.7)$$

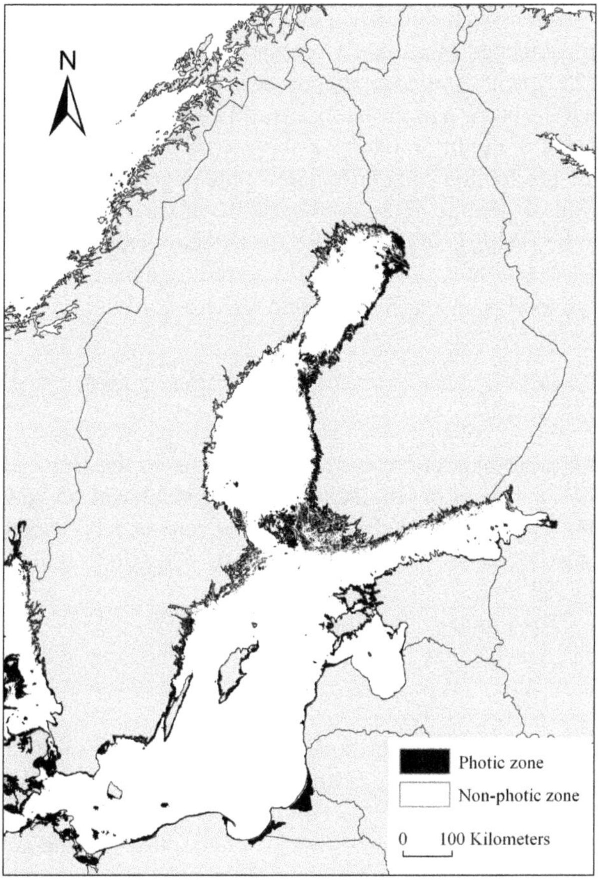

Fig. 4.10 Map showing the distribution of where at least 1% of the light reaches the *bottom*, i.e., the photic zone (corresponding to about 2·Secchi depth), of the Baltic Sea. Modified from Al-Hamdani and Reker (2007)

BM_{BA} = Biomass of benthic algae (kg ww); the initial BM_{BA}-value is set equal to the norm-value (NBM_{BA}), which is calculated from the normal production of benthic algae (NPR_{BA} from Eq. 4.8) times the turnover time for benthic algae (T_{BA}), i.e.: $NPR_{BA} \cdot T_{BA}$.

IPR_{BA} = Initial production of benthic algae (kg ww/month; see Eq. 4.9); note that the "production" of benthic algae, PR_{BA}, is given by the ratio between the calculated biomass, BM_{BA}, and the turnover time, T_{BA}.

CON_{BAZB} = Consumption of benthic algae by zoobenthos (kg ww/month).

EL_{BA} = Elimination (= turnover) of benthic algae (kg ww/month).

ER_{BA} = Physical erosion of benthic algae due to, e.g., wind/wave action. The same erosion rate is used for benthic algae and macrophytes. It is defined in Eq. (4.17). The erosion rate is a function of the macrophyte cover and the form factor (V_d) of the basin.

Intensive turbulence, generally predominating in shallow waters, is not beneficial for attached algae so the depth-production curve given by Håkanson and Boulion (2002a) in not linear but curved (see also Wetzel, 1983). This means that, on average in lakes and probably also in smaller coastal areas, the production of benthic algae in the littoral zone is about 40% (or, on average, 37%) of the total production (= production of benthic algae plus phytoplankton). It also means that the production of benthic algae is about 63% of the phytoplankton production in the littoral zone (PR_{PH}). The area above the depth given by the total depth of the photic zone (= 2·Secchi depth) in relation to the total area is A_{2Sec}/A is used as an estimate of the total production of benthic algae accordingly:

$$NPR_{BA} = 0.63 \cdot (A_{2Sec}/A) \cdot PR_{PH} \tag{4.8}$$

A_{2Sec} is calculated from the modeled monthly values of Secchi depth and the hypsographic curve for each sub-basin in the Baltic Sea and PR_{PH} is the primary production of phytoplankton in the given basin (in kg ww/month calculated according to the CoastWeb-model just given) and NPR_{BA} is the requested norm-value, which will be used to get order of magnitude reference values for the following model tests.

The initial production is given by:

$$IPR_{BA} = R_{BA} \cdot A_{2Sec} \cdot 2 \cdot Sec \cdot Y_{DayL} \cdot Y_{Temp} \cdot Y_{TP} \tag{4.9}$$

In the CoastWeb-model, we calculate the initial production of benthic algae, IPR_{BA}, based on the results from Håkanson and Boulion (2002a), i.e., by accounting for variations in Secchi depth (2·Sec in m), the morphometry of the basin (the area shallower than the two Secchi depths, i.e., the area of the littoral zone, A_{2Sec}, in m^2), the number of hours with daylight each month (as given by the dimensionless moderator Y_{DayL}; see Eq. (2.15); Y_{DayL} = HDL/12), water temperatures (as given by Y_{Temp} = SWT/9, where 9 is a reference SWT-value related to the length of the growing season; see Håkanson and Boulion, 2002a) and supply of phosphorus (as given by Y_{TP}; see Eq. 4.10). R_{BA} is the initial production rate constant for benthic algae, which gives production (kg ww/month) per volume unit (m^3). The initial production of benthic algae, IPR_{BA}, has the dimension kg ww/month (for the entire basins). The difference between initial production and production has already been discussed. One can note that the initial production is higher than the production because production also accounts for elimination, physical erosion and grazing by zoobenthos.

So, Eq. (4.9) gives that:

1. There is a benthic algae production rate constant, R_{BA}, which drives the production. The default value for R_{BA} is 0.01 kg ww/(month·m^3) in all sub-basins.
2. The morphometry of the basin will influence the production. This is expressed by the total area of the photic zone, i.e., for areas shallower then 2·Secchi depth (A_{2Sec} in m^2; as calculated from Secchi depth and the hypsographic curve). The

larger the A_{2Sec}-area, the higher the potential production of benthic algae, if all else is constant.

3. Secchi depth (in m), the more transparent the water, the higher the production of benthic algae. We assume a linear relationship. This gives the requested dimension for IPR_{BA} (kg ww/month).
4. Evidently, the production of benthic algae is also temperature and light dependent, and in this approach we use the same dimensionless moderators for temperature and light as in similar situations in the CoastWeb-model.
5. We also assume that the production of benthic algae depends on the availability of phosphorus, but, as already stressed, less so than phytoplankton. This dependency is given by the following dimensionless moderator:

$$Y_{TP} = (1 + 0.75 \cdot (TP/10 - 1)) \tag{4.10}$$

Where the TP-concentration is given in μg/l. The normal value is 10 μg/l in this dimensionless moderator. The amplitude value is set to 0.75, which means that Y_{TP} is 0.625 if TP = 5 μg/l (for an oligotrophic system such as the Bothnian Bay) and $Y_{TP} = 3.25$ for a system with $C_{TP} = 40$ (a eutrophic coastal area such as the Oder estuary in the Baltic Sea), if all else is constant. We will soon demonstrate how the model works along the three given gradients (reductions in TP-loading, increased salinity and increased water temperatures).

The consumption of benthic algae, CON_{BAZB}, by zoobenthos is calculated from:

$$CON_{BAZB} = BM_{BA} \cdot CR_{BAZB} \tag{4.11}$$

Where CR_{BAZB} is the actual consumption rate (dimension 1/month); BM_{BA} is the biomass of benthic algae (kg ww).

CR_{BAZB} is given by:

$$CR_{BAZB} = (NCR_{ZB} + NCR_{ZB} \cdot (BM_{ZB}/NBM_{ZB} - 1)) \tag{4.12}$$

NCR_{ZB} = The normal consumption rate for zoobenthos is defined by the ratio NR_{ZB}/T_{ZB}; T_{ZB} is the turnover time of zoobenthos (= 128 days, see Table 2.9); the number of first order food choices for zoobenthos, NR_{ZB}, is 2.

BM_{ZB} = The actual biomass of zoobenthos (kg ww); the higher the actual biomass of zoobenthos relative to the normal biomass of zoobenthos (NBM_{ZB}), the higher the grazing pressure exerted by zoobenthos on benthic algae.

The normal biomass of zoobenthos is given by (see Table 3.2):

$$NBM_{ZB} = Y_{Litt} \cdot (10^{-6}) \cdot 810 \cdot (TP^{0.71}) \cdot Area \tag{4.13}$$

The basic empirical regression ($810 \cdot TP^{0.71}$) is derived for lakes with mean TP-concentrations smaller than 100 μg/l and should not be used for coastal systems with TP-concentrations higher than that. It is unrealistic to assume that it could be used for hypertrophic systems with TP-concentrations higher than 100 μg/l, which are likely to have significant areas with anoxic sediments where zoobenthos cannot survive. The dimensionless moderator related to the difference between coastal areas and lakes related to the littoral zone, Y_{Litt}, is simply calculated from the ratio between the littoral zone in a coastal area $Area_{2Sec}$ (m^2) and the corresponding zone in a lake $Area_{2Seclake}$, i.e.:

$$Y_{Litt} = Area_{2Sec}/Area_{2Seclake} \tag{4.14}$$

The elimination of benthic algae in (kg ww/month) is given by:

$$EL_{BA} = BM_{BA} \cdot 1/T_{BA} \tag{4.15}$$

T_{BA} is the characteristic turnover time of benthic algae, set according to Table 2.9 to 4 days.

Physical erosion of benthic algae and macrophytes is a complicated process (see, e.g., Leclerc et al., 2000) involving wind speed, duration, fetch, wave characteristics, slope processes, erosion related to boating, etc. Basically, erosion of benthic algae (ER_{BA}) is given by:

$$ER_{BA} = BM_{BA} \cdot R_{Er} \tag{4.16}$$

The question is: Which value should be given to R_{Er} and what factors govern R_{Er}? According to Håkanson and Boulion (2002a), 60% of the variability in the values for the erosion rates can be statistically explained by differences among basins in macrophyte cover (Mac_{cov}) and the form factor (V_d; $V_d = 3 \cdot D_{MV}/D_{Max}$; D_{MV} = the mean depth in m; D_{Max} = the maximum depth in m). The larger the macrophyte cover, the smaller the erosion rate. This is logical since the exposed parts of the macrophyte cover could be regarded as a wave protection for the sheltered, inner parts. The next most important factor is the form factor. Basins with large shallow areas (V-shaped systems) logically have higher erosion rates than U-shaped basins. The algorithm for R_{Er} is:

$$R_{Er} = (0.1186 - 0.1338 \cdot \log(MA_{cov}) + 0.0769 \cdot V_d) \tag{4.17}$$

With this, we have given all processes regulating the production and biomass of benthic algae in the CoastWeb-model. The most important part for the model calibration, and actually the only calibration constant in the sub-model for benthic algae which is not mechanistically motivated, is the value used for the benthic algae production rate constant, R_{BA}. It has been calibrated to 0.01 kg ww/$m^3 \cdot$month (i.e., production per volume unit of the basin). In a following section, we will illustrate the correspondence between modeled values and norm-values for benthic algae.

4.2.4.2 Testing Modeling Results

This section will give results for the three scenarios. The background information related to scenario 1, reductions in TP-inflow to the Baltic Proper, is given in Fig. 4.2 for changes in TP-concentrations, chlorophyll, SPM and Secchi depths.

Reductions in TP-Loading

Figure 4.11 gives the monthly variations in the biomass of benthic algae in all five sub-basins. One can first note that the correspondence between dynamically modeled values and norm-values is good in term of order-of-magnitude values for the entire 40-year period. Figure 4.12 gives the relationships between dynamically modeled values and norm-values for the default conditions in the Baltic Sea. The correspondence between the two curves is quite good. Reductions in TP-loading will cause marked increases in Secchi depths (see Fig. 4.2d), which will influence the production and biomasses of benthic algae considerably. Also the changes in

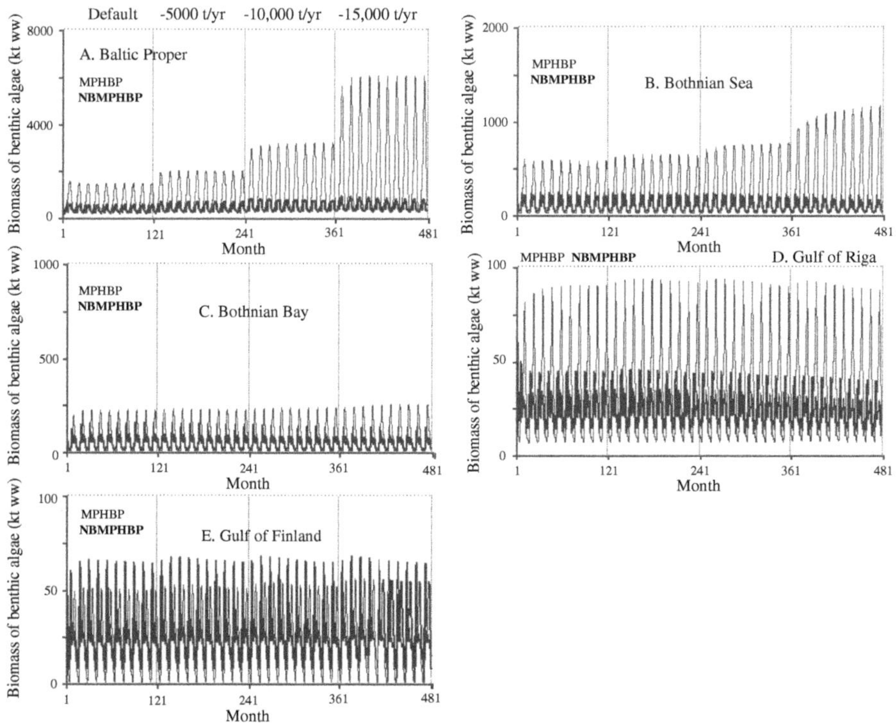

Fig. 4.11 A comparison between dynamically modeled biomasses of benthic algae and values calculated from the empirically-based norms for the 5 sub-basins in the Baltic Sea in a situation where the default conditions are shown for the first period (months 1–61), for the second period the annual TP-inflow via tributaries to the Baltic Proper has been reduced by 5,000 t/year, in the third period by 10,000 t/year and in the fourth period (months 361–481) by 15,000 t/year

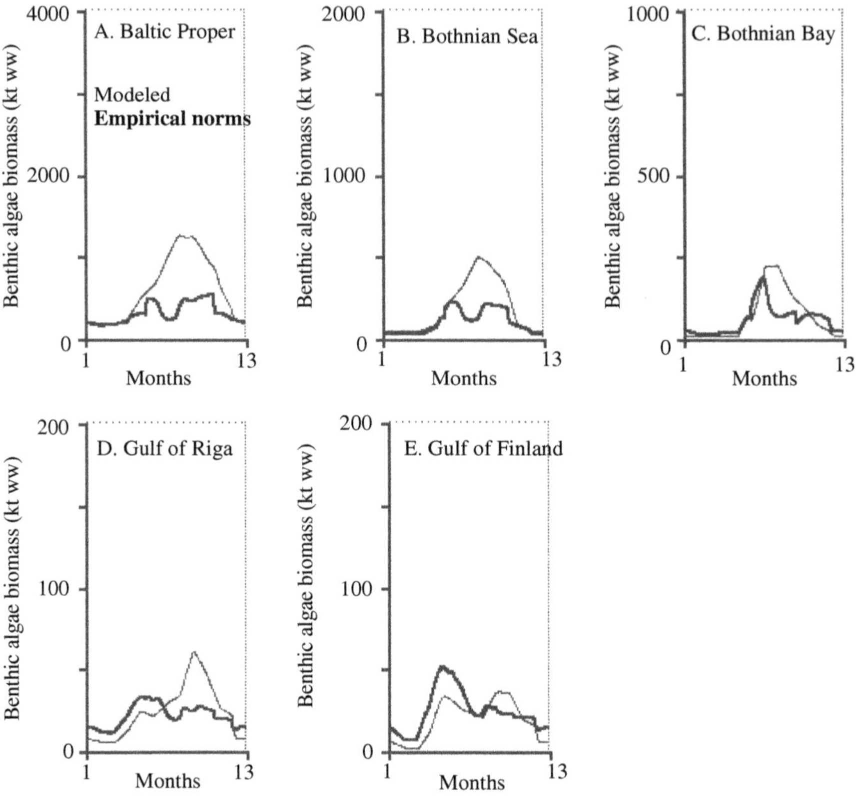

Fig. 4.12 A comparison between modeled empirically-based norm-values (*bolded* reference values for mainly the growing seasons) and dynamically modeled values of the biomass of benthic algae at steady-state (simulation time 241 months) under default conditions in the five basins of the Baltic Sea

TP-concentrations (Fig. 4.2a) and in primary phytoplankton production and biomass (see Fig. 4.3) will influence the benthic algae since the normal production values for benthic algae are related to the production of phytoplankton (see Eq. 4.10). One can also note from Fig. 4.11 that the increases in the biomass of benthic algae in this scenario are greatest in the Baltic Proper, but clear also in the Bothnian Sea.

Increasing Salt-Water Inflow

Increasing the salt-water inflow from the Kattegat to the Baltic Sea (see Fig. 4.5) will cause several changes to the structure and function of the Baltic Sea ecosystem. The Secchi depth will increase (see Fig. 4.5e), TP-concentrations will decrease (Fig. 4.5b) and primary phytoplankton production and biomass will decrease (Figs. 4.5c and 4.6). The changes in benthic algae are given in Fig. 4.13. There is a good correspondence between model-predicted values for benthic algae biomass and the norm-values (bolded in Fig. 4.12) in all five sub-basins for the 40-year

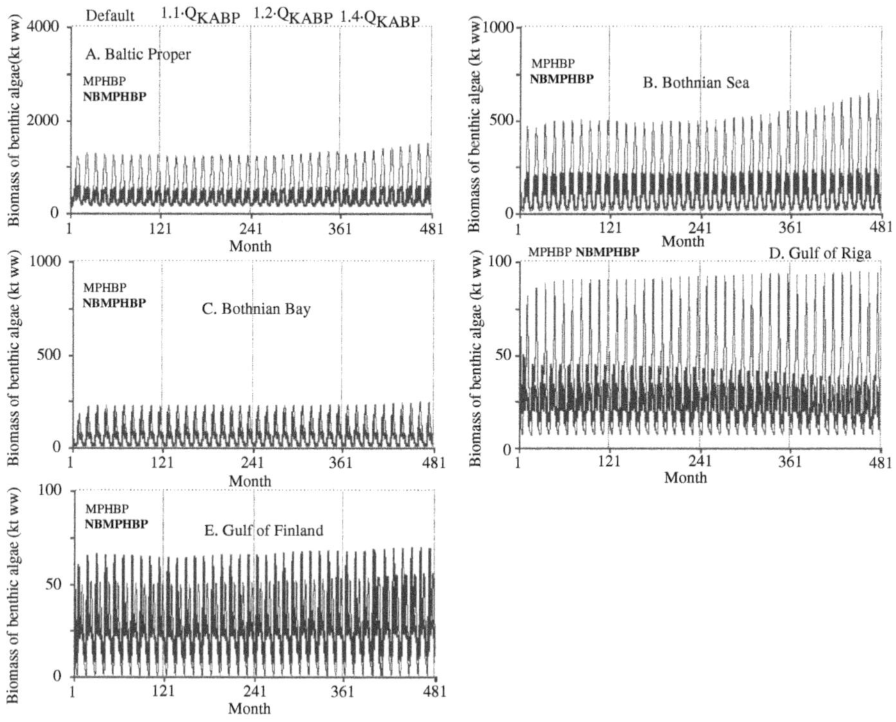

Fig. 4.13 A comparison (a sensitivity test) between dynamically modeled biomasses of benthic algae and values calculated from the empirically-based norms for the 5 sub-basins in the Baltic Sea in a situation where the default conditions are shown for the first period (months 1–61), for the second period the monthly saltwater inflow from the Kattegat has been increased by a factor of 1.1, in the third period 1.2 and in the fourth period (months 361–481) by 1.4

period. The increases in Secchi depth will mainly govern the relatively small increases in the biomasses of benthic algae in the five sub-basins seen in Fig. 4.13.

Temperature Scenario

Figure 4.7 gave the initial results from the "temperature scenario" when the water temperatures have been increased for a 40-year period. This means that the TP-concentrations (Fig. 4.7b) would likely increase very much and the Secchi depth decrease significantly (Fig. 4.7e).

The corresponding changes in the biomasses of benthic algae are given in Fig. 4.14. The correspondence between model-predicted values for benthic algae biomass and the norm-values (bolded in Fig. 4.14) is again good in all five sub-basins for the entire period.

Increasing the water temperatures may create rather severe eutrophication of the Baltic Sea system. This will reduce Secchi depths, increase TP-concentrations and there will be reductions in the production and biomasses of benthic algae, especially in the Bothnian Bay and the Bothnian Sea. These areas have the greatest land uplift

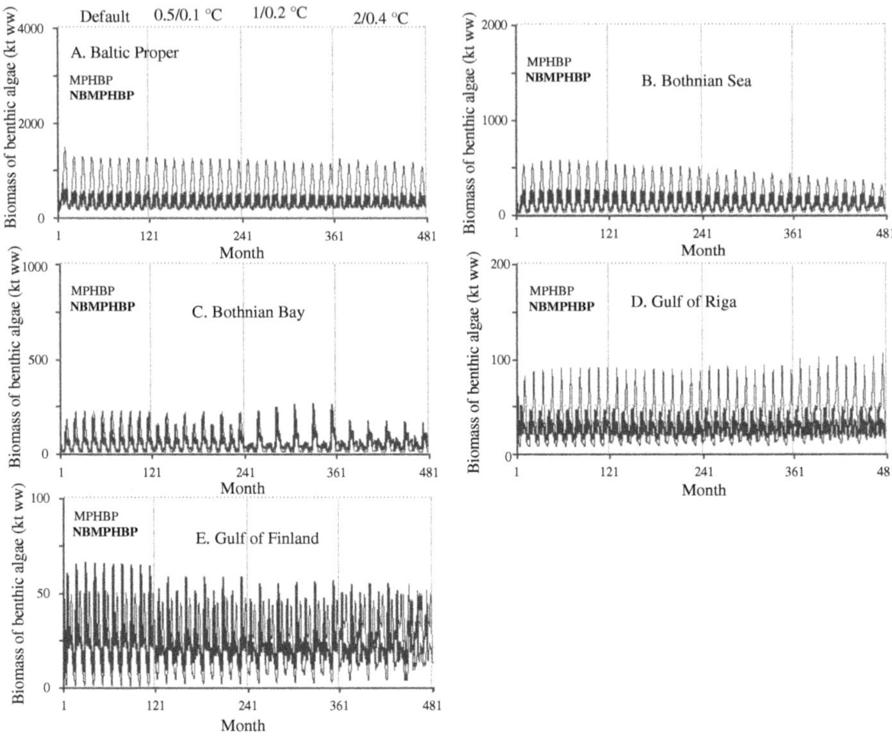

Fig. 4.14 A comparison (a sensitivity test) between dynamically modeled biomasses of benthic algae and values calculated from the empirically-based norms for the 5 sub-basins in the Baltic Sea in a situation where the default conditions are shown for the first period (months 1–61), for the second period the monthly surface-water temperatures have been increased by 0.5 and 0.1°C every second year, for the third period by 1 and 0.2°C every second year and in the fourth period by 2 and 0.4°C every second year. Middle- and deep-water temperatures have been increased by 0.75 and 0.5 of the increases given for the surface-water temperatures

and, in addition, the increased winter temperatures will cause the most dramatic reductions in ice cover and hence also in turbulence and resuspension.

4.2.5 *Macrophytes*

4.2.5.1 Background on Macrophytes

Macrophytes play several important roles in aquatic ecosystems, e.g., in proving shelter for small fish, binding nutrients and influencing secondary production by creating habitats for bacteria, benthic algae and zooplankton (Sandström et al., 2005). However, the quantitative role of macrophytes in aquatic systems is relatively poorly known because few general, validated models yielding high predictive power for macrophyte production, cover and biomass for marine areas have been presented. There are probably many reasons for this, e.g., related to the costs and efforts required to obtain relevant data (see Håkanson and Boulion, 2002a).

For the determination of one of the most fundamental properties of aquatic systems, the trophic status (see Table 4.2), the basic attention is generally given to phytoplankton production and biomass. However, the macrophytes can make a significant contribution to the total primary production, especially in small and shallow systems. Sometimes the macrophyte production exceeds the phytoplankton production in such systems (Wetzel, 1983). The macrophytes keep the nutrients bound for long periods. Consequently, they can help to improve of water quality. It is also important to emphasize that the evolution of any lake or smaller coastal system is closely connected with overgrowing by root plants (Beeton and Edmondson, 1972). Macrophytes also provide an important protective environment for small fish. Although they may not be especially important as a source of food for the fish, macrophytes can still influence fish production.

To determinate the relative role of macrophytes and phytoplankton in primary productivity, it is necessary to study the development of these plant groups relative to the morphometric and optical properties of the water. The utilization of the macrophyte biomass in the aquatic food chain is, as we understand it, poorly investigated. Vorobev (1977) analyzed data from 229 lakes of the Vologda district (Russia). He showed that the areal cover by macrophytes (Mac_{cov}, % of the area) is related to the ratio between the Secchi depth (Sec in m) and the mean depth (D_{MV} in m).

4.2.5.2 Modeling of Macrophytes

Figure 4.15 gives a general outline of the dynamic model for macrophytes and the panel of driving variables. This is a box model based on one ordinary differential equation. Subsequently, we will treat all the fluxes regulating macrophyte biomass and production. It should also be stressed that the model is meant to predict monthly, basin-characteristic values, and it is driven by parameters easily accessed from standard monitoring programs and maps. The driving variables are: Total phosphorus, Secchi depth, latitude, maximum depth, mean depth, coastal area and the hypsographic curve for the given basin. These factors are important for the prediction of macrophyte biomass and production, but not equally important. The model also uses data on daylight and surface-water temperatures, which could either be measured or predicted by a model driven by data on latitude, altitude and continentality (see Ottosson and Abrahamsson, 1998). For the following simulations, we have used empirical mean monthly temperatures for all layers in all Baltic Sea basins and mean monthly number of hours of daylight from Table 2.3. The model equations are complied in Table A.3 and will be explained in the following text.

The differential equation for macrophyte biomass (BM; values in kg ww) is given by:

$$BM_{MA}(t) = BM_{MA}(t - dt) + (IPR_{MA} - CON_{MA} - EL_{MA} - ER_{MA})dt \qquad (4.18)$$

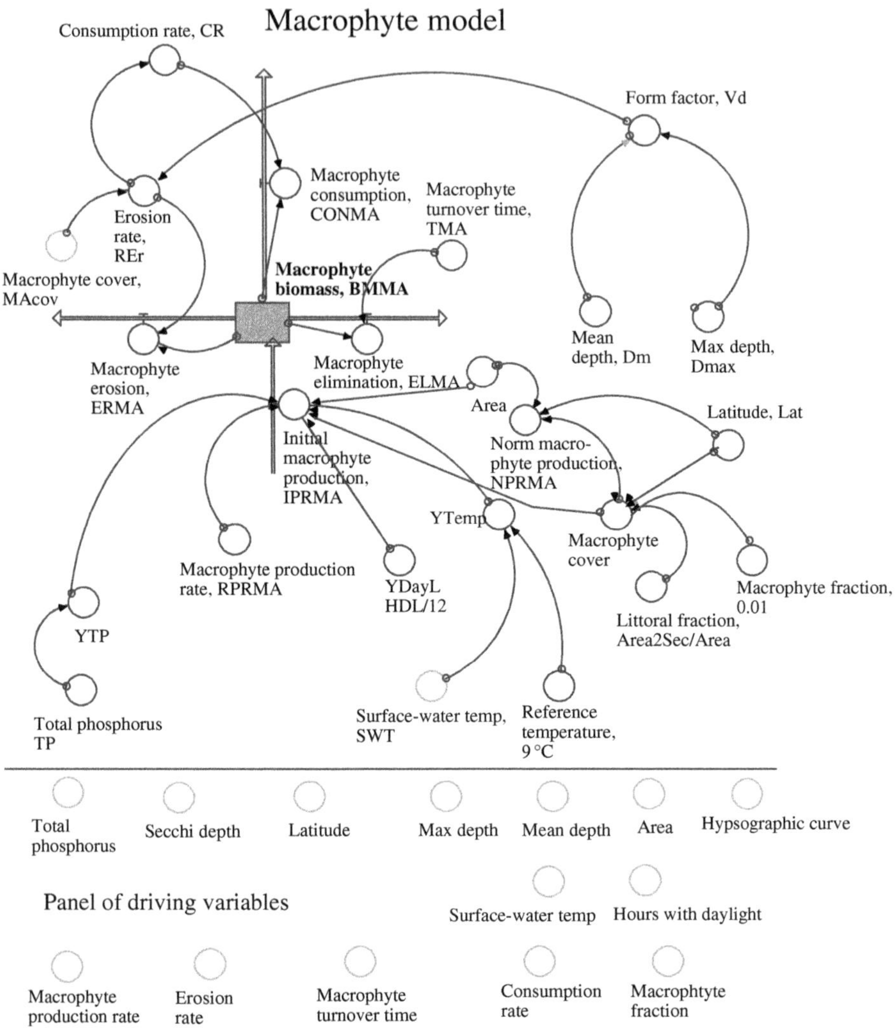

Fig. 4.15 Illustration of the sub-model for macrophytes

IPR_{MA} = Intial macrophyte production (kg ww/month).
CON_{MA} = Consumption of macrophytes by zoobenthos (kg ww/month).
EL_{MA} = Elimination of macrophytes related to macrophyte turnover (kg ww/month).
ER_{MA} = Physical erosion of macrophytes related to, e.g., wave action (kg ww/month).

We will treat these processes one by one in the following parts, starting with initial macrophyte production.

Initial Production

The initial macrophyte biomass [BM(0) in kg ww] is basically given by the empirical norm, i.e.;

$$BM(0) = NBM_{MA} = NPR_{MA}/T_{MA} \tag{4.19}$$

The normal production of macrophytes (kg ww/month) is given by (from Håkanson and Boulion, 2002a):

$$\begin{aligned} NPR_{MA} = {} & 0.001 \cdot Area \cdot 1/12 \cdot 10 \wedge (2.472 + 1.028 \cdot \log(MA_{cov}) \\ & -0.516 \cdot 90/(90 - Lat)) \end{aligned} \tag{4.20}$$

0.001 = Dimensional adjustment of g ww to kg ww.
Area = Coastal area in m^2.
1/12 = Dimensional adjustment of annual data to monthly data.
MA_{cov} = Macrophyte cover in %; from Eq. (4.21).
Lat = Latitude in °N.

The macrophyte cover is estimated here from the fraction of the littoral zone assuming that 1% of this area in the relatively large Baltic Sea basins is occupied by macrophytes. That is:

$$MA_{cov} = 0.01 \cdot Area_{2Sec}/Area \tag{4.21}$$

The initial macrophyte production (IPR_{MA} in kg ww/month) is calculated from the macrophyte cover and gives the theoretical maximum value for the macrophyte production without any losses. It is given by:

$$IPR_{MA} = R_{PRMA} \cdot Area \cdot (Mac_{cov} \cdot 0.01) \cdot Y_{DayL} \cdot Y_{Temp} \tag{4.22}$$

R_{PRMA} is the initial macrophyte production rate (kg ww/m^2·month). The influences of daylight and water temperature on macrophyte production are given by Y_{DayL} and Y_{Temp} in the same manner as for the initial production of benthic algae (see Eq. 4.9).

The model calibrations have focused on the value for the R_{PRMA}, the initial macrophyte production rate. The aim of the calibrations has been to seek a general, default value for R_{PRMA}, which could be used as a model constant for all sub-basins in the Baltic Sea and the default value is set to 1.

Consumption

The section addresses the question: How much of the macrophyte biomass is lost per month due to consumption by zoobenthos? This is given by:

$$CON_{MAZB} = BM_{MA} \cdot CR_{MAZB} \tag{4.23}$$

Where CR_{MAZB} (dimension 1/month) is calculated from:

$$CR_{MAZB} = (NCR_{ZB} + NCR_{ZB} \cdot (BM_{ZB}/NBM_{ZB} - 1)) \tag{4.24}$$

NCR_{ZB} = The normal consumption rate for zoobenthos as defined by the ratio NR_{ZB}/T_{ZB}; T_{ZB} is 128 days, see Table 2.9; the number of first order food choices for zoobenthos, NR_{ZB}, is 2.

BM_{ZB} = The actual biomass of zoobenthos (kg ww).

The normal biomass of zoobenthos is given by Eq. (4.13) (see Table 3.2).

Elimination

The elimination loss is given by:

$$EL_{MA} = BM \cdot 1/T_{MA} \tag{4.25}$$

T_{MA} is set to 300 days (i.e., 300/12 months, see Table 2.9).

Erosion

Physical erosion is calculated in the same way for macrophytes as for benthic algae from Eqs. (4.16) and (4.17) (as $ER_{MA} = BM_{MA} \cdot R_{Er}$).

In a following section, we will illustrate the correspondence between modeled values and norm-values for macrophytes.

4.2.5.3 Testing Modeling Results

Reductions in TP-Loading

Figure 4.16 gives the monthly variations in the biomass of macrophytes in all five sub-basins for the 40-year period. One can first from Fig. 4.17 note that the correspondence between dynamically modeled values and norm-values is good under default conditions. From Fig. 4.16, one can also see that the correspondence is fine for the entire 40-year period. Reductions in TP-loading will cause marked increases in Secchi depths (see Fig. 4.2d), which will also increase the production and biomass of macrophytes. One can also conclude from Fig. 4.16 that the increases in the biomass of macrophytes in this scenario are greatest in the Baltic Proper, but evident in all sub-basins.

Increasing Salt-Water Inflow

When the salt-water inflow from the Kattegat to the Baltic Sea has been increased (see Fig. 4.5), this will cause several changes to the system. The Secchi depth will increase (see Fig. 4.5e), TP-concentrations will decrease (Fig. 4.5b) and primary phytoplankton production will decrease (Figs. 4.5c and 4.6). The changes in macrophytes are given in Fig. 4.18. Again we can note the good correspondence between

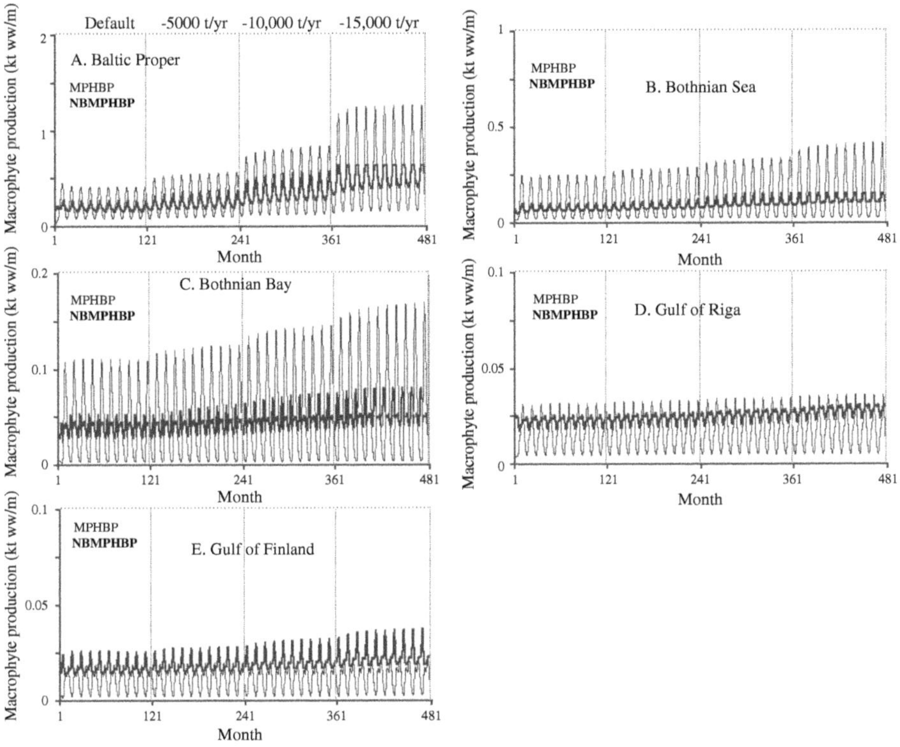

Fig. 4.16 A comparison between dynamically modeled biomasses of macrophytes and values calculated from the empirically-based norms for the 5 sub-basins in the Baltic Sea in a situation where the default conditions are shown for the first period (months 1–61), for the second period the annual TP-inflow via tributaries to the Baltic Proper has been reduced by 5,000 t/year, in the third period by 10,000 t/year and in the fourth period (months 361–481) by 15,000 t/year

model-predicted values for macrophyte production (not biomass in this example) and the corresponding norm-values (bolded in Fig. 4.18). The increases in Secchi depth will mainly explain the increases in macrophytes in all five sub-basins.

Temperature Scenario

Figure 4.7 gave the results from the "temperature scenario" where the water temperatures have been successively higher for a 40-year period. This means that the TP-concentrations (Fig. 4.7b) would likely go up and the Secchi depth decrease significantly (Fig. 4.7e). The corresponding decreases in the production of macrophytes are given in Fig. 4.19. The correspondence between model-predicted values and the norm-values is again good in all five sub-basins.

So, increasing the water temperatures would reduce Secchi depths and there will be reductions in the production of macrophytes, especially in the Bothnian Bay

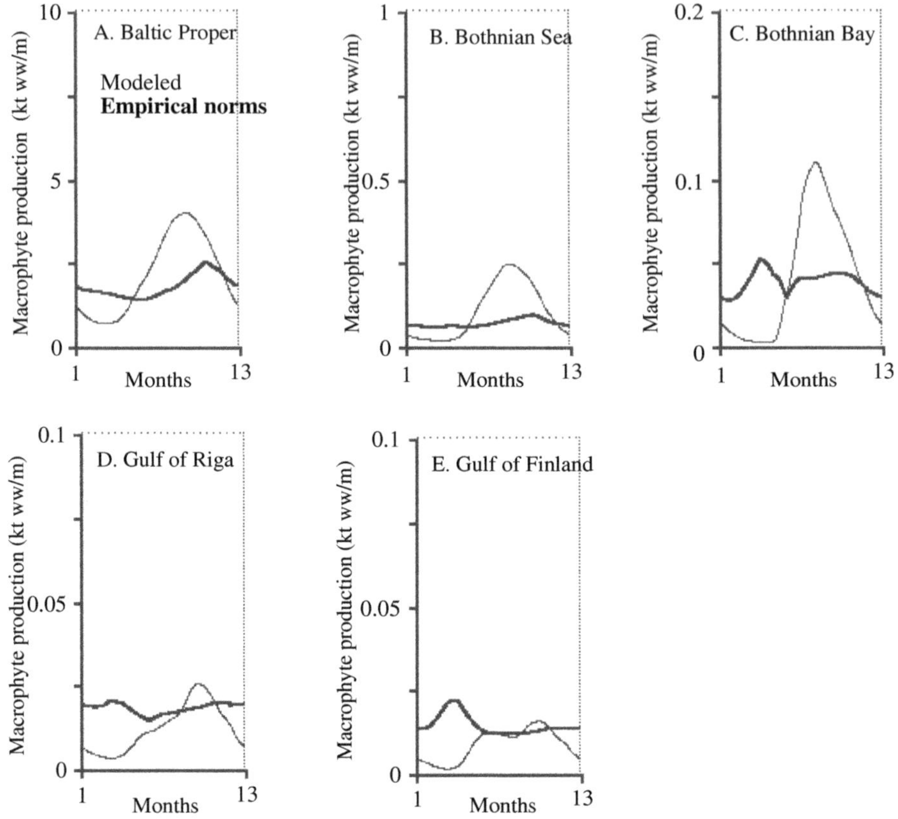

Fig. 4.17 A comparison between modeled empirically-based norm-values (*bolded* reference values for mainly the growing seasons) and dynamically modeled values of macrophyte production at steady-state (simulation time 241 months) under default conditions in the five basins of the Baltic Sea

where the higher winter temperatures will cause the most pronounced reductions in ice cover and hence also increase turbulence and resuspension the most.

4.3 The Reducer, Bacterioplankton

4.3.1 Background on Bacterioplankton

Many studies have shown that there exists a direct dependence between the biomass and/or growth of bacteria and the biomass and/or production of phytoplankton (Overbeck, 1972; Rai, 1978; Aizaki et al., 1981; Bird and Kalff, 1984; Currie, 1990; Conan et al., 1999). Nevertheless, there are few quantitative dynamic models capturing the most important factors and processes regulating bacterioplankton production

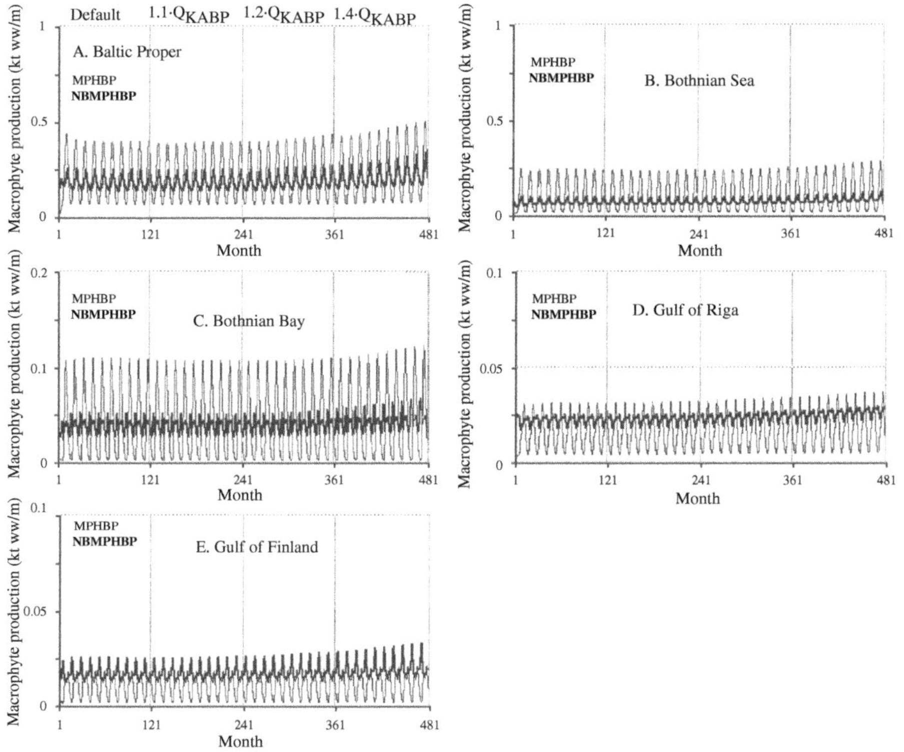

Fig. 4.18 A comparison (a sensitivity test) between dynamically modeled biomasses of macrophytes and values calculated from the empirically-based norms for the 5 sub-basins in the Baltic Sea in a situation where the default conditions are shown for the first period (months 1–61), for the second period the monthly saltwater inflow from the Kattegat has been increased by a factor of 1.1, in the third period 1.2 and in the fourth period (months 361–481) by 1.4

and no such models have, as far as we know, yielded good predictive power over a wide domain of marine coastal systems from just a few readily accessible driving variables. The basic aim of this section is to present such a dynamic model for the Baltic Sea system.

By analyzing data from the International Biological Program (IBP) and other literature data, Håkanson and Boulion (2002a) showed a logarithmic relationship between the number (NB in mill. cells per ml) and biomass of bacterioplankton (BM_{BP} in mg ww/l). This relationship is strong ($r^2 = 0.78$) and almost linear:

$$\log(BM_{BP}) = 0.973 \cdot \log(NB) - 0.438 \tag{4.26}$$

$$(r^2 = 0.78;\ n = 72;\ p < 0.001)$$

or

$$BM_{BP} = 0.365 \cdot NB^{0.973} \tag{4.27}$$

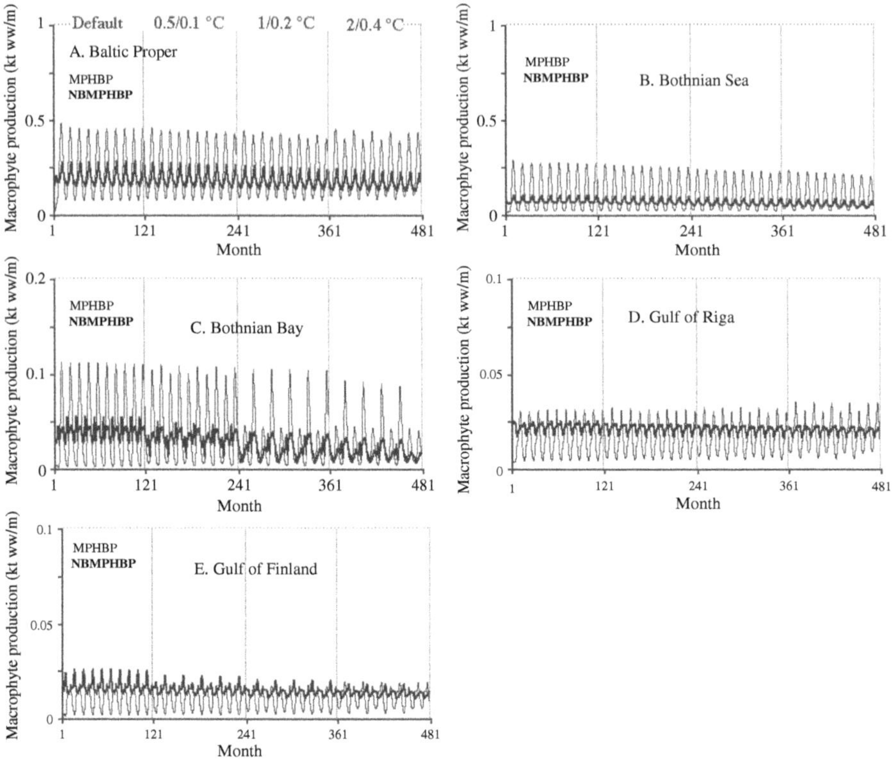

Fig. 4.19 A comparison (a sensitivity test) between dynamically modeled biomasses of macrophytes and values calculated from the empirically-based norms for the 5 sub-basins in the Baltic Sea in a situation where the default conditions are shown for the first period (months 1–61), for the second period the monthly surface-water temperatures have been increased by 0.5 and 0.1°C every second year, for the third period by 1 and 0.2°C every second year and in the fourth period by 2 and 0.4°C every second year. Middle- and deep-water temperatures have been increased by 0.75 and 0.5 of the increases given for the surface-water temperatures

The exponent is close to 1. It means that if the number of bacteria increases two orders of magnitude, from 0.1 to 10 million cells/ml, the bacterial biomass is also likely to increase by two orders of magnitude (or rather by a factor of 89 and not by 100). One can also conclude from Håkanson and Boulion (2002a) that the higher the productivity of the system (as given by concentrations of chlorophyll-a, Chl), the lower the expected ratio between BM_{BP} and Chl. This also implies that the ratio between bacterial production and biomass (the PR/BM-ratio) would grow more progressively than the bacterial biomass. In other words, the production of bacteria (but not the number and biomass of bacteria) is proportional to the chlorophyll-a concentration and the phytoplankton production.

An essential element for bacterial production is dissolved carbon produced by phytoplankton. The release of dissolved organic matter (DOM) by phytoplankton is confirmed by a number of experimental studies. The transformation of DOM to bacterial production is in fact a basic flow of organic carbon in fresh and seawaters (Derenbach et al., 1974; Harrison et al., 1977; Larson and Hagström, 1979; Cole et al., 1982; Wolter, 1982; Sondergaard et al., 1985; Baines and Pace, 1991). DOM production can vary from 1 to 50% of the total phytoplankton production. It has been argued that the percentage of DOM released by phytoplankton is reduced at higher phytoplankton production (Watt, 1966; Fogg, 1966; Anderson and Zeutschel, 1970; Thomas, 1971). However, a study by Baines and Pace (1991) did not confirm this, and they based their conclusions on 225 observations from fresh and marine systems. DOM may be released by living phytoplankton cells (e.g., Sadchikov and Frenkel, 1990; Sadchikov and Makarov, 1997), but also by dying cells (Marker, 1965; Cole et al., 1982).

4.3.2 Modeling of Bacterioplankton

The dynamic model for bacterioplankton and its panel of driving variables are illustrated in Fig. 4.20. The model has two targets, bacterioplankton biomass and production. The following ordinary differential equation gives the monthly production and biomass of bacterioplankton:

$$BM_{BP}(t) = BM_{BP}(t - dt) + (IPR_{BP} + MIG_{InBP} - MIG_{OutBPZH} - CON_{BPZH} - EL_{BP})dt \tag{4.28}$$

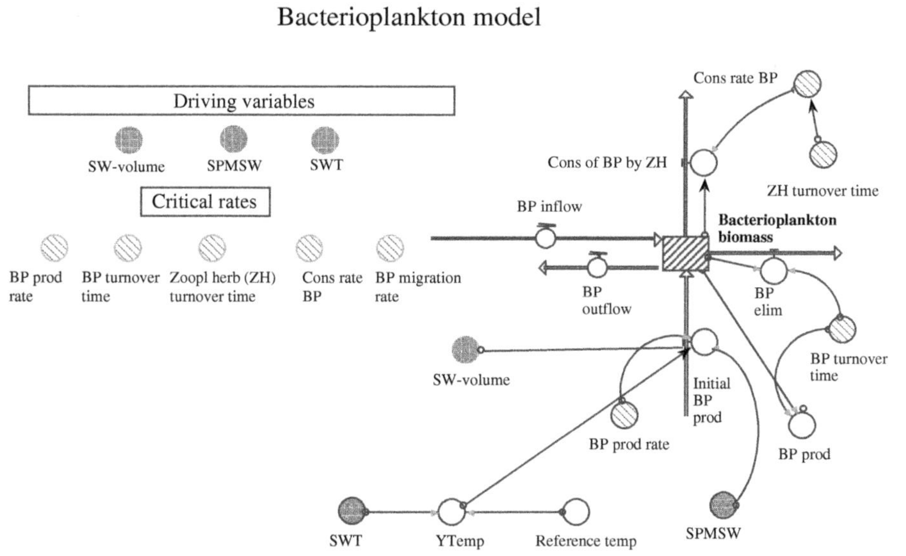

Fig. 4.20 The bacterioplankton sub-model in CoastWeb. The panel of driving variables gives the obligatory driving variables. The panel also lists the six critical rates. Migration (or transport by water currents) is not shown in this figure

BM_{BP} = Bacterioplankton biomass (kg ww).
IPR_{BP} = Initial bacterioplankton production (kg ww/month).
MIG_{InBP} = Inflow of bacterioplankton (kg ww/month).
MIG_{OutBP} = Outflow of bacterioplankton (kg ww/month).
CON_{BPZH} = Bacterioplankton consumption by herbivorous zooplankton (kg ww/month).
EL_{BP} = Bacterioplankton elimination (or turnover) (kg ww/month).

The initial bacterioplankton biomass is set equal to the norm-value (NBM_{BP} in kg ww), which is calculated from the empirical norm (see Table 3.2; from chlorophyll, SPM and SW-volume). Bacteria can generally be found in the entire water mass, although the highest bacterial biomasses often appear close to the bottom and near the water surface (see Wetzel, 1983 and Fig. 2.18).

NBM_{BP} is given by:

$$NBM_{BP} = Y_{SPM} \cdot 0.001 \cdot V_{SW} \cdot 10\hat{}(0.973 \cdot (0.27 \cdot \log(Chl) + 0.19) - 0.438) \quad (4.29)$$

Chl = Modeled concentration of chlorohyll-a (μg/l).
Vol_{SW} = The surface-water volume (m^3).
Y_{SPM} = A dimensionless moderator expressing the relationship between SPM in a given coastal area and in a lake with similar characteristics, defined from the ratio SPM_{SWlake}/SPM_{SW}. SPM_{SW} is modeled dynamically by CoastMab and SPM_{SWlake} is calculated from the regression based on TP (see Eq. 2.5).

The initial production of bacterioplankton (IPR_{BP}) is related to the available amount of SPM (= seston) from all sources and including dissolved organic matter (DOM), the basic fuel for bacterioplankton production, and other factors known to influence bacterioplankton production, e.g., temperature. IPR_{BP} is given by:

$$IPR_{BP} = Y_{Temp} \cdot R_{BP} \cdot SPM_{SW} \cdot Vol_{SW} \cdot (1/1,000) \quad (4.30)$$

Where Y_{Temp} is the same dimensionless moderator generally used in the CoastWeb-model for expressing the influence of water temperature variations on bioproduction – the higher the temperature the more bacterioplankton will be produced. Y_{Temp} is defined by the ratio between SW-temperatures and a reference temperature of 9°C (related to the duration of the growing season). The 1/1,000 term in Eq. (4.30) is a conversion constant from mg/l to kg/m^3 and R_{BP} is the rate of initial bacterioplankton production (in 1/month). The question is: Which R_{BP}-value gives the best correspondence between modeled values and the empirical norm-values? Many calibrations are been done to find a generic value for R_{BP} and the result is to set R_{BP} to 12 (1/month).

Multiplication with Vol_{SW} in Eq. (4.30) gives the requested dimension (values in kg ww/month).

The bacterioplankton consumption by herbivorous zooplankton in Eq. (4.28) is given by:

$$CON_{BPZH} = BM_{BP} \cdot CR_{BPZH} \tag{4.31}$$

Where BM_{BP} is the bacterioplankton biomass (kg ww) and CR_{BPZH} the actual consumption rate (1/month) expressing the loss of bacterioplankton (BP) from predation by herbivorous zooplankton (ZH). CR_{BPZH} is defined from the turnover time of herbivorous zooplankton (T_{ZH}; in months), which is set to 6.0 days (see Table 2.9). That is:

$$CR_{BPZH} = (NCR_{BPZH} + NCR_{BPZH} \cdot (BM_{ZH}/NBM_{ZH} - 1)) \tag{4.32}$$

NCR_{BPZH} is the normal consumption rate, as given by NR_{ZH}/T_{ZH} (1/months) for $NR_{ZH} = 2$; NR_{ZH} is the number of first order food choices for herbivorous zooplankton, i.e., phytoplankton and bacterioplankton. CR_{BPZH} is the actual consumption rate (1/months). NBM_{ZH} is the normal biomass of herbivorous zooplankton (kg ww); when the actual biomass of herbivorous zooplankton (BM_{ZH}) is equal to the normal biomass of herbivorous zooplankton (NBM_{ZH}), the actual consumption rate (CR_{BPZH}) is equal to the normal consumption rate (NCR_{BPZH}).

Migration (kg ww/month), or rather inflow, of bacterioplankton from the outside sea (in this example bacterioplankton in the Bothnian Sea, BS, transported by water currents to the Bothnian Bay, BB):

$$MIG_{InBPBB} = R_{MigBPBB} \cdot BM_{BPBS} \tag{4.33}$$

Where $R_{MigBPBB}$ is the migration rate for bacterioplankton ($= 1/T_{SWBB}$, T_{SWBB} is the water turnover time for the SW-water in BB in months) and BM_{PHBS} is the actual biomass of bacterioplankton in BS (modeled using CoastWeb). The outflow of bacterioplankton is calculated in the same manner using the modeled biomass of bacterioplankton in the given system (here BM_{PHBB}).

The last part of Eq. (4.28) concerns bacterioplankton elimination (or turnover). EL_{BP} is given by:

$$EL_{BP} = BM_{BP} \cdot 1/T_{BP} \tag{4.34}$$

T_{BP} is the turnover time of bacterioplankton, which is set to 2.8 days (Table 2.9). This value is used as a general value for all types of bacterioplankton just like 6.0 days is used for all types of herbivorous zooplankton.

The bacterioplankton production (in kg ww/month) is then given by $PR_{BP} = BM_{BP}/T_{BP}$.

4.3.3 Testing Modeling Results

4.3.3.1 Reductions in TP-Loading

Figure 4.21 gives the results for the biomass of bacterioplankton in the five sub-basins of the Baltic Sea. One can first note from Fig. 4.22 that the correspondence between dynamically modeled values and norm-values is good under default conditions and from Fig. 4.21, one can also see that the correspondence is relatively good for the entire 40-year period. Reductions in TP-loading will cause clear decreases in SPM (see Fig. 4.2c), which will reduce the production and biomasses of bacterioplankton. One can also note from Fig. 4.21 that the reductions in bacterioplankton are logically greatest in the Baltic Proper, but evident in all basins.

4.3.3.2 Increasing Salt-Water Inflow

When the salt-water inflow from the Kattegat is increased (see Fig. 4.5), this will cause several changes to the system. SPM will decrease (see Fig. 4.5d),

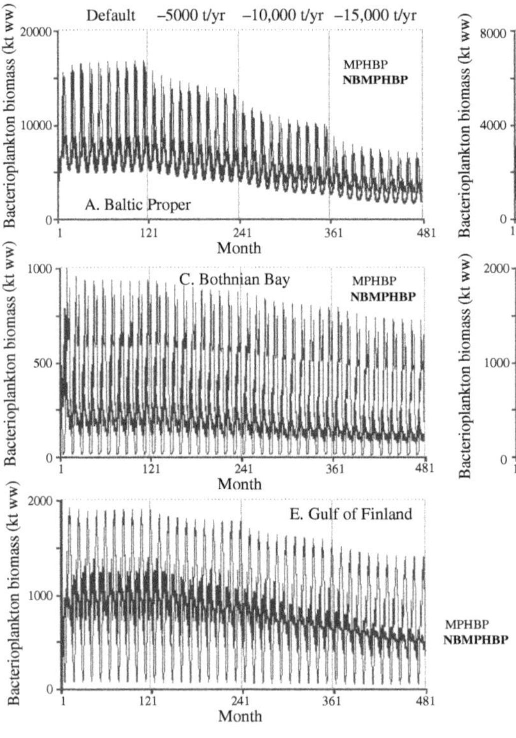

Fig. 4.21 A comparison between dynamically modeled biomasses of bacterioplankton and values calculated from the empirically-based norms for the 5 sub-basins in the Baltic Sea in a situation where the default conditions are shown for the first period (months 1–61), for the second period the annual TP-inflow via tributaries to the Baltic Proper has been reduced by 5,000 t/year, in the third period by 10,000 t/year and in the fourth period (months 361–481) by 15,000 t/year

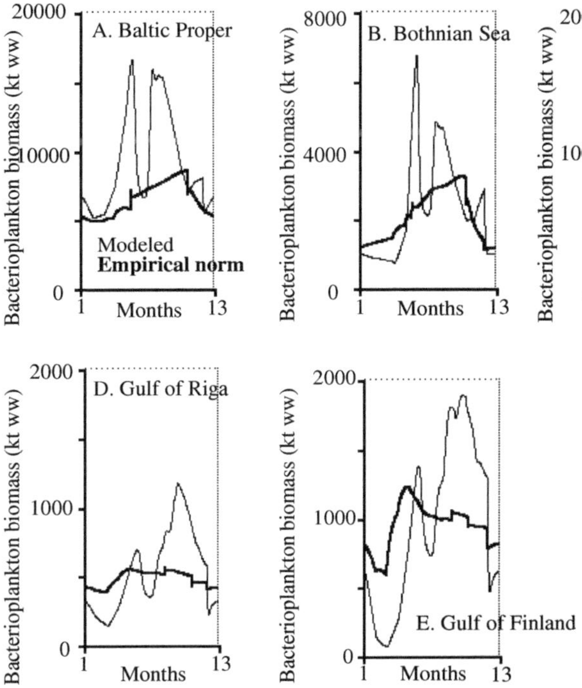

Fig. 4.22 A comparison between modeled empirically-based norm-values (*bolded* reference values for mainly the growing seasons) and dynamically modeled values of bacterioplankton biomass at steady-state (simulation time 241 months) under default conditions in the five basins of the Baltic Sea

TP-concentrations will also decrease (Fig. 4.5b) and primary phytoplankton production will decrease (Figs. 4.5c and 4.6). The changes in bacterioplankton biomass are given in Fig. 4.23. The correspondence between model-predicted values and the norm-values (bolded in Fig. 4.23) is relatively good the entire period. The decrease in SPM mainly explains the decreases in bacterioplankton biomass in the five sub-basins.

4.3.3.3 Temperature Scenario

Figure 4.7 gave the results when the water temperatures were increased for a 40-year period. This means that the TP-concentrations (Fig. 4.7b) would likely increase very much, the SPM-values increase (Fig. 4.7d) and the Secchi depth decrease (Fig. 4.7e).

The corresponding changes in the biomasses of bacterioplankton are logical and shown in Fig. 4.24. The correspondence between model-predicted values for benthic algae biomass and the norm-values is again relatively good in all five sub-basins for the entire 40-year period.

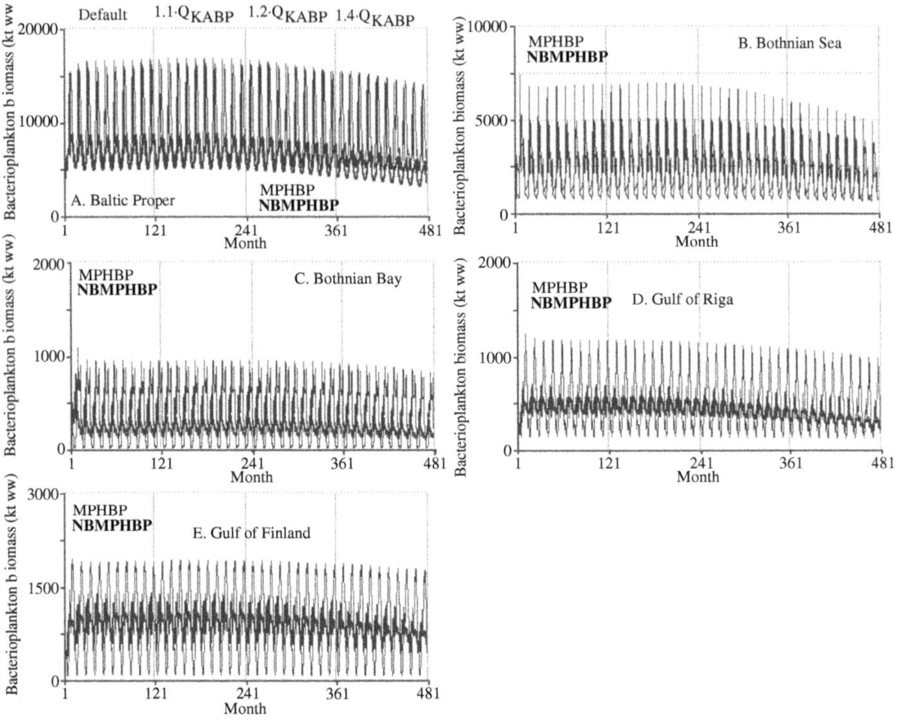

Fig. 4.23 A comparison (a sensitivity test) between dynamically modeled biomasses of bacterioplankton and values calculated from the empirically-based norms for the 5 sub-basins in the Baltic Sea in a situation where the default conditions are shown for the first period (months 1–61), for the second period the monthly saltwater inflow from the Kattegat has been increased by a factor of 1.1, in the third period 1.2 and in the fourth period (months 361–481) by 1.4

Increasing the water temperatures would increase SPM-concentrations and there will be marked increases also in the production and biomasses of bacterioplankton, especially in the Bothnian Bay and the Bothnian Sea where the higher winter temperatures will cause the most pronounced reductions in ice cover (and hence also increase turbulence and resuspension).

4.4 Secondary Producers

4.4.1 Zoobenthos

To predict the production and biomass of zoobenthos is important in water management, fish biology, aquatic ecology and in many types of studies concerned with the sediment habitat. As far as the authors are aware, there are no models for zoobenthos of the type presented here for marine ecosystems accounting to production, grazing,

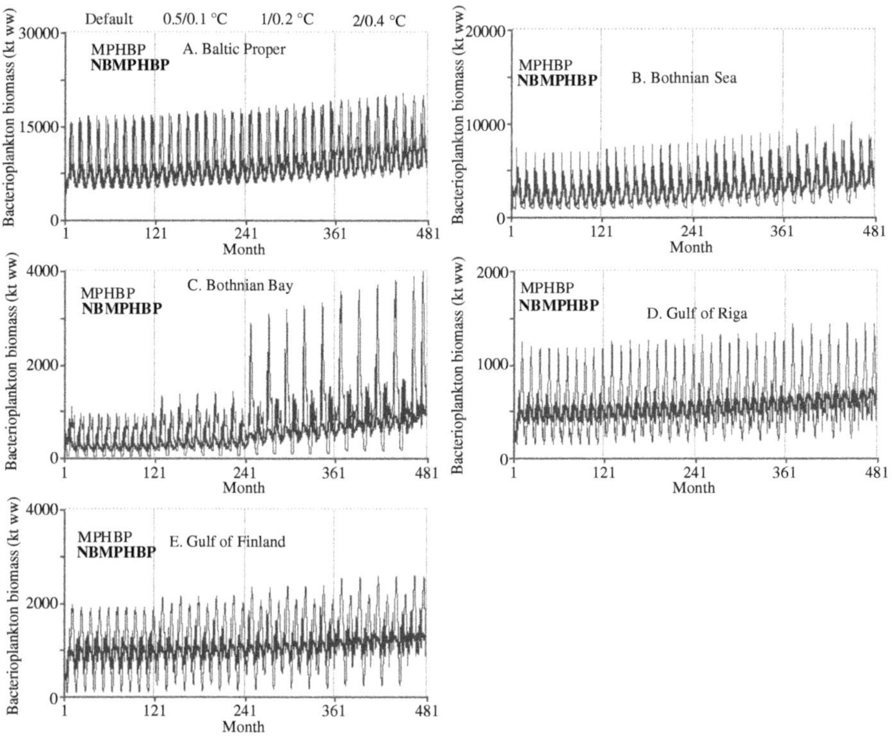

Fig. 4.24 A comparison (a sensitivity test) between dynamically modeled biomasses of bacterioplankton and values calculated from the empirically-based norms for the 5 sub-basins in the Baltic Sea in a situation where the default conditions are shown for the first period (months 1–61), for the second period the monthly surface-water temperatures have been increased by 0.5 and 0.1°C every second year, for the third period by 1 and 0.2°C every second year and in the fourth period by 2 and 0.4°C every second year. Middle- and deep-water temperatures have been increased by 0.75 and 0.5 of the increases given for the surface-water temperatures

growth, elimination and food choices in a general, holistic ecosystem framework designed to achieve practical utility.

Zoobenthos, benthic algae and macrophytes are the three categories of functional organisms related to sediments incorporated in the CoastWeb-model. The aim of this section is to present, motivate and test the dynamic model for zoobenthos. First, we will give an introductory section on this group of organisms and also define the empirical reference models used to test the dynamic model.

4.4.1.1 Background on Zoobenthos

Alimov (1982, 1989) emphasized that zoobenthos consists of non-predatory and predatory animals. However, from an energy point of view, this community can be considered as one. The food consumed by the non-predatory animals is the energy input to the system. The output from the system is the production of the community as a whole. Thus, the production of zoobenthos is the sum of production of non-predatory and predatory animals minus consumption (outflow) by predatory animals: $PR_{Com} = PR_{Non\text{-}pred} + PR_{Pred} - CON_{Pred.}$ PR_{Com} is a useful production measure of how much of the zoobenthos community that can be used by fish. It is necessary to note that the biomass of predatory zoobenthos is often relatively small (see Håkanson and Boulion, 2002a). The ratio between the biomasses of predatory and non-predatory zoobenthos also generally decreases with increasing production values (see Håkanson and Boulion, 2002a)

Håkanson and Boulion (2002a) gave the following relationship between production and biomass of zoobenthos:

$$PR_{ZB} = 1.62 \cdot {BM_{ZB}}^{0.98} \tag{4.35}$$

$$(n = 20;\ r^2 = 0.80)$$

Where PR_{ZB} is production in kg ww/day·m^2 during the growing season; BM_{ZB} is the mean biomass of zoobenthos in kg ww/m^2 during the growing season. By accounting for the duration of growing season for the investigated systems (using the model given by Håkanson and Boulion, 2002a), it turned out that the mean duration of growing season for these systems was 207 days. Hence, the average turnover time for zoobenthos biomass is 207/1.62 = 128 days. This is the value used in the CoastWeb-model (see Table 2.9).

Other functional parameters for zoobenthos (consumption and respiration) were also discussed by Alimov (1982) and Håkanson and Boulion (2002a), who gave the relationship between zoobenthos production (PR_{ZB} in kg ww/m^2 during the growing season) and the amount of food consumed (CON_{ZB} in kg ww/m^2 during the growing season) as:

$$\log(PR_{ZB}) = 1.10 \cdot \log(CON_{ZB}) - 0.95 \tag{4.36}$$

$$(n = 20;\ r^2 = 0.91)$$

or

$$PR_{ZB} = 0.11 \cdot {CON_{ZB}}^{1.1} \tag{4.37}$$

These data suggest that the metabolic efficiency ratio (MER) for zoobenthos growth (MER = PR/CON) averages 0.11. But since the exponent for CON is higher than 1, there is actually very good agreement between these results and the results given by Winberg (1985; see Table 2.9). So, the CoastWeb-model uses a MER-value of 0.15 for zoobenthos feeding on benthic algae and macrophytes, but a lower MER-value for zoobenthos feeding on sediments (25% lower).

It is important to stress that living macrophytes is not an important food sources for zoobenthos. According to Ostapenia (1989), only about 10% of the macrophyte biomass is utilized in the total lake foodweb. Epipelic and sedimenting planktonic algae, detritus and bacteria are the main foodstuffs for zoobenthos (Morgan, 1980).

Detritus is not a valuable food in comparison to, e.g., bacteria attached to detrital particles. According to Kuznetsov (1970), bacteria constitute on average 5% of the organic substances in the fine sediments. Obviously macrophytes, being the substrate for periphyton and a source of detritus, indirectly stimulate the development of zoobenthos. Taking into account that phytoplankton (alive and/or in the detrital form) is important food for zooplankton and that the macrophyte material (in transformed form) is food for zoobenthos, Winberg et al. (1986) came to the following conclusion: The higher the ratio of phytoplankton production to macrophyte production, the higher the ratio of zooplankton production to zoobenthos production.

4.4.1.2 Modeling of Zoobenthos

From this background, the following section presents the dynamic model for zoobenthos (see Fig. 4.25 for an overview of the model and Table 2.11 for an overview of abbreviations). The differential equation (Eq. 4.40) gives the fluxes (kg ww/month) to and from the compartment, which includes all types of non-predatory and predatory benthic animals.

$$BM_{ZB}(t) = BM_{ZB}(t-dt)+(IPR_{MAZB}+IPR_{BAZB}+IPR_{SedZB}-CON_{ZBPY}-EL_{ZB})\cdot dt \quad (4.38)$$

BM_{ZB} = The biomass of zoobenthos (kg ww); the initial BM_{ZB}-value is set to the norm-value, which is given by Eq. (4.13).
IPR_{MAZB} = The initial production of zoobenthos from macrophytes (kg ww/month).
IPR_{BAZB} = The initial production of zoobenthos from benthic algae (kg ww/month).
IPR_{SedZB} = The initial production of zoobenthos from all other sediment sources (kg ww/month).
CON_{ZBPY} = Consumption of zoobenthos by prey fish (kg ww/month); note that consumption means a loss from one compartment and production means that matter is added to a compartment. These two fluxes may be very similar numerically, but they are still differentiated to emphasize this functional difference.
EL_{ZB} = Elimination (turnover) of zoobenthos (kg ww/month).

A certain part of zoobenthos is insect larvae. However, for simplicity the model does not account for the flight of imago from a given system during the growing season. Another question related to zoobenthos production concerns the contribution from epiphytes? If epiphyte biomass is about 1.5% of macrophyte biomass (see Håkanson and Boulion, 2002a) in lakes and small coastal areas; the PR/BM-coefficient for epiphytes is between the values for phytoplankton and phytobenthos. This, however, is not accounted for in this modeling for entire large basins.

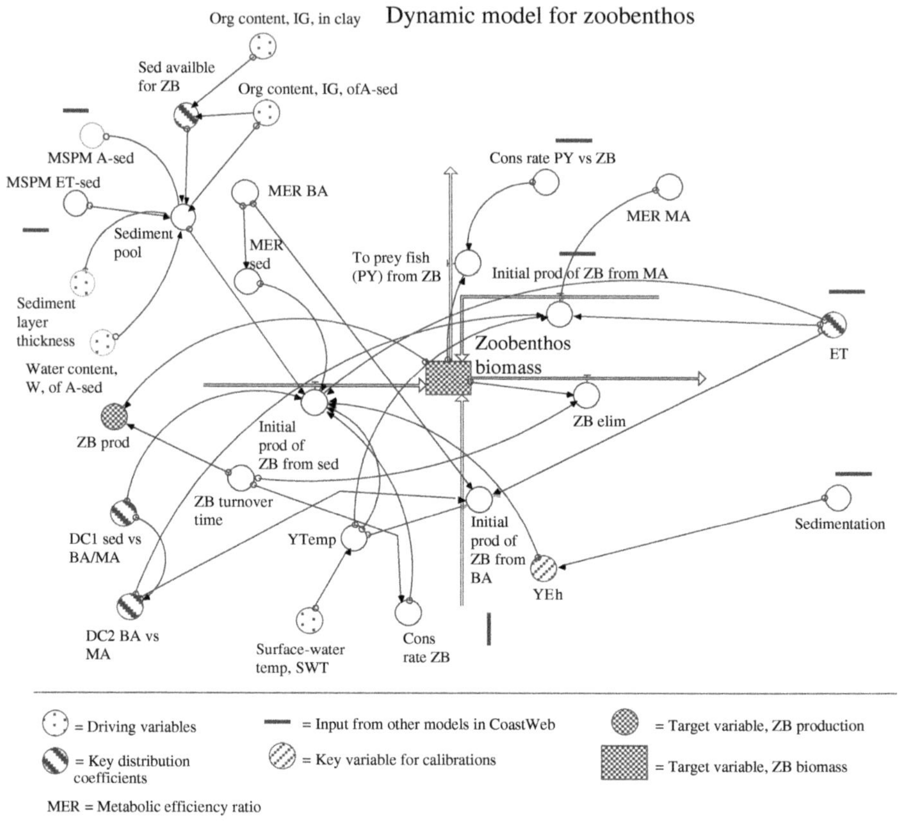

Fig. 4.25 Outline of the sub-model for zoobenthos in CoastWeb. The panel of driving variables gives information on important model variables

There are three food choices for zoobenthos and Fig. 4.25 gives the panel of food choices. There are two first order food choices ($NR_{ZB} = 2$), sediments (= all types of sediments) and "the rest", i.e., benthic algae and macrophytes. The default value for the first order distribution coefficient (DC_{ZB1}) between sediments and "the rest" is set to 0.75. This is based on the previous discussion concerning the relative role of various food choices for zoobenthos. The second order distribution coefficient, DC_{ZB2}, differentiates between zoobenthos eating either benthic algae or macrophytes. The default value for DC_{ZB2} is also set to 0.75 as in the basic LakeWeb-model (Fig. 4.26).

The initial production of zoobenthos from consumption of macrophytes is generally small and calculated from:

$$IPR_{MAZB} = CON_{MAZB} \cdot (1 - DC_{ZB1}) \cdot (1 - DC_{ZB2}) \cdot Y_{Temp}{}^{0.25} \cdot MER_{MAZB} \quad (4.39)$$

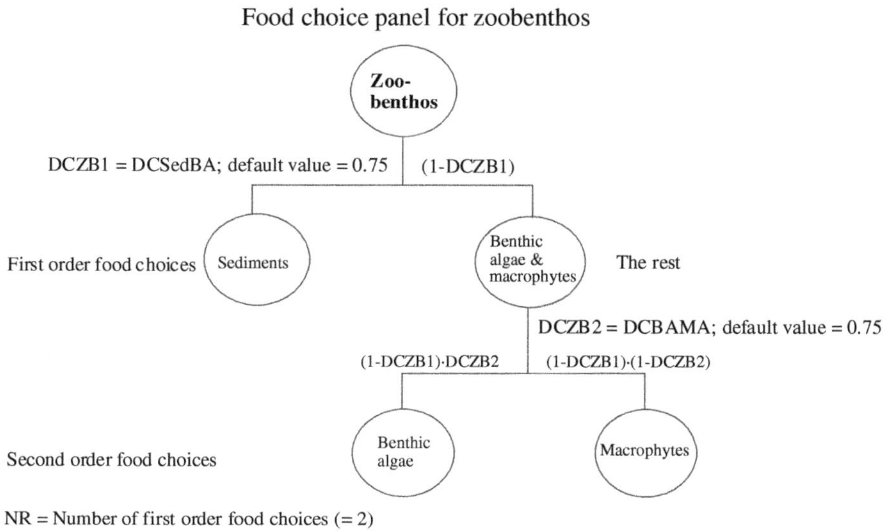

Fig. 4.26 The food-choice panel for zoobenthos

CON_{MAZB} = The consumption of macrophytes by zoobenthos (kg ww/month).
DC_{ZB1} = The default value for this first order distribution coefficient is set to 0.75. That is, the flow from all sediment sources is multiplied by 0.75 and the flow from benthic algae plus macrophytes by (1–0.75). Note that this does not mean that 75% of the fluxes to zoobenthos will emanate from sediments sources and 25% from benthic algae plus macrophytes. This would be the case only if these two fluxes are identical, which they rarely are.
DC_{ZB2} = The second order distribution coefficient regulating how much of either benthic algae or macrophytes that zoobenthos eat. The default value is 0.75, which does NOT mean that 25% of the food consumed by zoobenthos is macrophytes, but that the factor 0.25 times the given flux from macrophytes will go into zoobenthos.
Y_{Temp} = The ratio between SWT (i.e., monthly surface-water temperatures in °C) and the reference temperature of 9°C (related to the duration of the growing season, see Håkanson and Boulion, 2002a). The exponent 0.25 is used as a default value for zoobenthos in the CoastWeb-model. For phytoplankton and bacterioplankton living in the water the exponent is 1, for herbivorous and predatory zooplankton 0.5 and for fish 0.25 in the CoastWeb-model.
MER_{MAZB} = The metabolic efficiency ratio for zoobenthos eating macrophytes. It is set to 0.15 (see Table 2.9).

The bottom fauna will die if the sediments turn anoxic (see Fig. 3.4). In the algorithm for the initial production of zoobenthos from consumption of sediments, we will also account for this by means of a moderator called Y_{Eh}. But for zoobenthos eating macrophytes and benthic algae, which constitute only a small part of the diet for zoobenthos, and since macrophytes live in shallow water where the oxygenations is often good, this moderator is not used. Basically Y_{Eh} is related to sedimentation (see Eq. 4.42).

The initial production of zoobenthos from benthic algae is given in a similar way by:

$$IPR_{BAZB} = CON_{BAZB} \cdot (1 - DC_{ZB1}) \cdot DC_{ZB2} \cdot {Y_{Temp}}^{0.25} \cdot MER_{BAZB} \quad (4.40)$$

MER_{BAZB} is the metabolic efficiency ratio for zoobenthos eating benthic algae. It is also set to 0.15.

The third and most important production factor for zoobenthos is the consumptions of sediments (detritus) and sedimented matter from all sources (except those given by benthic algae and macrophytes). This initial production is given by:

$$IPR_{SedZB} = M_{Sed} \cdot DC_{ZB1} \cdot NCR_{ZB} \cdot MER_{SedZB} \cdot {Y_{Temp}}^{0.25} \cdot (ET + (1 - ET) \cdot Y_{Eh}) \quad (4.41)$$

M_{Sed} = The sediment pool, i.e., the mass (kg ww) of sediments in the given basin.
NCR_{ZB} = The normal consumption rate for zoobenthos, which is defined from the turnover time of zoobenthos (T_{ZB}) and the number of first order food choices for zoobenthos ($NR_{ZB} = 2$), i.e., $NCR_{ZB} = NB_{ZB}/T_{ZB}$.
MER_{SedZB} = The metabolic efficiency ratio for zoobenthos eating sediments, which is set to be 25% of the MER-value for zoobenthos eating benthic algae, i.e., $MER_{SedZB} = 0.25 \cdot 0.15 = 0.0375$ (dim. less).
Y_{Temp} = The dimensionless temperature moderator (already defined).
ET = The fraction of the bottom area in the given basins above the theoretical wave base where processes of fine sediment erosion and transportation dominate the bottom dynamic conditions and where relatively high O_2-concentrations are likely to prevail (see Fig. 2.1). The areas with continuous fine sediment accumulations (A = (1 − ET)) are more likely to have a higher oxygen consumption, lower oxygen concentrations and lower redox potential (Eh).

This dimensionless moderator, Y_{Eh}, concerns the potential redox-conditions in the sediments. An increasing sedimentation will lower the redox potential in the A-sediments more than in the ET-sediments, since, by definition, the latter are influenced by wind-/wave action and generally well oxygenated. In this modeling, Y_{Eh} is defined in a simple way (see Eq. 3.54). Sed (sedimentation in cm/year) is calculated in a standard way from gross sedimentation of SPM using the dynamic SPM-model (GS in μg dw/cm^2·d), sediment water contant (W in % ws; set to 70% ww as a default value for all A-sediments in the Baltic Sea) and sediment bulk density (d in g dw/cm^3) as:

$$Sed = 10^{-6} \cdot GS \cdot 365/((1 - W/100) \cdot d) \quad (4.42)$$

This means that one should expect relatively high oxygen concentrations and favorable conditions for the zoobenthos in areas with sedimentation lower than 0.75 cm/year. So, if sedimentation increases to values higher than 0.75 cm/year, it means a lower redox-potential (Eh) and a lower production of zoobenthos. Since one should also expect a more or less exponential decrease in zoobenthos biomass with increasing water depth and relatively low biomasses below the photic zone (see Fig. 4.27 and Håkanson and Rosenberg, 1985; Rosenberg, 1985). Since the mean monthly or annual sedimentation (see Table 2.14) in the Baltic Sea basins is generally much lower than 0.75, this Y_{Eh}-factor would not normally affect the production of zoobenthos very much. As discussed in Chap. 2, one should also stress that the oxygen concentrations below the halocline at 75 m in the Baltic Proper would imply that relatively few zoobenthos would be produced in these areas.

The amount (= mass in kg ww) of sediments available for zoobenthos from ET- and A-areas, M_{Sed}, has been calculated in the following manner from dynamically modeled amounts of SPM (in basins such as the Baltic Proper and the Gulf of Finland with three sediment compartments, M_{SPMAMW}, $M_{SPMETSW}$ and M_{SPMADW} in g dw):

$$M_{Sed} = DC_{IG} \cdot (IG/100) \cdot (1/(1 - W/100)) \cdot (M_{SPMAMW} + M_{SPMETSW} + M_{SPMADW}) \cdot 0.001 \quad (4.43)$$

Where IG is the organic content (loss on ignition), which is set to 12% dw (dry weight) for all Baltic Sea basins for the sediment interval 0–10 cm (see Chap. 2) and W is the water content of these sediments in percent wet weight (ww). DC_{IG} is simple a dimensionless moderator for the fraction of organic matter potentially available for the zoobenthos since all of the organic matter is not available, which is

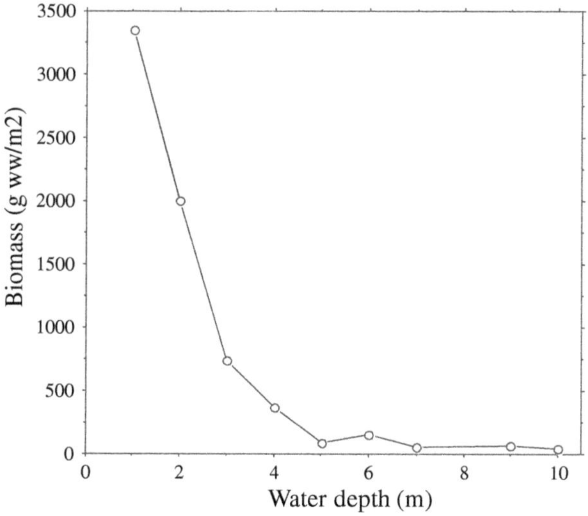

Fig. 4.27 The relationship between the biomass of the mobile epifauna (>0.5 mm) in g/m^2 and the water depth at Tylösand (on the Swedish west coast; data from Möller et al., 1985)

shown from the fact that the glacial clays in the Baltic Sea sediment generally have an organic content of about 5% dw. So DC_{IG} is defined as (12 – 5)/12.

Also note that ET-areas vary between 41% in the Bothnian Sea to 79% in the Gulf of Riga (see Table 2.1).

So, the amount of sediments available for zoobenthos may be calculated in this manner from the amount of SPM, which is already available in the CoastWeb-model for several other purposes. The water content (W = 70% ww) is used to transform values in gram dry weight to gram wet weight and the factor 0.001 gives the requested dimension in kg ww. Evidently, sediments also contain substances that cannot be utilised by zoobenthos. This is why the MER_{SEZB}-value is set to 0.25 of MER_{BAZB}.

The default thickness of the A-sediments (D_{AS}) is set to 10 cm. The decomposition of organic matter in sediments from zoobenthos, bacteria, etc. implies that organic matter is lost via metabolic activities. This will reduce the organic content (Håkanson and Jansson, 1983), and hence also influence the TP-content of the sediments. This process is accounted for in the dynamic SPM-model, which is linked to the dynamic phosphorus model.

The two losses from the compartment "zoobenthos" are consumption of zoobenthos by prey fish, CON_{ZBPY}, and elimination, EL_{ZB}. Consumption of prey fish is given by:

$$CON_{ZBPY} = CR_{ZBPY} \cdot BM_{ZB} \tag{4.44}$$

Where BM_{ZB} is the actual biomass of zoobenthos (kg ww). CR_{ZBPY} is the actual consumption rate (zoobenthos to prey fish, dimension 1/month). The normal consumption rate for prey fish eating zoobenthos (NCR_{PY}) is assumed to be in the range between the values for predatory zooplankton eating herbivorous zooplankton and predatory fish eating prey fish. From calibrations (see Håkanson and Boulion, 2002a), the normal consumption rate for prey fish, NCR_{PY}, is defined by:

$$NCR_{PY} = NR_{PY} \cdot (NCR_{ZP} \cdot 0.15 + NCR_{PD} \cdot 0.85) \tag{4.45}$$

NR_{PY} = The number of first order food choices for prey fish (NR_{PY} is 2 in the model).

NCR_{ZP} = The normal consumption rate for predatory zooplankton is defined from the ratio $1/T_{ZP}$, where T_{ZP} is the turnover time for predatory zooplankton (which is 11 days, see Table 2.9).

NCR_{PD} = The normal consumption rate for predatory fish, as defined from the ratio $1/T_{PD}$; the turnover time for predatory fish is 900 days (see Table 2.9).

This means that the normal consumption rate for prey fish is about 0.016 (1/month; see Table 2.9). Given the complex role of prey fish in the aquatic systems and in the CoastWeb-model, empirical calibrations (see Håkanson and Boulion, 2002a) have demonstrated that unrealistically low production values for prey fish would be obtained if the NCR_{PY} would be calculated directly from the turnover time of prey fish (300 days). With this approach, the normal consumption rate for prey fish is about 5 times higher.

The actual consumption rate for prey fish eating zoobenthos, CR_{ZBPY}, is defined by:

$$CR_{ZBPY} = (NCR_{PY} + NCR_{PY} \cdot (BM_{PY}/(NBM_{PY}) - 1)) \tag{4.46}$$

If the actual biomass of prey fish (BM_{PY}) increases relative to the normal biomass of prey fish (NBM_{PY}), the grazing pressure on zoobenthos will increase and this will give higher values on the actual consumption rate, CR_{ZBPY}.

Kitaev (1984) has presented data comparing the fraction of predatory fish to the total fish biomass for 122 lakes in Europe and North America. His results, and those of Swingle (1950, 1952), emphasize that on the basis of the ratio between prey and predatory fish, it is possible to estimate a characteristic ratio for a balanced or "normal" fish community. If the portion of predatory fish is 13–42% of the total fish biomass, then it is possible to consider the community as "normal". From this, one can assume that (13 + 42)/2 $\approx$ 27% could be regarded as a normal portion of predatory fish in balanced lake systems. This is elaborated in greater detail by Håkanson and Boulion (2002a) in discussing the fish model.

The normal fish biomass of prey fish (NB_{MPY}) is calculated in steps. First, we calculate the normal total fish biomass (NBM_{Fish}) from Eq. (3.22), then the normal biomass for prey fish from Eq. (3.23) and the normal biomass for predatory fish (NBM_{PD}) from Eq. (3.24) (see Tables 3.1 and 3.2).

The smoothing function in these equations will smooth the input (NBM_{Fish}) using the turnover time of the prey fish (T_{PY}) as an averaging time. The reason for this is simply that the seasonal variation in prey fish biomass should generally be smaller than the seasonal variations in TP-concentrations. The initial value is set equal to NBM_{Fish}. The distribution coefficient used in CoastWeb is the same as the one used in LakeWeb (i.e., DC_{PYPD}) and for these calculations for the Baltic Sea we have modified the boundary conditions (see Eq. 3.25).

The last term in the basic equation for zoobenthos, elimination (kg ww/month), is given in the same way as for other functional groups in the CoastWeb-model. That is:

$$EL_{ZB} = BM_{ZB} \cdot 1/T_{ZB} \tag{4.47}$$

T_{ZB} is the characteristic turnover time zoobenthos (128 days; see Table 2.9).

With this, we have presented the dynamic models for zoobenthos.

4.4.1.3 Testing Modeling Results

Reductions in TP-Loading

Figure 4.28 gives the results for the biomass of zoobenthos in the five sub-basins of the Baltic Sea and Fig. 4.29 shows the correspondence between dynamically

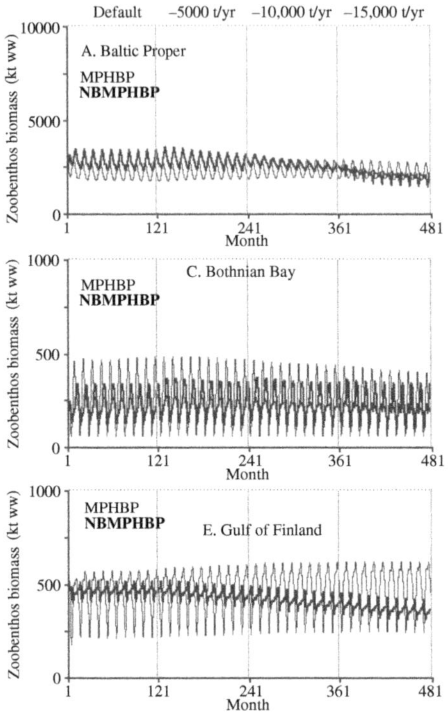

Fig. 4.28 A comparison between dynamically modeled biomasses of zoobenthos and values calculated from the empirically-based norms for the 5 sub-basins in the Baltic Sea in a situation where the default conditions are shown for the first period (months 1–61), for the second period the annual TP-inflow via tributaries to the Baltic Proper has been reduced by 5,000 t/year, in the third period by 10,000 t/year and in the fourth period (months 361–481) by 15,000 t/year

modeled values and norm-values for zoobenthos biomasses under default conditions in better detail. The correspondence is quite good under default condition, and also for the entire 40-year period. Reductions in TP-loading will cause reductions in SPM-concentrations (see Fig. 4.2c) and hence also in sedimentation, which will reduce the amount of food available for zoobenthos (the sediment pool). This would lead to minor reductions in zoobenthos production and biomasses, as shown in Fig. 4.28. Note that we do not have access to any compatible dynamic mass-balance model for oxygen which would account for the major processes regulating the oxygen concentrations in the given compartments in the basins forming the Baltic Sea from quantifications of (1) oxygen inflow from the Kattegat and tributaries, (2) oxygen input from wave action, (3) oxygen production related to primary production and (4) oxygen consumption by bacteria in water and sediments. It is likely that the oxygen consumption would be reduced when the amount of degradable organic matter in water and sediments goes down and that this would favor zoobenthos production. That process is not accounted for in this scenario, but it will be discussed more in Chap. 5.

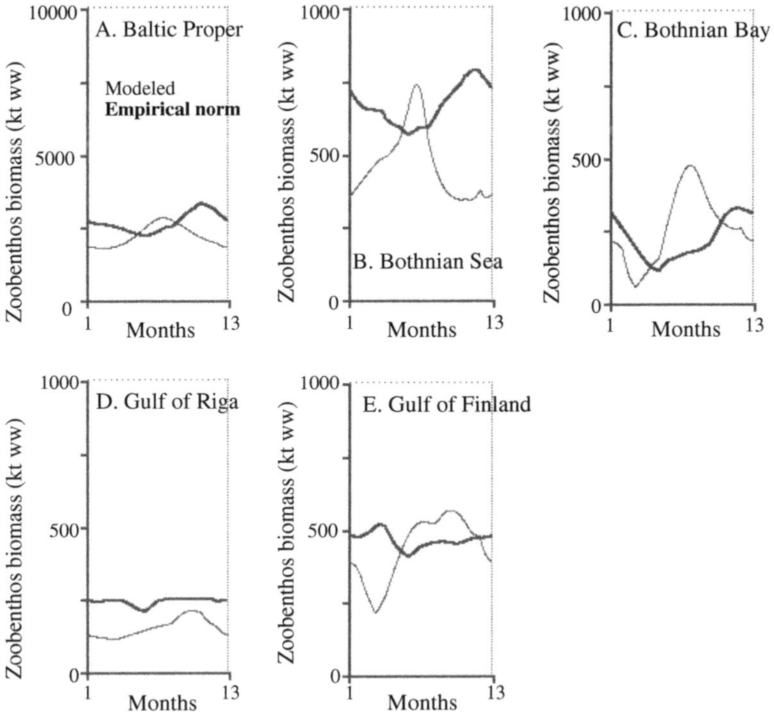

Fig. 4.29 A comparison between modeled empirically-based norm-values (*bolded* reference values for mainly the growing seasons) and dynamically modeled values of zoobenthos biomass at steady-state (simulation time 241 months) under default conditions in the five basins of the Baltic Sea

Increasing Salt-Water Inflow

Increasing the salt-water inflow from the Kattegat (see Fig. 4.5) will cause several changes to the system. SPM will decrease (see Fig. 4.5d) and primary phytoplankton production will decrease (Figs. 4.5c and 4.6). The changes for zoobenthos in this scenario are given in Fig. 4.30 and they are small. As just mentioned, we will discuss this in a more comprehensive manner in Chap. 5.

Temperature Scenario

Figure 4.7 gave the results from the "temperature scenario". This scenario gives marked increases in TP-concentrations (Fig. 4.7b) and SPM-values (Fig. 4.7d).

Increasing the water temperatures would also increase the production and biomasses of zoobenthos (see Fig. 4.31), especially in the Bothnian Bay and the Bothnian Sea, where the higher winter temperatures will cause the most pronounced reductions in ice cover and hence also increase turbulence and resuspension. One should note that higher water temperatures also imply higher mineralization, which means reduced sedimentation. The reductions in sedimentation are not great (see

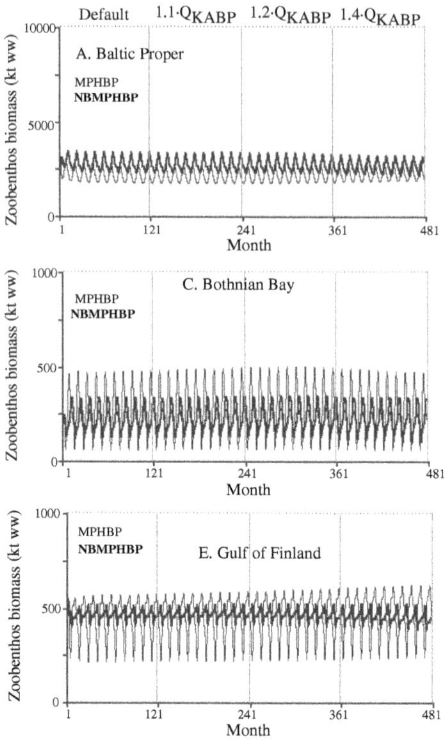

Fig. 4.30 A comparison (a sensitivity test) between dynamically modeled biomasses of zoobenthos and values calculated from the empirically-based norms for the 5 sub-basins in the Baltic Sea in a situation where the default conditions are shown for the first period (months 1–61), for the second period the monthly saltwater inflow from the Kattegat has been increased by a factor of 1.1, in the third period 1.2 and in the fourth period (months 361–481) by 1.4

Fig. 4.32) but more evident in the Bothian Bay than in the other basins. One can also see that sedimentation is lower than the critical limit for zoobenthos used in this modeling of 0.75 cm/year most months in all basins.

4.4.2 *Zooplankton*

4.4.2.1 Background on Zooplankton

Zooplankton play several fundamental roles in aquatic ecosystems. For example, they are important in the energy transfer from primary producers to pelagic fish, especially in systems that have a large pelagic zone (Möllmann et al., 2000). There are also different types of models and discussions about modeling approaches for zooplankton (see, e.g., Peters, 1991; Jørgensen, 1998; Scheffer, 1998). However, for zooplankton, as for most types of organisms, there are major differences

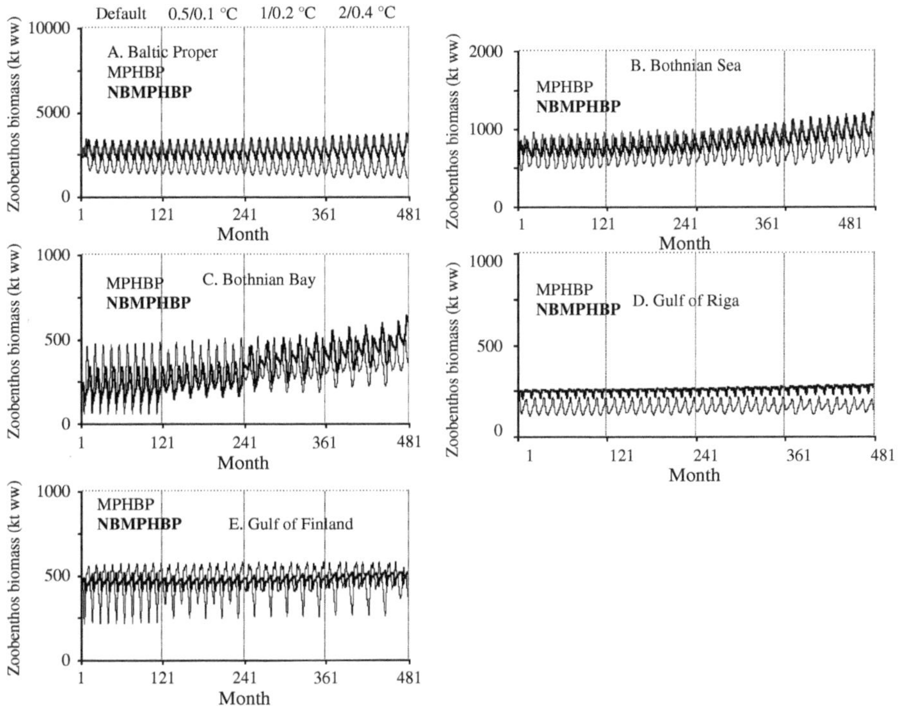

Fig. 4.31 A comparison (a sensitivity test) between dynamically modeled biomasses of zoobenthos and values calculated from the empirically-based norms for the 5 sub-basins in the Baltic Sea in a situation where the default conditions are shown for the first period (months 1–61), for the second period the monthly surface-water temperatures have been increased by 0.5 and 0.1°C every second year, for the third period by 1 and 0.2°C every second year and in the fourth period by 2 and 0.4°C every second year. Middle- and deep-water temperatures have been increased by 0.75 and 0.5 of the increases given for the surface-water temperatures

among models related to differences in target variables (from individual species to total biomass), modeling scales (daily to annual predictions), modeling structures (from empirical/regression models to differential equations) and driving variables (whether accessed from standard monitoring programs, climatological measurements or specific studies). So, to make meaningful model comparisons is not a simple matter, and this is not the focus of this section. As far as the present authors are aware, there are no coastal models for zooplankton of the type presented here accounting to production, grazing, growth, migration, elimination and food choices in a general, holistic ecosystem framework designed to achieve practical utility for marine ecosystems.

The aim here is first to give a brief background on zooplankton and motivate the features included in the dynamic (time-dependent) model, such as distribution coefficients (food choices), consumption rates, metabolic efficiency ratios and turnover times. Then, we will present the empirical models (norms) used to test

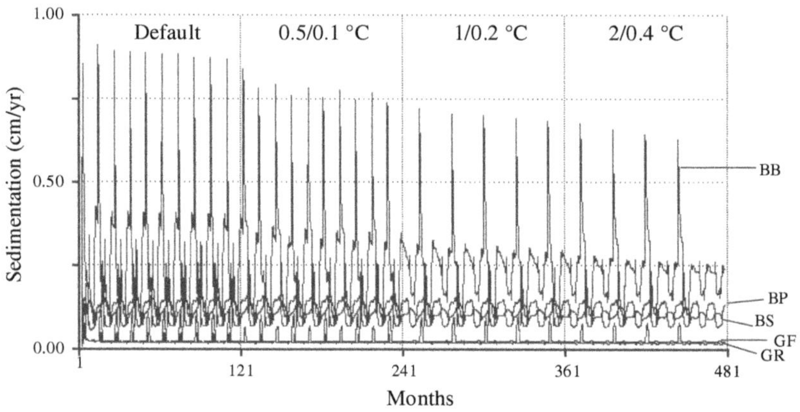

Fig. 4.32 A comparison (a sensitivity test) between dynamically modeled values of sedimentation (= deposition) of matter (in cm/year) on areas of accumulatiom nm in the five sub-basins in the Baltic Sea in a temperature gradient. The default conditions are shown for the first period (months 1 to 61), for the second period the monthly surface-water temperatures have been increased by 0.5 and 0.1 °C every second year, for the third period by 1 and 0.2 °C every second year and in the fourth period by 2 and 0.4 °C every second year. Middle-and deep-water temperatures have been increased by 0.75 and 0.5 of the increases given for the surface-water temperatures

the dynamic CoastWeb-model. After that, the dynamic model will be introduced, motivated and tested.

The primary aim of the CoastWeb-model is not to give good predictions for certain species or coast types. The aim is to quantitatively describe typical, characteristic foodweb interactions so that production, biomasses and predation can be determined for the functional groups of organisms included in the model. Other groups of organisms, such as benthic bacteria and fungi are not treated as individual groups but are accounted for in the sense that they are included in the flux to zoobenthos called "zoobenthos production from other sediment sources". Jellyfish is a separate unit described in Chap. 2. It should also be emphasized that the CoastWeb-model primarily handles feedbacks among these functional groups. Because biotic/abiotic feedbacks are also of great interest, such interrelationships are also included in the model. The first concerns the influence of material produced in the coastal area (autochthonous materials) on the depth of the photic zone, the second reductions in TP-concentrations related to biouptake of phosphorus, and the third suspended particulate matter (= SPM = seston), a key factor for bacterial production, sedimentation and other important functions. The LakeWeb-model has been tested (Håkanson and Boulion, 2002a) along many limnological gradients and against empirical models and data. Those tests have demonstrated that LakeWeb can capture fundamental foodweb interactions and abiotic/biotic interactions very accurately. Those tests will not be repeated here.

The dynamic model for zooplankton presented in this work is an important part of CoastWeb, and it must also be compatible with the larger entity, and this puts specific demands on this model:

1. It should give seasonal (monthly) variations.
2. It should give good predictions when tested against the empirical norms, and it must be driven by variables that are readily accessed from standard monitoring programs and maps.
3. It should capture processes and mechanisms regulating production and biomasses of the two functional groups of zooplankton in a general manner.

Ivanova (1985) has analyzed the relationships between production and biomass of zooplankton using materials collected by the International Biological Program (IBP). The wet weight of crustaceans, rotifers and protozoans often contain about 10% of dry substance. On a wet weight basis, zooplankton contains about 5% organic carbon and 10% of dry matter, while the wet weight of phytoplankton contains about 10% organic carbon and 20% of dry matter. This distinction is necessary in calculations of biomasses for zooplankton.

The relative biomass of protozoans (ciliates) is generally small in most systems (see Håkanson and Boulion, 2002a). However, protozoans can contribute with a significant part of the secondary production. Using data from Ivanova (1985), Håkanson and Boulion (2002a) presented the following regression on the relationship between production (PR) and biomass (BM) of herbivorous zooplankton:

$$\log(\mathrm{PR_{ZH}}) = 0.925 \cdot \log(\mathrm{BM_{ZH}}) + 1.37 \text{or} \, \mathrm{PR_{ZH}} = 23.4 \cdot \mathrm{BM_{ZH}}^{0.925} \tag{4.48}$$

($r^2 = 0.77$; n = 54)

Where the production is given in g ww/m^2·year and the mean biomass of herbivorous zooplankton in g ww/m^2. For predatory zooplankton, they gave the following analogous relationship using the same dimensions:

$$\log(\mathrm{PR_{ZP}}) = 0.916 \cdot \log(\mathrm{BM_{ZP}}) + 1.08 \text{or} \, \mathrm{PR_{ZP}} = 12.0 \cdot \mathrm{BM_{ZP}}^{0.916} \tag{4.49}$$

($r^2 = 0.71$; n = 54)

The exponents are close to 1 in both regressions. One can see that the annual BM/PR-ratio (as a measure of the turnover time) for predatory zooplankton is about 2 times longer than for herbivorous zooplankton. The latter group includes not only crustaceans but also protozoans and small rotifers with very specific production patterns. The scatter around the regression lines for the two zooplankton groups is determined by, (1) the structural heterogeneity of the herbivores (i.e., the differences between crustaceans, rotifers and protozoans), (2) the absence of precise criteria to distinguish between the herbivores and the predators (some abundant species, such *Asplanchna priodonta*, are facultative predators), (3) the dependence on temperature and (4) uncertainties in the empirical data.

To get a first estimate of the appropriate turnover times for herbivorous and predatory zooplankton, one can use the results from Eqs. (4.48) and (4.49). Both exponents are close to 1, and if one sets the duration of the growing season to 150 days, the turnover time for predators is about 12 days and for herbivores about 6 days. This is in good general agreement with literature data (see Håkanson and

Boulion, 2002a). For the dynamic model, it is essential to have reliable general default values of the turnover times for these two groups of zooplankton, and knowledge about the factors influencing the uncertainty of these default values. In the CoastWeb-model, we will use 6.0 days for herbivorous zooplankton (T_{ZH}) and 11 days for predatory zooplankton (T_{ZP}; see Table 2.9).

In many freshwater systems herbivorous zooplankton mainly consists of cladocerans, on average about 40% (from 20 to 60%), copepods make up about 30% (from 10 up to 50%), protozoans about 19% and rotifers about 7%. Predatory zooplankton consists of more copepods (about 60%) and rotifers (about 20–25%) but less cladocerans (about 11%) and very little protozoans. The protozoans together with bacteria are generally categorized as microbial organisms, which create a so-called microbial "loop" or microbial foodweb (Porter et al., 1988; Stockner and Porter, 1988; Stockner and Shortreed, 1988; Pace et al., 1990; Stone and Weisburd, 1992; Pace, 1993). Due to the small sizes and the high metabolic rates, microbial organisms increase the nutrient cycling in aquatic systems. To gain simplicity, we treat the protozoans as herbivorous zooplankton in the dynamic model.

Most herbivorous zooplankton are filter feeders, such as crustaceans, copepods and cladocerans. Phytoplankton, bacterioplankton and detritus are the energy source for planktonic crustaceans (Gutelmakher, 1986; Monakov, 1998). Mechanical filtration of water is the main way of feeding for many planktonic crustaceans, which consume particles of a certain size range, which corresponds to length of the animal and the structure of their filtration apparatus (Morgan, 1980; Gutelmakher, 1986). The size range of the food particles for filter feeders is generally about 1–100 μm. Effective filter feeders like cladocerans consume particles in the size range 1–50 μm (Pace et al., 1990; Monakov, 1998). Hence, many cladocerans can consume single bacterial cells. However, in natural aquatic systems there are many types of aggregated bacterial cells including bacteria attached to detrital particles. Evidently, bacteria in such forms can also be consumed by zooplankton (Knoechel and Holtby, 1986; Olsen et al., 1986; Pace et al., 1990; Hairston and Hairston, 1993).

Due to this mechanical filtration, zooplankton consume different fractions of suspended organic particles and generally in direct proportion to their respective concentrations in the water. If the detrital particles are in the suitable size range of food particles, they can be consumed by zooplankton. According to Gutelmakher (1986), detritus amount to 60–70% of the total food intake, while phytoplankton and bacterioplankton make up 10–22% and 18–24%, respectively. However, the relative amounts of phytoplankton, bacterioplankton and detritus as food objects for herbivorous zooplankton differ among systems of different trophic status. In the trophic range from oligotrophy to eutrophy, the portion of bacteria decreases relatively to chlorophyll (see Boulion and Paveljeva, 1998). It is also necessary to underline also another regularity: the higher the chlorophyll-a concentration, the higher the particulate organic matter concentration and, simultaneously, the higher the relative content of Chl in POM (particulate organic matter; see Boulion, 1997).

So, the role of phytoplankton as food for herbivorous zooplankton increases from oligotrophic to eutrophic systems. On the contrary, the role of bacteria and detritus grows from eutrophic to oligotrophic waters (Dillon and Rigler, 1974; Jones and Bachmann, 1975; Carlson, 1977; Schindler, 1978).

Unfortunately, we do not have access to similar information as this (based on many and extensive empirical studies) concerning the food choices of herbivorous zooplankton for the Baltic Sea system. This means that that the underlying assumptions in the dynamic model in these respects are relatively uncertain.

4.4.2.2 Empirical Reference Models for Zooplankton

The empirical regressions (here called "norms"; see Table 3.2) play an important role in the testing of the dynamic model. For this purpose, we would like to have as reliable empirical reference equations as possible for biomasses and production values of herbivorous and predatory zooplankton and empirically based criteria to differentiate between the two groups, and this is the focus of this section. Peters (1986) have presented the basic empirical norm for total zooplankton biomass (BM_{ZT}; see Table 3.1):

$$BM_{ZT} = 38 \cdot {TP_{SW}}^{0.64} \tag{4.50}$$

$$(r^2 = 0.86;\ n = 12;\ \text{TP-range} = 3\text{–}80\ \mu g/l)$$

Lacking better information for marine ecosystems, we will, after multiplication with the correction factor Y_{Chl} (= Chl_{Coast}/Chl_{Lake}), which transforms the regression for a lake to a norm (reference) for coastal areas, use this relationship also for the Baltic Sea basins, which all have mean TP-concentrations lower than 80 μg/l. To differentiate between herbivorous and predatory zooplankton, we need a distribution coefficient DC_{ZHZP}, which is 0.77 for lakes (see Table 3.1; i.e., 77% ZH and 23% ZP). For the Baltic Sea basins, we have for simplicity changed this DC-value to 0.8. Evidently, the biomass of the herbivores is generally bigger than that of the predatory zooplankton.

With this, we have presented the two empirical reference norms and can turn to the dynamic model for zooplankton.

4.4.2.3 Modeling of Herbivorous Zooplankton

This section will introduce the two differential equations for zooplankton. Figure 4.33 gives an overview of the model. In the following parts, we will first present the two equations and then model tests. The target variables are the biomasses and production values of the two groups of zooplankton.

The following ordinary differential equation gives the fluxes (kg ww/month) to and from this compartment.

$$\begin{aligned} BM_{ZH}(t) = BM_{ZH}(t - dt) + (&IPR_{PHZH} + IPR_{BPZH} + MIG_{InZH} - CON_{ZHZP} \\ &- CON_{ZHPY} - MIG_{OutZH} - EL_{ZH}) \cdot dt \end{aligned} \tag{4.51}$$

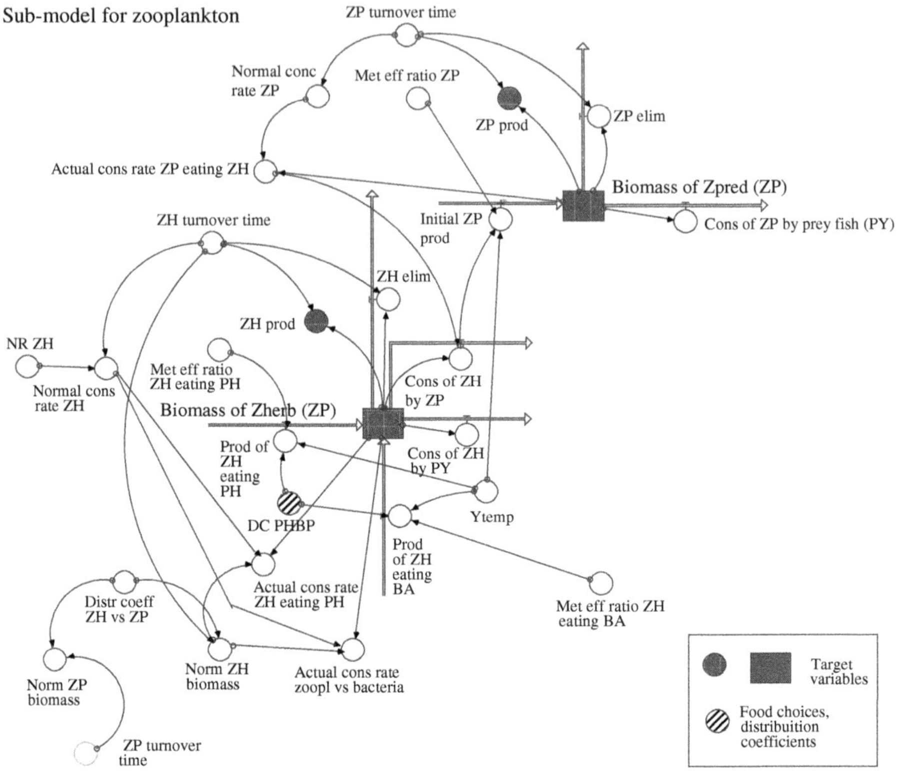

Fig. 4.33 An overview of the sub-model to calculate production and biomasses of herbivorous and predatory zooplankton

BM_{ZH} = Biomass of herbivorous zooplankton (kg ww); the initial BM_{ZH}-value is set equal to the norm-value, which is given in Table 3.2.
IPR_{PHZH} = Initial production of herbivorous zooplankton from eating phytoplankton (kg ww/month).
IPR_{BPZH} = Initial production of herbivorous zooplankton from consumption of bacterioplankton (kg ww/month).
MIG_{InZH} = Inflow of herbivorous zooplankton (kg ww/month).
MIG_{OutBP} = Outflow of herbivorous zooplankton (kg ww/month).
CON_{ZHZP} = Consumption of herbivorous zooplankton by predatory zooplankton (kg ww/month).
CON_{ZHPY} = Consumption of herbivorous zooplankton by prey fish (kg ww/month).
EL_{ZH} = Elimination (= turnover) of herbivorous zooplankton (kg ww/month).

There is a food choice for herbivorous zooplankton between phytoplankton and bacterioplankton; the default value of the distribution coefficient regulating the food choice, DC_{PHBP}, is 0.5. This means that herbivorous zooplankton have no preference for either phytoplankton or bacterioplankton. If there are equal amounts of phytoplankton and bacterioplankton available for consumption, the two fluxes regulating the production of herbivorous zooplankton are of equal size. The number of first order food choices for herbivorous zooplankton, NR_{ZH}, is 2. Figure 4.34 gives the food choice panel for zooplankton, both the input to herbivorous zooplankton and the output to prey fish. For the prey fish, the first order food choice is between zooplankton and zoobenthos. The default value of this distribution coefficient (DC_{PY1}) is 0.75. That is, the flux from zooplankton (both herbivorous and predatory) to prey fish (e.g., mainly herring, sprat and small cod in the Baltic Proper) is multiplied by 0.75 and the rest (i.e., prey fish eating zoobenthos) is multiplied by (1 – 0.75). The second order distribution coefficient directly affecting zooplankton is DC_{PY2} (= 0.2), which regulates how much predatory or herbivorous zooplankton prey fish eat. Note again that $DC_{PHBP} = 0.5$ does not mean that 50% of the fluxes to herbivorous zooplankton will emanate from phytoplankton and 50% from bacterioplankton. It would be so only if these two fluxes are identical.

The initial production of herbivorous zooplankton from phytoplankton is calculated from:

$$IPR_{PHZH} = CON_{PHZH} \cdot DC_{PHBP} \cdot {Y_{Temp}}^{0.5} \cdot MER_{PH} \qquad (4.52)$$

CON_{PHZH} is the consumption of phytoplankton by herbivorous zooplankton (kg ww/month). CON_{PHZH} is given by:

$$CON_{PHZH} = CR_{PHZH} \cdot BM_{PH} \qquad (4.53)$$

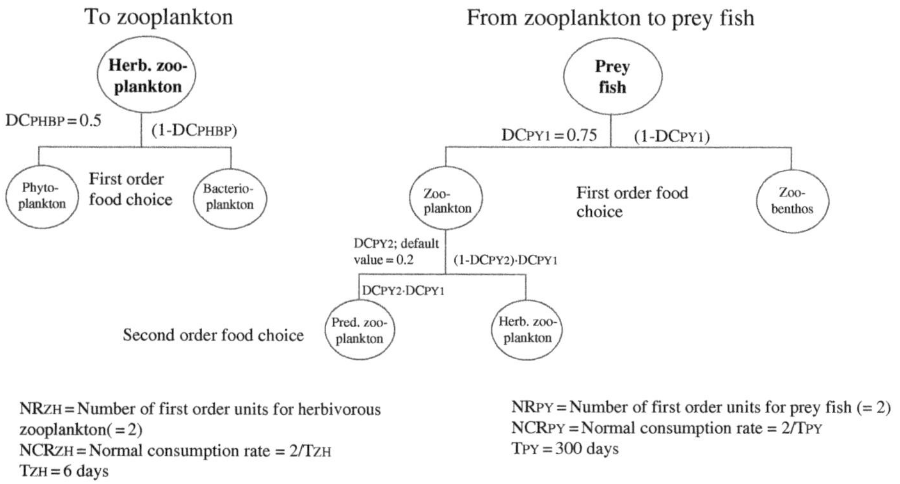

Fig. 4.34 Food-choice panel for zooplankton

Where the actual biomass of phytoplankton (BM_{PH} in kg ww) is calculated by the CoastWeb-model and the actual consumption rate for herbivorous zooplankton feeding on phytoplankton, CR_{PHZH}, is given by:

$$CR_{PHZH} = (NCR_{ZH} + NCR_{ZH} \cdot (BM_{ZH}/NBM_{ZH} - 1)) \quad (4.54)$$

BM_{ZH} is the calculated actual biomass of herbivorous zooplankton (kg ww) and NBM_{ZH} is the normal biomass of herbivorous zooplankton calculated from Table 3.2 (see Eq. 3.20). The normal consumption rate for herbivorous zooplankton eating phytoplankton, NCR_{ZH}, is set equal to NR_{ZH}/T_{ZH}.

$Y_{Temp}{}^{0.5}$ is the ratio between the monthly surface-water temperatures (°C) and the reference temperature of 9°C (related to the duration of the growing season, see Håkanson and Boulion, 2001b). The exponent 0.5 is used for herbivorous zooplankton. This will give realistic seasonal patterns in the biomasses of these organisms and hence also in the food supply for herbivorous zooplankton; so, the consumption is higher at higher temperatures if the amount of food is constant.

MER_{PH} is the metabolic efficiency ratio for herbivorous zooplankton eating phytoplankton. It is set to 0.24 (see Table 2.9).

The initial production of herbivorous zooplankton from bacterioplankton is given in the same manner by:

$$IPR_{BPZH} = CON_{BPZH} \cdot (1 - DC_{PHBP}) \cdot Y_{Temp}{}^{0.5} \cdot MER_{BP} \quad (4.55)$$

CON_{BPZH} is the consumption of bacterioplankton by herbivorous zooplankton, which is given by:

$$CON_{BPZH} = CR_{BPZH} \cdot BM_{BP} \quad (4.56)$$

The actual biomass of bacterioplankton (BM_{BP} in kg ww) is calculated by the model and the actual consumption rate for herbivorous zooplankton feeding on bacterioplankton, CR_{BPZH}, is:

$$CR_{BPZH} = (NCR_{ZH} + NCR_{ZH} \cdot (BM_{ZH}/NBM_{ZH} - 1)) \quad (4.57)$$

The normal consumption rate for herbivorous zooplankton eating bacterioplankton, NCR_{ZH}, is set equal to $2/T_{ZH}$, since there are two first order food choices for herbivorous zooplankton ($NR_{ZH} = 2$; see Fig. 4.34).

MER_{BP} is the metabolic efficiency ratio for herbivorous zooplankton eating bacterioplankton. It is also set to 0.24 (see Table 2.9).

There are four fluxes of matter from the compartment "Herbivorous zooplankton", (1) to predatory zooplankton, (2) to prey fish, (3) elimination and (4) outflow or outmigration. The flux (CON_{ZHZP} in kg ww/month) from herbivorous to predatory zooplankton is given by:

$$CON_{ZHZP} = CR_{ZHZP} \cdot BM_{ZH} \quad (4.58)$$

The actual consumption rate for predatory zooplankton feeding on herbivorous zooplankton, CR_{ZHZP} (1/month), is given by:

$$CR_{ZHZP} = (NCR_{ZP} + NCR_{ZP} \cdot (BM_{ZP}/NBM_{ZP} - 1)) \tag{4.59}$$

BM_{ZP} is the calculated actual biomass of predatory zooplankton (kg ww) and NBM_{ZP} is the normal biomass of predatory zooplankton, which is calculated from the equation in Table 3.2 (see also Eq. 3.21). The normal consumption rate for predatory zooplankton eating herbivorous zooplankton, NCR_{ZP}, is equal to $1/T_{ZP}$, since there is only one food choice for predatory zooplankton ($NR_{ZP} = 1$). The turnover time for predatory zooplankton is set to 11 days (see Table 2.9).

The flux from herbivorous zooplankton to prey fish is given by (CON_{ZHPY} in kg ww/month):

$$CON_{ZHPY} = CR_{PY} \cdot BM_{ZH} \tag{4.60}$$

The actual consumption rate for prey fish feeding on herbivorous zooplankton (as well as on predatory zooplankton and zoobenthos is abbreviated $CR_{PY} = CR_{ZHPY} = CR_{ZPPY} = CR_{BEPY}$), CR_{PY} (1/month) is:

$$CR_{PY} = (NCR_{PY} + NCR_{PY} \cdot (BM_{PY}/NBM_{PY} - 1)) \tag{4.61}$$

BM_{PY} is the calculated actual biomass of prey fish (kg ww), NBM_{PY} is the normal biomass of prey fish, NCR_{PY}, is the normal consumption rate for prey fish (calculated by the CoastWeb-model).

The initial production of prey fish from consumption of herbivorous and predatory zooplankton (IPR_{ZHPY} and IPR_{ZPPY}) will be given in the next section (the fish model).

Migration (kg ww/month), or rather inflow of herbivorous zooplankton from the outside sea by currents (in this example the transport is from the Bothnian Sea to the Bothnian Bay, BB), is calculated from:

$$MIG_{InZHBB} = R_{MigZHBB} \cdot BM_{ZHBS} \tag{4.62}$$

Where $R_{MigZHBB}$ is the migration rate for herbivorous zooplankton ($= 1/T_{SWBB}$, T_{SWBB} is the water turnover time for the SW-water in BB in months) and BM_{ZHBS} is the actual biomass of herbivorous zooplankton in BS (modeled using CoastWeb). The outflow of herbivorous zooplankton is calculated in the same manner using the modeled biomass of phytoplankton in the given system (here BM_{ZHBB}).

The last term is elimination or mortality (kg ww/month), which is given by:

$$EL_{ZH} = BM_{ZH} \cdot 1/T_{ZH} \tag{4.63}$$

T_{ZH} the characteristic turnover time for herbivorous zooplankton (6 days; see Table 2.9).

With this, we have presented the model for herbivorous zooplankton.

Testing Modeling Results

Reductions in TP-Loading

The results for the biomass of herbivorous zooplankton in the five sub-basins of the Baltic Sea in this gradient are given in Fig. 4.35. Figure 4.36 shown in more detail the correspondence between dynamically modeled values and norm-values for herbivorous zooplankton under default conditions. The correspondence is quite good. This is also the case for the entire 40-year period (Fig. 4.35). Reductions in TP-loading will cause reductions in TP-concentrations (see Fig. 4.2b). This would lead to reductions also in production and biomasses of herbivorous zooplankton, as shown in Fig. 4.35.

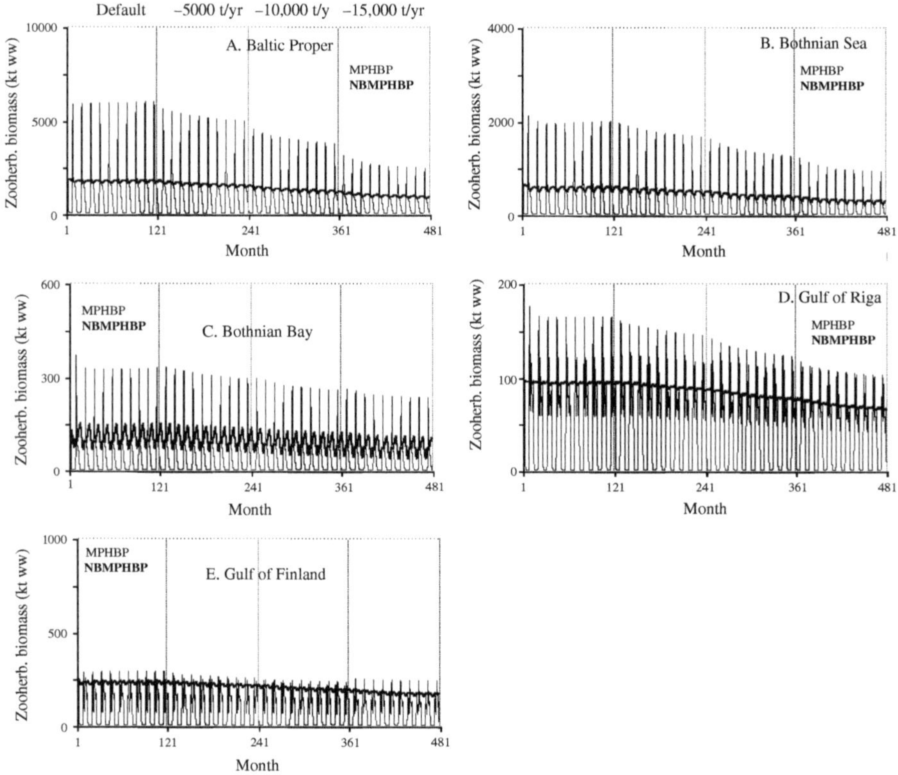

Fig. 4.35 A comparison between dynamically modeled biomasses of herbivorous zooplankton and values calculated from the empirically-based norms for the 5 sub-basins in the Baltic Sea in a situation where the default conditions are shown for the first period (months 1–61), for the second period the annual TP-inflow via tributaries to the Baltic Proper has been reduced by 5,000 t/year, in the third period by 10,000 t/year and in the fourth period (months 361–481) by 15,000 t/year

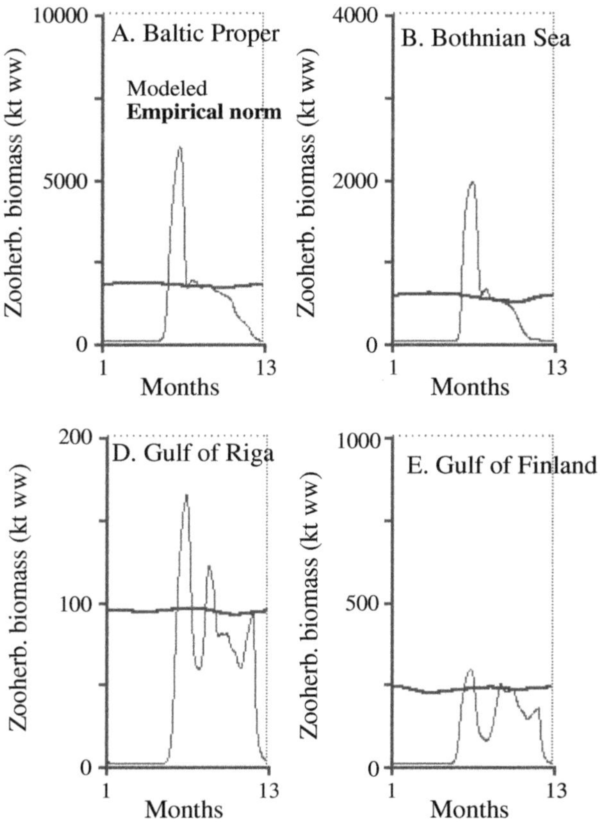

Fig. 4.36 A comparison between modeled empirically-based norm-values (*bolded* reference values for mainly the growing seasons) and dynamically modeled values of herbivorous zooplankton biomass at steady-state (simulation time 241 months) under default conditions in the five basins of the Baltic Sea

Increasing Salt-Water Inflow

When the salt-water inflow from the Kattegat is being been increased (see Fig. 4.5), this will cause several changes to the system. TP and SPM will decrease and primary phytoplankton production will decrease (Fig. 4.6). The changes for herbivorous zooplankton in this scenario are given in Fig. 4.37 and they are relatively small.

Temperature Scenario

Figure 4.7 gave the results from the "temperature scenario". This scenario gives marked increases in TP-concentrations (Fig. 4.7b) and SPM-values (Fig. 4.7d). There will also be increases in the production and biomasses of herbivorous zooplankton (see Fig. 4.38), especially in the Bothnian Bay, where the increased winter temperatures will cause the most pronounced reductions in ice cover and hence also increase turbulence and resuspension.

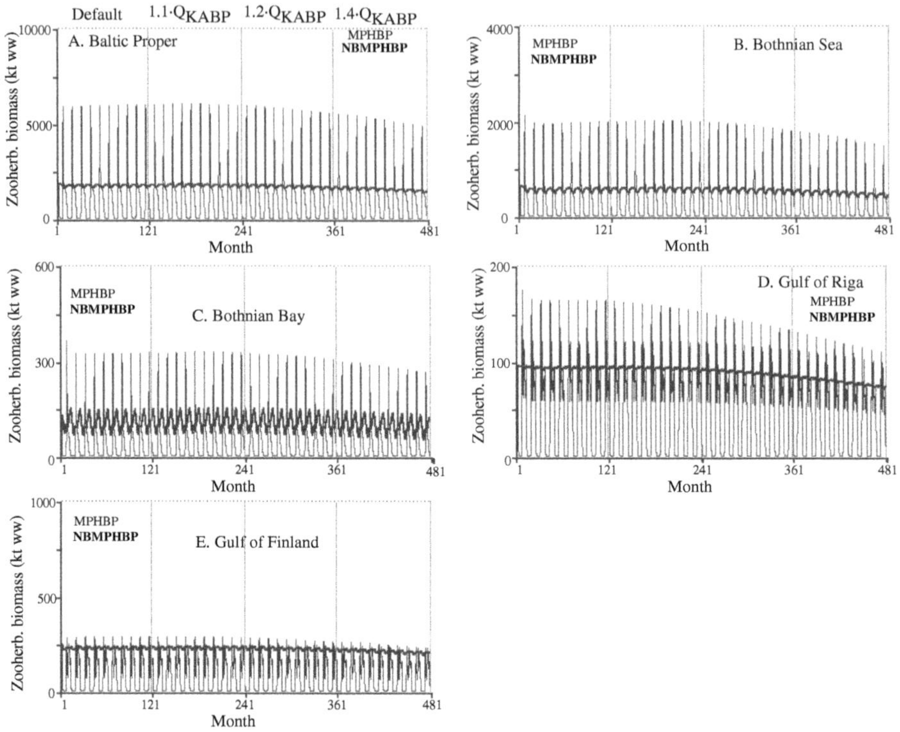

Fig. 4.37 A comparison (a sensitivity test) between dynamically modeled biomasses of herbivorous zooplankton and values calculated from the empirically-based norms for the 5 sub-basins in the Baltic Sea in a situation where the default conditions are shown for the first period (months 1–61), for the second period the monthly saltwater inflow from the Kattegat has been increased by a factor of 1.1, in the third period 1.2 and in the fourth period (months 361–481) by 1.4

4.4.2.4 Modeling of Predatory Zooplankton

This is a simple model with only two in-flows, initial production of predatory zooplankton from consumption of herbivorous zooplankton and inmigration, and three outflows, elimination, outmigration and losses from prey fish feeding on predatory zooplankton. The following differential equation gives the fluxes:

$$BM_{ZP}(t) = BM_{ZP}(t-dt)+(IPR_{ZHZP}+MIG_{InZP}-MIG_{OutZP}-CON_{ZPPY}-EL_{ZP})\cdot dt \quad (4.64)$$

BM_{ZP} = Biomass of predatory zooplankton (kg ww); the initial BM_{ZP}-value is set equal to the norm-value, which is given in Table 3.2.

IPR_{ZHZP} = Initial production of predatory zooplankton from feeding on herbivorous zooplankton (kg ww/month).

CON_{ZPPY} = Consumption of predatory zooplankton by prey fish (kg ww/month).

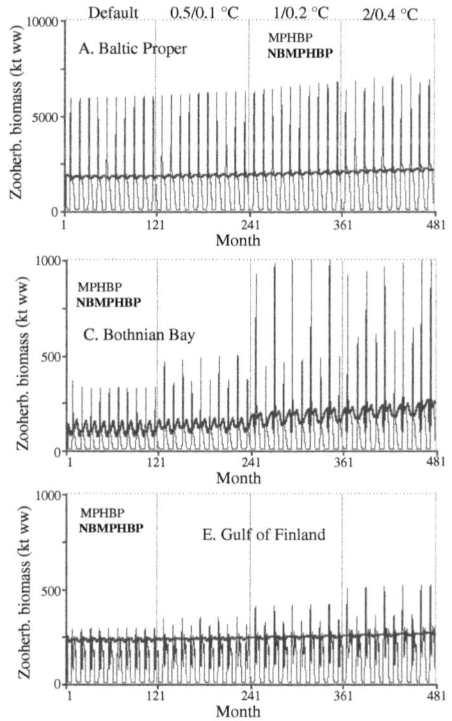

Fig. 4.38 A comparison (a sensitivity test) between dynamically modeled biomasses of herbivorous zooplankton and values calculated from the empirically-based norms for the 5 sub-basins in the Baltic Sea in a situation where the default conditions are shown for the first period (months 1–61), for the second period the monthly surface-water temperatures have been increased by 0.5 and 0.1°C every second year, for the third period by 1 and 0.2°C every second year and in the fourth period by 2 and 0.4°C every second year. Middle- and deep-water temperatures have been increased by 0.75 and 0.5 of the increases given for the surface-water temperatures

EL_{ZP} = Elimination (turnover) of predatory zooplankton (kg ww/month).

The turnover time of predatory zooplankton, T_{ZP}, is, as already explained, set to 11 days (see Table 2.9).

The initial production of predatory zooplankton from eating herbivorous zooplankton is given by:

$$IPR_{ZHZP} = CON_{ZHZP} \cdot MER_{ZH} \cdot {Y_{Temp}}^{0.5} \tag{4.65}$$

The consumption term, CON_{ZHZP}, was given in Eq. (4.60) and the Y_{Temp}-moderator has also been defined before. Note that the exponent is set to 0.5 for predatory zooplankton (and 0.25 fish). MER_{ZP} is the metabolic efficiency ratio for predatory zooplankton eating herbivorous zooplankton. It is set to 0.32 (see Table 2.9).

The first loss from this compartment (consumption of ZP by prey fish) is given by:

$$CON_{ZPPY} = CR_{PY} \cdot BM_{ZP} \tag{4.66}$$

The actual consumption rate for prey fish feeding on predatory zooplankton (1/month) is:

$$CR_{ZPPY} = (NCR_{PY} + NCR_{PY} \cdot (BM_{PY}/NBM_{PY} - 1)) \tag{4.67}$$

BM_{PY} is the calculated actual biomass of prey fish (kg ww), NBM_{PY} is the normal biomass of prey fish, NCR_{PY}, is the normal consumption rate for prey fish (calculated by the CoastWeb-model).

Migration (kg ww/month), or inflow from the outside sea by currents (in this example the transport is from the Bothnian Sea, BS, to the Bothnian Bay, BB) is calculated from:

$$MIG_{InZPBB} = R_{MigZPBB} \cdot BM_{ZPBS} \tag{4.68}$$

Where $R_{MigZPBB}$ is the migration rate for predatory zooplankton (= $1/T_{SWBB}$, T_{SWBB} is the water turnover time for the SW-water in BB in months) and BM_{ZPBS} is the actual biomass of predatory zooplankton in BS (modeled with CoastWeb). The outflow of predatory zooplankton is calculated in the same manner using the modeled biomass of phytoplankton in the given system (here BM_{ZPBB}).

Elimination of predatory zooplankton is:

$$EL_{ZP} = BM_{ZP} \cdot 1/T_{ZP} \tag{4.69}$$

T_{ZP} is the characteristic turnover time for predatory zooplankton (11 days, see Table 2.9).

With this, we have presented also the model for predatory zooplankton.

4.4.2.5 Testing Modeling Results

Reductions in TP-Loading

The results for the biomass of predatory zooplankton in the TP-gradient are given in Figs. 4.39 and 4.40 shows in more detail the correspondence between dynamically modeled values and norm-values for herbivorous zooplankton under default conditions. The correspondence is quite good for all the sub-basins. It should be stressed that for all these tests there has been no "tuning" or change in any of the model constants – the same model set-up have been used for all the sub-basins. The empirical data to distinguish between herbivorous and predatory zooplankton in the Baltic Sea leaves a lot for improvements.

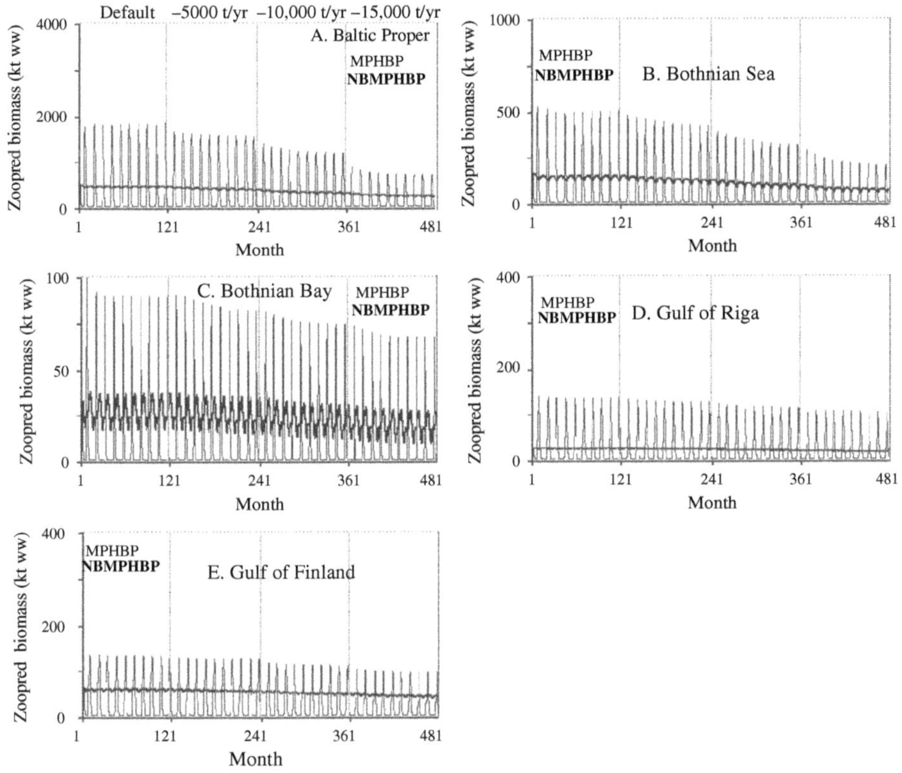

Fig. 4.39 A comparison between dynamically modeled biomasses of predatory zooplankton and values calculated from the empirically-based norms for the 5 sub-basins in the Baltic Sea in a situation where the default conditions are shown for the first period (months 1–61), for the second period the annual TP-inflow via tributaries to the Baltic Proper has been reduced by 5,000 t/year, in the third period by 10,000 t/year and in the fourth period (months 361–481) by 15,000 t/year

Reductions in TP-loading will cause reductions would lead to reductions also in production and biomasses of predatory zooplankton, as shown in Fig. 4.39.

Increasing Salt-Water Inflow

When the salt-water inflow from the Kattegat is being been increased (see Fig. 4.5), this will cause several changes to the system. The changes for predatory zooplankton are given in Fig. 4.41 and they are relatively small.

Temperature Scenario

Figure 4.7 gave the results from the "temperature scenario". This scenario gives marked increases in TP-concentrations (Fig. 4.7b) and there will also be increases in

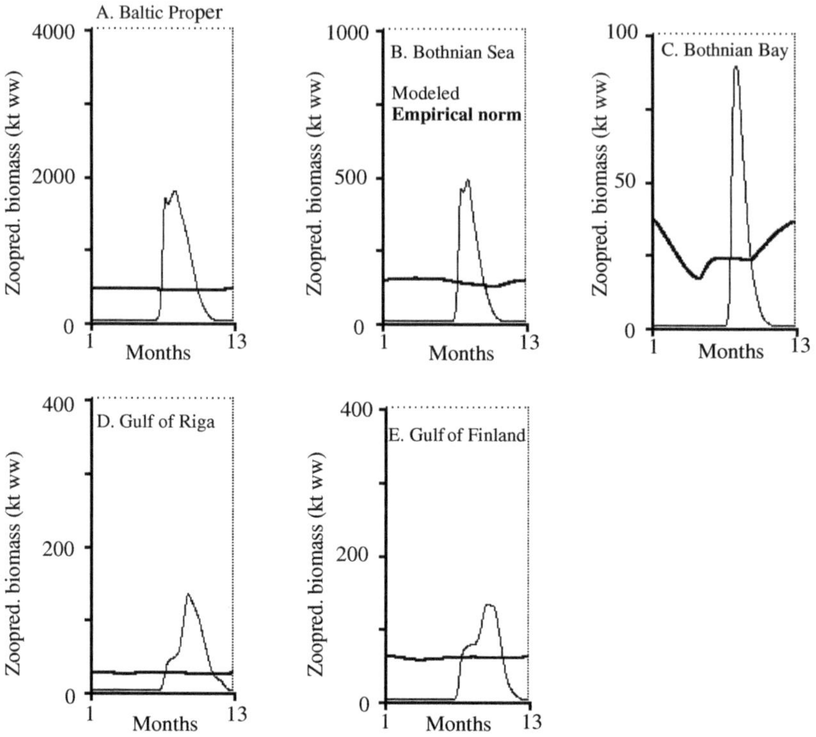

Fig. 4.40 A comparison between modeled empirically-based norm-values (*bolded* reference values for mainly the growing seasons) and dynamically modeled values of predatory zooplankton biomass at steady-state (simulation time 241 months) under default conditions in the five basins of the Baltic Sea

the production and biomasses of herbivorous zooplankton (see Fig. 4.42), especially in the Bothnian Bay.

4.4.3 Fish

4.4.3.1 A Short Background on Fish

To predict the production and biomass of fish in the basins of the Baltic Sea is, evidently, of fundamental importance for a sustainable Baltic Sea management, including how fish quotas should be determined in relation to variations in key abiotic variables, such as water temperature, salinity, oxygen conditions and nutrient concentrations. The aim of this section is not to give a review on fish biology, since much has already been written on that topic (see, e.g., Blanc et al., 1971; Maitland, 1977, 1978; Ojaveer, 2003). Chapter 2 presented results from our literature review on Baltic Sea fisheries. We will also present the empirical regressions (the norms)

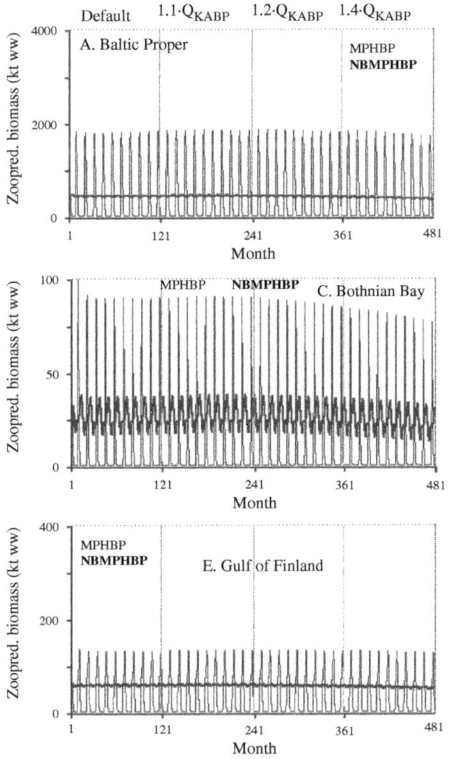

Fig. 4.41 A comparison (a sensitivity test) between dynamically modeled biomasses of predatory zooplankton and values calculated from the empirically-based norms for the 5 sub-basins in the Baltic Sea in a situation where the default conditions are shown for the first period (months 1–61), for the second period the monthly saltwater inflow from the Kattegat has been increased by a factor of 1.1, in the third period 1.2 and in the fourth period (months 361–481) by 1.4

used for the model testing, and discuss how and why we have accounted for the role of migration in the fish model and how fishing (by man, birds or animals) is handled. All equations in new fish model will be presented and motivated.

The first aim here is to focus on fundamental features used in the CoastWeb-model, such as empirical regressions to predict production and biomasses of the two categories of fish included in the model, i.e., prey fish and predatory fish, and on distribution coefficients regulating food choices for fish. In the following introduction, we give background information for the subsequent presentation of the dynamic model for fish.

Note that the primary aim of the CoastWeb-model is not to give predictions for certain species of fish or for specific coast types. The main aim of the model is to quantitatively describe typical, characteristic foodweb interactions so that production, biomasses and predation can be determined for the functional groups of organisms included in the model.

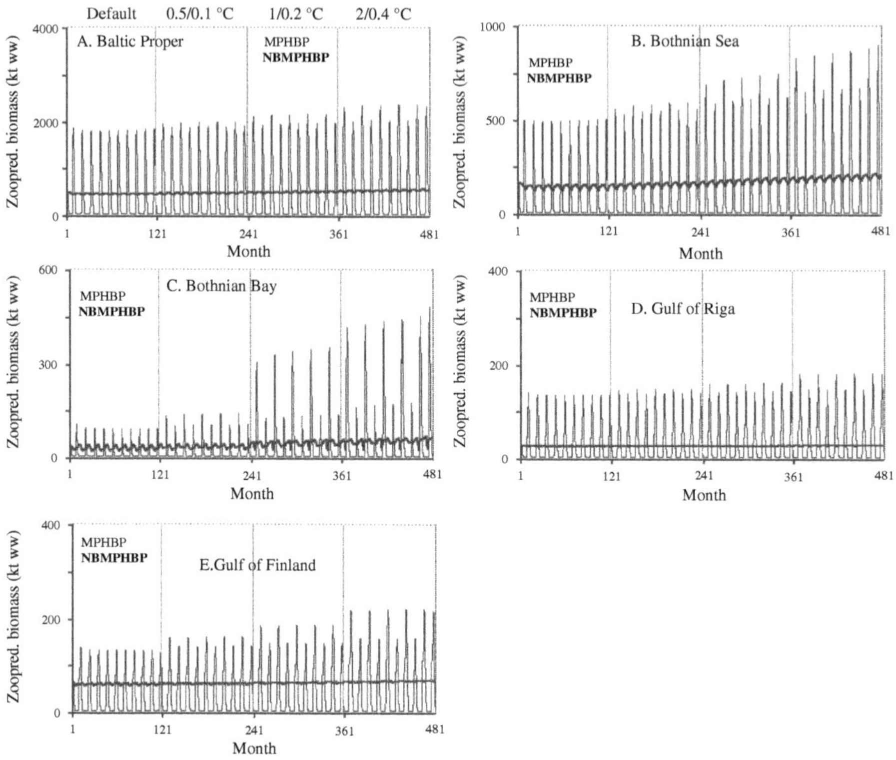

Fig. 4.42 A comparison (a sensitivity test) between dynamically modeled biomasses of predatory zooplankton and values calculated from the empirically-based norms for the 5 sub-basins in the Baltic Sea in a situation where the default conditions are shown for the first period (months 1–61), for the second period the monthly surface-water temperatures have been increased by 0.5 and 0.1°C every second year, for the third period by 1 and 0.2°C every second year and in the fourth period by 2 and 0.4°C every second year. Middle- and deep-water temperatures have been increased by 0.75 and 0.5 of the increases given for the surface-water temperatures

Table 4.3 provides background information on many common European fish species; Chap. 2 gave information on dominating fish species in the Baltic Sea. Table 4.3 also lists typical weight ranges for the given species, the trophic range where the given fish are likely to appear, characteristic habitats, main diets and very short notes on the significance of the given species.

An important question for the CoastWeb-model for fish concerns the distribution coefficient which regulate how much of the total fish biomass in a given lake that is prey fish and predatory (= piscivorous) fish. The way in which fish feed is certainly complicated because (see Kitaev, 1984; Jeppesen et al., 1990):

(1) Fish, e.g., cod in the Baltic Sea, change their feeding preferences, especially during their early development (see Fig. 2.47).

Table 4.3 Compilation of some European fish species (modified from Brittain, 1998). CM = consumed by man. I, II and II = stages

Species	Typical weight range (kg)	Target weight (kg)	Trophic range	Habitat	Food habit	Significance
Sturgeon	20–50	30	meso- to eutrophic	benthic. riverine	piscivore	CM, caviar
Brown trout. I	0.05–0.1	0.1	oligo- to mestotr.	benthic. littoral	planktivore	CM, early
Brown trout. II	0.1–0.3	0.2	oligo- to mestotr.	benthic. littoral	benthivore	CM
Brown trout. III	0.5–1.5	1	oligo- to mestotr.	benthic. littoral	piscivore	CM
Arctic char	0.01–0.2	0.1	oligo- to mestotr.	pelagic	planktivore	CM
Whitefish	0.1–1	0.3	oligo- to mestotr.	pelagic	planktivore	CM
Smelt	0.01–0.05	0.01	oligo- to eutrophic	pelagic	planktivore	prey
Pike	0.5–3	1	oligo- to mestotr.	benthic. littoral	piscivore	CM
Roach	0.05–0.2	0.1	oligo- to hypertr.	benthic/ pelagic. litt.	omnivore	prey
Minnow	0.001–0.01	0.01	oligo- to mesotr.	benthic. littoral	omnivore	prey species
Asp	0.5–3	1	meso- to eutrophic	pelagic. riverine	piscivore	CM
Nase	0.1–1	0.3	meso- to eutrophic	pelagic. riverine	planktivore	CM
Barbel	0.5–3	1	meso- to eutrophic	benthic. riverine	benthivore	CM, angling
Bream	0.5–2	1	meso- to eutrophic	benthic. littoral	benthi-/detrivore	CM
Carp	0.5–3	1	meso- to eutrophic	benthic. littoral	benthi-/detrivore	CM
Wels	2–20	5	meso- to eutrophic	benthic	piscivore	CM
Eel	0.1–1	0.5	oligo- to eutrophic	benthic. riverine	omnivore	CM
Burbot	0.1–1	0.5	oligotrophic	benthic. profun.	benthivore	CM
Perch. I	0.01–0.1	0.1	oligo- to eutrophic	benthic. littoral	planktivore	CM, early stage
Perch. II	0.1–0.3	0.2	oligo- to eutrophic	benthic. littoral	benthi-/omni.	CM
Perch. III	0.3–0.6	0.5	oligo- to eutrophic	benthic. littoral	piscivore	CM
Pike-perch	0.5–3	1	oligo- to eutrophic	pelagic	piscivore	CM
Ruffe	0.005–0.02	0.01	oligo- to eutrophic	benthic. littoral	benthivore	prey species

(2) As adults, they sometimes switch between organisms, like zooplankton and zoobenthos, and from small fish to larger fish.
(3) So, they transform and influence the foodweb system.
(4) They also influence abiotic conditions by binding nutrients, and this will influence the primary production, which in turn will influence the secondary production, including the fish production, and so on.

The basic aim of the CoastWeb-model is to quantify such interrelationships in a general manner.

It should also be noted that macrophytes constitute an excellent environment for several species of predatory fish, e.g., for pike to make an ambush. The area between the macrophyte beds and the shoreline is often a fine "nursery" for young fish, which help to sustain a high fish reproduction in sheltered coastal areas. Larger prey fish generally prefer to stay outside the macrophyte cover where zooplankton are more abundant. So, macrophytes are favorable for both prey fish and predatory fish, and it is important to account for this role, especially in lakes and smaller coastal areas. However, macrophytes are not an important food source for fish. They provide shelter and can thus reduce the predation pressure, especially on small fish targeted by piscivores.

Most fish are visual grabbers, who attack each prey individually according to which size of prey they prefer (Ivlev, 1961; Skoptsov et al., 1983). The total time required for consumption of one unit of prey often equals the time for search plus the "handling" time, i.e., the time to swallow and initially digest the food. The higher the biomass of the prey, the shorter the time of the search, but the time of "handling" remains more of less the same. The handling time is usually dependent on the size of the prey (e.g., Turesson et al., 2002).

Many fish are near-sighted, but fish generally have a maximal acute vision at about 35 lux, whereas for man it is at about 300 lux (Nikolskiy, 1974). Therefore, for the feeding of planktivorous fish, the water clarity is generally of less importance than the abundance of food organisms. Since most of the "favorite food" for planktivorous fish is transparent organisms (like *Leptodora* or larvae of *Chaoborus*; even *Daphnia* is transparent), the distance between predator and prey must be relatively short (<0.5 m).

Predatory fish generally head for the prey laterally. So, the eyesight is used mainly for correction of the rush against the prey. For example, a blind pike is able to hunt laterally. And benthivorous fish are able to feed in darkness due to a sense of touch, taste and smell. Some species have acute sense of touch due to sensitive barbels, whisker-like organs near the mouth, e.g., the carp (*carpio carpio*). Many fish species have very advanced senses of taste (see Kasumyan and Doving, 2003).

If the water is turbid, the chances for the predatory fish to find its prey may be lower than in clear-water conditions and the possibilities for the prey to escape may be higher because of many complicated features related to how signals between the prey and the predator travel in waters of different hydro-optical properties (see Eq. 3.10). In general, turbidity should affect piscivorous fish more, as they feed on larger and more visible prey, than the particle-feeding planktivorous fish (De Robertis et al., 2003).

4.4.3.2 Criteria to Differentiate Between Prey and Predatory Fish

To produce 1 kg of a predatory fish requires approximately 4 kg of prey fish. This is given by the metabolic efficiency ratio, which is set to 0.25 for predatory fish (see Table 2.9). So, if the biomass of predatory fish is large, the grazing pressure on the prey fish is also large.

Kitaev (1984) has presented a formula to estimate the predation pressure of predatory fish on the total fish biomass:

$$\mathrm{RED} = 1/(1 + (\mathrm{FCR} - 1) \cdot (1 - \mathrm{DC_{PYPD}})) \tag{4.70}$$

RED = A coefficient describing the reduction of the fish biomass (dimensionless, 1 means 100% reduction).
FCR = The feed conversion ratio (dimensionless) for predatory fish; i.e., the prey fish biomass (kg ww) needed to produce 1 kg ww of predatory fish.
DC_{PYPD} = The portion of prey fish; this is the requested distribution coefficient needed in the CoastWeb-model.

Assuming that for predatory fish FCR = 4, and the portion of predatory fish is 0.27 (i.e., $DC_{PYPD} = 0.73$), we get that the "normal" coefficient of reduction, RED, is 0.55.

An interesting work on the same subject has been presented by Jeppesen et al. (1990). Their study focused on the ratio between the number of planktivorous fish (fish smaller than 10 cm) to planktivorous plus piscivorous fish (>10 cm). This ratio, PLV/(PLV+PSV), increased in these studied systems in a significant manner with increasing TP-concentration up to a threshold level at about 100 μg TP/l. In hypertrophic systems, the ratio was relatively constant and the fish biomass dominated by planktivores. The data were fitted to a Monod equation were the half-saturation coefficient was 22 ± 4 μg TP/l, and the saturation coefficient is 1 ± 0.03. That is:

$$\mathrm{PLV}/(\mathrm{PLV} + \mathrm{PSV}) = \mathrm{C_{TP}}/(\mathrm{C_{TP}} + 22) \tag{4.71}$$

$(r^2 = 0.98)$

It should be noted that Eq. (4.71) is valid when PLV < 10 cm and PSV > 10 cm. So, increasing eutrophication is accompanied by a marked decrease in the relative abundance of piscivorous fish, to less than 20%, at TP levels above 100 μg/l. In this modeling for the Baltic Sea, and given the empirical data on the proportion of predatory fish in the system (see Fig. 2.46), we have used two boundary conditions for this distribution coefficient (DC_{PYPD}) – it should be larger than 0.9 and smaller than 0.99.

From this, it may be concluded that:

- The results presented by Jeppesen et al. demonstrate, that there is a marked relationship between the trophic state and the portion of predatory fish, which should decrease with increased TP-concentrations. Equation (4.71) is not directly valid for predatory fish and prey fish as these groups are defined in the CoastWeb-model. We have based the dynamic fish model on functional criteria and predatory fish includes all types of piscivores and prey fish includes "the rest", i.e., all other types of fish, such as zooplanktivores and benthivores.

Håkanson and Boulion (2002a) presented the algorithm, which is used in the LakeWeb/CoastWeb-model to estimate the portion of prey fish (DC_{PYPD}). That algorithm and the connected boundary conditions were discussed when Eq. (3.25) was presented.

4.4.3.3 Empirical Norms for Prey and Predatory Fish

First, we estimate the normal total fish biomass (NBM_{Fish} in kg ww) from Eq. (3.22). It should be noted, that the fish catch "useful for man" is generally significantly lower then the fish production calculated by the CoastWeb-model. Most of the fish are young fish and the age frequency distribution for fish age is positively skewed (the mean value higher than the median value), as it is for most animals. To catch bigger fish is more profitable so this is the exploited part of the actual fish production (Blazka et al., 1980). Much of the primary production (PrimP) is lost in the trophic chain, which is quite long in most aquatic ecosystems. The efficiency in the transition from level to level is traditionally set to about 10%. So, fish production is often about 0.5% of primary production (see Chap. 1).

The normal prey fish biomass is calculated from Eq. (3.23) and the normal predatory fish biomass from Eq. (3.24).

These empirical reference norms for production and biomasses of fish will subsequently be compared to values using the dynamic CoastWeb-model.

4.4.3.4 Fish Migration

Migration may be important for the production of fish, especially in systems with great temporal fluctuations or with marked spatial patterns in habitat quality (Blanc et al., 1971; Maitland, 1978; Northcote, 1978; Tonn et al., 1990; Busch and Sly, 1992). Some general characteristics of fish migrations are illustrated in Fig. 4.43.

Spawning migrations for salmonids have been the focus of many studies but both feeding and refuge or wintering migrations can be important in terms of survival and production of fish. The most common migration direction is probably along longitudinal routes up or down river systems, but many fish species move laterally to utilize the margins of rivers, wetlands and lakes for feeding and spawning. In large systems, some fish species undertake vertical migrations between the surface-water

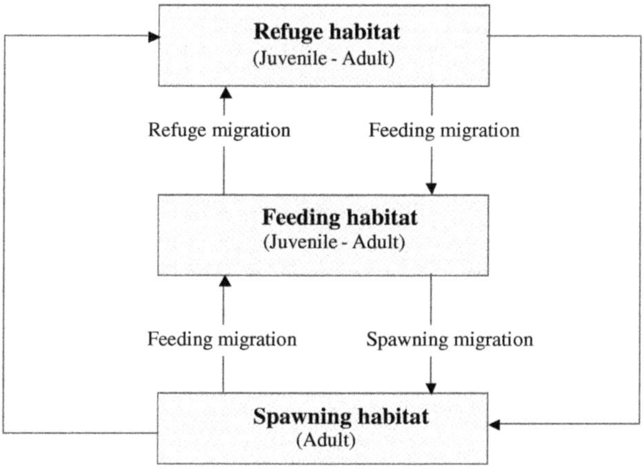

Fig. 4.43 Generalized patterns of migration between the three basic habitats utilized by many migratory fishes (from Brittain and Braband, 2001, as simplified after Northcote, 1978)

and deep-water layer, usually on a daily basis in search of food. Evidently, fish also move horizontally, e.g., in and out of littoral areas. Several species lay their eggs in the substrate where the fry may remain for some time before they move away from the substrate to the water.

Some fish species leave the area where they hatch and move to feeding habitats early in their life, e.g., smelt (*Osmerus eperlanus*) and pike-perch (*Stizostedion lucioperca*). Other species disperse more gradually from the spawning area and reach their feeding habitat after several months, e.g., grayling (*Thymallus thymallus*) and sturgeon (*Acipenser sturio*). Some populations of migratory whitefish (*Coregonus*) remain for a long time near the spawning area before migrating to a second feeding area. Survival is obviously enhanced if the fish attains a certain size before migration. Several species migrate in order to avoid unfavourable abiotic or biotic conditions, e.g., roach (*Rutilius rutilus*), bream (*Abramis brama*) and pike-perch (*Stizostedion lucioperca*) may leave feeding areas in brackish and marine environments and migrate to over-wintering areas in rivers or lakes.

Most fish undertake spawning migrations. In many cases, this involves a return from a habitat rich in food but unsuitable for spawning to an area, which can provide the requirements for reproduction, even though the possibilities for feeding may be sub-optimal. There are many examples of migrations from lacustrine or marine habitats to streams or rivers in order to spawn, e.g., for trout, salmon, char, lampreys, sturgeon, graylings and whitefish and even in some cyprinids.

There are major differences among different species and populations in the distance over which they migrate (Fig. 4.44). Migration of fish is a very complicated issue (Levinton, 2001; Lucas and Baras, 2001; Neuenfeldt et al., 2007; Saulamo and Neuman, 2002), but nevertheless it has to be quantified in the CoastWeb-model where the aim is to obtain realistic predictions of fish biomass in larger coastal areas.

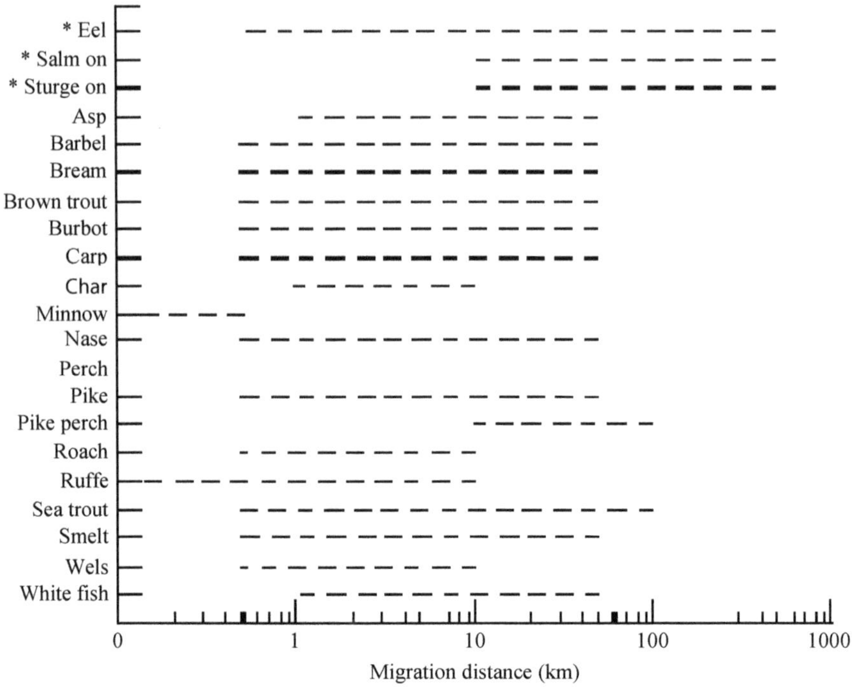

Fig. 4.44 Typical migration distances of some common European fish. The *asterisks* indicate that the species migrate to marine areas (from Brittain and Braband, 2001)

Not only prey and predatory fish migrate in and out of coastal areas, but also jellyfish, predatory and herbivorous zooplankton, bacterioplankton and phytoplankton are transported in and out of the Baltic Sea basins by water currents. The algorithms for in- and outmigration used for all these functional groups in the CoastWeb-model were compiled in Table 3.4. These algorithms are based on the following principles:

1. The migration rate (R_{Mig}, per month) is related to the surface water retention rate ($R_{Mig} = 1/T_{SW}$ in months). This is meant to account for the physical possibilities for organisms to migrate: if there is no inflow or outflow of water, no organisms will migrate in- and out of the system. This approach is used for smaller organisms that travel with the water rather than in the water, i.e., phytoplankton, bacterioplankton and zooplankton, but not for fish. Plankton (predatory zooplankton, herbivorous zooplankton, phytoplankton and bacterioplankton) are not likely to migrate in search for food, but are mainly transported by water currents, as given by the surface water exchange (T_{SW}). The deep-water exchange is generally smaller than the surface-water exchange and the focus here is on the water exchange for the productive surface-water layer.

2. For the interconnected system of basins in the Baltic Sea, we use the same migration rate for prey and predatory fish and the algorithms to quantify migration were presented and discussed in Sect. 3.4.2.

4.4.3.5 Fishing

Birds and other animals fish for feeding purposes, and so does man, but we also fish for fun. These activities create a predation pressure on the fish. Evidently, this pressure can be heavy during, e.g., intensive trawling. The most important criteria regulating fish production concern the abundance and size of the food organisms. For example, small planktivorous fish select, as a rule, objects smaller than 1 mm, young fish generally prefer organisms of about 1 mm, and larger fish eat larger objects (Salazkin and Ogorodnikova, 1984). For benthivorous fish, the large larvae of potential food organisms, such as *Chironomus,* generally live several centimeters down in the sediments and hence are not readily available for consumption, except for some fish species such as bream (*abramis brama*) and carp (*carpio carpio*) that are adapted to digging in sediments (Froese and Pauly, 2008). And, e.g., small molluscs inhabiting the surface of sediments and/or macrophytes are consumed by fish in spite of the fact that the food value of molluscs is less than that of *Chironomus*.

The food selection by fish may be estimated by the following index of selection (E_S, from Ivlev, 1961):

$$E_S = (r_i - p_i)/(r_i + p_i) \tag{4.72}$$

Where r_i is the relative content of a given food component in a fish stomach, p_i is the same in forage reserve. The value of E_S varies between -1 and $+1$. If $E_S = 0$, there is no food selection.

The algorithm to quantify fishing of both prey and predatory fish was given by Eq. (3.51).

The seasonal variations in fishing rate under default conditions in all the major sub-basins of the Baltic Sea are shown in Fig. 4.45. From this figure, we can note that, (1) the fishing rates in the Baltic Proper, the largest basin, are lowest (about 2% of the prey biomass per month) and in the summer time highest in the Gulf of Finland ($R_{Fish} \approx 0.05$) and (2) the seasonal variations are highest in the Bothnian Bay, the Gulf of Finland and the Gulf of Riga and lowest in the Baltic Proper.

Historically, the seal composition and the number of seals have varied due to large-scale changes in the Baltic Sea (Schmölcke, 2008). The biomass of seals in the Baltic Sea has decreased radically during the last 100 years mainly due to first hunting by man and then reproduction problems caused by toxic substances (Hårding and Härkönen, 1999). Seals consume significant amounts of mostly prey fish but also predatory fish (Lundström et al., 2007). We have not specifically addressed this in this model set-up, but the fishing rate is meant

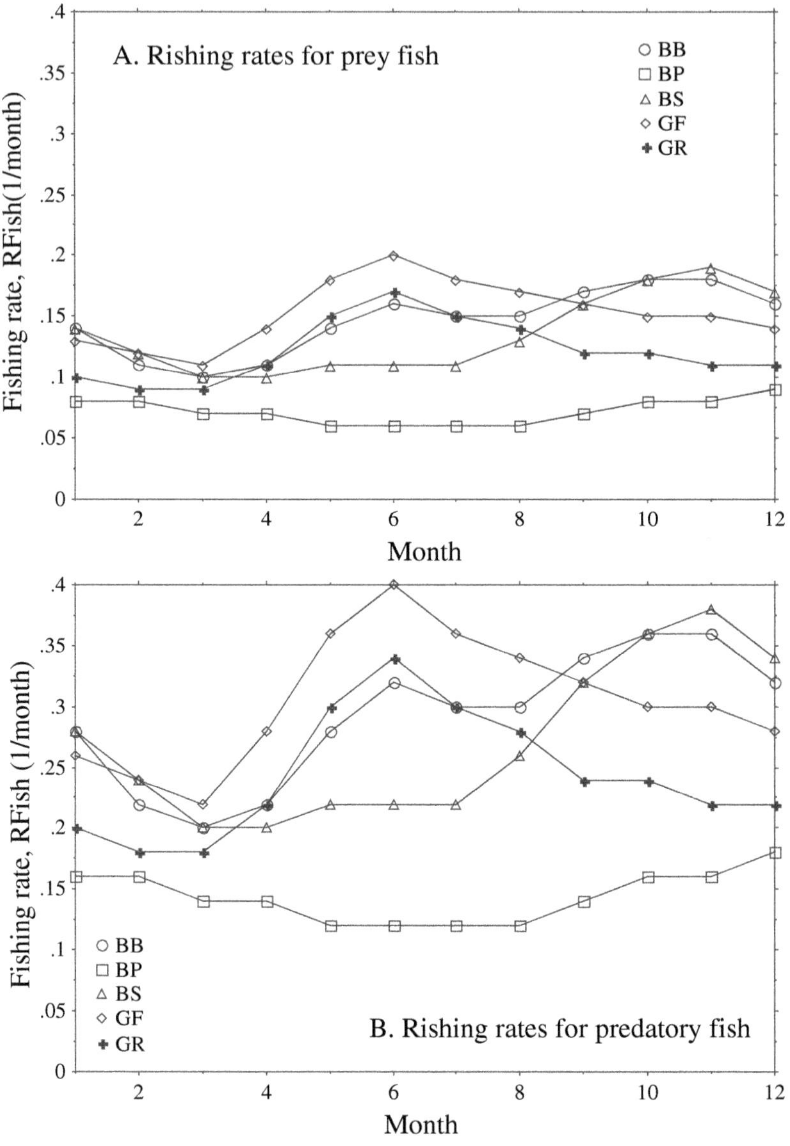

Fig. 4.45 Default fishing rates for prey and predatory fish using data for the five sub-basins in the Baltic Sea

to account for also the "seal factor". Without the seals, we would have set the fishing rate, R_{Fish}, to a lower value. In Chap. 5, we will discuss the problem of finding "optimal fish quotas" and then we will vary the default fishing rate.

4.4.3.6 The Dynamic Model for Fish

Figure 4.46 gives an overview of the fish model. Figure 4.34 shows the food choice panel for prey fish. It must be noted that there are many simplifications in the CoastWeb-model for fish. For example, the prey fish compartment is a simplification since three functionally different feeding patterns have been combined. So, instead of making a more complicated model with separate fluxes (and differential equations) for benthivores, planktivores (two varieties, one for herbivorous and one for predatory zooplankton) and prey fish consuming "everything" including small-sized fish (omnivores), the CoastWeb-model is simplified. Instead of expanding the model to account for these feeding patterns, we have tried to account for different feeding patterns by means of distribution coefficients and the number of food choices. The

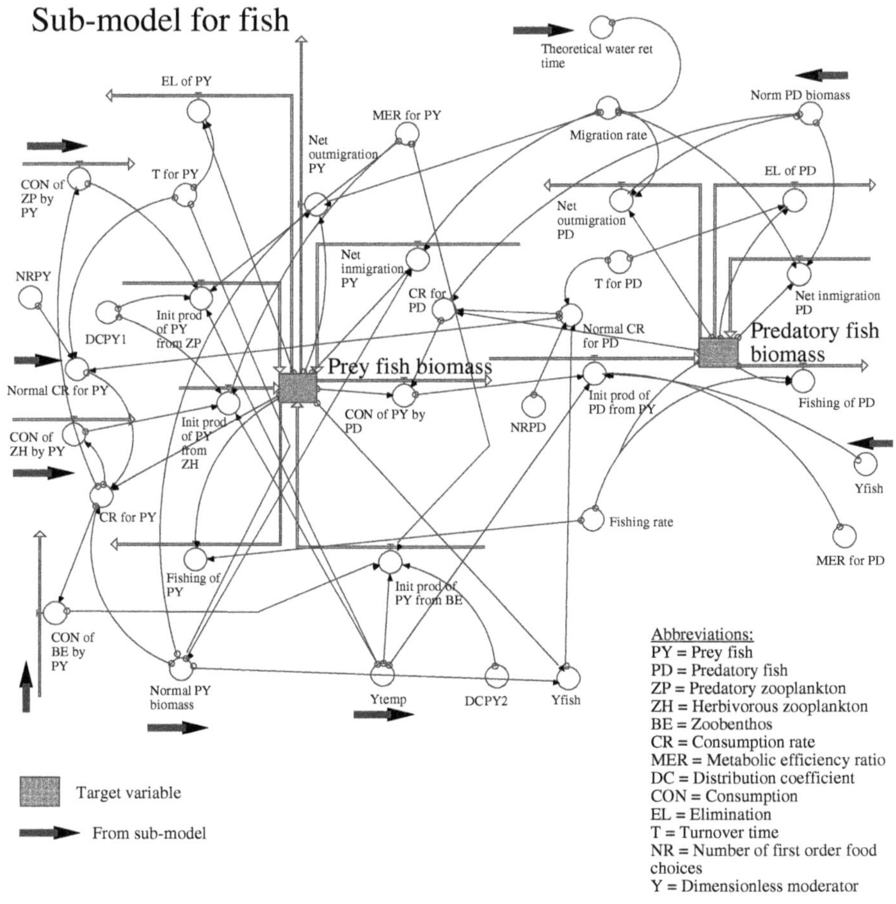

Fig. 4.46 Outline of the sub-model for fish

model calculates net fish production and fish consumption is restricted for predatory fish. This approach has several advantages for the practical use of the model, but it also entails problems. One such problem concerns the consumption rate and the number of food choices for predatory fish when the prey fish compartment is defined in this manner. We will address that problem in the following section.

Modeling of Prey Fish

The following differential equation gives the fluxes (kg ww/month) to and from the compartment "Prey fish".

$$\begin{aligned}BM_{PY}(t) = {} & BM_{PY}(t-dt) + (IPR_{ZHPY} + IPR_{ZPPY} + IPR_{ZBPY} + MIG_{InPY} \\ & - CON_{PYPD} - EL_{PY} - MIG_{OouPY} - FISH_{PY}) \cdot dt\end{aligned} \quad (4.73)$$

BM_{PY} = Biomass of prey fish (kg ww); the initial BM_{PY}-value is set equal to the normal prey fish biomass, NBM_{PY}, which is given by Table 3.2.
IPR_{ZHPY} = Initial production of prey fish from consumption of herbivorous zooplankton (kg ww/month).
IPR_{ZPPY} = Initial production of prey fish from consumption of predatory zooplankton (kg ww/month).
IPR_{ZBPY} = Initial production of prey fish from eating zoobenthos (kg ww/month).
MIG_{InPY} = Migration of prey fish into the basin (kg ww/month).
CON_{PYPD} = Consumption of prey fish by predatory fish (kg ww/month).
EL_{PY} = Elimination of prey fish (kg ww/month).
MIG_{OutPY} = Migration of prey fish out of the system (kg ww/month).
$FISH_{PY}$ = Loss of prey fish from all sorts of fishing (kg ww/month).

The initial production of prey fish from consumption of herbivorous zooplankton (IPR_{ZHPY} in kg/month) is given by:

$$IPR_{ZHPY} = CON_{ZHPY} \cdot Y_{Temp}{}^{0.25} \cdot MER_{PY} \cdot DC_{PY1} \cdot (1 - DC_{PY2}) \quad (4.74)$$

Where CON_{ZHPY} is the consumption of herbivorous zooplankton by prey fish (kg ww/month). The actual consumption rate for prey fish feeding on herbivorous zooplankton is the same as for prey fish eating zoobenthos. The initial production of prey fish from eating predatory zooplankton (IPR_{ZPPY} in kg ww/month) is given in an analogous manner by:

$$IPR_{ZPPY} = CON_{ZPPY} \cdot Y_{Temp}{}^{0.25} \cdot MER_{PY} \cdot DC_{PY1} \cdot DC_{PY2} \quad (4.75)$$

Y_{Temp} = The temperature moderator. The exponent is set to 0.25 as a default value for fish. There are many factors, beside mean monthly temperature influencing the feeding behavior of fish, such as access to food, solar light, water clarity and short-term variations in water temperature. So, fish production is less dependent on mean weekly surface water temperatures than plankton production. Numerous tests of this kind have been done for more exponents along different gradients (see Håkanson and Boulion, 2002a). Generally, the best correspondence to the norms and the most realistic seasonal patterns are obtained if the exponent is set to 0.25.

MER_{PY} = The metabolic efficiency ratio for prey fish eating zooplankton and zoobenthos ($MER_{PY} = MER_{ZHPY} = MER_{ZPPY} = MER_{BEPY}$). It is set to 0.16 (see Table 2.9).

DC_{PY1} = The first order distribution coefficient regulating how much of either zooplankton or "the rest" (= zoobenthos) prey fish eat. The default value is set to 0.75 (see Fig. 4.34).

DC_{PY2} = The second order distribution coefficient regulating how much of either predatory zooplankton (first) or herbivorous zooplankton that prey fish eat. The default value is set to 0.2.

The biomass of the prey fish also depends on migrations of fish to and from the given system.

In this context, it should be stressed that inherent instincts are of great importance for fish migration. In contrast to zooplankton, fish are highly organized vertebral animals with complex behavior. The relation between migration and the theoretical water renewal (or retention) time, T_{SW}, is a simple mechanistic approach which is meant to account for physical presuppositions related to fish migrations, i.e., whether there are physical possibilities for the fish to migrate and if these are limited or not. This approach does not address the fact that many species of fish prefer to migrate against the stream currents.

The quantification of inmigration and outmigration of fish were given in Sect. 3.4.2.

The definition of the normal consumption rate for predatory fish, NCR_{PD}, is related to the structuring of the model and the definition of the prey fish compartment. The default value for the number of first order food choices for predatory fish (NR_{PD}) is set to 1 because there is one food choice for predatory fish. They eat prey fish only. So, the normal consumption rate for predatory fish is first given by: $NCR_{PD} = NR_{PD} / T_{PD} = 1 / T_{PD}$, where the turnover time of predatory fish, T_{PD}, is set to 900 days for the Baltic Sea where large cod dominates among the predatory fish (2 times the value used for lakes with predatory fish such as pike and large perch; see Table 2.9).

Håkanson and Boulion (2002a) showed, however, that this approach may yield too low predatory fish consumption, and hence too low biomasses of predatory fish, too small predation pressure on prey fish and hence too high biomasses of prey fish. Given the complex nature of the prey fish compartment in nature and in this model, and the consequences for the predatory fish of the given simplifications for the prey fish compartment, it seems unrealistic to assume that the normal consumption rate for predatory fish should increase in a linear way with trophic state and prey fish biomass. In highly productive systems, the prey fish biomass will increase both in absolute numbers, and as already stressed, in relation to the biomass of predatory fish, so the normal consumption rate for predatory fish should not increase linearly with prey fish biomass. We have used a dimensionless moderator, Y_{Fish}, to handle these interactions. It was defined in Eq. (3.10). So, the normal consumption rate for predatory fish is given by: $NCR_{PD} = Y_{Fish}/T_{PD}$.

Equation (3.10) gives that for a system with a TP-concentration of 30 μg/l, Y_{Fish} is 1 and the normal consumption rate is given by the basic approach. For less productive systems, Y_{Fish} will increase; for a system with a TP-concentration of 10 μg/l, Y_{Fish} is 2.67, and the normal consumption rate will be so much higher. For more productive systems, Y_{Fish} will decrease but the value is never permitted to go below 0.2.

Note that many approaches have been tested in this context, e.g., to base the algorithm on variations in Secchi depth or on suspended particulate matter. There have been drawbacks with all those alternatives, but not (as we have experienced) with the given approach. The main reason for this is probably that since this approach is based on alterations in TP-concentrations, it accounts for changes in trophic state in a broad manner and hence incorporates several potential mechanisms (e.g., sight-feeders are less efficient in more turbid water) that may help to explain why the consumption rate for predatory fish should not be the same in the entire trophic range.

The consumption of prey fish by predatory fish is then given by:

$$CON_{PYPD} = BM_{PY} \cdot CR_{PD} \tag{4.76}$$

And the consumption rate, CR_{PD}, is:

$$CR_{PD} = YFish \cdot (NCR_{PD} + NCR_{PD} \cdot ((BM_{PD}/(NBM_{PD}) - 1))) \tag{4.77}$$

Elimination (kg ww/month) for prey fish is given by:

$$EL_{PY} = BM_{PY} \cdot 1/T_{PY} \tag{4.78}$$

The turnover time for prey fish, T_{PY}, is set to 300 days (see Table 2.9). It should be stressed that the mortality is high for prey fish even when there are few predator fish (Fig. 4.47). A large percentage of especially the young prey fish die from diseases and parasites and the basic principles behind the mortality curve for prey fish shown in Fig. 4.47 should apply to all functional animal groups. The mortality rate is defined from the relationship between the number of deaths at age t divided by the population size at age t.

One flow for prey fish remains to be discussed, the loss of prey fish from fishing, $FISH_{PY}$. This flow (kg ww/month) is given by:

$$FISH_{PY} = BM_{PY} \cdot R_{Fish} \tag{4.79}$$

The fishing rate, R_{Fish} (in month) is defined in Eq. (3.51).

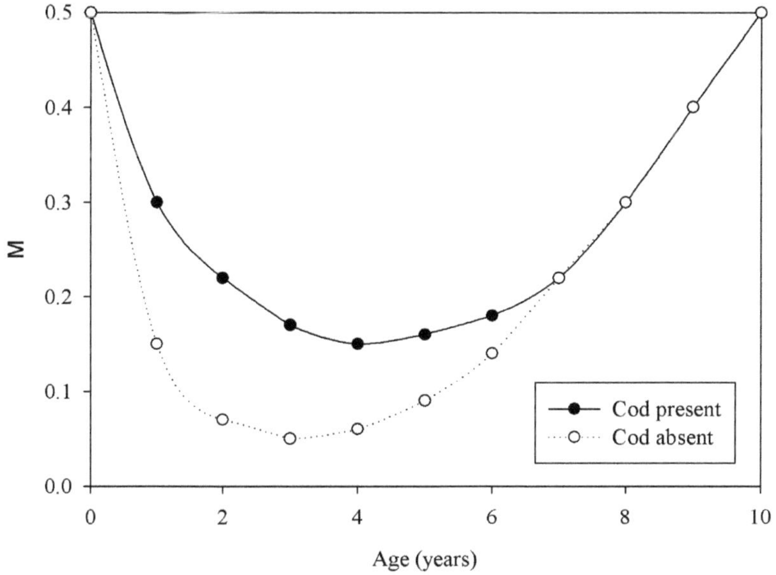

Fig. 4.47 The instantaneous natural mortality rate (M) of autumn spawning herring in the northeastern Baltic Sea with predatory cod present and absent. Modified from Ojaveer (2003)

With this, we have presented the sub-model for prey fish and can turn to the model for predatory fish.

Modeling of Predatory Fish

The following equation gives the fluxes (kg ww/month) to and from "Predatory fish".

$$BM_{PD}(t) = BM_{PD}(t-dt)+(IPR_{PYPD}+MIG_{InPD}-EL_{PD}-MIG_{OutPD}-FISH_{PD})\cdot dt \quad (4.80)$$

BM_{PD} = The biomass of predatory fish (piscivores; kg ww). The initial BM_{PD}-value is set equal to the norm-value, NBM_{PD}, which is given by Table 3.2.
IPR_{PYPD} = Initial production of predatory fish from eating prey fish (kg ww/month), as given by Eq. (4.81).
MIG_{InPD} = Migration of predatory fish into the basin (kg ww/month).
EL_{PD} = Elimination of predatory fish (kg ww/month).
MIG_{OutPD} = Migration of predatory fish out of the system (kg ww/month).
$FISH_{PD}$ = Loss of predatory fish due to all sorts of fishing (kg ww/month).

The initial production of predatory fish from eating prey fish, IPR_{PYPD}, is given by:

$$IPR_{PYPD} = CON_{PYPD} \cdot {Y_{Temp}}^{0.25} \cdot MER_{PD} \quad (4.81)$$

CON_{PYPD} = Consumption of prey fish by predatory fish (kg ww/month), see Eq. (4.76).

Y_{Temp} = The temperature moderator. The exponent for fish is, as motivated before, set to 0.25.

MER_{PD} = Metabolic efficiency ratio for predatory fish eating prey fish. It is set to 0.25 (see Table 2.9).

The migration of predatory fish in and out of the lake is handled in the same manner as for prey fish.

Elimination (kg ww/month) for predatory fish is:

$$EL_{PD} = BM_{PD} \cdot 1/T_{PD} \tag{4.82}$$

The turnover time for predatory fish, T_{PD}, is set to 900 days (see Table 2.9).

The last outflow for predatory fish is the loss from fishing, $FISH_{PD}$. This flow (kg ww/month) is given by:

$$FISH_{PD} = BM_{PD} \cdot 2 \cdot R_{Fish} \tag{4.83}$$

The fishing rate, R_{Fish} (in 1/month), for predatory fish is, as mentioned, set to 2 times the fishing rate for prey fish.

With this, we have presented also the sub-model predatory fish and can focus on critical model tests.

Testing Modeling Results

Reductions in TP-Loading

The results for the biomass of prey and predatory fish along the TP-gradient are given in Fig. 4.48. Figure 4.49 shows in more detail the good correspondence between dynamically modeled values and norm-values for prey and predatory fish under default conditions. Note that there has been no "tuning" to obtain these results. This means that the migration, fishing, predation, etc. is quantified in the standardized manner for all the five sub-basins. Reductions in TP-loading will cause lower TP- and SPM-concentrations in water and higher Secchi depths and this would lead to reductions also in production and biomasses of prey and predatory fish, as shown in Fig. 4.48.

Increasing Salt-Water Inflow

When the salt-water inflow from the Kattegat is being been increased (see Fig. 4.5), this will cause several changes to the system. The changes for prey and predatory fish are given in Fig. 4.50 and they are relatively small.

Temperature Scenario

Figure 4.7 gave the results from the "temperature scenario". This scenario gives marked increases in TP-concentrations (Fig. 4.7b) and there will also be clear

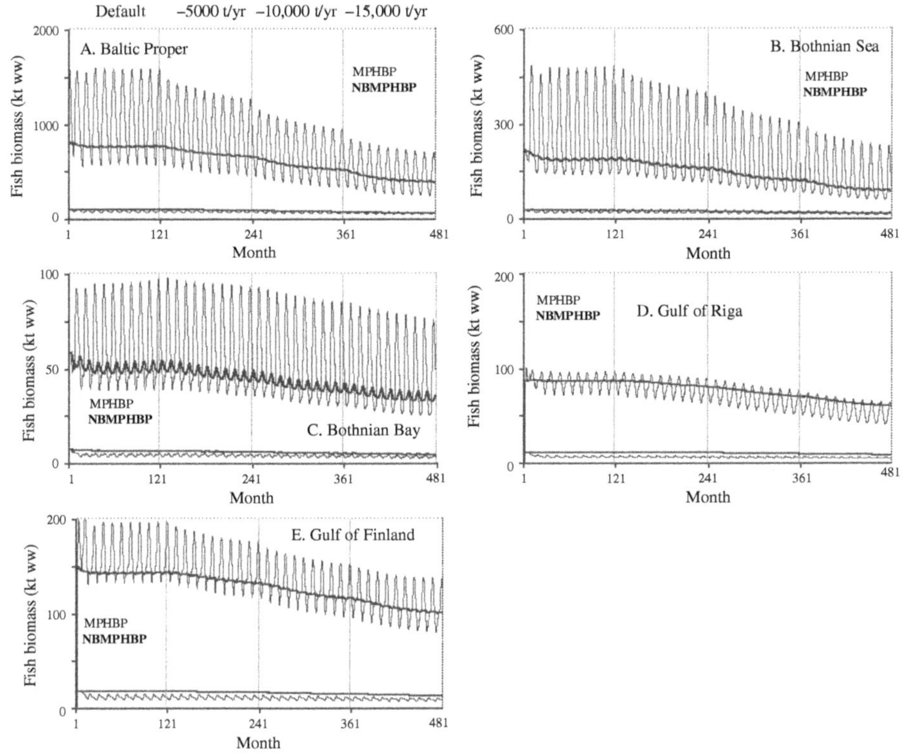

Fig. 4.48 A comparison between dynamically modeled biomasses of prey and predatory fish and values calculated from the empirically-based norms for the 5 sub-basins in the Baltic Sea in a situation where the default conditions are shown for the first period (months 1–61), for the second period the annual TP-inflow via tributaries to the Baltic Proper has been reduced by 5,000 t/year, in the third period by 10,000 t/year and in the fourth period (months 361–481) by 15,000 t/year

increases in the production and biomasses of, foremost prey (see Fig. 4.51) in the Bothnian Bay.

4.5 A Compilation of Modeled Concentrations of Biomasses for all Functional Organisms

Figures 4.52, 4.53 and 4.54 give a compilation of concentrations for all nine key functional organisms in the five sub-basins under default (present-day) conditions. To get comparability, we have defined the concentrations as the ratio between the calculated biomasses and the total area for each basin, and the dimension is g ww/m^2. From these figures, one can first note that there are:

1. Major differences in the seasonal variations among the functional groups. Predatory fish biomasses/concentrations vary the least, the zooplankton groups and the primary producers vary most, and zoobenthos and prey fish in-between.

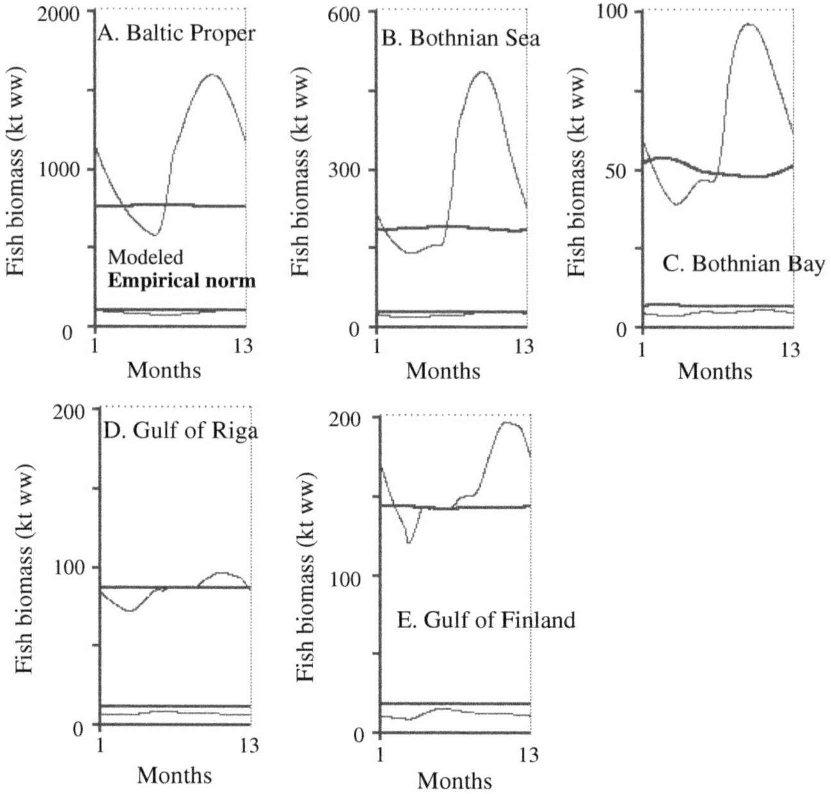

Fig. 4.49 A comparison between modeled empirically-based norm-values (*bolded* reference values for mainly the growing seasons) and dynamically modeled values of biomass of prey and predatory fish at steady-state (simulation time 241 months) under default conditions in the five basins of the Baltic Sea

2. There are major differences in concentrations among the functional groups in the different five basins. By looking at the scales of the y-axes, we can note that bacterioplankton, phytoplankton and zoobenthos appear with the highest concentrations, then herbivorous zooplankton, benthic algae, predatory zooplankton, prey fish, predatory fish and macrophytes. One can specifically note the difference between zoobenthos and zooplankton. It is, for example, evident that predatory zooplankton production and biomass cannot alone sustain a high fish production in the Baltic Sea basins. The difference in concentrations between predatory fish and prey fish is a factor of 6 (see Fig. 4.55); the difference in concentrations between prey fish and the food eaten by prey fish (zoobenthos, herbivorous and predatory zooplankton) is also a factor of about 6; and the difference in concentrations between predatory zooplankton and herbivorous zooplankton is a factor of 5; and between herbivorous zooplankton and their main food (phytoplankton and bacterioplankton) the difference is also about a factor of 6, although there are major variations among different basins and with season of the year.

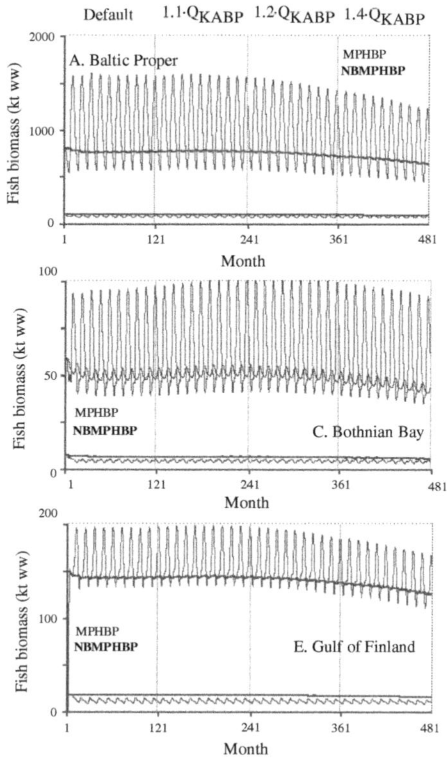

Fig. 4.50 A comparison (a sensitivity test) between dynamically modeled biomasses of prey and predatory fish and values calculated from the empirically-based norms for the 5 sub-basins in the Baltic Sea in a situation where the default conditions are shown for the first period (months 1–61), for the second period the monthly saltwater inflow from the Kattegat has been increased by a factor of 1.1, in the third period 1.2 and in the fourth period (months 361–481) by 1.4

Figure 4.55 compares how the modeled proportions of predatory fish to the total fish biomass under default conditions vary among the basins and seasonally and one can note that this ratio on average varies around 0.05–0.07 in the five basins, and this should be compared to the empirically-based mean value of 0.06 given in Fig. 2.46. The empirically-based biomass of predatory fish in the Baltic Sea around the year 2000 should (according to our results in Chap. 2) be about 100 kt ww, and our modeling gives (see Table 5.3):

BP	BS	BB	GR	GF
67	19	3.8	5.0	11 = 104.4 kt ww for the entire Baltic Sea

The total biomass of prey fish has fluctuated very much indeed during the last 4 decades. Typically, the annual biomasses for the dominating species of prey fish

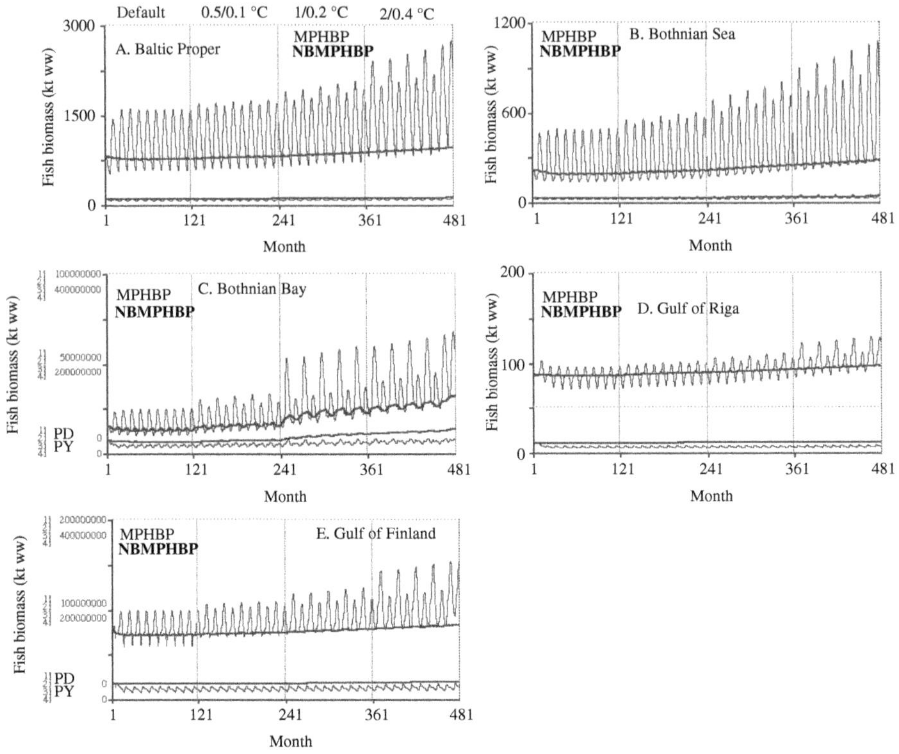

Fig. 4.51 A comparison (a sensitivity test) between dynamically modeled biomasses of prey and predatory fish and values calculated from the empirically-based norms for the 5 sub-basins in the Baltic Sea in a situation where the default conditions are shown for the first period (months 1–61), for the second period the monthly surface-water temperatures have been increased by 0.5 and 0.1°C every second year, for the third period by 1 and 0.2°C every second year and in the fourth period by 2 and 0.4°C every second year. Middle- and deep-water temperatures have been increased by 0.75 and 0.5 of the increases given for the surface-water temperatures

should vary around 2,500 and 5,000 kt ww. This is also what this modeling shows (see also Chap. 5).

4.6 Summary and Comments

This chapter has presented the details of the modeling of the key functional groups of organisms included in the CoastWeb-model, which is based on ordinary differential equations (the ecosystem scale) and gives seasonal variations (the calculation time, dt, is 1 month and Euler's integration method has been applied).

The sub-model for phytoplankton is meant to account for all fundamental abiotic/biotic interactions and feedbacks (including predation by herbivorous zooplankton). The model has not been tested in the traditional way using data from a

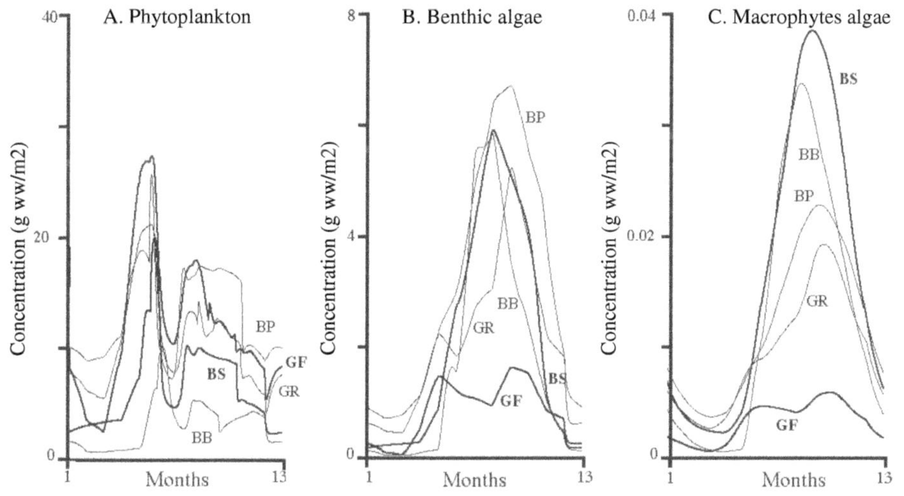

Fig. 4.52 Compilation and comparison of concentrations (g ww/m^2) for the three primary producers in the five basins of the Baltic Sea under default conditions. These concentrations have been defined from the total biomass divided with the total area of each basin

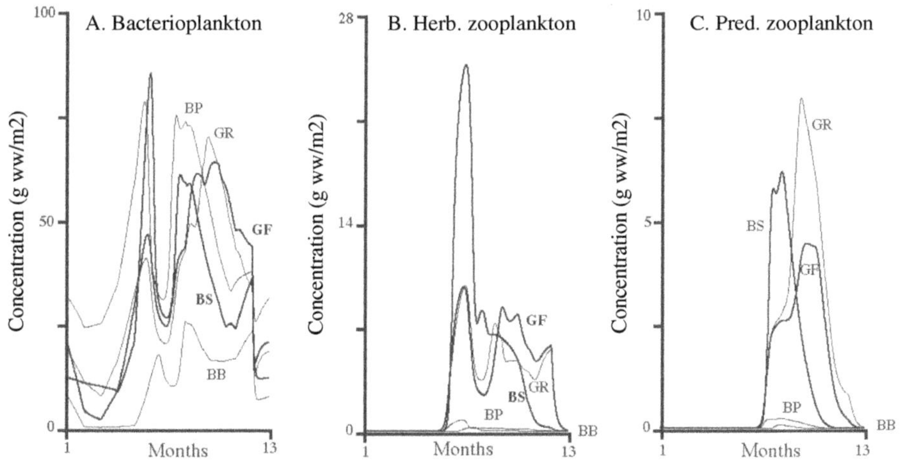

Fig. 4.53 Compilation and comparison of concentrations (g ww/m^2) for (**a**) bacterioplankton, (**b**) herbivorous zooplankton and (**c**) predatory zooplankton in the five basins of the Baltic Sea under default conditions. These concentrations have been defined from the total biomass divided with the total area of each basin

few well-investigated systems. Instead, it has been tested using empirical regressions, the norms (see Table 3.2). The basic aim of this dynamic sub-model is that it should capture typical functional and structural patterns in a general way. It accounts for how variations in (1) phosphorus concentrations, (2) water clarity, (3) basin morphometry, (4) light conditions, (5) salinity, migration or water transport between

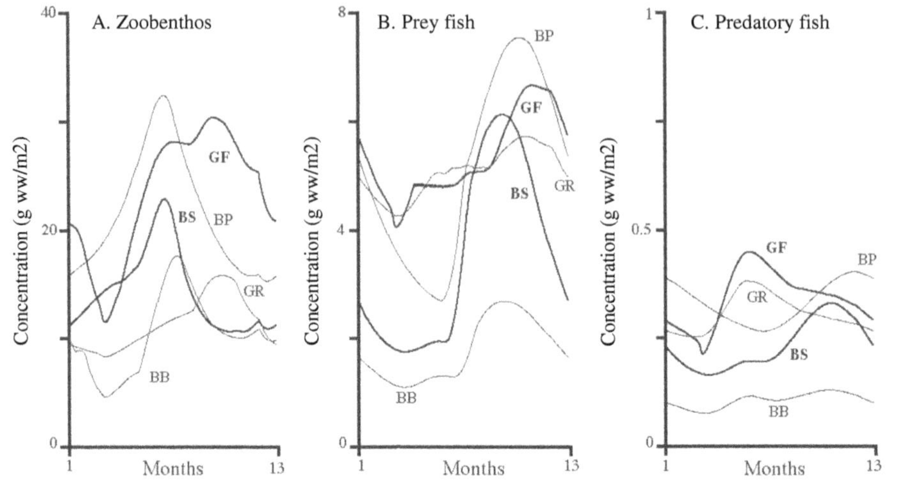

Fig. 4.54 Compilation and comparison of concentrations (g ww/m^2) for (**a**) zoobenthos, (**b**) prey fish and (**c**) predatory fish in the five basins of the Baltic Sea under default conditions. These concentrations have been defined from the total biomass divided with the total area of each basin

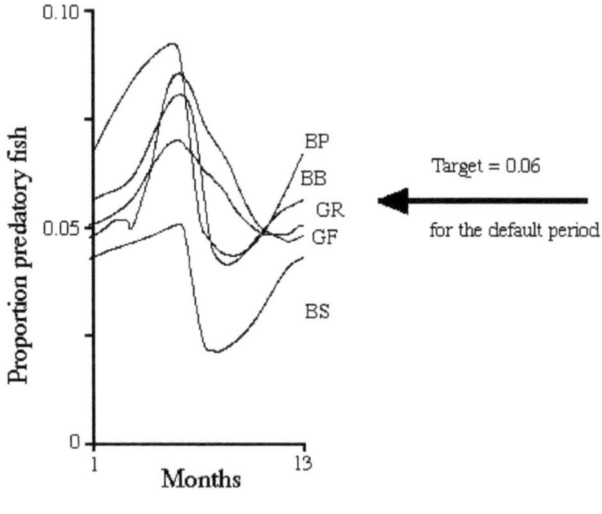

Fig. 4.55 Modeled values of the proportion of predatory fish during the default period (1997–2006) for the five basins. The empirical target value for the entire system is about 0.06 (or 6%; see Fig. 2.46)

adjacent basins and (6) predation by herbivorous zooplankton influence production and biomass of phytoplankton. An important prerequisite for this model is that it should be driven by variables easily accessed from standard monitoring programs and maps. We have demonstrated that the model gives predictions that agree well with the values given by the empirical norms, and also expected and requested divergences from these norms when they do not provide sufficient resolution.

The model has been have tested in three scenarios, (1) by lowering the tributary TP-inflow to the Baltic Proper in steps, (2) by increasing the salt-water inflow to the Baltic Sea system from the Kattegat in steps and (3) in a temperature scenario where water temperatures in all Baltic Sea basins have been increased in a systematic and structured manner. Phytoplankton production and biomass decrease in the TP-scenario and in the salinity scenario. The corresponding changes for all functional groups have been presented. There are major and opposite changes in the temperature scenario. With higher water temperatures the ice cover during winter would largely disappear, which leads to more turbulence and more resuspension, and thus also to significantly higher TP-concentrations and SPM-concentrations, and corresponding reductions in Secchi depths.

We have also presented empirical and dynamic models for benthic algae. The empirical model is based on the primary production of phytoplankton and the ratio between the area of the photic zone, i.e., where the depth is less than two Secchi depths, and the total area. The dynamical and empirical models give corresponding results over a wide domain created in the three scenarios. We have given algorithms for (1) the production rate of benthic algae (2) the elimination rate (related to the turnover time of benthic algae), (3) the rate of benthic algae consumption by zoobenthos, and (4) the rate of physical erosion of benthic algae. Our results indicate that the production of benthic algae is highly dependent on morphometry of the basins, as well as water clarity, and less dependent on nutrient (phosphorus) concentrations in water and sediments.

Macrophytes play several important roles in aquatic ecosystems, e.g., proving shelter for small fish, binding nutrients and influencing secondary production by creating habitats for bacteria, benthic algae and zooplankton. The main factors influencing macrophyte cover are the Secchi depth and the basin morphometry. We have presented a dynamic model for macrophyte production and biomass and several critical tests of the model. We have given algorithms for (1) the macrophyte production rate, (2) the elimination rate (related to the macrophyte turnover time), (3) the rate of macrophyte consumption by zoobenthos, and (4) the rate of macrophyte erosion.

Furthermore, we have presented a dynamic model to predict bacterioplankton production and biomass and given algorithms for (1) bacterioplankton production, (2) elimination (related to the turnover time of bacterioplankton), (3) bacterioplankton consumption by herbivorous zooplankton, (4) migration or rather physical transport by water current of bacterioplankton to and from a given basin, and the factors influencing these processes/rates. We have demonstrated that the dynamic model provides predictions that agree well with the values given by the empirical norms. The dynamic model is driven by data easily accessed from standard monitoring programs or maps and meant to be of practical used in Baltic Sea management.

In this chapter, we have also presented a dynamic model to predict zoobenthos in the Baltic Sea basins. The basic aim of the dynamic model is that it should capture general functional patterns for zoobenthos in the Baltic Sea basins. We have shown algorithms for (1) production of zoobenthos from eating macrophytes, benthic algae and sediments, (2) elimination (related to the turnover time of zoobenthos), and (3) zoobenthos consumption by prey fish, and the factors influencing these processes/rates. The model is driven by data easily accessed from standard monitoring programs or maps a prerequisite for practical utility in contexts of water management.

In addition, we have discussed the details regarding a dynamic model to predict two fundamental functional categories of zooplankton, herbivorous and predatory zooplankton. We have also presented the empirically-based norms which are used in the CoastWeb-model and we have provided algorithms for (1) production of herbivorous and predatory zooplankton, (2) elimination (related to the turnover time of zooplankton), (3) zooplankton consumption by prey fish, (4) physical transport of zooplankton to and from the given basins and the factors influencing these processes/rates.

This work presents a new dynamic model to predict two fundamental functional categories of fish in the Baltic Sea, and all other aquatic systems, prey and predatory fish. The dynamic fish model gives seasonal variations (the calculation time is 1 month). It is meant to account for all key factors regulating the production of fish in a general way. The basic aim of the fish model is that it should capture typical functional and structural patterns in all major Baltic Sea basins. It accounts in a relatively simple manner for many complicated processes, like fishing (by birds, animals and man) and fish migration to and from basins. Food choices are handled by distribution coefficients regulating how much of the different available food sources a given organism would consume. Beside these distribution coefficients, and the way the food choices are structured (the food choice panel), fundamental concepts in the fish model are: (1) metabolic efficiency ratios, which express how much of the food consumed by the predator that will increase the biomass of the predator and how much that will be lost by respiration and faeces, (2) actual consumption rates, which are defined from the ratio between the actual biomass of the predator and the normal biomass of the predator, and the normal consumption rates, which are related to the turnover time of the predator. We have demonstrated that the new model gives predictions which agree well with the values given by the empirical norms, and also expected and requested divergences from these regression lines when they do not provide sufficient resolution.

It should be stressed that the model presented in this book (CoastWeb) can provide good predictions of the biomasses for all functional groups of organisms when tested against empirical data (for fish), and against the empirical norms in all Baltic Sea basins without basin-specific tuning and without taking nitrogen concentrations into account.

In the next chapter, we will combine the scenarios discussed in this chapter, especially the increased eutrophication related to higher water temperatures and the oligotrophication from increased salt-water inflow, to see if these factors provide compensatory effects. However, to do that as realistically as possible, we will first modify and improve the first scenario presented in this chapter, the TP-scenario, and discuss a more optimal strategy to reduce eutrophication in the Baltic Sea basins, and use those results in the combined scenario in Chap. 5. We will also make the salinity scenario discussed in this chapter more realistic by accounting for changes in oxygen that would follow an increase in salt-water inflow from the Kattegat. The last scenario in Chap. 5 is meant as a comprehensive holistic approach to Baltic Sea management. We will use the CoastWeb-model including the CoastMab-model to discuss a management plan for the Baltic Sea system.

Chapter 5
Strategies for Remediation, Cost-Benefit Analyses and a Holistic Management Plan for the Baltic Sea

This chapter will present several scenarios, which are meant to focus on key problems related to sustainable Baltic Sea management. The last scenario will put many of the results in this book together and discuss a management plan for the Baltic Sea, which also includes a cost-benefit analysis. The first scenario is similar to the one presented by Håkanson and Bryhn (2008b). It concerns a strategy to find an "optimal" abatement plan to reduce the eutrophication of the Baltic Sea. Anthropogenic nutrient emissions have seriously altered the trophic state of the Baltic Sea, and our optimal strategy in Sect. 5.1 defines a goal for the remedial actions and presents a realistic avenue to reach that goal.

The second scenario was introduced already in Figs. 1.8, 1.9 and 1.10. The key question is: How can the very large fluctuations in biomass of fish in the Baltic Sea be explained? We will present an answer to that question using the CoastWeb-model and simulate the effects of changes/increases in the inflow of oxygenized salt-water from the Kattegat.

The third scenario concerns an approach to find a strategy to adjust the fish quota to variations in temperature, salinity, oxygen concentrations and nutrient concentrations. This scenario is related to the "optimal" fishing rate discussed in Chap. 3 (see Table 3.7): If too much fish are caught, there will remain too little fish to catch. One cannot fish more than a certain threshold value. The CoastWeb-model is a tool to calculate the total fish production potential of the Baltic Sea system, and we will present further simulations to try to find the optimal fishing rate and to demonstrate the consequences if too much fish are taken out of the system.

In Chap. 4, we presented a temperature scenario and a salinity scenario and we used CoastWeb to study the related consequences for the structure and function of the Baltic Sea ecosystem if there are changes in water temperature and salinity. Increased salt-water inflow from the Kattegat leads to an oligotrophication of the system whereas increased water temperatures lead to a higher eutrophication. The question here is that if there is a global temperature increase, as argued by IPCC (see IPCC, homepage), this could lead to both increased water temperatures and increased salt-water inflow from the Kattegat. In this scenario, we will combine these two aspects, and we will also relate this to the "optimal strategy" for nutrient emissions discussed in the first section of this chapter.

L. Håkanson et al., *The Fish Production Potential of the Baltic Sea*, Environmental Science and Engineering, DOI 10.1007/978-3-642-11562-2_5,

As mentioned, the last scenario is meant to give a more holistic management plan, and such a plan should also, we think, include some constructive actions, and not just costly nutrient reductions. So, we will introduce a scenario with fish cage farming and we will discuss criteria for when and where and how much fish (mainly rainbow trout) that may be produced in the Baltic Sea without clear negative ecosystem effects, but causing economic benefits, employment opportunities and more tax revenues.

The invasion of alien species and the extinction of traditional ones is, and should be, a major concern for Baltic Sea management. Invasions of jellyfish could potentially reduce Baltic Sea fishing, because jellyfish are not readily consumed by prey fish and the jellyfish would consume food (mainly zooplankton) that would otherwise sustain a higher fish production. We will use the jellyfish model discussed in Chap. 3 to address that problem.

The final scenario gives and motivates our suggestions to a strategy to find a solid scientific framework for a management plan for the Baltic Sea.

5.1 A Strategy for "Optimal" Nutrient Reductions

5.1.1 Introduction and Aim

Nutrient reductions are ultimately related to political decisions. One can safely assume that it is practically impossible to remediate all human emissions of nutrients to the Baltic Sea. The 15,000 t/year suggested by HELCOM (2007) represent a reduction of about 50% of the phosphorus transported via rivers/countries to the Baltic Sea. Only a relatively smaller part of the remaining anthropogenic nutrient fluxes can be reduced from countries that have already carried out costly measures to reduce nutrient discharges to the Baltic Sea. The scenario presented here is based on the same principles as the one given by Håkanson and Bryhn (2008b). This means that the "optimal" remedial strategy should account for the following arguments:

- It is sub-optimal to give reduction quotas to different countries (such a strategy is based on political considerations rather than science). Our strategy is based on the identified "hotspots", so the idea is to target on basins (generally estuaries) with a high eutrophication and reduce nutrient input to such systems. From the maps given in Chap. 2 (see Fig. 2.19 for chlorophyll and Fig. 2.20 for Secchi depth), one can identify the Gulf of Riga, the Gulf of Finland, the Oder and Vistula estuaries and the coastal area outside of Kaliningrad as hotspots.
- It is not realistic to carry out measures that would bring the Baltic Sea ecosystem including key structural and functional characteristics, functional groups and species back to the conditions as they were, say 100 years ago due to major changes in population structure, agriculture, species composition, fishing/trawling, etc. However, it would be possible to reduce nutrient inputs so that the Secchi depth in the Gulf of Finland could return to the values discussed in Chap. 4 and shown in Table 4.1, i.e., to about 7 m. To reach such a specific goal,

there must also be major phosphorus reductions not just in the rivers entering the Gulf of Finland, but also in the rivers entering the Baltic Proper, since the water and nutrient exchange between the Baltic Proper and the Gulf of Finland is intense (see Fig. 2.10).

In this scenario, we will present a realistic remedial scenario that would considerably improve the conditions in the Gulf of Riga, the Gulf of Finland as well as the Baltic Proper and the entire Baltic Sea.

The default conditions using the CoastWeb-model have already been described in detail for TP, SPM, chlorophyll, Secchi depth and all functional groups and organisms in Chap. 4 also showed the generally good correspondence between modeled values and empirical norms and empirical data.

Many tests have been carried out to find a strategy to reach the goal that the mean annual Secchi depth in the Gulf of Finland should be 7.0 m. The final results are given on a monthly basis in Fig. 5.1 for TP-concentrations in the surface-water layer

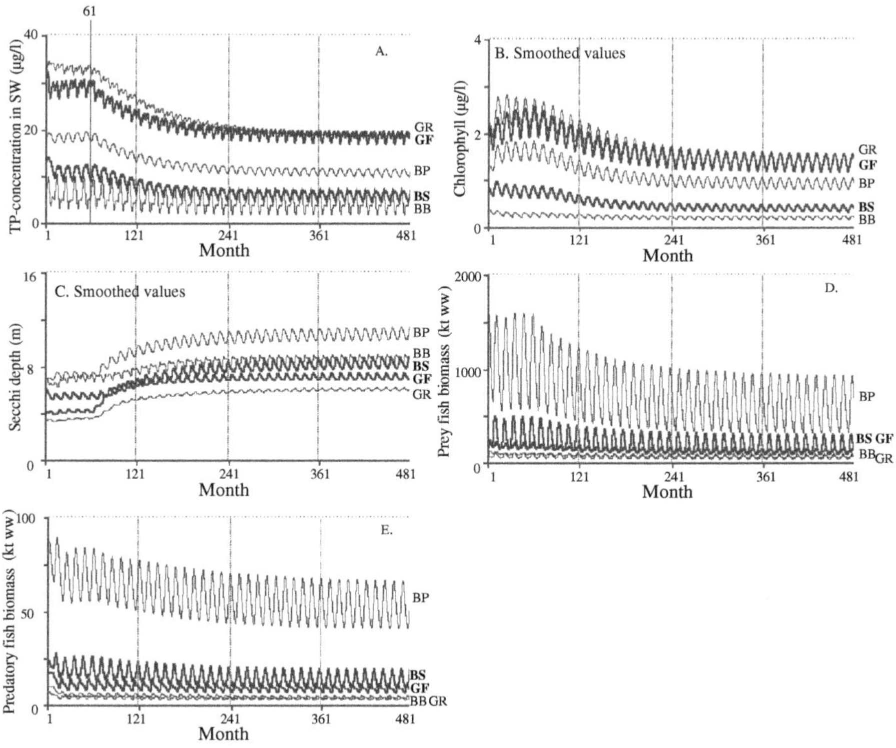

Fig. 5.1 "Optimal" TP-reductions. Modeled TP-concentrations (**a**), modeled smoothed values of chlorophyll (**b**) and Secchi depth (**c**), biomasses of prey fish (**d**) and predatory fish (**e**). The default conditions are shown for the first period (months 1–60), for the second period (61–481) the annual TP-inflow via tributaries to the Baltic Sea has been reduced by 8,730 t/year (according to Table 5.1). The goal is the get a Secchi depth of 7.0 m in the Gulf of Finland and this is achieved in this scenario

in all five Baltic Sea basins, for smoothed values of chlorophyll-a concentrations (B) and Secchi depths (C) and for the target biomasses of prey and predatory fish. The simulation period is 40 years and the reductions in TP-loading started month 61. Evidently, it is not realistic to implement such major TP-reduction suddenly, and these curves are meant to illustrate the dynamic response of the system in this hypothetical remediation. An important question is: How long would it take to reach a new stable situation? Figure 5.1 shows that it takes at least 20 years, and the factors regulating the response are essential parts of the CoastWeb-model.

One can note from these tests (see the compilation in Table 5.1), and also from Håkanson and Bryhn (2008b), that a reduction of 15,000 t/year of phosphorus to the Baltic Sea as suggested by HELCOM (see Table 2.6) and agreed upon by the Baltic Sea states in November 2007, would likely increase the Secchi depth in the Gulf of Finland beyond the mean or median values around the year 1900. This indicates that 15,000 t/year is likely an "overkill". A lowering of the primary production will imply also a reduction in the secondary production, including zooplankton and fish, and it would increase the concentration of organic toxins in fish – "in the clearest waters swim the most toxic fish". This is a well established fact called biological dilution (see Håkanson, 1999, 2000). It relates to the definition of the average concentration of toxins in fish, C = M/BM, where M is the total mass of a given toxin in fish (in g; e.g., total-PCB, total dioxins or methyl mercury) and BM is the total biomass of the fish (e.g., prey or predatory fish, or a given species of fish, such as cod; in kg). If BM decreases as it does in this oligotrophication scenario, C should increase if there are no simultaneous reductions in the loading (and mass) of toxins to the system. There is evidently no point in lowering the trophic status of the Baltic Sea system to levels where the environmental drawbacks become larger than the benefits, and every action could potentially imply benefits as well as drawbacks. So, one can conclude that 15,000 t/year of phosphorus may not be an optimal way to remediate the situation in the Baltic Sea.

Table 5.1 Mean annual Secchi depths (m) in the Bothnian Bay (BB), the Baltic Proper (BP), the Bothnian Sea (BS), the Gulf of Finland (GF) and the Gulf of Riga (GR) under default conditions and related to the optimal strategy, which means that there have been no reductions in the anthropogenic TP-loading via rivers to the Bothnian Bay and the Bothnian Sea, 38% reductions to BP, 60% reductions to GR and 70% reductions to GF. The goal is to achieved a mean Secchi depth of 7.0 m in the Gulf of Finland and to reach that goal a total of 8,730 t of TP should be reduced from the present TP-inflow (see Table 2.5)

	SecBB	SecBP	SecBS	SecGF	SecGR
Anthrop. TP-load (t), default	1,153	12,989	1,291	4,543	917
TP reduction (%/amount)	0	38/5,000	0	70/3,180	60/550
Default (m), modeled	6.3	6.7	4.8	4.0	3.5
Empirical, mean ± 1 SD	5.8±1.7	5.6±1.7	6.3±1.7	4.8±1.5	3.0±1.1
Optimal strategy (m)	8.6	10.7	8.1	7.0	5.9
Total reduction: 8,730 t TP					

The strategy we ask for should also concur with some evident practical constraints. For example, it is not really realistic to reduce all anthropogenic TP-discharges. Furthermore, it will become increasingly expensive to reduce the remaining tons for countries where major investments in nutrient reductions have already been made. So, in our search for an optimal strategy, we have set a limit to 70% TP-reductions in the anthropogenic emissions to the Gulf of Finland, the Gulf of Riga and from Poland, and a 30% reduction in the anthropogenic TP-emissions from Sweden, Finland and Germany. We will focus on the conditions in the Gulf of Finland and the Gulf of Riga, and not on smaller coastal areas, and not on the already low-productive basins (i.e., the Bothnian Bay and the Bothnian Sea).

From these presuppositions, we can present a more "optimal" strategy (see Table 5.1), which concerns a total reduction of 8,730 t/year of phosphorus (and no reductions in nitrogen emissions) and of these reductions 5,000 t/year (38% of anthropogenic emissions) are removed from the countries/rivers adding nutrients to the Baltic Proper, 3,180 t/year from the Gulf of Finland (corresponding to 70% of the anthropogenic input) and 550 t/year of TP to the Gulf of Riga (or 60% of the anthropogenic input to this basin). The results of this strategy largely corresponds with the results presented by Håkanson and Bryhn (2008b); the main difference is that is this scenario we reduce 1,600 kt less to the Baltic Proper, but 450 t TP more to the Gulf of Finland and 125 t more to the Gulf of Riga per year. This means that we achieve the same end result (7.0 m Secchi depth in the Gulf of Finland) but with 1,000 t less TP-reductions per year.

This would give a Secchi depth of 8.6 m in the Bothnian Bay, about 8.1 m in the Bothnian Sea, 5.9 m in the Gulf of Riga and 10.7 m in the Baltic Proper. The primary phytoplankton production would go down and the risks of harmful algal blooms (of cyanobacteria) would be significantly reduced, by a factor of 3–5 in GR, GF and BP (see Håkanson and Bryhn, 2008b). It is also interesting to note that the modeled Secchi depths are within 1 standard deviation of the empirical mean values (see Table 5.1) and from Fig. 5.1, one can see the dynamic response of the system. In this scenario with the hypothetical sudden reduction in a given month (month 61), one can identify two response phases, first an initial quick response of about 6 years and then a slower response. After about 20 years, the system has almost reached a new steady state. To reach a perfect steady-state takes longer in the sediments. The sediment compartments with the lowest deposition rates will have the slowest response to these changes, and vice versa.

It is interesting to see the results for prey and predatory fish in Fig. 5.1. One can note drastic reductions in prey fish biomass (Fig. 5.1d), but not in predatory fish biomass (Fig. 5.1e). Such a marked change in trophic state as this would be relatively favorable for the predatory fish. The general rule is that the more eutrophic the system becomes, relatively less of the more desirable predatory fish are produced and relatively more of the less desirable prey fish, such as roach, perch, etc. – and vice versa when the system is undergoing an oligotrophication such as this scenario.

5.1.2 Comments and Conclusions

In this section, we have challenged the wisdom of the HELCOM strategy to reduce the eutrophication in the Baltic Sea. Later in this chapter, we will discuss the costs gained in our scenario as compared to the HELCOM strategy. There are major differences if we just look at the figures. In our strategy, there are no nitrogen reductions at all and our strategy implies that 8,730 and not 15,000 t of TP should be reduced per year.

Nitrogen reductions may fail to give lower N/P-ratios in the water because of compensatory increases in the nitrogen fixation by cyanobacteria (see Håkanson and Bryhn, 2008b). If nitrogen reductions lower the N/P-ratios in the surface water, this could increase the competitiveness of cyanobacteria in relation to other algae even more, which is a clearly negative consequence of an expensive remedial strategy implemented to improve rather than worsen the conditions in the Baltic Sea. The results presented in this section indicate that a reduction of 15,000 t/year of phosphorus would likely create what may well be an undesired oligotrophication of the Baltic Sea system in the sense that the trophic status, as revealed by the operational bioindicators (Secchi depth and chlorophyll), would approach lower levels than Baltic Sea managers should seek, also considering the lower total fish production in the system and the higher probable levels of organic toxins in fish.

We have also presented and motivated an alternative remedial strategy to reduce the eutrophication in the Baltic Sea based on the following cornerstones:

1. Focus on the major hotspots, i.e., first improve the conditions in systems such as the Gulf of Finland and the Gulf of Riga.
2. Many remedial measures in agriculture, in urban areas or industry, would remove both nutrients and when substance-specific methods are available, they should target on phosphorus removal, because phosphorus has been shown to be the nutrient limiting primary production in the Baltic Sea over longer periods of time, such as the growing season. The effects of nitrogen reductions cannot be predicted with any certainty, but nitrogen reductions may increase the blooming of harmful algae (cyanobacteria), which should be avoided.
3. We have motivated a remedial strategy where 3,180 t/year (or 70% of the anthropogenic contributions) of the phosphorus to the Gulf of Finland, 550 t/year (or 60% of the anthropogenic contributions) to the Gulf of Riga and 5,000 t/year (or 38% of the anthropogenic contributions) to the Baltic Proper and no reductions at all to the Bothnian Sea and the Bothnian Bay would be carried out. Evidently, it would take a long time to implement such reductions and the Baltic Sea system could face several changes in that time (e.g., related to climatic variations). This means that these recommendations should be taken with due reservations and that they should be adjusted to such future changes. Such adjustments can also be elaborated using the CoastWeb-model.
4. The recommended remedial strategy would, under the given presuppositions, give an annual mean Secchi depth of about 7 m in the Gulf of Finland, of 8 m in the Bothnian Sea, of 8.5 m in the Bothnian Bay, 6 m in the Gulf of Riga and

10.5 m in the Baltic Proper. The large-scale annual prey fish production would go down (from about 625 kt ww in the Baltic Proper to about 420 kt ww and much less in the other Baltic Sea basins, see Sect. 5.3). The predatory fish production would go down from 75 to 72 kt/year in the Baltic Proper and relatively less in the other Baltic Sea basins.

5.2 How Can Variations in Fish Biomasses (Increases and Recovery After Heavy Fishing) Be Explained?

5.2.1 Introduction, Background and Aim

Relevant background information for the scenario presented in this section (the "oxygen" scenario) has been discussed in the following figures:

- Figure 1.9 gave information on the major intrusions of oxygen-rich salt-water from the Kattegat to the Baltic Proper. In this scenario, we will focus specifically on the period between 1970 and 2000 and one can note that there were massive salt-water intrusions between 1975 and 1983.
- Figure 1.10 shows the changes in the three dominating fish species, cod, sprat and herring in this period. Sprat, herring and small cod may be categorized as prey fish, while large cod is a predator. Key information from Fig. 1.10 for this scenario is that:

  1. In the Baltic Sea, the total prey fish biomass in the years around 1983 was about 2,500–5,000 kt ww but the fluctuations and uncertainties in the data are large.
  2. The total biomass of the predatory cod should be lower than the total cod biomass, by a factor of 2. This means that one should expect maximum biomasses of predatory fish in the Baltic Sea in the order-of-magnitude of 500 kt ww in the period around 1983.
  3. There is a very distinct pattern with maximum values for predatory cod around 1983; relatively low values from the mid-1990s and onwards and corresponding fluctuations for the prey fish eaten by predatory cod; the most evident changes are that the peak values for sprat logically correspond to the minimum values for cod.

  The aim of this scenario is to describe and explain these patterns and quantitative changes using the CoastWeb-model.
- Figure 3.17 gave the changes in total catch of cod (predatory cod plus cod in the category "prey fish") and in the permitted fish quota for cod in the Baltic Sea. From this figure, one can note that the maximum fish quota (TAC) in the Baltic Sea was about 210 kt ww in 1981 when the fish biomass reached very high values; the lowest fish quota up till the year 2003 was 45 kt ww. In the following scenario, we will test different realistic fish quota for the period and

stress the difference between the fishing rate and fish quota. One should also note the difference between total catch of landed fish and total fishing, which includes landed fish, illegal fishing, fishing by seals, birds, other animals and recreational fishing. The total fishing of predatory fish (including about 90% big cod) should be lower than the total catch of cod given in Fig. 3.17. If the maximum total catch of cod in the years around 1980 was about 1,000 kt ww/year, the model should predict values for total fishing of predatory fish in the Baltic Sea system that should agree with this amount accounting also for the fact that (1) not all cod are predatory, (2) that there are also significant amounts of unregistered and illegal fishing, (3) and also dumping of dead fish and that (4) there are other predatory fish in the system than big cod.

- Figure 2.35 gave the variations in empirical oxygen concentrations in the deep-water layer of Baltic Proper in the years between 1968 and 2008. One can note from this figure that:

1. Looking at the O_2-concentrations below the halocline at 75 m in the entire Baltic Proper, the values for the default period used in this modeling, i.e., for the year 1997–2006, are often relatively low. There are major variations around the mean value in the oxygen concentrations and it should be stressed that the zoobenthos will not survive if the oxygen level in the sediments (not the water) is lower than about 2 mg/l.
2. The oxygen consumption should be significantly higher in the sediments than in the water (see Fig. 2.18).
3. The oxygen concentration in the sediments will have a major impact on the production/survival of zoobenthos (see Fig. 3.4).
4. The intrusion of oxygenated high saline water from the Kattegat will mainly influence the conditions in the deep-water layer in the Baltic Proper (see Fig. 2.9) but also the entire system from mixing processes.

From this background, the aim of this scenario is to model the changes in oxygen concentrations in all five Baltic Sea basins related to the salt-water intrusions with a focus on the period around 1980 when very high fish biomasses have been documented. We will present a simple dynamic mass-balance model, which is meant to handle how these salt-water intrusions would affect first the oxygen concentrations, and then calculate how those changes would likely affect zoobenthos production, and also the production and biomasses of prey and predatory fish. We have already demonstrated that changes in salt-water inflow will also affect salinities, TP-concentrations, SPM-concentrations, Secchi depths and chlorophyll-a concentrations, and all other functional groups in the CoastWeb-model. The new part here concerns the oxygen sub-model and how it is linked to the CoastWeb-model. This will be explained in the following text.

So, the main aim of this section is to model and explain the very marked changes in fish biomasses and fishing, which took place in the years around 1980.

5.2.2 The Oxygen Sub-model

The model is shown in Fig. 5.2 and all equations are compiled in Table 5.2. Note that this is a simple mass-balance model (it is not a CoastMab-model for oxygen). It only calculates increases in O_2-concentrations relative to the default conditions. The oxygen model is based on the following presuppositions:

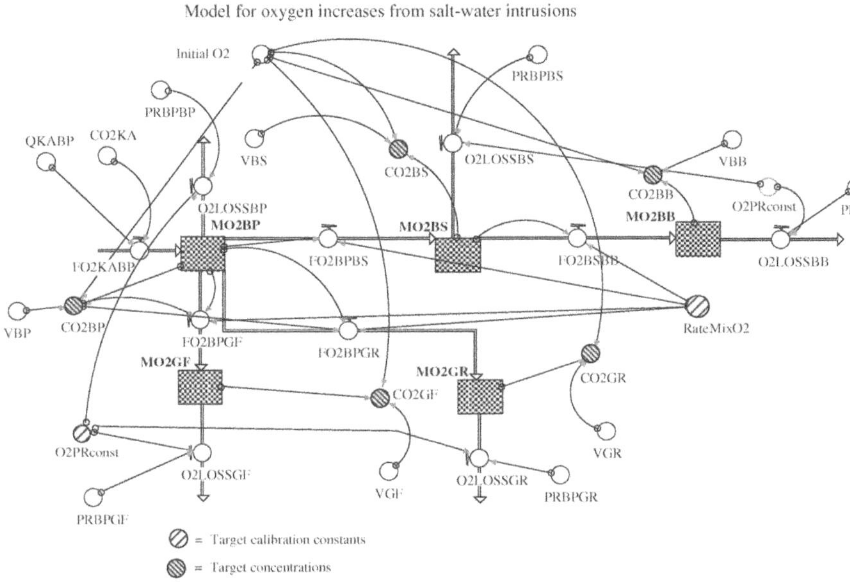

Fig. 5.2 Outline of the model to calculate oxygen increases in the five sub-basins of the Baltic Sea from salt-water intrusions from the Kattegat (KA)

1. It is basically meant to handle the salt-water intrusion for the period 1960–2010 shown in Fig. 5.3a (see also Fig. 1.9).
2. The water fluxes (e.g., Q_{DWBPGR} in m^3/month; i.e., water flow from DW in BP to GR) are calculated in the same way as used in the CoastMab-model for salt, TP and SPM.
3. The volumes of the five systems (e.g., V_{GR} in m^3) are also the same as used in the CoastWeb-model.
4. The oxygen consumption is calculated from the production of bacterioplankton (e.g., PR_{BPGR} in kg ww/month), which is calculated automatically in the CoastWeb-model.
5. The loss of oxygen related to mixing and all other processes (water-temperature changes, diffusion, etc.) is handled in a simplistic manner by means of the loss rate, R_{MixO2} (in 1/month). This is one of two calibration constants in the oxygen model. The higher the value of R_{MixO2}, the faster the loss of oxygen from any given compartment. The default value has been set to 0.001 (per month) to obtain realistic oxygen reductions in the five basins after a salt-water intrusion. The

Table 5.2 Compilation of equations for the mass-balance model for oxygen increases in the five basins in the Baltic Sea related to major intrusions of oxygenized salt water from the Kattegat (KA). The model calculates net transport between the compartments

Bothnian Bay (BB)

$M_{O2BB}(t) = M_{O2BB}(t-dt) + (F_{O2BSBB} - O2_{LOSSBB})dt$

$F_{O2BSBB} = M_{O2BS} \cdot Rate_{MixO2}$
$O2_{LOSSBB} = O2_{PRconst} \cdot PR_{BPBB}$

Baltic Proper (BP)

$M_{O2BP}(t) = M_{O2BP}(t-dt) + (F_{O2KABP} + F_{O2BPGF} - F_{O2BPGR} - F_{O2BPBS} - O2_{LOSSBP})dt$

$F_{O2KABP} = C_{O2KA} \cdot Q_{KABP}$
$F_{O2BPGF} = C_{O2BP} \cdot Q_{DWBPGF} + M_{O2BP} \cdot Rate_{MixO2}$
$F_{O2BPGR} = M_{O2BP} \cdot Rate_{MixO2}$
$F_{O2BPBS} = M_{O2BP} \cdot Rate_{MixO2}$
$O2_{LOSSBP} = O2_{PRconst} \cdot PR_{BPBP}$

Bothnian Sea (BS)

$M_{O2BS}(t) = M_{O2BS}(t-dt) + (F_{O2BPBS} - F_{O2BSBB} - O2_{LOSSBS})dt$

$F_{O2BPBS} = M_{O2BP} \cdot Rate_{MixO2}$
$F_{O2BSBB} = M_{O2BS} \cdot Rate_{MixO2}$
$O2_{LOSSBS} = O2_{PRconst} \cdot PR_{BPBS}$

Gulf of Finland (GF)

$M_{O2GF}(t) = M_{O2GF}(t-dt) + (F_{O2BPGF} - O2_{LOSSGF})dt$

$F_{O2BPGF} = C_{O2BP} \cdot Q_{DWBPGF} + M_{O2BP} \cdot Rate_{O2}$
$O2_{LOSSGF} = O2_{PRconst} \cdot PR_{BPGF}$

Gulf of Riga (GG)

$M_{O2GR}(t) = M_{O2GR}(t-dt) + (F_{O2BPGR} - O2_{LOSSGR})dt$

$F_{O2BPGR} = M_{O2BP} \cdot Rate_{MixO2}$
$O2_{LOSSGR} = O2_{PRconst} \cdot PR_{BPGR}$

Model variables

$C_{O2BB} = M_{O2BB}/V_{BB}$
$C_{O2BP} = M_{O2BP}/V_{BP}$
$C_{O2BS} = M_{O2BS}/V_{BS}$
$C_{O2GF} = M_{O2GF}/V_{GF}$
$C_{O2GR} = M_{O2GR}/V_{GR}$
$C_{O2KA} = 10$ (mg/l)
$O2_{PRconst} = 3.8$ (g O_2/kg ww)
$PR_{BPGF} = M_{BPGF}/T_{BP}$ (kg ww/month); $PR_{BPBB} = M_{BPBB}/T_{BP}$ (kg ww/month); $PR_{BPBP} = M_{BPBP}/T_{BP}$ (kg ww/month)
$PR_{BPBS} = M_{BPBS}/T_{BP}$ (kg ww/month); $PR_{BPGR} = M_{BPGR}/T_{BP}$ (kg ww/month)
Q_{KABP} = if Salscen = 0 then $(345/12) \times 10^9$ else $(\text{Salscen}+345/12) \times 10^9$ (m^3/month); Salscen gives salt-water intrusions from Fig. 1.9.

$Rate_{MixO2} = 0.001$ (1/month)
$V_{BB} = 1{,}500 \times 10^9$ (m^3); $V_{BP} = 13{,}055 \times 10^9$ (m^3); $V_{BS} = 4{,}889 \times 10^9$ (m^3); $V_{GF} = 1{,}073 \times 10^9$ (m^3); $V_{GR} = 409 \times 10^9$ (m^3)

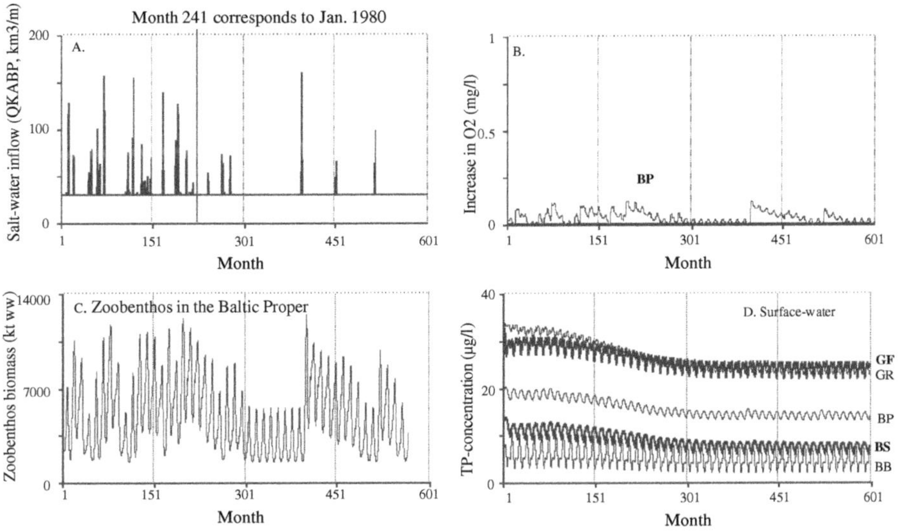

Fig. 5.3 Results from the "oxygen" scenario (600 months from 1960 to 2010). (**a**) Gives the major salt-water intrusions to the Baltic Proper from the Kattegat. This scenario focuses on explaining the drastic increases in predatory fish biomasses for the years around 1980 (month 241 in this scenario), (**b**) gives the corresponding increases in the oxygen concentrations in the Baltic Proper (there are only marginal changes in the Gulf of Riga and the Gulf of Finland), (**c**) the modeled increases in zoobenthos biomass in the Baltic Proper, and (**d**) gives the corresponding changes in TP-concentrations in the SW-layers in all five Baltic Sea basins

oxygen will decrease at a higher rate than the salinity after a given salt-water intrusion.

6. Under default conditions, when there are no salt-water intrusions but an annual inflow of salt water of 345 km^3/year, there are relatively low oxygen concentrations in all the five deep-water layers. Since the conditions in the Baltic Proper are most important in this scenario, since this is the largest basin, which also actually receives the salt-water from Kattegat, the following results focus on the conditions in the Baltic Proper. The oxygen model predicts the *added oxygen* from the salt-water intrusions.
7. The second calibration constant is related to the oxygen consumption from the production of bacterioplankton. During a month, there would be several generations of bacterioplankton consuming organic matter and using oxygen. The constant called oxygen consumption constant, $O2_{PRconst}$ (in mg/(kg ww·month)), is related to the bacterioplankton production and has been determined from calibrations. The higher the values for $O2_{PRconst}$, the greater the oxygen consumption by bacterioplankton. The default value for $O2_{PRconst}$ is 3.8 (g O_2/kg ww).
8. So, the target variable in this model is the added oxygen concentration in the Baltic Proper (C_{O2BP} in mg/l) related to the given salt-water intrusions.

Figure 5.3b gives the modeled changes in oxygen concentrations in the Baltic Proper. One can note that there are only very small changes in the other basins and that the salt-water intrusions given in Fig. 5.3a will increase the oxygen concentrations in the Baltic Proper by about 0.1 mg/l during the period around 1980, as compared to the default period.

The critical question now is: How would this influence the Baltic Sea fishery? Can the documented variations in fish biomass be explained? By accounting for this, would the total fishing of predatory fish agree with the actual documented fishing shown in Fig. 3.17 and the biomasses shown in Fig. 1.10?

The quantitative link between the modeled increases in oxygen concentrations (C_{O2}) and the initial production of zoobenthos has been modified by including the term, $Y(C_{O2})$, in Eq. (4.41). That gives:

$$IPR_{SedZB} = Y(C_{O2}) \cdot M_{Sed} \cdot DC_{ZB1} \cdot NCR_{ZB} \cdot MER_{SedZB} \cdot Y_{Temp}^{0.25} \cdot (ET + (1 - ET) \cdot Y_{Eh})) \qquad (5.1)$$

Where

$$Y(C_{O2}) = SMTH((0.01 + C_{O2BP})/0.01, T_{ZB}, 1) \qquad (5.2)$$

The average time in the smoothing function is set to the turnover time for zoobenthos (T_{ZB}); the initial value is 1 and the file to be smoothed gives the ratio between modeled O_2-concentrations and the boundary O_2-value of 0.01 (mg/l) for the default period ((0.01+ C_{O2BP})/0.01). So, this is dimensionless moderator meant to quantitatively describe variations in initial zoobenthos production from variations in C_{O2}.

5.2.3 Results

The results are shown in Fig. 5.3c for the zoobenthos biomass in the Baltic Proper. One can see a significant increase in zoobenthos biomass, which will continue to about 1990 (month 361) related to the salt-water intrusions in Fig. 5.3a and the corresponding changes in oxygen concentrations shown in Fig. 5.3b. Note that many changes might have happened to the Baltic Sea system during these decades related to variations in water temperatures, tributary water discharges, prevailing wind conditions and foodweb characteristics, which are not handled in this scenario. This is a sensitivity analysis where we have changed the salt-water inflow and everything else has been kept constant. Under these presuppositions, one can also note from Fig. 5.3d, that there would be clear reductions in TP-concentrations in all surface-water compartments in the Baltic Sea basins. This was discussed in Chap. 4 (the salinity scenario).

Figure 5.4 gives the results for prey and predatory fish. There are major increases in the prey fish biomass, mainly in the Baltic Proper (Fig. 5.4a), and also increases in predatory fish biomass, also primarily in the Baltic Proper (Fig. 5.4b). Figure 5.4c, d give the changes in prey fish biomass and the total fishing of predatory cod in all

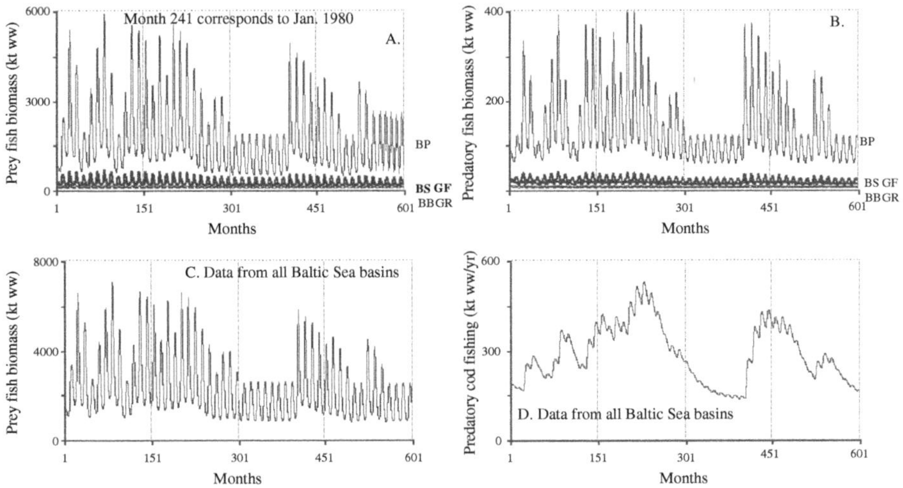

Fig. 5.4 More results from the "oxygen" scenario. (**a**) Gives the modeled increases in prey fish biomass in the Baltic Sea basins, (**b**) gives the corresponding information for predatory fish, (**c**) gives the changes in total prey fish biomass in all five Baltic Sea basins, and (**d**) gives the corresponding changes in total default fishing of predatory cod in all five Baltic Sea basins

five basins. These values can now be checked against the empirical values discussed in the introduction of this scenario.

1. One should expect that the maximum total prey fish biomass would be about 2,500–5,000 kt ww according to Fig. 1.10. This modeling gives values corresponding to that, which is good given the fluctuations and uncertainties in the empirical data.
2. One should expect maximum biomasses of predatory fish in the Baltic Proper in the order-of-magnitude of 400 kt ww and this is also what this modeling gives.
3. The official total maximum catch of cod in the Baltic Sea in the years around 1980 was 1,000 kt/year (see Fig. 3.17), and this is also what this modeling gives, as calculated from 0.86·0.5·0.1·TPDfish, where 0.86 is the average fraction of cod of the predatory fish, 0.5 is reduction in total predatory fishing (TPDfish) related to illegal fishing plus dumping of already caught fish onboard the fishing vessels and 0.1 is fishing by birds, seals, etc. All these figures have already been discussed and from Fig. 5.4d one can note that they are also well motivated by these CoastWeb-calculations.

There should also be a distinct pattern with maximum values for predatory cod around 1983 and minimum values from the late-1990s and onwards and corresponding fluctuations for the prey fish eaten by predatory cod (i.e., sprat, herring and small cod). To study this, we have in Fig. 5.5 also differentiated between the fishing rate and the fish quota. The results in Fig. 5.5 are based on the default conditions for the CoastWeb-model with salt-water intrusions and we have used the sub-model for

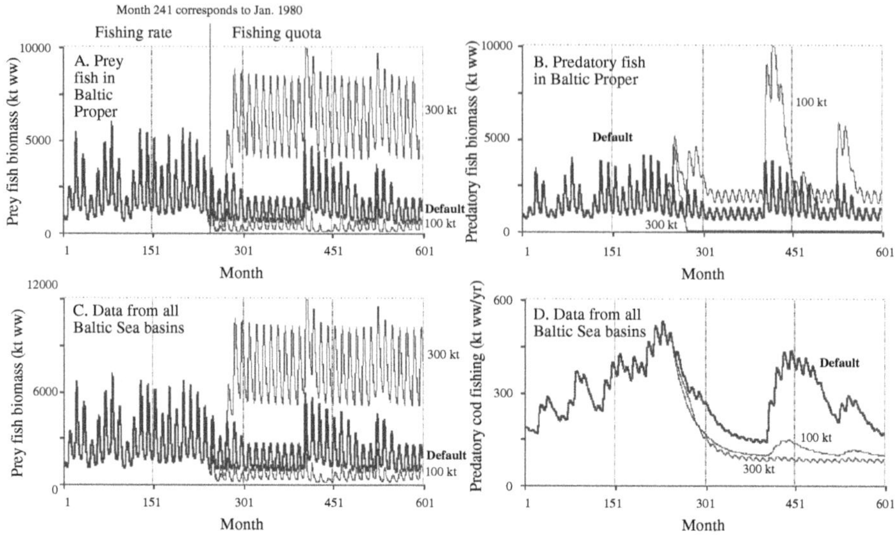

Fig. 5.5 More results from the "oxygen" scenario when also the fishing rate has been changed to a total fishing amount month 241 (100 and 300 kt ww of predatory fish being caught annually in the Baltic Proper alone; month 241 corresponds to January 1980). (**a**) Gives the modeled increases in prey fish biomass in the Baltic Proper, (**b**) gives the corresponding information for predatory fish in the Baltic Proper, (**c**) gives the changes in total prey fish biomass in all five Baltic Sea basins, and (**d**) gives the corresponding changes in total fishing of predatory cod in all five Baltic Sea basins

oxygen. This means that all initial values have been based on the default fishing rate (see Eq. 3.51). So, we have used the default fishing rate until month 241 (i.e., January of 1981) and then replaced the fishing rate by an annual total fishing in the Baltic Proper, first of 300 kt ww/year of predatory fish, and then of 100 kt ww/year of predatory fish in BP.

For comparative purposes, this figure also gives the default results for prey and predatory fish from Fig. 5.4 for the entire 50-year period (the bolded curves in Fig. 5.5). We can note major increases in the prey fish biomass in the Baltic Proper (Fig. 5.5a) when the predatory fish stock has collapsed due to too heavy fishing (Fig. 5.5b; 300 kt ww/year all years after month 241). The low predation pressure on prey fish explains the dramatic increase in prey fish biomass. When 100 kt ww of predatory fish are caught after month 241, the system is in better balance and the values are more close to those given by the default condition.

5.2.4 Comments

This scenario has demonstrated that the marked intrusions of oxygen-rich salt-water in the years prior to 1983 caused clear increases in oxygen concentrations in the deep-water compartments of mainly the Baltic Proper (these results may be compared with the empirical data given in Fig. 2.29b). The increased oxygen concentrations have been dynamically modeled using the new sub-model for oxygen presented in Fig. 5.2 and Table 5.2. The higher oxygen concentrations in the basins lead to significant increases in the production and biomass of zoobenthos (mainly in the Baltic Proper). The salt-water intrusions would also be beneficial for the survival of the cod roe and the increased biomasses of zoobenthos would mean more food for all types of benthivorous prey fish, including small cod. The zoobenthos biomass in the Baltic Proper has according to these calculations varied by a factor of 2, from about 3,500 kt ww under default condition to about 7,000 kt ww in the years around 1980 compared to the years around 2000. This means that the biomass of prey fish in the Baltic Proper has increased from about 2,000 kt ww to about 4,000 kt ww; the corresponding increases in predatory fish biomass in the Baltic Proper are from about 100 kt ww to about 200 kt ww. A part of the change in predatory fish biomass is connected to the lower trophic state related to the salt-water intrusions; the TP-concentration in the Baltic Proper has decreased from about 20 to 17 μg/l.

It is interesting to note the difference between a fishing rate and a fishing quota. The fishing quota represents a defined removal in kt of fish and if the fish biomass goes down due to changes in nutrient loading, salinity or water temperature variations, one can reach a state where the fishing pressure causes the system to collapse, as shown in Fig. 5.5 using a constant fishing of 300 kt ww/year of predatory fish. In the next section we will discuss how optimal fishing rates may be established so that such disasters may be avoided.

5.3 A Strategy to Set "Optimal" Fishing Rates and Fish Quota

5.3.1 Background and Aim of This Scenario

The results presented in Table 3.7 can be seen as an introduction to this scenario. The aim here is to discuss in more detail how an optimal fishing rate may be defined for prey and predatory fish in all Baltic Sea basins. First, under default conditions and then in the situation defined in the first section of this chapter (i.e., the "optimal" strategy related to phosphorus reductions).

The main point in this scenario is to stress by quantitative data that the fish quota should be adjusted to variations in abiotic variables regulating fish production and biomass, and that changes also in fishing pressure will have consequences for the fishing efforts, i.e., the time and money that mainly professional fishermen need to spend to get the amount of fish regulated by the fish quota.

5.3.2 Results Scenario 1 – Default Conditions

Table 5.3 gives the results for predatory fish in all Baltic Sea basins and Table 5.4 the corresponding results for prey fish under default condition. These tables give

Table 5.3 Calculated annual biomasses (mean values in kt ww) and production values of predatory fish (kt ww/year) and fluxes (initial production, elimination, fishing, in- and outmigration) of predatory fish in the Baltic Sea basins for different fishing rates under default conditions. The fishing rate constant has the dimension 1/month. Simulation time 481 months

Fishing rate constant (1/month)	Mean biomass BM_{PD} (kt ww)	IPR_{PD} (kt ww/year)	$Elim_{PD}$ (kt ww/year)	$Fishing_{PD}$ (kt ww/year)	MIG_{InPD} (kt ww/year)	MIG_{OutPD} (kt ww/year)	Production BM_{PD}/T_{PD} (kt ww/year)
A. Baltic Proper							
0.5 (default)	67	143	27	118	7.7	5.5	27
0.75	49	113	20	95	6.3	4.0	20
1	39	93	16	79	5.4	3.2	16
2	21	54	9	48	3.9	1.7	8.6
0.25	104	194	42	141	5.9	16	42
0.1	166	231	67	144	7.9	28	67
B. Bothnian Sea							
0.5 (default)	19	60	7.6	63	13.3	2.2	7.6
0.75	14	48	5.6	51	11	1.3	5.6
1	11	40	4.4	42	8.3	1.0	4.4
2	5.8	23	2.3	25	4.5	0.5	2.3
0.25	29	83	12	77	12	5.4	12
0.1	48	91	19	81	19	9.4	19
C. Bothnian Bay							
0.5 (default)	3.8	9.8	1.5	14	6.1	0.7	1.5
0.75	2.7	7.4	1.1	10	4.4	0.5	1.1
1	2.1	5.8	0.8	8.0	3.4	0.4	0.8
2	1.1	3.1	0.4	4.3	1.8	0.2	0.4
0.25	6.1	14	2.5	17	6.9	1.6	2.5
0.1	10	19	4.1	19	8.0	3.4	4.1
D. Gulf of Riga							
0.5 (default)	5.0	21	2.0	15	14	0.6	2.0
0.75	3.7	2.8	1.5	11	10	0.4	1.5
1	3.0	2.3	1.2	8.7	7.9	0.4	1.2
2	1.8	1.5	0.7	4.8	4.3	0.2	0.7
0.25	7.7	5.4	3.1	23	21	0.9	3.1
0.1	11	7.3	4.3	19	18	2.3	4.3
E. Gulf of Finland							
0.5 (default)	10	8	4.1	38	37	3.2	4.1
0.75	7.0	5.8	2.8	28	27	2.2	2.8
1	5.4	4.5	2.2	22	21	1.7	2.2
2	2.8	2.4	1.1	12	12	0.9	1.1
0.25	15	12	6.2	43	44	7.1	6.2
0.1	24	19	9.7	42	47	14	9.7

Table 5.4 Calculated annual biomasses (mean values in kt ww) and production values of prey fish (kt ww/year) and fluxes (initial production, elimination, fishing, in- and outmigration) of predatory fish in the Baltic Sea basins for different fishing rates under default conditions. The fishing rate constant has the dimension 1/month and the default value is 0.5

Fishing rate constant	Biomass BM_{PY}	IPR_{ZBPY}	IPR_{ZHPY}	IPR_{ZPPY}	From PY to PD	$Elim_{PY}$	$Fishing_{PY}$	MIG_{InPY}	MIG_{OutPY}	Production (BM_{PY}/T_{PY})
A. Baltic Proper										
0.5	1,082	2,940	1,747	181	605	1,316	956	73	134	1,317
0.75	1,166	3,109	1,901	186	479	1,418	1,136	74	149	1,418
1	1,213	3,204	1,988	188	394	1,475	1,247	74	158	1,476
2	1,286	3,347	2,120	191	229	1,564	1,451	73	173	1,565
0.25	954	2,675	1,512	172	822	1,160	649	72	115	1,161
0.1	714	2,170	1,084	147	980	868	309	68	79	869
B. Bothnian Sea										
0.5	285	984	653	52	259	348	495	154	36	347
0.75	309	1,054	720	52	207	377	594	164	39	376
1	324	1,096	760	52	171	395	655	169	42	395
2	349	1,160	822	52	100	426	761	174	46	425
0.25	258	918	590	52	362	315	346	137	32	314
0.1	179	687	382	46	400	218	153	105	21	217
C. Bothnian Bay										
0.5	67	206	74	5	50	81	122	61	13	81
0.75	71	212	79	5	38	87	137	64	14	87
1	73	214	81	5	30	89	144	64	15	89
2	76	218	84	5	16	93	155	64	17	93
0.25	62	199	69	5	74	76	90	53	11	76
0.1	49	176	51	5	97	60	47	37	8	60

Table 5.4 (continued)

Fishing rate constant	Biomass BM_{PY}	IPR_{ZBPY}	IPR_{ZHPY}	IPR_{ZPPY}	From PY to PD	$Elim_{PY}$	$Fishing_{PY}$	MIG_{InPY}	MIG_{OutPY}	Production (BM_{PY}/T_{PY})
D. Gulf of Riga										
0.5	85	110	48	7	17	103	124	148	13	103
0.75	88	112	49	7	13	106	126	148	14	106
1	89	113	50	8	11	109	125	148	15	109
2	94	117	53	8	7	114	122	143	17	114
0.25	80	107	45	7	26	98	115	143	11	98
0.1	80	106	45	7	35	97	72	109	10	97
E. Gulf of Finland										
0.5	161	283	102	9	42	195	293	327	79	196
0.75	165	288	106	9	30	200	314	339	83	200
1	167	290	108	9	23	202	326	346	84	203
2	169	294	111	9	12	206	345	357	87	206
0.25	164	285	104	10	66	199	229	288	80	199
0.1	159	277	96	10	100	194	139	229	73	194

mean annual biomasses and production values (initial production and production = PR/T), elimination (death), total fishing (i.e., all forms of fishing by man, birds, and animals, legal and illegal fishing) and in- and outmigration of fish. Five different fishing rates have been tested, 0.5 (the default value), 0.75, 1, 2, 0.25 and 0.1 (the dimension for the fishing rate is 1/month). The basic aim of this scenario is to highlight how different fishing rates influence the production, biomasses and amounts of fish that man can take out of the systems. It should be stressed that the legal fishing quota 2009 was 45 kt ww for the Baltic Sea, but that the permitted quota is only one part of the total fishing calculated in this modeling.

- There is also illegal fishing, which may be 20–50% of the permitted quota and in the following we will set this value to 35% as a default value.
- Dumping at sea of dead fish – this is, unfortunately, a well documented practice and it is done, for example if the quota is already reached or if the fish is deemed too small for the market. We will set the illegal dumping to 15% of the total fishing.
- Fishing by birds, seals and other animals, which may be up to about 10% of the total fishing.
- There are also uncertainties in all empirical data related to fishing and fish biomasses.

So, from this, it is evident that one cannot set the fish quota equal to the calculated total fishing using the CoastWeb-model. The fish quota should be somewhere around 25–40% of the total fishing.

From Tables 5.3 and 5.4, one may note:

- There are major differences among the main basins related to the production potential for fish depending on, e.g., the fishing rate. Most predatory fish are produced in the Baltic Proper – the annual production varies from 9 to 67 kt ww; the biomasses from 21 to 166 kt ww. The smallest basin, the Gulf of Riga, has a relatively low production potential for predatory fish – the production values vary from 0.7 to 4.3 kt ww/year; the biomasses from 1.8 to 11 kt ww. The Baltic Proper also has the highest production of prey fish (from 869 to 1,565 kt ww/year) and the highest biomasses of prey fish (714 to 1,290 kt ww). The lowest production for prey fish is found in the Bothnian Bay, which has the lowest trophic level (oligotrophic conditions favor production of predatory fish rather than prey fish). The production of prey fish in the Bothnian Bay is about 60–93 kt ww/year.
- As stressed in Chap. 3, the initial production is higher than the production because the initial production does not include predation.
- In- and outmigration are relatively important in all basins for both prey and predatory fish, but highest for the Gulf of Finland. There is a comparatively high net inmigration of prey and predatory fish to the Gulf of Riga.
- The elimination of predatory fish is by definition identical to production and may be higher or lower than the total fishing in all basins for both prey and predatory fish.

- The biomass of prey fish is a factor of 15–20 higher than for predatory fish in the basins with the highest trophic level (e.g., the Gulf of Riga), and a factor of 3–4 higher in the basin with the lowest trophic level (e.g., the Bothnian Bay) under default conditions.
- Zoobenthos is the most dominating food for prey fish in all Baltic Sea basins and predatory zooplankton the smallest contributor to prey fish production in all basins.

5.3.3 Results Scenario 2 – Conditions Defined by the "Optimal" Phosphorus Reduction Strategy

We have run the CoastWeb-model according to the scenario given in the first section of this chapter. The goal of that scenario was to find an approach so that the Secchi depth in the Gulf of Finland could reach 7.0 m, as it was between 1900 and 1920 before the more severe eutrophication started in the Baltic Sea. We have motivated a remedial strategy where 3,180 t/year of the phosphorus to the Gulf of Finland, 550 t/year to the Gulf of Riga and 5,000 t/year to the Baltic Proper and no reductions at all to the Bothnian Sea and the Bothnian Bay would be carried out.

Tables 5.5 and 5.6 give the results for predatory and prey fish in all Baltic Sea basins. These tables provide similar information as Tables 5.3 and 5.4, i.e., mean annual biomasses and production values, elimination, total fishing and in- and out-migration of fish. The same fishing rates as in scenario 1 have been tested also in this scenario. The aim is to study how the fish production potential of the Baltic Sea system would change if the system would be more oligotrophic. We can note that:

- The biomasses and production values would go down, as compared to the default conditions, by about 20% for predatory fish; the reductions in prey fish biomasses are higher, about 40%. As stressed, the general rule is that oligotrophic conditions favor the production of predatory fish, and vice versa.
- The same observations can be made concerning the role of the initial production relative to the production, in- and outmigration, elimination in relation to total fishing and the very important role of the zoobenthos for the fish production in all Baltic Sea basins.

5.3.4 Fishing Rate and Fish Quota

These results can also be used to discuss the setting of optimal fish quota in relation to the fish production potential of the system. We have shown that there is an optimal fishing rate. This has several implications:

1. There is a maximum amount of fish that can be removed from a given system corresponding to a given fishing rate. For example, under default condition this

Table 5.5 Calculated annual biomasses (mean values in kt ww) and production values of predatory fish (kt ww/year) and fluxes in kt (initial production, elimination, fishing, in- and outmigration) of predatory fish in the Baltic Sea basins for different fishing rates if the conditions change according to the optimal tributary nutrient reductions given in Sect. 5.1 (a reduction in riverine TP-inflow with 8,730 t/year). The fishing rate constant has the dimension 1/month (default value 0.5)

Fishing rate constant (1/month)	Mean biomass BM_{PD} (kt ww)	IPR_{PD} (kt ww/year)	$Elim_{PD}$ (kt ww/year)	$Fishing_{PD}$ (kt ww/year)	$MIGIn_{PD}$ (kt ww/year)	$MIGOut_{PD}$ (kt ww/year)	Production BM_{PD}/T_{PD} (kt ww/year)
A. Baltic Proper							
0.5	55	138	22	116	5.6	5.7	22
0.75	41	113	17	99	5.2	3.3	17
1	33	96	13	84	4.6	2.6	13
2	18	58	7.4	53	3.5	1.5	7.4
0.25	83	176	34	133	4.2	14	34
0.1	130	198	53	130	6.1	22	53
B. Bothnian Sea							
0.5	13	47	5.4	50	10	1.6	5.4
0.75	9.7	38	3.9	41	8.0	1.1	3.9
1	7.7	32	3.1	35	6.7	0.7	3.1
2	4.2	20	1.7	21	3.7	0.4	1.7
0.25	20	63	8.2	60	8.8	3.7	8.2
0.1	33	67	13	61	14	6.4	13
C. Bothnian Bay							
0.5	2.9	8.2	1.2	11	4.2	0.5	1.2
0.75	2.1	6.3	0.8	8.2	3.1	0.4	0.8
1	1.6	5.1	0.6	6.6	2.4	0.3	0.6
2	0.8	2.8	0.3	3.6	1.3	0.1	0.3
0.25	4.6	12	1.9	14	4.8	1.2	1.9
0.1	7.7	15	3.1	15	5.4	2.6	3.1
D. Gulf of Riga							
0.5	3.8	4.4	1.6	14	11	0.4	1.6
0.75	2.9	3.5	1.2	10	8.2	0.3	1.2
1	2.4	3.0	1.0	8.3	6.5	0.3	1.0
2	1.5	1.9	0.6	4.8	3.6	0.2	0.6
0.25	5.8	6.2	2.4	19	16	0.8	2.4
0.1	8.0	8.0	3.2	17	14	1.8	3.2
E. Gulf of Finland							
0.5	8.1	10	3.3	34	30	2.6	3.3
0.75	5.7	7.5	2.3	25	22	1.8	2.3
1	4.4	5.9	1.8	20	18	1.4	1.8
2	2.3	3.2	0.9	11	9.8	0.7	0.9
0.25	11	15	4.6	33	29	6.1	4.6
0.1	20	23	8.0	40	36	12	8.0

value is about 140 kt/year of predatory fish in the Baltic Proper at a fishing rate of 0.1–0.25; 43 kt/year of predatory fish in the Gulf of Finland at a fishing rate of 0.25. If the system would become more oligotrophic (as defined by the given scenario in Table 5.5), the maximum amount of fish that can be taken out of the

Table 5.6 Calculated annual biomasses (mean values in kt ww) and production values of prey fish (kt ww/year) and fluxes in kt (initial production, elimination, fishing, in- and outmigration) of predatory fish in the Baltic Sea basins for different fishing rates if the conditions change according to the optimal tributary nutrient reductions given in Sect. 5.1. The fishing rate constant has the dimension 1/month (default value 0.5)

Fishing rate constant	Biomass BM_{PY}	IPR_{ZBPY}	IPR_{ZHPY}	IPR_{ZPPY}	From PY to PD	$Elim_{PY}$	$Fishing_{PY}$	MIG_{InPY}	MIG_{OutPY}	Production (BM_{PY}/T_{PY})
A. Baltic Proper										
0.5	634	2,050	1,022	124	582	770	678	52	76	771
0.75	689	2,172	1,127	129	476	837	832	54	84	838
1	726	2,252	1,197	132	402	881	937	54	90	883
2	788	2,386	1,316	136	246	957	1,144	56	100	959
0.25	537	1,834	842	114	742	652	432	51	62	653
0.1	387	1,486	575	93	833	470	194	49	39	471
B. Bothnian Sea										
0.5	162	647	371	35	201	197	318	89	20	197
0.75	177	694	415	35	164	216	389	96	23	216
1	186	723	443	35	138	227	436	101	24	227
2	204	776	493	35	83	249	526	108	27	249
0.25	143	598	326	34	270	173	213	76	18	174
0.1	97	452	202	29	289	117	91	56	11	117
C. Bothnian Bay										
0.5	50	172	59	5	42	61	95	34	9	61
0.75	54	179	64	5	32	65	109	37	10	65
1	56	183	67	5	26	68	117	39	10	68
2	58	186	70	5	14	71	128	39	12	71
0.25	46	165	52	4	61	55	69	28	8	55
0.1	36	147	39	4	80	44	36	19	6	44

Table 5.6 (continued)

Fishing rate constant	Biomass BM_{PY}	IPR_{ZBPY}	IPR_{ZHPY}	IPR_{ZPPY}	From PY to PD	$Elim_{PY}$	$Fishing_{PY}$	MIG_{InPY}	MIG_{OutPY}	Production (BM_{PY}/T_{PY})
D. Gulf of Riga										
0.5	53	88	31	5	21	65	92	97	7	65
0.75	56	90	33	5	17	68	96	99	8	68
1	57	91	34	5	14	70	98	99	8	70
2	61	94	36	5	9	74	99	98	10	74
0.25	49	84	28	5	29	60	79	90	6	60
0.1	47	80	26	5	38	57	48	68	5	57
E. Gulf of Finland										
0.5	109	242	68	6	53	132	223	214	47	132
0.75	112	246	72	6	39	137	244	224	51	137
1	114	249	74	6	31	139	256	230	53	139
2	118	253	77	6	17	143	278	241	57	143
0.25	113	247	71	6	77	137	162	178	49	137
0.1	100	233	61	6	119	121	100	145	37	121

system would be lower – 133 kt/year of predatory fish in the Baltic Proper at a fishing rate of 0.25 and 19 kt/year of predatory fish in the Gulf of Riga at a fishing rate of 0.25.

2. The corresponding values for prey fish are: (1) under default condition 956 kt/year in the Baltic Proper at a fishing rate of 0.5 and as little as 309 kt/year at a fishing rate of 0.1; and 293 kt/year of prey fish in the Gulf of Finland at a default fishing rate of 0.5; (2) under more oligotrophic conditions as little as 194 kt/year of prey fish in the Baltic Proper at a fishing rate of 0.1 because then the biomass of predatory fish would be very high (130 kt ww), and hence also the predation pressure on the prey fish; and, for example, only 91 kt/year of prey fish in the Bothnian Sea at a fishing rate of 0.1.

5.3.5 Summary and Comments

Intensified fishing of predatory fish means a lower remaining biomass of predatory fish and a lower predation pressure on the prey fish. This is easy to state and understand, but it is more important and crucial for Baltic Sea management to dispose of a model, which enables realistic and dynamic quantitative predictions of these complicated relationships. The CoatsWeb-model is intended as a tool for such calculations.

We suggest the following simple formula to estimate a realistic fish quota from the fishing rate regulating the total fishing. The same approach may be used for all basins and both for prey and predatory fish. The focus in this context would, however, generally be set on the predatory fish, which should be the most important functional group of fish in contexts of professional fishing, but the same principle applies also to prey fish. For example for fish in the Baltic Proper ($\mathrm{FISH_{PDBP}}$), we have:

$$\begin{aligned}&\text{Fish quota for predatory fish in the Baltic Proper}\\&\quad = \mathrm{FISH_{PDBP}} \cdot (1 - 0.35 - 0.15 - 0.1 - 0.1) = 0.3 \cdot \mathrm{FISH_{PDBP}}\end{aligned} \tag{5.3}$$

The factor 0.35 in Eq. (5.3) is a default reduction related to illegal fishing; 0.15 is a reduction related to fish caught in the net and dumped dead at sea because the quota for the given boat is reached and the fisherman may expect inspection of the load. Today these values may be even higher, but no one knows for sure. The mass-media debate, several debate books and the intensive discussions have meant that the instruments to combat illegal fishing are probably sharper today than a few years ago. One can presuppose that there will always be illegal fishing in the Baltic Sea; 0.1 in Eq. (5.3) is fishing from all other causes, such as seals and birds; the last 0.1 factor has been added to stress the precautionary principle – there are always uncertainties in these complicated ecosystem calculations.

From the regular fish statistics for the Baltic Sea kept by ICES, one can see that for the given period of time about 86% of the predatory fish in the Baltic Sea would be large cod. This would imply that the fish quota for large predatory cod would be 0.86·0.3·$FISH_{PDBP}$ (= 0.26 or ≈ 0.25 of $FISH_{PDBP}$) and $FISH_{PDBP}$ is given by the CoastWeb-model. This would be a complementary method to set a fish quota, as compared to the standard approach based on fish statistics used today. Both these methods have inherent uncertainties but together they would provide a better base for determining a more reliable fish quota, which would also take into account the fish production potential of the system and the factors included in the CoastWeb-model on that matter.

We have shown that there are major differences among the basins in their fish production potential due to variations in size, nutrient loading, water temperatures, oxygen status and salt-water intrusions. Such variations in abiotic governing factors should warrant ongoing changes in the fishing quota. The CoastWeb-model can handle such relationships, but it should be stressed that the CoastWeb-model does not quantify different species, only different functional groups. This means that CoastWeb quantifies the production potential of the system under different abiotic conditions, and that adjustments to different species should be done using complementary, traditional methods based on fish statistics (ICES). The CoastWeb-model can, however, provide a framework also for such analyses, but it gives data for entire basins and not for sites. This means that one can not directly compare our results with empirical data based on information from individual sites since the areal, vertical and temporal variations in all ecosystem variables are great in the Baltic Sea.

5.4 Expanded Temperature-Increase Scenario

5.4.1 Background and Aim

This scenario is related to the temperature scenario discussed in Chap. 4, where we stressed that the given scenario was not a global change scenario because future possible global climatic changes might affect the Baltic Sea system in many different ways besides causing increased water temperatures. It is also probable that the water level of the sea may rise and that this would influence the inflow of oxygen-rich salt-water from the Kattegat, as discussed in the second scenario in Chap. 4. In this section, we first combine these two aspects in the first round of simulations. We will utilize the sub-model for oxygen presented in this chapter. This means an expanded temperature scenario, but it is still not a climate change scenario, just one form of sensitivity analysis where we have accounted for two effects related to possible future climate changes. These two aspects are interesting in the sense that increased water temperatures would likely cause eutrophication of the Baltic Sea system, as shown in Chap. 4, and increased salt-water fluxes from Kattegat would likely cause oligotrophication. So, one should expect compensatory effects when these two aspects are combined.

In the second round of simulations, we will also add the results related to the "optimal" phosphorus reductions discussed in the first section of this chapter. The idea is that within 40 years there may be increases in water temperatures, increases in salt-water fluxes to the Baltic Sea and also further nutrient reductions to combat the eutrophication of the Baltic Sea.

5.4.2 Higher Water Temperatures and More Salt-Water Inflow

The results are given in Fig. 5.6. We have used the same stepwise increases in salt-water fluxes from Kattegat to the Baltic Proper (Q_{KABP}) as before and they are shown in Fig. 5.6a (default, 1.1, 1.2 and 1.4 times the default value of 345 km^3/year for the given 40-year period) and also the same stepwise increases in water temperatures as before (default; 0.5 and 0.1; 1 and 0.2 and finally for the last 10-year period, 2 and 0.4°C).

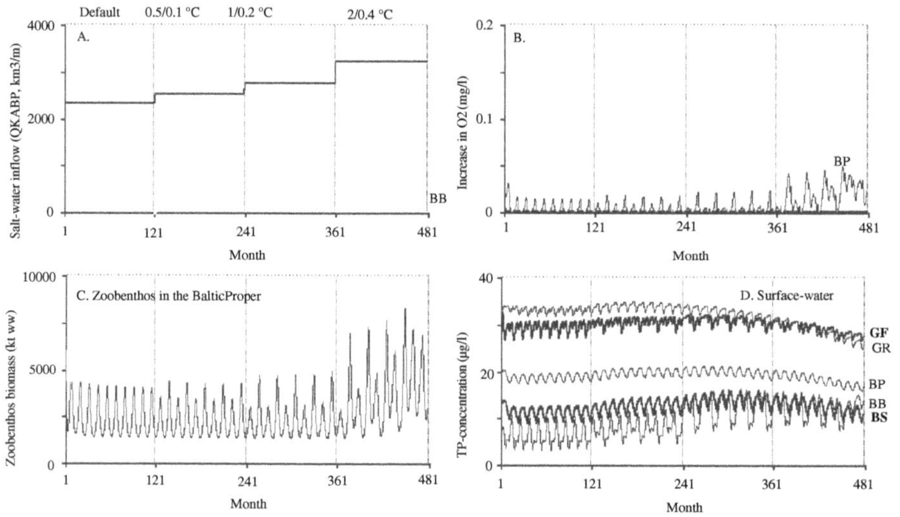

Fig. 5.6 Results from the scenario with increased water temperatures (**a**), increased salt-water fluxes (**a**) and the CoastWeb-model including the sub-model of oxygen concentrations, (**b**) gives the corresponding increases in the oxygen concentration in the Baltic Proper, (**c**) gives the modeled increases in zoobenthos biomass in the Baltic Proper, and (**d**) shows the corresponding changes in TP-concentrations in the SW-layers in all five Baltic Sea basins

Figure 5.6b gives the modeled changes in oxygen concentrations in the Baltic Proper. Figure 5.6c gives the corresponding increases for zoobenthos biomass in the Baltic Proper and Fig. 5.6d the changes in TP-concentrations in all surface-water compartments in the Baltic Sea basins – higher TP-concentrations in the Bothnian Bay and the Bothnian Sea and lower TP-concentrations in the Gulf of Riga, the Gulf of Finland and the Baltic Proper. From this background, one can interpret the changes for fish shown in Fig. 5.7.

There are only relatively small changes until the last 10-year period. So, for the first 20 years there are evident compensatory effects. The increases in TP-concentrations and the corresponding changes in chlorophyll, SPM, Secchi depth,

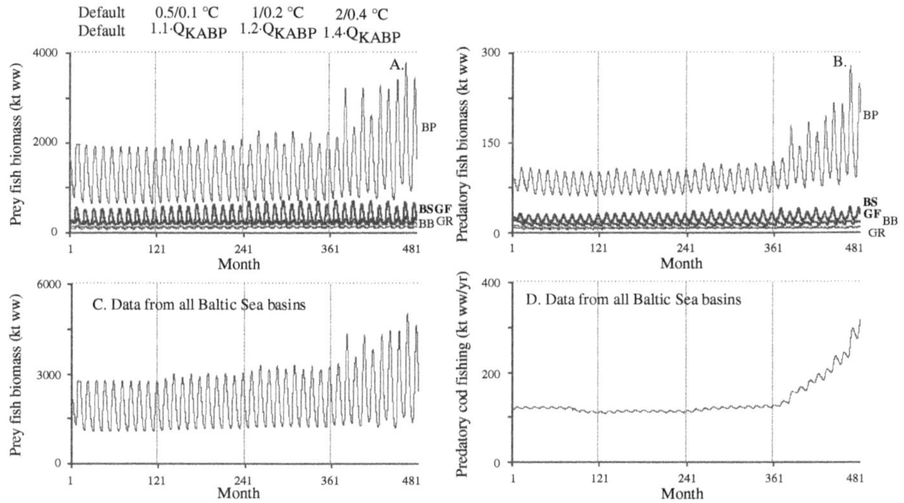

Fig. 5.7 More results from the scenario with increased water temperatures and increased salt-water fluxes. (**a**) Gives the modeled increases in prey fish biomass in the Baltic Sea basins, (**b**) gives the corresponding information for predatory fish, (**c**) gives the changes in total prey fish biomass in all five Baltic Sea basins, and (**d**) gives the corresponding changes in total default fishing of predatory cod in all five Baltic Sea basins

and all other foodweb characteristic were demonstrated in the temperature scenario in Chap. 4 and the decreases in TP-concentrations (and the corresponding changes in all other variables) related to the higher salt-water inflows more or less take out one another until the last two periods when the higher oxygen concentrations causes increased production of zoobenthos, which will trigger the given changes in fish biomasses and in default fishing. Note that the changes are much more pronounced for predatory fish than for prey fish and that the changes are small in all basins except for the Baltic Proper. The increased possibilities for fishing of predatory fish are again explained by increased oxygen concentrations.

5.4.3 Higher Water Temperatures, More Salt-Water and Reduction in TP-Loading

In the following simulations, we will reduce the tributary inflow of phosphorus in steps corresponding to the steps in water-temperature increases and salt-water inflows just given. We will use the default data for the first 10-year period, then for the second 10-year period we will remove 1,000 t TP/year from the Baltic Proper, 800 t/year from the Gulf of Finland and 150 t/year from the Gulf of Riga (i.e., a total of 1,950 t/year), twice that for the next 10-year period and three times that for the last 10-year period. The total "optimal" TP-reduction was 8,730 kt/year so this means that we have reached a significant part of those reductions in this scenario. This stepwise reduction is evidently more realistic than the sudden reduction simulated before, but more realistic does not necessarily mean very realistic. These reductions in tributary phosphorus loading would be costly (see last section of this chapter) and difficult to implement.

The results are shown in Fig. 5.8 for:

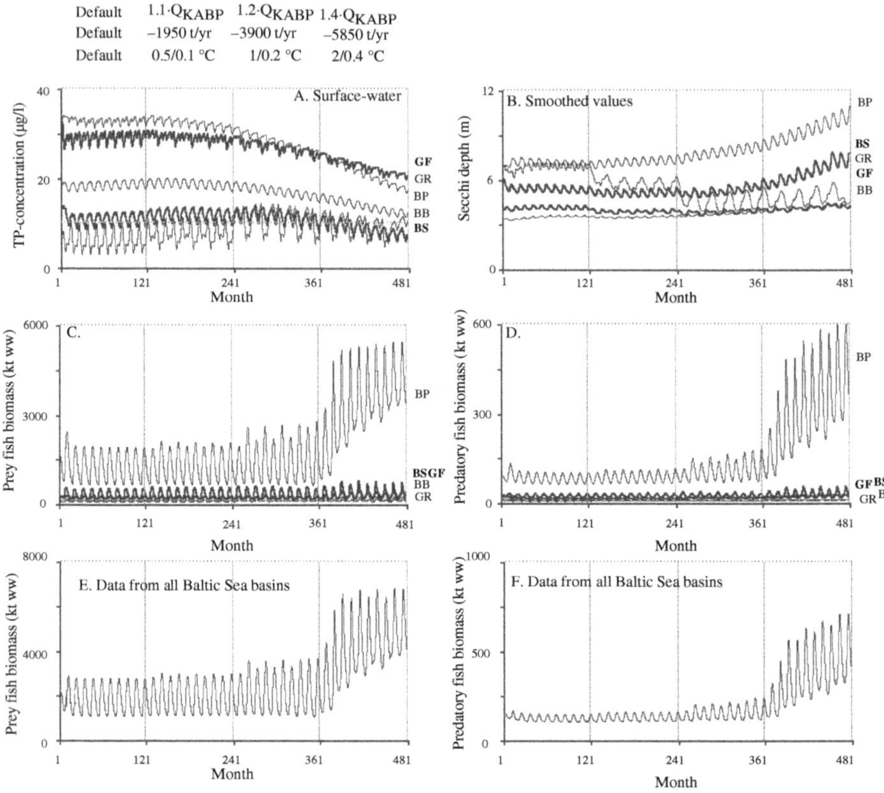

Fig. 5.8 Results from the scenario with steps in increased water temperatures, increased salt-water fluxes and increased phosphorus reductions to the Baltic Proper (1,950, 3,900 and 5,850 t/year) using the CoastWeb-model including the sub-model for oxygen. (**a**) Shows the changes in TP-concentrations in the SW-layers in all five Baltic Sea basins, (**b**) gives the changes in Secchi depth, (**c**) gives the modeled increases in prey fish biomass in the Baltic Sea basins, (**d**) gives the corresponding information for predatory fish, (**e**) gives the changes in total prey fish biomass in all five Baltic Sea basins, and (**f**) gives the corresponding changes in total biomass of predatory fish in all five Baltic Sea basins

A. TP-concentrations in surface water in all five Baltic Sea basins; the TP-concentrations clearly decrease in four basins, but not in the Bothnian Bay, i.e., in the basin with the highest land uplift and the coldest water, which will be influenced the most by changes in water temperature.

B. The Secchi depths, the water clarity, will increase in all basins except for the Bothnian Bay.
C. The prey fish biomasses increase significantly in the Baltic Proper mainly as a response to the changes in oxygen concentrations and the corresponding increases in zoobenthos biomass; in the other four basins, the prey fish biomass stay fairly stable throughout the 40 year period in this simulation.
D. The corresponding changes in predatory fish biomasses are also relatively small, except for the Baltic Proper.
E. Gives the changes in prey fish biomass for all five basins and one can note the increase for the last 10-year period.
F. Gives the results related to the default fish of predatory fish and one can note the significant increase in the production potential and hence also fishing potential in the last period.

Figure 5.9 explains these results in more detail:

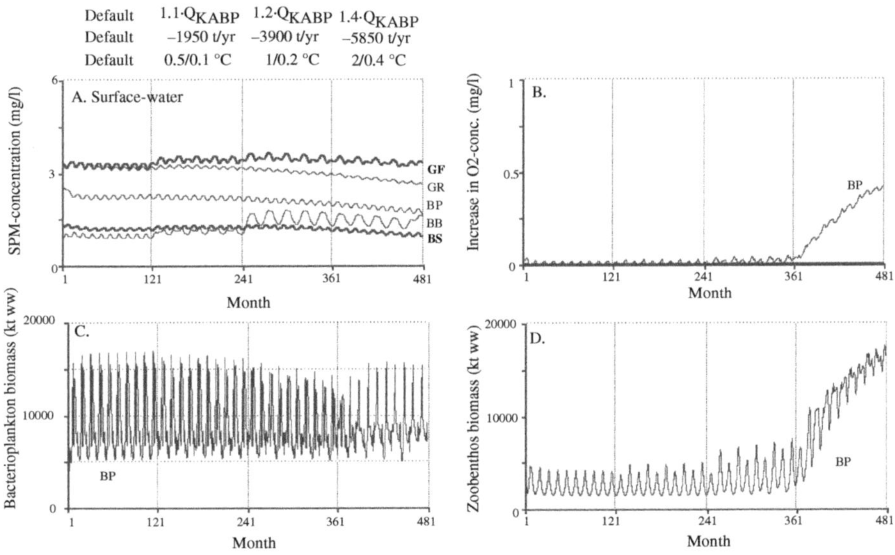

Fig. 5.9 Results from the scenario with steps in increased water temperatures, increased salt-water fluxes and increased phosphorus reductions to the Baltic Proper (1,950, 3,900 and 5,850 t/year) using the CoastWeb-model including the sub-model for oxygen. (**a**) Shows the changes in SPM-concentrations in the SW-layers in all five Baltic Sea basins, (**b**) gives the changes in oxygen concentrations in the Baltic Proper, (**c**) gives the modeled increases in bacterioplankton biomass in the Baltic Sea basins, and (**d**) gives the corresponding information for zoobenthos in the Baltic Proper

- The changes in SPM-concentrations (smoothed values) are first shown (in Fig. 5.9a) and they regulate the changes in Secchi depths shown in Fig. 5.8b.
- There are marked increases in the oxygen concentrations in the Baltic Proper (Fig. 5.9b) because of (1) the stepwise increases in salt-water inflow, (2) the

lower bacterioplankton biomasses in the Baltic Proper (Fig. 5.9c) from the stepwise reductions in TP-loading and SPM-concentrations and (3) the increased biomasses of zooplankton (Fig. 5.9d) from the improved oxygen concentrations.

5.4.4 Comments

Baltic Sea management has for a long time been dominated by remedial measures based on relatively vague claims and suggestions rather than mass-balance calculations using validated models which have been tested for several independent coastal areas and shown to predict well (see, e.g., the website of the National Swedish Environmental Protection agency; http://www.naturvardsverket.se/sv/), e.g., suggestions that nitrogen reductions should be carried out, that cultivation of mussels may be useful method to improve the eutrophication in the Baltic Sea, that invasions of jellyfish may be a serious threat to Baltic Sea fishery, that extensive fishing of sprat may stop an ongoing regime shift in the Baltic Proper, that oxygenation of deep-water layers by large pumps would reduce the diffusion of phosphorus from deep-water sediments and thereby reduce the eutrophication in the Baltic Proper and that addition of chemicals to sediments would also reduce the diffusion of phosphorus and reduce the eutrophication. These claims and suggestions are, however, neither based on validated mass-balance models nor on well tested foodweb models, which is a scientific requirement if one has scientific claims related to changes in nutrient fluxes. If the claim is that one particular measure would reduce a particular nitrogen or phosphorus flux, e.g., diffusion from sediments, it is a scientific demand to have all other major processes under control so that the favorite process can be put into its proper context. We would argue that the CoastWeb-model including the CoastMab-model is the first more holistic tool to address questions regarding remedial measures to reduce Baltic Sea eutrophication and associated foodweb characteristics, including the fish production potential of the system. In the next section in this chapter, we will address how predation of fish by seals and porpoise, invasion of jellyfish and cultivation of mussels would likely affect Baltic Sea eutrophication. In a previous section of this chapter, we demonstrated an approach to find an optimal reduction in the tributary loading of nutrients (phosphorus). We do not advocate any reductions in nitrogen loading, since such reductions, as stressed in Chap. 2, can not be scientifically evaluated and since they would possibly do more harm than good in triggering the blooming of cyanobacteria.

We have demonstrated that if 8,730 t/year of phosphorus would be removed from the present tributary loading, this would mean that a defined goal would be reached – that the Secchi depth in the Gulf of Finland would return to what it was between 1900 and 1920, i.e., to 7 m. We have also presented several scenarios related to possible future possible climate changes, namely increases in water temperatures, which would imply a loss in ice cover and connected increases in turbulence (and hence also much lower sedimentation of particulate forms of nutrients and SPM) and resuspension, and rather drastic changes in internal phosphorus loading, especially the Bothnian Bay and the Bothnian Sea. We have also demonstrated probable consequences of higher salt-water inflows, which would create an oligotrophication of the Baltic Sea system, and also, if the salt-water intrusions are large enough, increase deep-water oxygen concentrations, which in turn would increase zoobenthos production, which would be beneficial for Baltic Sea fish production. We have shown results where these aspects have been combined, i.e., higher water temperatures, more salt-water inflow and reduced tributary TP-loading. The point is that our results are quantitative, not qualitative, they are based on a validated mass-balance model which has been tested for more than 20 coastal areas and more than 50 lakes, and shown very good results; the CoastWeb-model has been extensively tested for the Baltic Sea basins and it does not include any tuning for different sub-basins in the Baltic Sea. Tuning is a practice used by some modelers – if the model does not produce correct results, it can always be tuned to do so by changing model constants. In fact, one can get almost any model to describe empirical data well, but models based on tuning are largely useless for predictive purposes and should not be used as tools for Baltic Sea management.

5.5 Possible Consequences of Invasions of Jellyfish, Cultivation of Mussels and Increased Number of Seals for the Fish Production

This scenario will discuss possible effects both on fish production and eutrophcation that may be derived from invasions of jellyfish, cultivation of mussels and increased number of seals.

5.5.1 Background, Presuppositions and Aim

In the default version of the CoastWeb-model, we use a threshold salinity value of 10 psu for jellyfish, i.e., if the salinity is lower than that, there will not be any jellyfish in the system, and since the salinity is lower than 10 in all surface-water compartments in the Baltic Sea, there is no jellyfish modeled in the system. We will change that by lowering the threshold salinity to 6.7 psu, which means that there will be an invasion of jellyfish to the Baltic Proper. Jellyfish will eat both herbivorous and predatory zooplankton and the default value for the distribution coefficient on the food choice menu is set to 0.5, which means that jellyfish would consume both types of zooplankton without preference for any of them. The default value for the jellyfish production rate is 8.5 (1/month), which is derived after calibrations so that the modeled actual biomass of jellyfish in the Baltic Proper corresponds to the normal biomass of jellyfish.

The key question is: How would a possible invasion of jellyfish influence the fish production?

An increase in the production (= cultivation) and harvesting of mussels (blue mussel, *Mytilus edulis*; see Lindahl, 2008) and (www.miljomusslor.loven.gu.se) will also be tested. This simulation is rather similar to the simulation for jellyfish except that the mussels are assumed to eat both types of zooplankton and also phytoplankton and bacterioplankton (see Davenport et al., 2000; Arifin and Bendell-Young, 2001; Lehane and Davenport, 2002). The distribution coefficients have been set in a simple manner. The first distribution coefficient on the food choice menu between both types of zooplankton and bacterioplankton plus phytoplankton has been set to 0.5 and this value has also been used for the second distribution coefficient between bacterioplankton and phytoplankton. This means that the mussels in this scenario do not have any favorite food.

The turnover time for the blue mussels in the Baltic Proper has been set to 2 years and the harvesting is done on a monthly basis when the mussels are 3 years old. The normal biomass for mussels has been set to be a factor of 4 lower than the mean value for the four normal biomasses of their prey and we have also used a smoothing function using the harvesting time (36 months) as an averaging function to get a realistic seasonal production pattern – the seasonal variations in the biomass of the mussels should be smaller than the seasonal variations in the biomasses of zooplankton, phytoplankton and bacterioplankton (which has much shorter turnover times). The initial production rate has been set to 1.51 (per month), which gives modeled biomasses of mussels close to the normal values in the Baltic Proper. The reductions in TP-concentrations in the Baltic Proper have been calculated directly from the modeled biomasses of mussels. That is, the reductions concern mussels consuming zooplankton, phytoplankton and bacterioplankton included in the compartment called biota with short turnover times (see section "Calculations of Biouptake and Retention in Biota" in Chap. 2). From this compartment, we have accounted for a TP-flux related to the harvesting of the mussels, which is given by:

$$F_{TPmussels} = SMTH(M_{MUBP}/36, 36, M_{MUBP}/36) \cdot 0.7 \tag{5.4}$$

Where 0.7 is the average phosphorus concentration in blue mussels (Lindahl, 2008) and 36 is the average number of months before harvesting the mussels (i.e., 1/36 of the total biomass of mussels are harvested any given month).

This gives that the average biomass of the blue mussels in the Baltic Proper would be about 2.97×10^9 kg ww (see Table 5.7) and that the total annual elimination of phosphorus from the Baltic Proper when the mussels are harvested would be 650 t TP, a small figure in this context where 158,400 t are added to the Baltic Proper from all sources each year (see Table 2.7). It should be stressed that to cultivate 3×10^9 kg ww of blue mussels would require an area of about 180 km^2 within the coastal zone (calculated from background data supplied by Lindahl, 2008). The point here is that it would be unrealistic to cultivate such large amounts of mussels, and if such large amounts of mussels were to be produced, the effects on the bioindicators, nutrient concentrations and functional groups would nevertheless be very small indeed, as this scenario will demonstrate (see Table 5.7)

This set-up implies that there will be reductions in the food available for fish at two lower trophic levels – phytoplankton plus bacterioplankton and zooplankton, respectively. The question is: How will this influence the fish production?

Table 5.7 Compilation of results related to (1) default conditions, (2) intrusion of jellyfish to the Baltic Proper (but not to the other basins) under default conditions by setting the threshold value for the salinity to 6.7 psu (the default value is 10 psu), (3) cultivation and harvesting of mussels in the Baltic Proper under default conditions, (4) by increasing the number of seals with 10,000 animals in the Baltic Proper and thereby increase the predation pressure of prey fish (but not predatory fish) and (5) by increasing the number of seals from 10,000 to 100,000 animals and by 25,000 porpoises in the Baltic Proper to mimic the situation in the years 1900–1920 and also reduce the tributary phosphorus loading according to the "optimal" strategy

	1. Default	2. Jellyfish	3. Mussels	4. Seals	5. Seals and porpoises 1900
BA	631	618	636	630	1,044
BP	9,084	9,199	9,013	9,088	6,381
MA	2.4	2.4	2.4	2.4	3.8
PD	67	68	68	66	44
PH	2,591	2,608	2,565	2,592	2,085
PY	1,082	1,116	1,088	1,073	505
ZB	1,814	1,778	1,806	1,827	2,003
ZH	1,216	1,256	1,230	1,214	653
ZP	348	309	323	350	337
JE	0	1,887	0	0	0
Mussels	0	0	2,970	0	0
Sec	6.9	6.8	6.8	6.8	10.8
TP_{SW}	19.2	19.7	19.1	19.2	10.4

The table gives the mean annual biomasses (kt ww) for benthic algae (BA), bacterioplankton (BP), macrophytes (MA), predatory fish (PD), phytoplankton (PH), prey fish (PY), zoobenthos (ZB), herbivorous zooplankton (ZH), predatory zooplankton (ZP), jellyfish (JE), mussles in kt wet weight; and also mean annual values of Secchi depths (in m) and TP-concentrations (μg/l) in surface-water layer in the Baltic Proper. Simulation time 241 months.

In the next tests, we will set the threshold value for the jellyfish back to the default value of 10, which means that there will be no jellyfish and no cultivation of mussels in the system. Instead, we will increase the biomass and number of seals. Today that value is about 10,000–15,000 animals and this is included in the normal fishing rate. We will change this in two steps. First, we will increase the number of seals in the Baltic Proper by adding 10,000 animals under default conditions and see how this would likely influence the fish production in the Baltic Proper. Each seal will eat about 5.5 kg/day (see Hårding and Härkönen, 1999) and this means that 10,000 more seals will eat 5.5·30·10,000 ($=1.65 \times 10^6$) kg ww of prey fish per month. Will this seriously affect the fish production?

Finally, we will return to the condition as there probably were 100 years ago when there were about 200,000 seals in the Baltic Sea and of these there were about 100,000 in the Baltic Proper and also about 25,000 porpoise in the Baltic Proper. We will allow these animals to eat 5.5 kg prey fish per day each. And 100 years ago the Baltic Sea was much more oligotrophic and we will use the results from the first scenario in this chapter to mimic the trophic status of the system (i.e., the results from the optimal phosphorus reduction scenario). It is probable that the inflow of oxygenated saline water from the Kattegat 100 years ago was more like the situation in the 1970s and 1980s, but in this scenario, we will test and see if it is possible to have so many seals and porpoise in the system during such oligotrophic conditions and still have a good and sustainable fishery at a default salt-water inflow of 345 km^3/year, as we have today.

5.5.2 Results

The results are given in Table 5.7 for the Baltic Proper. We give the mean annual biomasses in kt ww for benthic algae (BA; bacterioplankton (BP), macrophytes (MA), predatory fish (PD), phytoplankton (PH), prey fish (PY), zoobenthos (ZB), herbivorous zooplankton (ZH), predatory zooplankton (ZP), jellyfish (JE) and mussels in kt wet weight; and also mean annual Secchi depths and TP-concentrations in surface-water layer in the Baltic Proper. From Table 5.7, we can note:

- Compared to the results under default conditions (column 1), there are no or very small changes for all variables, except for zooplankton, and especially for zooplankton in the "Jellyfish", "Mussels" and "Seals" scenarios. In the "Jellyfish" scenario, the biomass of predatory zooplankton is reduced from 348 to 309 kt ww and given the related lowering in predation pressure on herbivorous zooplankton they will increase in biomass from 1,216 to 1,256 kt ww. There are also small reductions in prey and predatory fish biomass and in TP-concentrations and phytoplankton biomass. From this one can conclude that under these presuppositions an invasion of jellyfish and/or a cultivation of mussels would not influence the system very much, not the abiotic factors such as the TP-concentration and the Secchi depth and not the functional groups of organisms.
- One hundred years ago the system was more oligotrophic. The TP-concentration was about 11 μg/l in the surface-water of the Baltic Proper, which is about half the value today; the Secchi depth about 11 m, as compared to about 7 m today. It is very interesting to note that given the higher fishing of prey fish by so many seals and porpoises, the biomass of predatory fish would logically decrease, from 67 to 44 kg ww. The biomass of the prey fish would also be significantly lower, 505 kt ww compared to 1,982 kt ww; also the biomass of herbivorous zooplankton would be clearly lower in the "Seals and porpoises" scenario.

5.5.3 An Extended Scenario Related to Jellyfish

There were no invasions of jellyfish or cultivation of mussels in the years around 1980 when the intrusions of salt water from the Kattegat were frequent. This indicates that the threshold salinity for jellyfish is not likely as low as 6.7 psu used in the previous scenario. One would probably need larger salt-water intrusions that would cause higher salinities in the Baltic Sea. In this extended and more realistic scenario, we have accounted for the actual salt-water intrusions (as given in Fig. 5.3a)

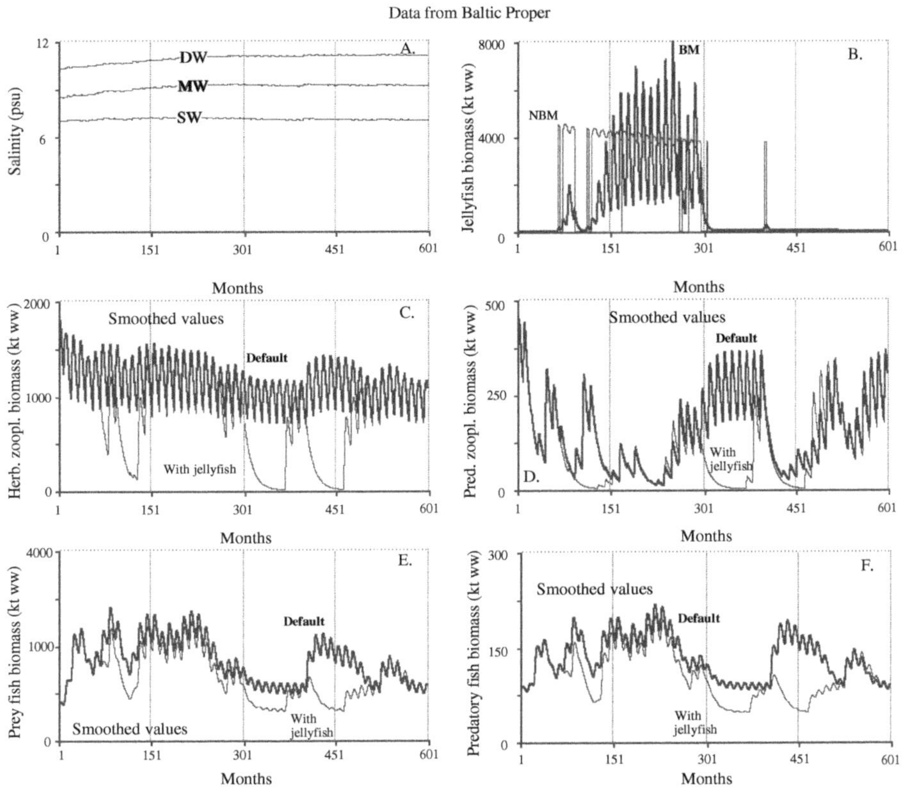

Fig. 5.10 Results from the scenario with the actual salt-water fluxes are accounted for and when there are invasions of jellyfish. (**a**) Gives the modeled resulting surface-water salinities in the three layers in the Baltic Proper, (**b**) the modeled increases in biomasses of jellyfish. This figure also gives the normal biomasses for jellyfish using the CoastWeb-model including the sub-model for oxygen, (**c**) compares the smoothed default values (the values have been smoothed to get a better overview) with the smoothed values in this jellyfish scenario for the biomass of herbivorous zooplankton, (**d**) shows the corresponding changes in biomass of predatory zooplankton, (**e**) and (**f**) informs about the corresponding values for prey and predatory fish in the Baltic Proper (the values for predatory fish have not been smoothed). Note that the threshold salinity for the invasion of jellyfish has been set to 7.1 psu (the default salinity is 10 psu)

and we have lowered the threshold salinity from 10 to 7.1 psu. This leads to higher salinities, as shown in Fig. 5.10a for the three water layers in the Baltic Proper. The corresponding changes in the jellyfish biomass are given in Fig. 5.10b. This figure also gives the normal biomasses for jellyfish and one can see that there is a good correspondence between norm-values and dynamically modeled values. Figure 5.10c–f give the changes in biomasses for herbivorous and predatory zooplankton and prey and predatory fish and also the corresponding data when the threshold salinity has not been lowered (but set to 10 psu) to prevent jellyfish in the system.

From Fig. 5.10, one can see that in a situation with jellyfish potentially all herbivorous and predatory zooplankton could be lost (see, e.g., the results around month 361). An important point here is that the consequences for single species living only on zooplankton, such as sprat, which is a staple food for predatory fish, and one of the most dominating species of prey fish in the Baltic Proper (see Chap. 2), could very well collapse under these conditions. This collapse is not seen in Fig. 5.10, but clearly indicated. In this modeling with the CoastWeb-model, we do not calculate changes for individual species only for prey fish as a functional group. The prey fish group as a whole would not collapse even though there are marked reductions in prey fish biomass in connection to the reduction in zooplankton biomass (Fig. 5.10e), and the corresponding reductions also in predatory fish biomass (Fig. 5.10f).

This example illustrates the need for complementary information for important species. There may be serious effects for the Baltic Sea system from invasions of jellyfish. This has been demonstrated for the Black Sea (see Zaitsev and Mamaev, 1997). It is also probably that such invasions are likely to be triggered by more massive salt-water intrusions than we have seen so far. This may be a cause for alarm related to possible future climate changes and it is interesting to note that on one hand such intrusions of oxygenated salt-water would likely be beneficial for the Baltic Sea system since this would increase the oxygen concentrations, increase the production of zoobenthos, which in turn would increase the fish production. On the negative side, we have connected possible invasions of jellyfish, which could potentially consume large amounts of zooplankton, and thereby threaten important species of prey fish (such as sprat in the Baltic Proper), which mainly feed on zooplankton. Such changes in the inflow of salt-water have caused significant and well documented changes in the structure and function of other marine systems, such as the Ringkobing Fjord (see Håkanson and Bryhn, 2008c).

5.5.4 Comments

Literature data demonstrate that the Baltic Sea system 100 years ago had a very large population of seals and porpoise. Under default conditions (with a salt-water inflow of about 345 km^3/year) neither invasions of jellyfish, large-scale cultivation of mussels nor clear increases of seals and porpoise would probably influence the system very much.

There may, however, be serious effects for the Baltic Sea system from invasions of jellyfish during periods with large salt-water intrusions. Such salt-water intrusions of oxygenated could cause both beneficial effects since this would increase oxygen concentrations, which would trigger the production of zoobenthos and also of prey fish feeding on zoobenthos and lower the eutrophication in the system. But, on the other hand, massive invasions of jellyfish could potentially lead to severe reductions of zooplankton, and thereby threaten important species of prey fish (such as sprat in the Baltic Proper feeding on zooplankton).

In this modeling, we differentiate between predatory and herbivorous zooplankton. From our literature survey (in Chap. 2), it seems like this differentiation is not always done by Baltic Sea ecologists. If we, e.g., omit predatory zooplankton in this modeling, it means a lower predation pressure on herbivorous zooplankton and increased biomasses of herbivorous zooplankton. At the end it leads to more food for prey fish and increased biomasses of both prey and predatory fish.

5.6 A Strategy to Increase Fish Cage Production in the Baltic Sea in a Sustainable Manner

5.6.1 Background and Aim

Fish cage farming is very important in many countries (e.g., Norway and Chile) and an expanding industry in many parts of the world (see Mäkinen, 1991; Ackefors and Olburns, 1996; FAO, 2001; Subasinghe, 2004). From a world production of about 100 kt in the early 1950s, the total fish farm production has increased to about 1,500 kt in 2002 and it may reach 2,000 kt in 2010 (Subasinghe, 2004). Fish farming is a significant part of the total aquaculture production, which means that it is very important to develop methods and strategies to evaluate the effects of fish farm emissions. This includes the consequences for the marine ecosystems if more and more prey fish are caught and used for fish meal production to feed cultivated fish, and also genetic consequences for the wild fish from escaping cultivated fish and the spread of fish diseases (e.g., Hutchinson, 1997; Youngson et al., 2001; Järvi, 2002). It is important to look at fish cage farming from a holistic perspective (e.g., Read and Fernandez, 2003), which could improve decision support and lead to more accurate environmental regulations and policies (see Ackefors and Olburs, 1996; Burbridge et al., 2001). This section, however, will focus on eutrophication effects related to fish farm emissions (i.e., emissions of TP and SPM), see Fig. 5.11.

There may be positive as well as negative effects of fish cage farming. Positive effects are, e.g., increased catches of wild fish in areas with fish farms (see Håkanson and Boulion, 2002a). Negative effects are, e.g., decreased caches for certain species of fish and altered fish communities (see, e.g., Kautsky and Kautsky, 1989; Elmgren, 1989a, b; Hansson and Rudstam, 1990; Kautsky, 1991; Tucker, 1999; Holmer et al., 2008), changes in the structure and composition of other species, e.g., large perennial algae (e.g., *Fucus vesiculosus*). Håkanson and Gyllenhammar (2005) have discussed the eutrophication effects for the Baltic Sea related to emissions from fish farms at four different scales:

1. The site scale at and around the farm. At this scale (< 1 ha), the fish farmer would benefit the most from emission reductions of food spill and faeces. The "footprint" expressing the impact areas of fish cage farm often corresponds to the size of a "football field" (50–100 m) if the annual fish production is about 50 t.

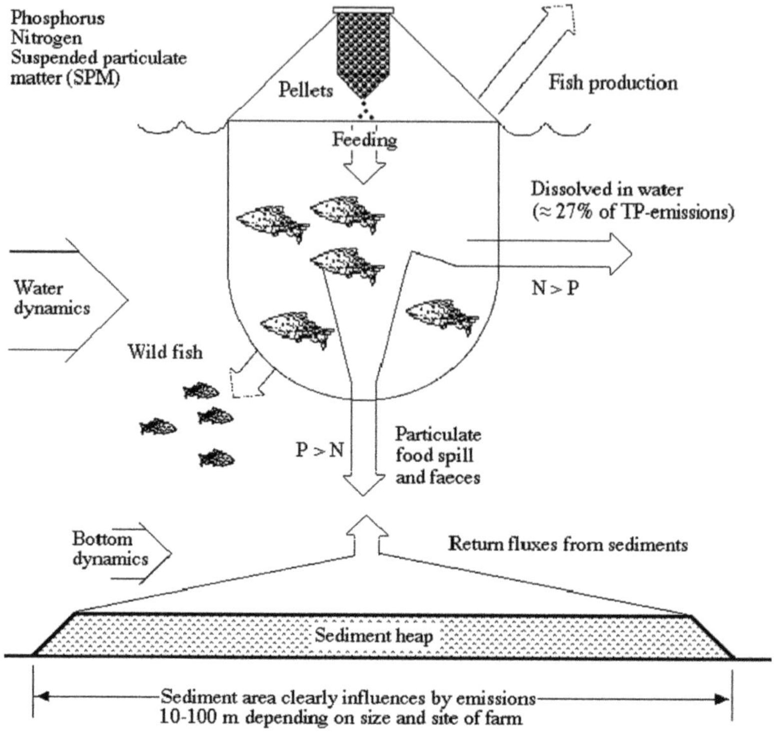

Fig. 5.11 Key components in budgets for phosphorus, nitrogen and suspended particular matter related to emissions from fish cage farms in coastal areas (modified from Håkanson and Gyllenhammar, 2005)

2. At the local scale (1 ha to 100 km^2), i.e., in the coastal area where the farm is situated, there exists a load diagram (effect-load-sensitivity; see Håkanson and Gyllenhammar, 2005) to relate the environmental effects from a specific fish farm production. This makes it possible to obtain a first estimate of the maximum allowable fish production in a specific coastal area. At this scale, local and regional authorities would be the main interesting party, as well as the farmer and people interested in recreational or other activities in the given coastal area.

3. At the regional scale (100–10,000 km^2), it has been shown (see Gyllenhammar et al., 2008) that it is possible to create negative nutrient fluxes, i.e., use fish farming as a method to reduce the nutrient loading to the sea. The breaking point is to use more than about 1.3 g ww wild fish per gram feed dry weight for the cultivated fish. Table 5.8 shows nutrient fluxes to Finnish Archipelago Sea (size 25,000 km^2). Typical current velocities are about 2–4 cm/s in the section areas defining the borders of this coastal areas (Håkanson et al., 1986; Persson et al., 1994). The net current velocity may, however, be a factor of 5–10 lower than 2.5 cm/s, and the related nutrient fluxes smaller. The values given in Table 5.8

Table 5.8 Estimated nitrogen and phosphorus fluxes to the Archipelago Sea (modified from Håkanson and Gyllenhammar, 2005)

	Nitrogen (t/year)	Phosphorus (t/year)
Rivers	6,000	400
Industries + urban emissions	1,400	110
Atmospheric deposition	7,200	180
Nitrogen fixation	900	–
Fish farm emissions	400	55
Water transport from BP, velocity, u = 2.5 cm/s	13,000,000	930,000
More probable, u = 0.25 cm/s	1,500,000	110,000

are meant to put the nutrient emission from the fish farming in this coastal area into a realistic context. For this calculation, it was assumed that about 7,000 t/year of rainbow trout were cultivated in the Finnish Archipelago Sea. The fish farm emissions can evidently not influence the nutrient status of the Finnish Archipelago Sea very much.

4. At the international scale, the contribution from fish farms to the overall nutrient fluxes are generally very small. The following data on fluxes of nitrogen and phosphorus (river data, atmospheric deposition, point source data, etc.) for 1990 from different sources, including fish farms, concern the entire Baltic Sea (for background information, see Ambio, 1990; EPA, 1994; Lozán et al., 1996; Wallin et al., 1992); data from Table 2.8:

	Nitrogen	Phosphorus
Total input to the Baltic Sea (t/year)	1,400,000–2,950,000	120,000–190,000
Total emissions from fish farms (t/year)	1,500–2,000	200–500

Also note that since the fish farm emissions play such a relatively small role in the overall nutrient budgets for the Finnish Archipelago Sea and the Baltic Sea, the ecosystem benefits in terms of reducing these fluxes are also generally small. An important motive for the fish cage farming in the Baltic Sea is to create more jobs for pellet producers, fishermen and fish farmers. It would also mean that the import of salmonids from mainly Canada and Norway to meet markets demands in the Baltic Sea countries can be reduced. This in turn means positive environmental effects related to transport and storage and a stronger local trade and a better regional economy. For a country like Finland, which has a net import of fish for consumption to a value of 79,000,000 USD (1996; from SVT, 1997), an increase in fish farming would meet present market demands and be beneficial from many other aspects.

Many studies have discussed the factors regulating the growth of cultivated fish (see Azevedo et al., 1998; Ruohonen et al., 1998; Alanärä, 2000; and Gyllenhammar and Håkanson, 2005). Figure 5.11 gives an illustration of basic concepts related to the emissions, the spread of the emissions and the "footprint", i.e., the area affected by the emissions. Several papers and books treat this subject (see, e.g., Håkanson et al., 1988; Holby and Hall, 1991; Mäkinen, 1991; Enell, 1994; Wahlström, 2000). Evidently, the "footprint" depends on the load of nutrients and suspended particulate matter (SPM) – the larger the emissions, the larger the "footprint".

About 27% (see Johansson et al., 1998) of the phosphorus emissions from a farm are in dissolved form, the rest is particulate phosphorus. Many papers have discussed the drastic reductions in FCR-value (feed conversion rate) during the last 30 years, from about 2 to about 1 (Enell, 1994; Alanärä, 2000). The mean FCR-value for all fish farm in Sweden today is about 1.2, according to official statistics (Statistiska centralbyrån, SCB, Stockholm, Sweden) and we will use that value in the following simulations. This means that the effluents from modern farms and the heaps below the farms are generally much smaller than they were 20–30 years ago. The feed conversion ratio may be as low as 1.0 in a well run farm. Then there may be small or no (see Håkanson, 1995) heaps of faeces and food spill beneath the cages. The distribution of the emissions from the farm depends very much on the prevailing water and sediment dynamics of the area around the cages.

There are several evident negative impacts of fish farms emissions, e.g., formation of toxic hydrogen sulfide in gas bubbles from the heaps beneath the cage, which may spread up to 5 m from the bottom. If H_2S reaches the cage, it may kill the fish (see Håkanson et al., 1988). Mäkinen (1991) gives a compilation of several studies related to the distribution of heavy metals, impact on zoobenthos, sedimentation and oxygen condition at sites at various distances from fish farms. Generally, such studies are difficult to evaluate because the results depend very much on season of the year, water temperature and prevailing water and bottom dynamics conditions during and before the study.

There are also effects on the wild fish biomass in the area where the fish farm is located. Johansson et al. (1998) have presented an experimental study with marked food (pellets) to see how much marked pellets there were in wild fish around a fish farm and caught at different distances from the farm. Fish (mainly roach and bream) caught up to 1,000 m from the farm had significant amounts of marked pellets in their guts. Fish farm emissions often cause significant increases in the biomasses of wild fish without corresponding increases in algal volume. So, the fish farm emissions influence the secondary production more markedly than the primary production. This may seem like a paradox, but it is related to the fact the wild fish consumes food spill and faeces from the fish farm and this creates a specific foodweb pathway described by Håkanson and Boulion (2002a).

An important presupposition for this scenario is that a significant fraction (at least 20%) of the food eaten by the cultivated fish should be wild fish from the Baltic Sea, such as sprat, i.e., commercially less attractive and more abundant prey fish that could be caught without negative consequences for the Baltic Sea foodweb and fisheries.

From this background, the aim of this section is to add a fish farm sub-model to the CoastWeb-model and see how large fish cage farm production of rainbow trout that the Baltic Sea system could sustain without causing clear ecosystem effects, and those effects will be defined and quantified using the CoastWeb-model.

5.6.2 Quantifying TP and SPM Emissions from Fish Cage Farms and Effects on the Baltic Sea Ecosystem

The emissions from the farm (see the model in Fig. 5.12; from Håkanson, 2006) may be quantified in the standard way from (1) annual fish production (AFP), (2) feed conversion ratio, FCR (i.e., the amount of food added to the farm in kg divided by the amount of fish produced in the farm in kg), (3) TP-concentration in feed (TP_{Feed}; this value depends on the type of feed used; in these simulations, we will use a value of 0.9%) and (4) TP-concentration in fish (TP_{Fish}; for rainbow trout from this farm; 0.42%) and (5) a seasonal moderator, which accounts for the fact that there is generally a typical seasonal pattern in fish growth, fish feeding and emissions of TP such that high emissions generally occurs in the fall, just before the harvest. The TP and SPM emissions are added to the surface-water compartments in the Baltic Sea basins.

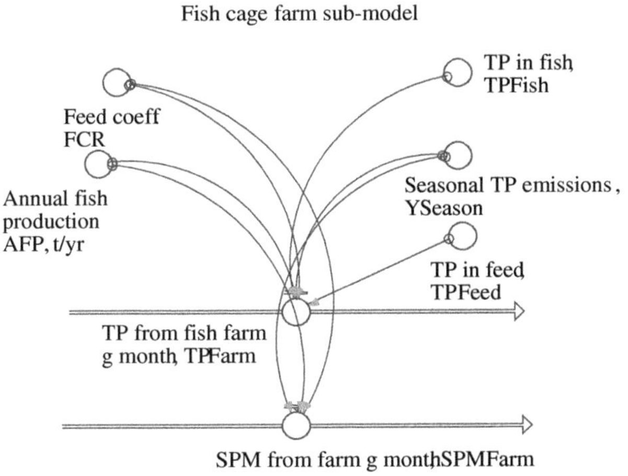

Seasonal TP emissions, YSeason

Jan	Feb	Mar	Apr	May	Jun	jul	Aug	Sep	Oct	Nov	Dec
0.034	0.038	0.044	0.053	0.075	0.107	0.131	0.146	0.15	0.086	0.048	0.044

<u>Equations:</u>
AFP = e.g., 30,000 t/yr
FCR = 1.2 dim. less
TPFeed = 0.9 %
TPFish = 0.42 %
TPFarm = YSeason·10^6·0.01·AFP·(FCR·TPFeed-TP Fish); g TP/month
SPMFarm = 0.23·AFP·FCR·YSeason·10^6; g SPM/month

Fig. 5.12 The sub-model to calculate fish farm emissions of TP and SPM to the surface-water layers in the basins of the Baltic Sea

5.6.3 Results

5.6.3.1 Default Conditions

Table 5.9 first gives the results related to the default conditions in the five basins for mean Secchi depths, TP-concentrations, biomasses of phytoplankton (PH), herbivorous zooplankton (ZH), prey fish (PY) and predatory fish (PD). The Secchi depths vary from 3.5 m in the Gulf of Riga (GR) to 6.9 m in the Baltic Proper. One can then ask: How would these conditions change if:

Table 5.9 Results under default conditions related to changes in fish cage farm production of rainbow trout (1) cultivation of a total of 10,000 t/year, 1/3 of this is BP, BS and BB, respectively, (2) cultivation of a total of 20,000 t/year, 1/3 of this is BP, BS and BB, respectively, (3) cultivation of a total of 30,000 t/year, 5,000 in BB, 10,000 in BS and 15,000 in BP, and (4) cultivation of a total of 60,000 t/year, 10,000 in BB, 20,000 in BS and 30,000 in BP

	Default	10,000 (1/3)	20,000 (1/3)	5/15/10	10/30/20
Sec_{BB}	6.6	6.6	6.5	6.5	6.4
Sec_{BP}	6.9	6.9	6.9	6.8	6.8
Sec_{BS}	5.0	5.0	5.0	4.9	4.9
Sec_{GF}	4.0	4.0	4.0	4.0	4.0
Sec_{GR}	3.5	3.5	3.5	3.5	3.5
TP_{BB}	6.4	6.4	6.5	6.5	6.7
TP_{BP}	19.2	19.2	19.3	19.4	19.5
TP_{BS}	11.5	11.6	11.6	11.7	11.9
TP_{GF}	29.8	29.8	29.9	29.9	30.1
TP_{GR}	33.3	33.3	33.4	33.5	33.7
PH_{BB}	108	109	109	109	111
PH_{BP}	2,591	2,594	2,596	2,598	2,605
PH_{BS}	524	525	526	527	530
PH_{GF}	345	345	346	346	347
PH_{GR}	170	170	170	170	171
ZH_{BB}	44	44	45	45	45
ZH_{BP}	1,216	1,219	1,221	1,225	1,234
ZH_{BS}	322	324	325	327	332
ZH_{GF}	101	101	101	101	102
ZH_{GR}	50	50	51	51	51
PY_{BB}	67	67	68	68	69
PY_{BP}	1,082	1,086	1,089	1,094	1,106
PY_{BS}	285	287	289	290	296
PY_{GF}	161	161	161	161	162
PY_{GR}	85	85	85	85	86
PD_{BB}	4	4	4	4	4
PD_{BP}	67	67	67	67	67
PD_{BS}	19	19	19	19	19
PD_{GF}	10	10	10	10	10
PD_{GR}	5	5	5	5	5

The table gives the mean biomasses in kt ww for predatory fish (PD), phytoplankton (PH), prey fish (PY), herbivorous zooplankton (ZH) in kt wet weight; and also mean Secchi depths (in m) and TP-concentrations (μg/l) in surface-water layers in all five Baltic Sea basins. Simulation time 241 months.

1. 10,000 t/year of rainbow trout would be produced in the Baltic Sea (1/3 of this in the Bothnian Bay, the Bothnian Sea and the Baltic Proper each)?
2. 20,000 t/year of rainbow trout would be produced (1/3 in each basin, BB, BS and BP)?
3. 30,000 t/year of rainbow trout would be; 5,000 t/year in the Bothnian Bay, 10,000 t/year in the Bothnian Sea and 15,000 t/year in the Baltic Proper?
4. 60,000 t/year of rainbow trout would be; 10,000 t/year in the Bothnian Bay, 20,000 t/year in the Bothnian Sea and 30,000 t/year in the Baltic Proper?

From Table 5.9, one can note that such high fish cage production as 60,000 t/year would reduce the Secchi depth; from 6.9 to 6.8 m in BP, from 6.6 to 6.4 m in BB and from 5.0 to 4.9 m in BS; the changes in the Gulf of Riga (GR) and the Gulf of Finland (GF) would be even smaller. There are also corresponding changes in phytoplankton biomasses, but very small changes in fish biomasses.

From this, we can conclude that a production of 60,000 t/year, a factor of 10 higher than today, would influence the Baltic Sea system and increase the eutrophication, and that a production of 30,000 t rainbow trout per year would cause only very marginal changes.

In the next scenario, we will simulate the effects on the Baltic Sea system in a similar way if 8,730 t/year of phosphorus would be removed from the present tributary loading, since this would mean that the goal would be reached, that the Secchi depth in the Gulf of Finland would return to what it was between 1900 and 1920, i.e., to 7 m. So, how large fish cage production could be accepted if that goal remains?

5.6.3.2 "Optimal" Conditions

Table 5.10 gives the results for mean Secchi depth, TP-concentration and biomasses of phytoplankton (PH), herbivorous zooplankton (ZH), prey fish (PY) and predatory fish (PD). The modeled Secchi depths under default conditions (Table 5.9) vary from 3.5 m in the Gulf of Riga (GR) to 6.9 m in the Baltic Proper. How would these values change if 10,000, 20,000, 30,000 and 60,000 t/year of rainbow trout would be produced in the Baltic Sea if the "optimal!" scenario would be implemented?

One can see from Table 5.10 that under these conditions the goal would not be met with a total fish cage production of 60,000 t/year because the mean annual Secchi depth in the Gulf of Finland would then decrease from 7 to 6.9 m.

5.6.3.3 Comment

With a fish cage production of 30,000 t/year (5,000 in the Bothnian Bay, 10,000 in the Bothnian Sea and 15,000 in the Baltic Proper), the goal that the mean annual Secchi depth in the Gulf of Finland should be 7 m could be met. In the next section where we present our management plan for the Baltic Sea, we will also include a total fish cage production of 30,000 t/year, as motivated in this scenario.

Table 5.10 Results for conditions prevailing 100 years ago (as defined by the "optimal" reductions in tributary TP-loading) related to changes in fish cage farm production of rainbow trout (1) cultivation of a total of 10,000 t/year, 1/3 of this is BP, BS and BB, respectively, (2) cultivation of a total of 20,000 t/year, 1/3 of this is BP, BS and BB, respectively, (3) cultivation of a total of 30,000 t/year, 5,000 in BB, 10,000 in BS and 15,000 in BP, and (4) cultivation of a total of 60,000 t/year, 10,000 in BB, 20,000 in BS and 30,000 in BP

	Optimal	10,000 (1/3)	20,000 (1/3)	5/10/15	10/20/30
Sec_{BB}	8.6	8.5	8.5	8.5	8.3
Sec_{BP}	10.7	10.6	10.6	10.6	10.5
Sec_{BS}	8.1	8.0	8.0	7.9	7.7
Sec_{GF}	7.0	7.0	7.0	7.0	6.9
Sec_{GR}	5.9	5.9	5.9	5.9	5.9
TP_{BB}	4.1	4.2	4.2	4.2	4.3
TP_{BP}	10.9	10.9	10.9	11.0	11.1
TP_{BS}	5.8	5.8	5.9	5.9	6.0
TP_{GF}	18.6	18.7	18.7	18.8	18.9
TP_{GR}	19.3	19.4	19.4	19.5	19.6
PH_{BB}	86	86	86	86	87
PH_{BP}	2,104	2,108	2,111	2,116	2,127
PH_{BS}	357	358	360	361	365
PH_{GF}	363	364	364	365	366
PH_{GR}	163	164	164	164	165
ZH_{BB}	36	37	37	37	37
ZH_{BP}	709	711	713	716	723
ZH_{BS}	202	204	205	206	210
ZH_{GF}	70	70	70	71	71
ZH_{GR}	35	35	36	36	36
PY_{BB}	50	50	51	51	51
PY_{BP}	634	635	637	640	645
PY_{BS}	162	163	164	165	168
PY_{GF}	109	109	109	109	110
PY_{GR}	53	54	54	54	54
PD_{BB}	3	3	3	3	3
PD_{BP}	55	55	55	55	55
PD_{BS}	13	13	13	13	14
PD_{GF}	8	8	8	8	8
PD_{GR}	4	4	4	4	4

The table gives the mean annual biomasses in kt ww for predatory fish (PD), phytoplankton (PH), prey fish (PY), herbivorous zooplankton (ZH) in kt wet weight; and also Secchi depths (in m) and TP-concentrations (μg/l) in surface-water layers in all five Baltic Sea basins. Simulation time 241 months.

5.7 A Holistic Management Plan for the Baltic Sea, Including Cost-Benefit Analyses

5.7.1 Background and Aim

This section will present an economic evaluation of Baltic Sea fishery so that an overall cost-benefit analysis can be made (see Sect. 5.10). We will present three options for Baltic Sea management:

1. Default conditions and no further remedial actions. This would imply no further costs and no further improvements of the conditions in the Baltic Sea. We are not advocating this as a desired option, just as a hypothetical one to be used as a reference. This also means that we will not calculate possible future changes in water temperatures, salt-water intrusions, invasions of jellyfish, cultivation of mussels or fish cage farming. Since the Baltic Sea has undergone major changes in the last 100 years, and especially between 1920 and 1980, all/most groups interested in the management of the Baltic Sea probably agree on at least this particular point, namely that we need a cost-efficient management plan for the Baltic Sea to combat the present eutrophication and that doing nothing is not a viable option.
2. The next option is our "optimal" strategy to reduce tributary phosphorus loading presented in the first section of this chapter. We have also motivated why we will include a fish cage production of 30,000 t rainbow trout in the management plan and in this scenario we will add also economic aspects related to the fish cage production. Also in this scenario, we will not calculate possible future changes in water temperatures, salt-water intrusions, invasions of jellyfish and cultivation of mussels.
3. The third option is given by the HELCOM strategy (see Table 2.6) already agreed upon by the Baltic Sea government but not yet implemented. This means that a total of 15,000 t/year of phosphorus and about 133,000 t of nitrogen should be removed from the present tributary nutrient loading. We will discuss the environmental consequences and costs of this option, but we will not calculate possible future changes in water temperatures, salt-water intrusions, invasions of jellyfish, cultivation of mussels and fish farming.

The differences related to these three options will be related to the three key bioindicators for coastal management, (1) Secchi depth, (2) chlorophyll-a concentrations and (3) concentrations of cyanobacteria. This means that we will also present the model used to calculate concentrations of cyanobacteria. We will also calculate changes in prey fish and predatory fish biomasses, the economic value of the fish produced in the Baltic Sea, and also include the production and economic value related to the fish cage farming. So, the focus in this evaluation and management plan is on the three bioindicators and on fish. First, however, we will present the economic criteria.

5.7.2 Economic Criteria and Tools for Optimizing Baltic Sea Fisheries Management

Fish provide a large number of market and non-market valued services, such as affecting the function of the foodweb, and providing medicine, recreational opportunities and healthy food to humans (Holmlund and Hammer, 1999). This section will exemplify how conflicting societal interests around fisheries can be treated by applying economics concepts such as utility optimization and cost-effectiveness. Some readers may frown at the thought of making economic analyses on fish stocks and other types of wildlife. Against that standpoint we would argue that many fish stocks have already become severely threatened due to improved catching techniques during recent decades (Myers and Worm, 2005; van Densen and McCay, 2007) and it may indeed have been the failure to assign proper monetary values to fish stocks which has left the ground open for treating fish as a free and endlessly exploitable resource. Healthy fish communities are good for both ecosystems and economies. Although this book mainly focuses on the natural sciences, we will now briefly discuss the core parts of the economics of fish stock management; the maximum utility of various strategies regarding fishing quotas and phosphorus emissions, and also how phosphorus abatement can be undertaken in a cost-effective manner.

5.7.2.1 Cost-Effective Phosphorus Abatement

Cost-effective environmental management is the art of reaching environmental targets at the minimal cost (Turner et al., 1999). Such management may bring substantial benefits, and not only to those who care about monetary resources. Cost-effectiveness also means that the environmental "job" can get done as quickly as possible – i.e., that the time period during which considerable environmental harm is allowed is shortened to a minimum. Cost-effectiveness should therefore be of major concern for most people who are interested in a clean environment and in well-functioning ecosystems.

One of the scenario types for fisheries management in this book concerns phosphorus abatement. Ideally, P-abatement measures should be, (1) effective (decrease the loading with a large number of tons per year), (2) cost-effective (have low marginal costs for each abated kilogram or ton) and (3) politically feasible to ensure that measures are quickly adopted into legislation and policy. How much phosphorus can be removed from the loadings to the Baltic Sea through various measures and some examples from the literature is displayed in Table 5.11. One relevant target according to Håkanson and Bryhn (2008b) would be to remove 9,775 t TP/year to restore the trophic state to conditions prevailing 100 years ago and in this chapter we have shown that the same target may be reach if 8,730 t TP/year could be reduced if a higher P-reduction to the Gulf of Finland could be achieved. Table 5.11 shows that upgrading the sewage treatment would be inevitable to meet this goal. Agricultural measures and banning detergents containing phosphates could be two other options with substantial, but insufficient, effects on the TP-loading. Some other measures, such as cultivating mussels/clams would make only very small differences to the conditions in the Baltic Sea (as we have shown) and this is valid also for the construction of wetlands, since they have a low efficiency to reduce nutrient flows when the fluxes are high and a somewhat higher capacity to reduce the nutrient fluxes when little water is transported in the rivers/brooks.

Secondly, the marginal phosphorus abatement cost should be studied. This is the cost for removing one kg (or another mass unit) of phosphorus from the loading to

Table 5.11 Possible measures to decrease the phosphorus input to the Baltic Sea. Swedish examples from Swedish EPA (2008a) and other examples from HELCOM (2007)

Measure	Area	Tons/year
Sewage treatment	Poland	5,332
Sewage treatment	Russia	3,844
Sewage treatment	Belarus	1,984
Sewage treatment	Baltic States	992
Sewage treatment	Czech R.	372
Phosphate-free detergents[a]	All	3,100
Agriculture	All	5,600
Rural sewage treatment	Sweden	175
Clam cultivation	Sweden	35
Dams	Sweden	10
Protective zones	Sweden	7.5
Wetlands	Sweden	4.3
Soil drainage	Sweden	2.6

[a]In combination with sewage treatment.

the Baltic Sea. Table 5.12 gives some marginal abatement costs in euro (calculated from Swedish crowns, adjusted to 2008 prices, and converted at a rate of 1 SEK = 0.10 EUR ≈ 0.13 USD) per kg phosphorus in connection with measures in agriculture, sewage treatment and wetland construction. These estimates are from Turner et al. (1999) and are thus to some extent outdated since some of the measures may already have been implemented, but the important message from this table regards which marginal costs these three different abatement strategies commonly have in relation to each other.

It is evident that marginal abatement costs in sewage treatment plants (STPs) are always lower in all regions listed in Table 5.12 than any of the costs for measures in agriculture or wetland construction. Agricultural measures are commonly a factor of

Table 5.12 Marginal abatement costs (euro/kg; 2008 prices) for phosphorus to the Baltic Sea

Region	Agriculture	STPs	Wetlands
Sweden	18–772	4.8–6.1	2,133
Finland	26–711	4.8–6.1	204
Denmark	17–305	4.8–8.0	141
Germany	22–347	4.8–8.0	105
Poland	13–238	2.3–12	71
Estonia	33–658	2.3–12	712
Latvia	27–662	2.3–12	144
Lithuania	22–783	2.3–12	113
St. Petersburg	27–505	2.3–12	96
Kaliningrad	40–502	2.3–12	64

STPs, sewage treatment plants. From Turner et al. (1999).

5–50 more expensive and costs for removing phosphorus in wetlands are commonly a factor of 25 more costly than sewage treatment. Thus, if the environmental goal can be achieved by means of sewage treatment, then the cost-effective strategy should focus on TP-reductions in STPs. Removing phosphorus in Swedish wetlands is the least cost-effective option according to Table 5.12.

An additional source of marginal cost data is HELCOM and NEFCO (2007) and the most relevant data are listed in Table 5.13. Wetland construction generally has low cost-effectiveness (just as in Table 5.12), with the exception of Lithuanian wetlands, although there are no data on how many tons of phosphorus that could be abated by constructing new Lithuanian wetlands. More cost-effective than wetlands is sewage treatment, and particularly in urban settlements in the eastern basin of the Baltic Sea when no additional pipes need to be constructed (19 euro/kg according to Table 5.13). Some examples of urban STPs in the eastern basin are specified in Kiirikki et al. (2003), where chemical P-treatment in functioning STPs could remove 520 t phosphorus per year at a marginal cost of 8 euro/kg; constructing the STP in St. Petersburg (Russia) would remove 200 t/year at 35 euro/kg while constructing the 12 km long northern sewage collector in the same area would remove 220 t/year at 43–88 euro/kg P. These figures were compared with much less cost-effective measures in Finnish agriculture at an average marginal cost of 220 euro/kg P (Kiirikki et al., 2003; 2008 prices).

Banning phosphates in detergents seems quite cost-effective according to Table 5.13, although attention must be paid to the regional differences. In Sweden, where sewage treatment has been implemented with relatively ambitious standards, marginal costs are actually higher than most sewage treatment alternatives in Table 5.13 and this would probably be the case for more countries if they would first upgrade their sewage treatment to Swedish standards. Vice versa, if phosphates in detergents have already been banned, then constructing STPs would be less cost-effective than if phosphates are allowed in detergents. This means that if the goal is to remove 3,100 t of P per year, a phosphate ban in detergents would be a cost-effective option according to Tables 5.11 and 5.13. If more than 12,400 t phosphorus should be removed from yearly loadings, as stipulated in the Baltic Sea Action Plan (HELCOM, 2007), sewage treatment together with a phosphate ban could also be cost-effective. However, if the goal lies well above 3,100 but below 12,400 t/year, as in Håkanson and Bryhn (2008b) or in the "optimal" scenario of this book, upgrading sewage treatment to Swedish standards without banning phosphates in detergents could be both sufficient and the most cost-effective option according to Tables 5.11 and 5.13.

It should also be noted that alternatives to reducing phosphorus in detergents may have adverse environmental effects which do not necessarily affect the Baltic Sea. One of the most viable alternatives to reducing P is to use Zeolite A, which produces greater volumes of sludge, which cannot be recycled in the same manner as phosphorus in sewage sludge but can be used as a fertilizer in agriculture. Thus, with effective sewage treatment in place, phosphorus in detergents may actually be the most environmentally friendly option in a life-cycle perspective (Köhler, 2006),

Table 5.13 Marginal abatement costs (euro/kg; 2008 prices) for phosphorus to the Baltic Sea (from HELCOM and NEFCO, 2007)

Measure	Area	Marginal abatement cost
Wetlands	Germany	103
Wetlands	Denmark	170
Wetlands	Estonia	153
Wetlands	Finland	92
Wetlands	Lithuania	35
Wetlands	Latvia	142
Wetlands	Poland	73
Wetlands	Russia	643
Wetlands	Sweden	163
Urban sewage treatment, no pipes	Eastern basin	20
Urban sewage treatment, including pipes	Eastern basin	42
Rural sewage treatment, no pipes	Eastern basin	43
Rural sewage treatment, including pipes	Eastern basin	92
Urban sewage treatment, no pipes	Western basin	30
Urban sewage treatment, including pipes	Western basin	63
Rural sewage treatment, no pipes	Western basin	65
Rural sewage treatment, including pipes	Western basin	138
Phosphate-free detergents	Denmark	44
Phosphate-free detergents	Estonia	19
Phosphate-free detergents	Finland	39
Phosphate-free detergents	Lithuania	14
Phosphate-free detergents	Latvia	19
Phosphate-free detergents	Poland	18
Phosphate-free detergents	Russia	13
Phosphate-free detergents	Sweden	53
Less milk cows	Germany	14
Less milk cows	Denmark	20
Less milk cows	Estonia	13
Less milk cows	Finland	19
Less milk cows	Lithuania	4.1
Less milk cows	Latvia	12
Less milk cows	Poland	10
Less milk cows	Russia	13
Less milk cows	Sweden	12
Less pigs	Germany	15
Less pigs	Denmark	18
Less pigs	Estonia	22
Less pigs	Finland	18
Less pigs	Lithuania	6.2
Less pigs	Latvia	19
Less pigs	Poland	11
Less pigs	Russia	12
Less pigs	Sweden	14

which would imply lower cost-effectiveness for a phosphate ban than what can be extracted from Table 5.13.

The effects on the TP-loading and the marginal cost of most available agricultural measures are not specified in HELCOM and NEFCO (2007) due to a reported lack of data. One exception is the possibility to reduce the livestock (milk cows and/or pigs), which would have comparatively low marginal costs in most countries (Table 5.13). This measure includes replacing the manure from the animals with artificial fertilizer on the fields. How many tons of phosphorus would then actually be prevented from reaching the Baltic Sea through this measure is unfortunately not specified (directly or through proper references) in HELCOM and NEFCO (2007). Relationships between changes in land use and TP-fluxes to the sea are generally poorly researched and sometimes unclear or even contradictory (Bergström et al., 2007). In other words, it is unclear whether there would be any effects at all from this measure, or whether clear effects would require such radical measures (e.g., removing 80% of the livestock which was the upper limit for model simulations for nitrogen abatement in Schou et al., 2006) that no action would be a preferable option for policymakers who wish to keep and extend their contracts. In addition, decreasing meat production without enforcing any changes in eating habits among the population could have the effect that local meat production would be substituted with imported meats, so that the eutrophication problem would instead be exported to other waters outside the Baltic Sea drainage basin. Increased imports could also bring other adverse environmental effects such as increased greenhouse gas emissions during transportation. Thus, although marginal costs may be low, the feasibility of eliminating meat production to decrease the P-loading seems low nonetheless when reliable alternatives such as improving sewage treatment are readily at hand. Livestock reductions and most other agricultural measures investigated by Gren and Elofsson (2008) except for modest (no amount specified) reductions in fertilizer application, were less or much less cost-effective than urban sewage treatment and phosphate-free detergents.

Most of the Swedish measures in Table 5.11 have low or very low cost-effectiveness compared to sewage treatment in Table 5.13. For instance, rural sewage treatment in Sweden has a marginal cost of 770–3,000 euro/kg (Swedish EPA, 2008b). The only exception of Swedish measures in Table 5.11 is clam/mussel cultivation whose marginal costs would be 35 euro/kg, although since the potential impact is very small (35 t/year; Table 5.13) it is disputable whether this option is worthwhile to develop beyond its present experimental stage.

Instead, the most cost-effective P-abatement plan with the ambition stated in this book (8,730 t/year, see Table 5.17 later) would primarily include sewage treatment and Table 5.13 gives at hand that the yearly cost of such a plan would probably not exceed 0.37 billion euro/year (8,730 t/year 42 euro/kg 1,000 kg/t). As a comparison, the recently adopted Baltic Sea Action Plan has an estimated yearly cost of 3 billion euro/year (HELCOM and NEFCO, 2007), and includes many abatement options with low cost-effectiveness.

5.7.2.2 Maximum Utility of Fish Stock and Nutrient Management in the Baltic Sea

After having motivated a cost-effective P-abatement plan, we will now use the CoastWeb-model to motivate a new sustainability plan for the Baltic Sea. This includes searching for the maximum total utility of the total fish stock, the fish yield (catches plus fish cage farming of rainbow trout), and the algal biomass by means of combining economic, social and environmental utilities from these variables. Such a plan will have to take into account that environmental management also concerns managing societal conflicts. One of society's many desires, to catch and consume as much fish as possible, already contains a conflict in itself, between the short-term and the long-term benefit. Too intensive fishing has in many well-documented cases caused fish stocks to collapse or decrease below threshold levels, which have severely hampered the prospects for future yields (Jackson et al., 2001; Myers and Worm, 2005; van Densen and McCay, 2007). Furthermore, the desire to maximize the yield may constrain the desire to maximize the size of the fish stock, which would be beneficial for recreational fishing. Finally, the societal desire to decrease the level of eutrophication may affect the production potential for fish, as this book demonstrates.

Table 5.14 describes the total number of people employed in the fish industry (including aquaculture and food processing) in Baltic Sea countries belonging to the European Union. It should be noted that Russia is not included in this table and that parts of the numbers in the table refer to people who work in the North Sea and other waters. In 2004, the total volume of commercial landings in the Baltic Sea countries (including catches outside of the Baltic Sea) was 4.8 million tons, with an average value of 1.8 euro/kg, at an average value of 1.6 euro/kg. Aquaculture in these countries produced 0.16 million tons in 2003, at a value of 2.4 euro/kg (data from HELCOM and NEFCO, 2007; 2008 prices). In Sweden and Denmark, aquaculture accounts for less than 20% of the total fish production while this figure exceeds

Table 5.14 Number of employees in the fisheries sector in 2004 (including aquaculture and food processing), total production (2003), production value (catches 2004 plus aquaculture 2003) exports and imports of fisheries products (2003) and the total mean yearly public aid to the fisheries sector 2000–2006 (from European Communities, 2006)

Country	Employed in fisheries	Production, kt	Exports, kt	Imports, kt	Production value, million euro	Total public aid, million euro
Denmark	4,490	1,063	1,198	1,618	439	44.0
Estonia	7,954	79	119	44		2.6
Finland	4,762	135	11	93	56	12.7
Germany	4,358	335	550	1,076	223	41.3
Latvia	4,115	115	109	35		4.6
Lithuania	3,030	160	67	73		2.5
Poland	6,300	215	111	251		40.3
Sweden	1,913	293	311	288	323	16.3

60% in Denmark and Germany (European Communities, 2006). The food processing industry roughly tripled the value of the total commercial fisheries production (European Communities, 2006; HELCOM and NEFCO, 2007). Table 5.14 shows the size of subsidies from the public sector, which in Finland, Sweden, Germany and Denmark accounted for 114 million euro/year in 2000–2006, which was about 11% of the production value from fisheries and aquaculture (1,040 million euro/year), or about 4% of the total value of fisheries sector production including food processing.

Table 5.14 also shows that fisheries products are intensively traded in the Baltic Sea region and that exports from some countries even exceed catches plus aquaculture production. Poland, Finland, Denmark, Germany and Lithuania are net importers of fisheries products, and the total imports of such goods to the countries in Table 5.14 exceeded exports in 2003 by 1 million tons. This means that if the production of fish and other seafood would increase in this area, up to about 1 million tons of new products could potentially be consumed locally at the disadvantage of producers in other parts of the world.

Different fish species attract human consumers in different ways. The price differences between different fish species in Sweden and Estonia is given in Table 5.15 expressed as the "herring factor"; the median price of each species divided by the median price of herring. Herring prices and overall price levels were about 35% higher in Sweden than in Estonia, but the relative prices were conspicuously similar in the two countries; sprat and herring were the two least expensive species while salmon, zander and eel cost the most. Swedish prices on herring and vendace

Table 5.15 Market prices for various fish species, expressed in factors of the median herring price (the "herring factor")

English name	Scientific name	CV SWE	Herring factor SWE	Herring factor EST	Mean herring factor
Sprat	*Sprattus sprattus*	0.18	0.75	1.1	0.91
Herring	*Clupea harengus*	0.50	1	1	1
Flounder	*Platichthys flesus*	0.11	5.7	3.5	4.6
Burbot	*Lota lota*	0.17	5.5	4.1	4.8
Pike	*Esox lucius*	0.11	7.0	6.8	6.9
Cod	*Gadus morhua*	0.24	7.8	8.6	8.2
Vendace	*Coregonus alburnus*	0.62	8.8	n.a.	8.8
Plaice	*Pleuronectes platessa*	0.23	11	n.a.	11
Perch	*Perca fluviatilis*	0.21	12	14	13
Whitefish	*Coregonus lavaretus*	n.a.	n.a.	13	13
Salmon	*Salmo salar*	0.27	20	16	18
Zander	*Sander lucioperca*	0.11	28	19	24
Eel	*Anguilla anguilla*	0.21	34	44	39

Swedish (SWE) median prices and CV-values, January–September 2008 from www.fiskeriverket.se, and Estonian (EST) mean prices 2006–2007 from www.agri.ee

fluctuated strongly in 2008, perhaps because of strong catching seasons, while prices for flounder, pike and zander were relatively stable, as reflected by the CV-value (coefficient of variation) in Table 5.15.

Table 5.15 gives at hand that the "herring factor" for cod is about 8.2 in the Baltic Sea. Cod comprises a major part of the predatory fish caught. Our management scenarios in this chapter will include oligotrophication and be particularly beneficial for predatory fish, and in order not to exaggerate the benefits of oligotrophication we will use the ratio 8:1 for the price of predatory fish to the price of prey fish. According to our calculations, the typical catch of predatory fish to the catch of prey fish during recent years should have been about 1:3.5, as estimated from the relationship between the cod, sprat and herring that was actually caught in the Baltic Sea (see Fig. 1.10). This gives a typical price for predatory fish at 5.0 euro/kg and a price of prey fish at 0.62 euro/kg given that the average price of fish is 1.6 euro/kg. It should be noted that these estimates are rather conservative, and another conservative estimate used in this chapter is that cultivated rainbow trout should yield a typical price of 6.0 euro/kg, slightly higher than the price of wild predatory fish (mainly cod).

Recreational fishing adds substantial additional economic value to that of the total ecosystems services of the Baltic Sea. In the case of Sweden, recreational fishing is, according to the Fisheries Board, worth about 250 million euro/year, out of which 175 million is actual costs and 75 million is the value of the consumer surplus (the willingness-to-pay in addition to actual costs; Fiskeriverket, 2008) A pan-Nordic study performed at the turn of the millennium found a higher sum for Sweden; about 470 million euro (Toivonen et al., 2004; converted to 2008 prices). The willingness-to-pay (wtp) for doubling the fish availability has been estimated at 57 million euro (Fiskeriverket, 2008). Such surveys over the whole Baltic Sea basin are unfortunately not available, although according to Toivonen et al. (2004), the willingness-to-pay was higher in Denmark than in Sweden, but lower in Finland. To roughly estimate the value of recreational fishing in the Baltic Sea, we will use the data range from Fiskeriverket (2008) and Toivonen et al. (2004) and the following assumptions (1) that half of the willingness-to-pay concerns freshwater (lake, stream and river) fishing and that the rest concerns fishing at sea and (2) that the differences between Baltic Sea countries regarding willingness-to-pay for recreational fishing are similar to the differences in willingness-to-pay for combating eutrophication as stated in HELCOM and NEFCO (2007). Then, the Swedish benefit of recreational fishing in the Baltic Sea would be worth 49–92 million euro/year, and the corresponding figures would be 20–37 million for Germany, 29–55 million for Denmark, 0.9–1.7 million for Estonia, 25–48 million for Finland, 2.0–3.7 million for Lithuania, 1.2–2.3 million for Latvia, 25–47 million for Poland, and 7.5–14 million for Russia. The total value would be 160–300 million euro/year and the willingness-to-pay for doubling the fish availability would be about 36 million euro/year. It is worth noting that while recreational fishing may be more worth than commercial fishing in many lakes and streams, this may not be the case for the Baltic Sea.

The willingness-to-pay for restoring the environmental state of the Baltic Sea to conditions preceding the 1960s has been estimated (Gren, 2001) at 3.6 million euro/year expressed in 2008 prices. Eutrophication is commonly treated as one of the most serious environmental threats to the Baltic Sea, and we will assume that roughly one fourth of the total willingness to pay in Gren (2001), or 1,000 million euro/year, can be attributed to restoring the trophic state.

Thus, four types of ecosystem services that the Baltic Sea may deliver and their yearly market plus non-market values are (1) catches from the fish fleet (2) water with algal blooms at the pre-1960s level (3) fish cage farm products and (4) recreational fishing. Service number 2 is not produced at all today while the other three services are produced to a certain extent. We will now investigate whether this is optimal utility-wise, or whether it would be possible to improve the overall utility of these four services taken together by producing more of some services and less of others. We will also discuss the extent to which uncertainties in the presented market and non-market values may lead to uncertain conclusions.

5.7.3 Summary

This section has discussed the core part of the economics of fisheries management in the Baltic Sea. This part includes cost-effective phosphorus abatement since the trophic state of aquatic systems also determines the production potential of fish. Phosohorus abatement measures in agriculture and wetland construction often have low cost-effectiveness compared to other options. Agricultural measures, which imply decreased production should be avoided because it could be difficult to raise public and political support for such actions, and they could also export the eutrophication problem to other regions of the world on behalf of increasing food transports. There are, however, two main actions which may (1) lead to substantial decreases in P-loading, (2) be cost-effective and (3) be politically feasible; upgrading sewage treatment in the Eastern Baltic basins, and banning phosphates in detergents. The extent to which these two alternatives should be used depends on the size of the abatement target. If it is estimated that this target requires Swedish standards on sewage treatment in all countries, then this alternative could decrease the cost-effectiveness of banning phosphate in detergents to the point where sewage treatment and no phosphate ban is an optimal option.

Increasing the fish catch can be done using economic incentives to fisheries such as subsidies or tax breaks, or by allowing larger fish quotas. Decreasing the catch may be more difficult and action alternatives could include:

- Fiscal measures (raising taxes or cutting subsidies). Table 5.14 shows that subsidies to the fisheries sector are modest (about 4%) compared to the total revenue. Thus, it is unlikely that cutting subsidies will have a strong impact on the total catch. However, the present subsidies may indeed create ecologically unsustainable incentives for those fishermen who are particularly successful at applying for aid, although a detailed review of the current subsidy policy is beyond the scope of this book. Taxing fisheries with the aim to decrease the catch of certain species is also an option, although taxes may actually increase the relative profitability of illegal fishing, which should be avoided.

- Decreasing fish quota. Our results show quantitatively for different environmental conditions how lower fish quota means a smaller predation pressure from man on fish and higher fish biomasses and vice versa. If the fish quota is very high, there is less fish to catch, and the costs and efforts to catch the remaining fish increase. This means that there is an optimal fish quota, which should be adjusted to variations in the environmental conditions regulating the fish production potential of the given system. The CoastWeb-model is meant as a tool for such analyses to optimize the fishing efforts.

Aquaculture carried out as suggested in this chapter appears to have modest impacts on phosphorus concentrations, Secchi depths and the key functional groups (see Sect. 5.6). These emissions can be counteracted by applying a stricter policy on P-emissions elsewhere; i.e., by banning phosphates in detergents in different countries or by spending more money on upgrading sewage treatment. Such measures can be financed by taxing the polluters within the aquaculture business. Table 5.14 shows that the EU-countries around the Baltic Sea have an import surplus of about 1 million tons of fisheries products per year and increased aquaculture production could potentially decrease this number.

5.8 The Model for Cyanobacteria

One of the target bioindicators in the final evaluation presented later in this chapter, and certainly a very important bioindicator for Baltic Sea management, is the concentration of cyanobacteria. Photolithoautotrophic bacteria (sometimes called bluegreen algae; here referred to as cyanobacteria), play two key roles in eutrophication contexts: they can form extensive nuisance blooms that may be toxic (Smith, 2003), and many cyanobacterial species can fix large amounts of dissolved gaseous nitrogen of atmospheric origin (Rahm et al., 2000; Tõnno, 2004). In the Baltic Sea, they constitute the dominating form of harmful algal blooms. Other harmful algae may be more important in other seas and coastal areas.

Håkanson and Bryhn (2008a) have demonstrated a highly significant and mechanistically understandable strong positive co-variation between chlorophyll-a concentrations and concentrations of cyanobacteria. Both are measures of algal biomass in water. Håkanson et al. (2007) have presented the general model to predict cyanobacteria (CB), which we will use here for the Baltic Sea. It should be stressed that this model does not concern cyanobacteria produced in the benthic zone and it does not differentiate between cyanobacteria fixing atmospheric nitrogen and non-fixing species. This model has been tested using data from more systems than the model presented by Smith (1985). It also gives a higher r^2-value so it is more general and provides better predictive power. Given the inherent uncertainties in the empirical CB-data, one should not expect to derive predictive models for cyanobacteria that give much higher r^2-values than 0.76–0.8. It should be noted that:

- The most important factor regulating cyanobacterial growth is the TP-concentration in the surface-water (see Eq. 5.4).
- High CB-values only appear in systems with TN/TP-ratios (TN = total–N) lower than 15.
- In the literature, temperatures between 15 and 17°C have been reported as the minimum for cyanobacteria blooms in freshwater systems and in the Baltic Sea (Reynolds and Walsby, 1975; Edler, 1979; Wasmund, 1997). Laboratory experiments on cyanobacteria also support this conclusion (Konopka and Brock, 1978; Lehtimäki et al., 1994, 1997) since many species of cyanobacteria have a slow growth rate below 15°C. This modeling sets the threshold value for the temperature influences at 15°C (see Fig. 5.13). So, e.g., when the water temperature is

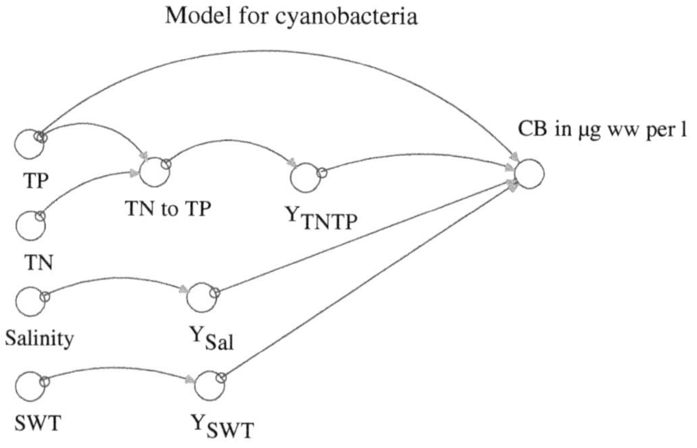

$$CB = ((5.85 \cdot \log(TP) - 4.01)^4) \cdot Y_{TNTP} \cdot Y_{SWT} \cdot Y_{Sal}$$

CB in µg ww/l
Sal_{SW} = Surface-water salinity in psu
SWT= Surface-water temperature in °C
Total-N (TN) in µg/l
Total-P (TP) in µg/l

Y_{TNTP} = if TN/TP < 15 then (1-3·(TN/TP/15-1)) else 1
Y_{SWT} = if SWT ≥15 then (0.86+0.63·((SWT/15)^1.5-1)) else (1+1·((SWT/15)^3-1))
Y_{Sal} = if Sal_{SW} < 10 then (2.1+1.1·((Sal_{SW}/10)^2-1)) else (2.1-115·((Sal_{SW}/10)^0.01-1))

Model domain: 4 < TP < 1300; 165 < TN < 6830; 0 < Sal_{SW} < 40; 8 < SWT < 25

Fig. 5.13 Outline of the model to predict cyanobacteria (from Håkanson et al., 2007)

25°C, the risks of getting high concentrations of cyanobacteria a factor of 1.48 higher than at 17.5°C, if all else is constant, using this approach.

- The salinity of the water also influences the production of cyanobacteria. The salinity moderator (Y_{Sal}) shown in Fig. 5.13 accounts for this. The maximum expected CB-values should be a found if the median salinity for the growing season is 10 psu.

In many lakes, the biomass of cyanobacteria can be very high (Reynolds, 1987). In brackish systems, the situation is probably slightly different. Howarth et al. (1988a, b) found no data on N_2-fixing planktonic species in estuaries and coastal seas, except for the Baltic Sea and Pell-Harvey estuary, Australia. Cyanobacteria of picophytoplankton size have, in addition, been found to form blooms in the Adriatic Sea (Lučić et al., 2003; Totti et al., 2005; Bernardi Aubry et al., 2006) and at least one of the subgroups in these blooms, *Synechococcus*, can fix atmospheric nitrogen (Paoli et al., 2007). Also results from Marino et al. (2006) support this general lack of N_2-fixing cyanobacteria in some estuaries. A field study from the Baltic Sea (Wasmund, 1997) indicates that in this brackish environment species of cyanobacteria have, interestingly, the highest biomass at 7–8 psu and that the blooms in Kattegatt and Belt Sea are more frequent if the salinity is below 11.5 psu. A laboratory experiment with cyanobacteria from the Baltic Sea supports the results that the highest growth rate was at salinities in the range between 5 and 10 psu (Lehtimäki et al., 1997). Water blooms of cyanobacteria in marine environments may not be as common as in freshwater systems but according to Sellner (1997) they can be the dominating factor in carbon and nutrient fluxes in some saline systems. In marine systems, there are just a few dominant genera. In a field study in the Pacific Ocean (Marumo and Asaoka, 1974), there was no correlation between the salinity and the cyanobacteria abundance and no cyanobacteria were found in the cooler, less saline subartic waters. Those marine cyanobacteria species are mainly found in high-saline conditions. In the data discussed by Marumo and Asaokas (1974), the salinity was around 32–36 psu.

The CB-model used here within the CoastWeb-model is summarized in Fig. 5.13. In the derivation of the model, data from many databases were used (see Håkanson et al., 2007). The basic regression between TP and CB is given by:

$$CB^{0.25} = 5.85 \cdot \log(TP) - 4.01 \tag{5.5}$$

$(r^2 = 0.76; n = 86)$

The general model for cyanobacteria may give rather uncertain predictions for systems with high TN/TP and low temperatures. However, during such conditions, the N-fixation should be small. Predicting conditions with high CB is evidently most important. Measured N-fixation tends to follow a similar pattern as the prevalence of cyanobacteria (Howarth et al., 1988a, b; Wasmund et al., 2001). Analyses using modern gene sequencing techniques have suggested that more organisms than we currently know may fix nitrogen in both lakes and marine systems (Zehr et al., 2003).

5.8.1 Applying the Model for the Baltic Sea

The calculated lowest and highest values for the maximum annual N-fixation in the Baltic Proper using this approach for the period 1997–2005 are given in Table 5.16. One can note that the variation among the years is very high – the smallest value is 45 kt, the highest 908 kt and that the average annual value is 191 kt/year. The order-of-magnitude is about the same in these measurements as those given by Larsson et al. (2001) and Wasmund et al. (2001) for 1997 and 1998.

Table 5.16 Compilation of calculated data on annual maximum nitrogen fixation in the Baltic Proper between 1997 and 2005

Year	N-fix. (t/year, max.)
1997	58,300
1998	70,960
1999	60,440
2000	116,920
2001	45,180
2002	131,898
2003	73,640
2004	254,980
2005	908,090
Min.	45,180
Max.	908,090
Mean	191,160
Median	73,640
Stand. Dev.	276,450
Coeff. of var.	1.45
n	9

Note that the results from 2005 are based on interpolated data for the missing months (from Håkanson and Bryhn, 2008b).

A key message in Table 5.16 related to the costly nitrogen reductions suggested by HELCOM (Table 2.7) and agreed upon by the Baltic States, that 133 kt TN/year should be reduced from the Baltic countries may more than well be compensated for by nitrogen fixation by cyanobacteria, which was over 900 kt in 2005.

In the following scenario, we will use this model for cyanobacteria, and run it with empirical water temperature data for the default period, with mean empirical TN-concentrations for the period, since the these have not varied very much during the last decades (see Fig. 2.30) and with dynamically modeled TP-concentrations.

5.9 Useless or Sub-optimal Remedial Measures for the Baltic Sea

In this section, we will briefly discuss several methods that have been used or suggested to be used to reduce the eutrophication in the Baltic Sea and mainly in the Baltic Proper: (1) measures to lower the tributary nitrogen loading, as advocated by HELCOM (see Table 2.7), (2) oxygenation of the deep-water layer in the Baltic Proper by means of large pumps (as proposed by Stigebrandt and Gustafsson,

2007; Stigebrandt, 2008), (3) chemical treatment to reduce diffusion of phosphorus from deep-water sediments (as suggested by Blomqvist and Rydin, 2009) and (4) cultivation of mussels (as proposed by Lindahl, 2008).

5.9.1 Reduced Nitrogen Loading

This has been the main strategy for a long time advocated in many papers and reports not just to mitigate the eutrophication in the Baltic Sea but also in many other coastal areas (see, e.g., Boesch et al., 2008 and many references given in that report). We have summarized the critical aspects related to this strategy (see Håkanson and Bryhn, 2008b) and they are:

1. It is not possible to predict how the Baltic Sea system would respond to reductions in nitrogen loading since there are major uncertainties related to the quantification of nitrogen fixation, wet and dry deposition of nitrogen, the algorithm regulating the particulate fraction for nitrogen and hence also sedimentation of particulate nitrogen and denitrification. Only the uncertainties in the annual nitrogen fluxes to the Baltic Proper amount to 1,400–2,950 kt TN/year (see Table 2.8). The reduction suggested by HELCOM is 133 kt TN/year (see Table 2.6)!
2. Nitrogen reductions are likely to favor the blooming of harmful algae (cyanobacteria), and such events should be avoided. Algae need both nitrogen and phosphorus. If the TN/TP-ratio is lower than 7.2, the conditions would favor phytoplankton species which can take up dissolved nitrogen gas of atmospheric origin. Empirical data show that for the growth of cyanobacteria, there is a threshold limit for the TN/TP-ratio not at 7.2 but rather at 15 (see Håkanson et al., 2007). Figure 5.14 shows variations in median monthly TN/TP-ratios in

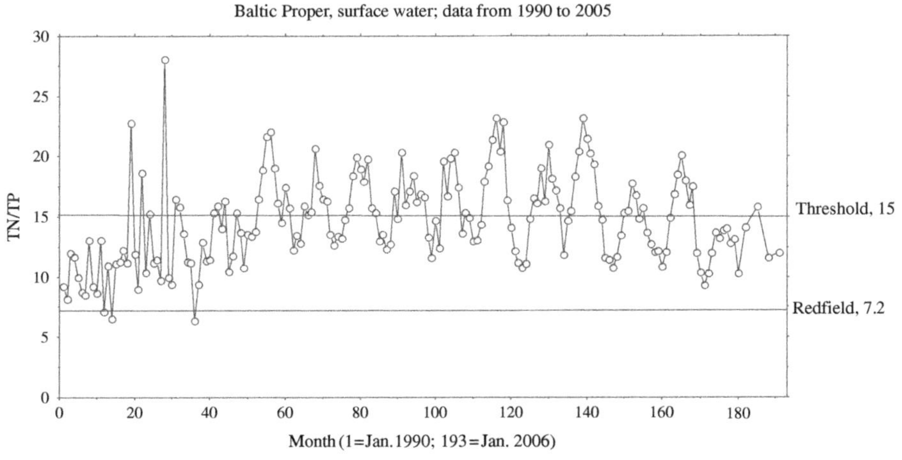

Fig. 5.14 Variations in mean monthly TN/TP-ratios based 24,048 data from surface-water samples in the Baltic Proper from 1990 to 2005 in relation to the threshold ratio of 15 and the Redfield ratio of 7.2 (modified from Håkanson and Bryhn, 2008b)

relation to the Redfield ratio of 7.2 and the threshold ratio of 15. Less than 7% of the empirical data are lower than 7.2 and between 30 and 50% of the data are lower than 15. When the TN/TP-ratio is higher than 15, nitrogen reductions are generally useless because the primary production is then limited by phosphorus; when the TN/TP-ratio is lower than 15, nitrogen reductions would favor cyanobacteria, which should be avoided. So, in the Baltic Proper, costly nitrogen reductions are never motivated.

3. There are no validated mass-balance models for nitrogen which have been tested for independent coastal systems and been demonstrated to yield good predictive power. If the model does not predict well because of inherent errors in the structure and the equations, any nitrogen model can be tuned/calibrated to give good descriptive power, but that should be an unacceptable practice in science. In spite of this, it is common practice.
4. The general phosphorus model presented in this book (CoastMab within CoastWeb) can provide good predictions ($r^2 = 0.98$; see Fig. 2.15a) when modeled annual values are compared to empirical data in all Baltic Sea basins without basin-specific tuning and without taking nitrogen concentrations into account. This fundamentally contradicts the popular "vicious circle theory" (see Vahtera et al., 2007), which argues that phosphorus diffusion from sediments is driven by nitrogen limited diatoms.
5. Phosphorus rather than nitrogen seems to limit the long-term (growing season period) primary production in the Baltic Sea (see Fig. 5.14).
6. In spite of the fact that costly measures have been implemented to reduce nitrogen transport from agriculture, urban areas (e.g., from water purification plants) and industries, the nitrogen concentrations in the surface-water in the main basins in the Baltic Sea, the Baltic Proper, have remain almost constant for the since 1965 (see Fig. 2.30).
7. Nitrogen abatement efforts in practice have been poorly correlated to chlorophyll concentrations in many areas in and outside the Baltic Sea (Duarte et al., 2009). The oligotrophication of several Danish coastal waters reported by Carstensen et al. (2006) was preceded by phosphorus reductions in addition to nitrogen reductions, so one cannot rule out that phosphorus reductions alone has been the only effective measure in this case.

5.9.2 Oxygenation of the Deep-Water Compartment in the Baltic Proper by Large Pumps

The following motivation for this method is given by Stigebrandt (2008):

> Symptoms of eutrophication – blooming of cyanobacteria and dead bottom areas – should be reduced by increased oxygenation of the deep-water layer (in the Baltic Proper). Proof of this is that both symptoms disappeared a few years in the 1990s when natural variations caused an oxygenation of the deep-water and the phosphorus content in the Baltic Proper nearly was halved.

Stigebrandt's group has received considerable funding from the National Swedish Environment Agency to test this hypothesis, which is motivated by using data from a site in the Baltic Proper, the Gotland Deep. We have shown in Chap. 2 that this site is not representative for the conditions in the Baltic Proper (see Figs. 2.26, 2.27, 2.28, and 2.29). The seasonal variations are higher both for phosphorus (TP) and oxygen and the oxygen concentrations significantly higher in the period after year 2000 than in the entire Baltic Proper. Our key points against this method are:

(1) The TP-concentrations in the deep-water layer in the entire Baltic Proper do not fluctuate in any clear way with the oxygen concentrations and neither does the ratio between phosphate and TP (see Figs. 2.31, 2.32, 2.33, 2.34, 2.35, and 2.36).
(2) The diffusion from the deep-water sediments in the Baltic Proper (which the artificial oxygenation is meant to reduce) is already very small indeed (0.05 kt TP/year in BP; see Fig. 2.16). In the ranking of the annual abiotic TP-fluxes in the Baltic Proper, it is evident that the most dominating one is from land uplift (107 kt/year), followed by tributary inflow (16.4 kt/year), inflow from the Bothnian Sea (13 kt/year) and total inflow from Kattegat (10.4 kt/year). Sedimentation in the SW-layer is also important with fluxes in the range between 70 and 100 kt/year. Sedimentation in the DW-layer is very small since most (99%; see also Fig. 2.33 and Tables 2.10 and 2.14) of the phosphorus in the DW-layer is in dissolved form. This also implies that the amount of phosphorus in the sediments below the halocline at 75 m in the Baltic Proper available for diffusion is small. The largest diffusive flux is from water from the DW-compartment to the MW-compartment (138 kt/year). The flux related to resuspension is 94 kt/year from ET-areas to the SW-layer, whereas diffusion from AMW-sediments is 0.6 kt/year in the Baltic Proper. Burial, i.e., the transport of TP from the sediment biosphere to the sediment geosphere is 113 and 14 kt/year, respectively, from the MW and DW-zones in the Baltic Proper.
(3) The low phosphorus content in the sediments below the theoretical wave base in the Baltic Proper and the reasons for these low TP-concentrations have been documented in extensive Soviet sedimentological research during the 1970s and 1980s. Less than 3% of the organic material in the SW-layer reaches the deep-water sediments while the rest is mineralized in the water column (Emelyanov, 1988). The major part of the particulate matter in the DW-sediments is instead of terrigenic origin (Blazhchishin and Shuyskiy, 1973; Blazhchishin, 1984). This means that the difference between the phosphorus concentration on particles in sediments and the TP-concentrations on settling particles in the deep-water layer is small; in the southern part of the Baltic Proper, empirical TP-concentrations on particles of deep-water areas were actually slightly lower than in corresponding surface sediments (Emelyanov, 1986).
(4) More evidence to clarify why there are so small amounts of phosphorus in these DW-sediments in the Baltic Proper is that most of the phosphorus (99%; see Table 2.10) in the DW-layer is in dissolved form (more than 90% is phosphate,

see Fig. 2.34), so the sedimentation of particulate phosphorus must of necessity be very small (see Fig. 2.16).

(5) Artificial oxygenation by large pumps does not address the basic cause of the eutrophication in the Baltic Proper, it would be symptom treatment. A much more efficient way to reduce the eutrophication in the Baltic Proper would be to reduce the TP-transport from the main sources in the catchment.

There might, however, be another motive for the oxygenation and that is that this would favor the production of zoobenthos, which in turn would favor the fish production. Our simulations indicate that the total annual inflow would then have to be increased from 345 km^3/year (today) to about 1.5–2 times that value.

So, the method proposed by Stigebrandt and financed by the Swedish Environmental Protection Agency addresses the smallest of all TP-fluxes in the Baltic Proper.

5.9.3 Chemical Treatment to Reduce Diffusion of Phosphorus from Sediments

Blomqvist and Rydin (2009) write:

> If the anoxic areas (70,000 km^2 below the halocline at 75 m in the Baltic Proper) is to be treated to reduce the mobile phosphorus the costs would be between 0.2 and 3 billion euro....

The same arguments can be given against this method (treatment of the sediments in the deep-water layer in the Baltic Proper with aluminum) as already given against the oxygenation: the diffusion from the deep-water sediments in the Baltic Proper is the smaller flux of all phosphorus fluxes because the sedimentation of particulate phosphorus is small, because most (more than 90%) phosphorus in the deep-water layer appear in dissolved form; the TP-concentrations in the DW-sediments are low, about 0.5 mg/g dw according to our calculations, which are supported by data not just from Emelyanov (2001) but also by data from Jonsson (1992). The concentration of hard-bound TP in the sediments is about 0.36 mg/g dw (this is a typical TP-concentration in glacial clays) and only the difference between 0.5 and 0.36 is in practice available for diffusion. Chemical sediment treatment is also symptom treatment and it would be much more efficient to use the money to reduce the TP-inflow to the system.

5.9.4 Cultivation of Mussels

We have already discussed cultivation of mussels as a method to reduce eutrophication and shown that even with an unrealistically high production of mussels, the effects on the bioindicators (e.g., Secchi depth), phosphorus concentrations and functional groups or organisms would be very small. There may, however, be other arguments in favor of mussel cultivation in the Baltic Sea and those arguments are also given by Lindahl (2008), e.g., to use mussel meal in feed for chicken.

5.10 The Management Plan for the Baltic Sea

5.10.1 Background and Aim

The aim of this final section is to put our key results into an overall and holistic framework for Baltic Sea management. Our aim is to motivate and present a management plan, which includes both environmental and economic aspects. So, the idea is to create a link between Baltic Sea ecology and economics. Similar compilations have been done in aquatic radioecology, e.g., in the decision-support system MOIRA (see Monte et al., 2000; Gallego et al., 2004), which is used in many European countries as a stand-by method if there is a new accident such as the one in Chernobyl in 1986. That system includes models of the type discussed in this book and an overall multi-attribute analysis (MAA) including environmental, economical and social aspects. We will not carry out any multi-attribute analysis in this chapter, but one form of cost-benefit analysis, where the environmental attributes are derived from the CoastWeb-model, the costs related to different remedial actions and the values of fish are taken from the previous compilations given in this chapter.

We will, as already mentioned, give three scenarios:

1. Default conditions and "business as usual", i.e., no actions and no investments in remedial measures to reduce eutrophication. We would like to stress again that we are not advocating this as an option, just as a hypothetical example to be used as a reference. In this scenario, we will not calculate possible future changes in water temperatures, salt-water intrusions, invasions of jellyfish, cultivating mussels or fish cage farming.
2. The next option is our "optimal" strategy to reduce tributary phosphorus loading presented in the first section of this chapter. We have also motivated why we will include a fish cage production of 30,000 t/year rainbow trout in this management plan and in this scenario we will add also economic aspects related to the fish cage production. Also in this scenario, we will not calculate possible future changes in water temperatures, salt-water intrusions, invasions of jellyfish and cultivation of mussels.
3. The third option is given by the HELCOM strategy (see Table 2.6) already agreed upon by the Baltic Sea governments but not yet implemented. This means that about 15,000 t/year of phosphorus and about 133,000 t/year of nitrogen should be removed from the present tributary nutrient loading. We will discuss the environmental consequences and costs related to this strategy.

The environmental attributes connected to these options will be related to Secchi depths, chlorophyll-a concentrations and concentrations of cyanobacteria. We will also calculate changes and values (in euro) in prey fish and predatory fish biomasses and also include the production and economic value of the fish cage farming. So, the focus in this evaluation and management plan is on the three bioindicators and on fish.

Our aim is to put all pertinent aspects into one sheet of paper (the calculations are done using data produced by the CoastWeb-model and economic data given previously in this chapter) to get an overview of the options, benefits and drawbacks of the different strategies.

5.10.2 The Management Plan

This section may also be regarded as a summary and conclusion of this chapter. The results are given in Table 5.17. Some background information for Table 5.17 motivated earlier in this chapter include:

Table 5.17 Our management plan for the Baltic Sea is given under "Optimal" scenario, the first column gives "no action", the third column the HELCOM strategy (BSAP; by setting BP – 13,206; GF – 1,510; GR – 300; BS = BB – 0 according to contributing countries in Table 2.6; t P per year), which has been agreed upon by the Baltic States

		Default, no action	"Optimal" scenario+fish farm	HELCOM
Nutrient reductions	TP	0	8,730	15,016
(t/year)	TN	0	0	133,170
Costs, nutrient reductions	TP	0	367	TP+TP =
(million euros)	TN	0	0	3,100
Bioindicators	SecBB	6.3	8.4	11.7
(m)	SecBP	6.7	10.7	16.8
	SecBS	4.8	8.0	14.2
	SecGF	4.0	7.0	8.5
	SecGR	3.5	6.0	7.3
(μg/l), for growing season	ChlBB	0.3	0.3	0.1
	ChlBP	1.7	1.3	0.5
	ChlBS	0.8	0.6	0.2
	ChlGF	2.4	1.9	1.0
	ChlGR	2.4	1.9	1.0
(μg/l), for growing season	CBBB	0.03	0.0	0.6
	CBBP	62.2	9.5	0.1
	CBBS	7.0	0.0	2.5
	CBGF	41.5	8.8	4.0
	CBGR	98.5	22.6	6.2
Professional large-scale	PYBB	85	61	45
fishing (0.65 of total)	PYBP	626	418	307
(kt ww/year)	PYBS	343	203	130
	PYGF	194	143	117
	PYGR	81	58	45
	PDBB	9.3	6.8	5.0
	PDBP	75	72	62
	PDBS	42	32	23

Table 5.17 (continued)

		Default, no action	"Optimal" scenario+fish farm	HELCOM
	PDGF	25	21	18
	PDGR	10	8.5	7.0
Value	PYBB	53	38	28
(million euros)	PYBP	388	259	190
	PYBS	212	126	81
	PYGF	120	89	72
	PYGR	50	36	28
	PDBB	47	34	25
	PDBP	375	360	312
	PDBS	211	160	116
	PDGF	123	107	88
	PDGR	48	42	35
Small-scale and recreational mainly coastal fishing	BB	14	10	7.6
	BP	108	75	57
(0.1 of total fishing)	BS	59	36	24
(kt ww/year)	GF	34	25	21
	GR	14	10	8.1
Added value small-scale fishing	BB	15	11	8.1
(million euros)	BP	117	95	77
	BS	65	44	30
	GF	37	30	25
	GR	15	12	9.7
Added fish cage production	BB	0	15	0
(rainbow trout)	BP	0	10	0
(kt ww/year)	BS	0	5	0
(million euros)	BB	0	90	0
	BP	0	60	0
	BS	0	30	0
Total value (million euros)		1,876	1,624	1,127
Willingness-to-ap (million euros)			1,000	

The table gives mean annual values for Secchi depth and the fish biomasses and mean values for the growing season for chlorophyll and cyanobacteria. Simulation time 481 months.

- The abatement cost for 8,730 t/year of phosphorus is 42 euro/kg or 367 million euro/year.
- The cost for nutrient reductions in the Baltic Sea Action Plan is 3,100 million euro/year (this cost is for both N+P abatement and has not been specified substance-wise).
- The value of wild prey fish, wild predatory fish and cultivated rainbow trout is 0.62, 5.0 and 6.0 euro/kg, respectively.

- If one calculates the value of the small-scale coastal and recreational fishing only based on the value of the fish, 0.62 for prey fish and 5.0 euro/kg for predatory fish, and if one sets this fishing to be 0.1 of the total fishing, the result is 249 million euro/year under default conditions, 192 million euro/year in the "optimal" scenario and 150 million euro/year according to the HELCOM strategy.

Further explanation of the results in Table 5.17:

- No action under default conditions means no further nutrient reductions and no costs (lines 1 and 2 in Table 5.17). In the "optimal" scenario, we reduce 8,730 t/year of phosphorus and no nitrogen. The costs for this would be about 367 million euro if this is done in a cost-effective manner, which means a focus on phosphorus reductions connected to the most polluted estuaries and coastal areas (here the Gulf of Finland and the Gulf of Riga). Evidently, there would be major differences in these costs depending on the country, the method to reduce phosphorus, etc. We assume that most of this would go to the building of water treatment plants in the Baltic countries and Poland. The costs to reduce 15,016 t/year of TP and 133,170 t/year of nitrogen according to the HELCOM strategy would be 3,100 million euro/year. That is, 2,733 million euro/year higher than our "optimal" strategy!

In addition, the HELCOM strategy has thus far been difficult to implement in practical terms. Its cost of 3,100 million euro/year is much higher than the estimated willingness-to-pay for eutrophication measures of 1,000 euro/year. For instance, Sweden's nitrogen goal according to the HELCOM plan may be impossible to reach without closing down a substantial part of Swedish agricultural production (Swedish EPA, 2008b). Since the societal cost for the HELCOM strategy is very high, we would argue that this strategy is more likely to fail than the "optimal" scenario in Table 5.17.

- How would the Baltic Sea ecosystem react to these two options? This is shown in the next lines first for the three bioindicators, Secchi depth, chlorophyll-a concentrations and concentrations of cyanobacteria. It is interesting to note that according to the HELCOM strategy, we would create a situation where the Baltic Sea would be more oligotrophic than before (maybe for several 100 years). The fish production would be much lower than today and about half of what one would get in the "optimal" scenario.
- The CoastWeb-model calculates total fishing of prey and predatory fish. About 10% if this concerns fishing by birds, seals, etc. and value of this is not included in these calculations; about 15% of the fishing would be dead fish dumped at sea and the value of those fish is also disregarded. This means that 75% of the fishing is accounted for, including legal, illegal and recreational

fishing. We have assumed that 10% of the total fishing is recreational fishing and the rest (65%) is professional fishing by bigger boats including big trawlers.

- This means that the total value of the fish would be 1,130 million euro/year according to the HELCOM scenario, 1,880 million euro/year under default conditions and 1,620 million euro/year in the "optimal" scenario. However, looking at the willingness-to-pay, one should note that about 1,000 million euro/year should be added to the "optimal" scenario because this is the sum that people would be ready to pay to do something about the present eutrophication in the Baltic Sea.
- The risks of getting blooms of harmful algae (cyanobacteria, CB) would be significantly lower than today both in the "optimal" scenario and the costly and unrealistic HELCOM scenario. About 30% of the anthropogenic emissions would be removed each year from the Bothnian Sea (390 t TP/year) and the Bothnian Bay (345 t TP/year), 70% from the Gulf of Finland (3,180 t TP/year), 60% from the Gulf of Riga (550 t TP/year) and 81% from the Baltic Proper (10,551 t TP/year) to be able to reach the 15,016 t TP/year requested in the HELCOM strategy. It should be noted that increased water temperatures would increase the risks of getting high concentrations of cyanobacteria and in these simulations we have used the default empirical nitrogen concentrations and default water temperatures to obtain these results using the model for cyanobacteria.
- The Secchi depths inform about water clarity and the depth of the photic zone, which is important for the primary production, and hence also secondary production of zooplankton and fish. For the general public, the Secchi depth is one of the most informative bioindicators. Most people are probably interested in clear waters with little turbidity and no visible cyanobacteria. The target variable for the "optimal" scenario is that the Secchi depth in the Gulf of Finland could return to 7 m, which is was 100 years ago. This concrete goal is met in the optimal scenario, but not in the HELCOM scenario, which would give a Secchi depth higher than most knowledgeable Baltic managers should ask for, 8.5 m, as compared to about 4 m today.
- The concentrations of chlorophyll informs about the phytoplankton biomass. This means that we would advocate that a mean summer chlorophyll-a concentration for the Baltic Proper of about 1 μg/l would be a target for Baltic Sea management. The present value is about two times higher than that.
- The economic value of prey and predatory fish are evidently quite uncertain and vary among the Baltic States and with time depending on supply and demand. In these calculations, we have used a general value of 5 euro/kg for predatory fish and 0.62 euro/kg for prey fish and 6 euro/kg for the cultivated rainbow trout from fish farming. This gives for the "optimal" scenario a total value of the cultivated fish (30,000 t/year) of 180 million euro/year. The added direct value related to small-scale coastal and recreational fishing of prey and predatory has been set to be 10% of the total fishing. This gives for the default conditions:

Basin	BP	BS	BB	GF	GR	Sum
Amount of fish caught	108	59	14	34	14	229 kt ww/year
Value	117	65	15	37	15	249 million euro/year

It should be stressed that there are many other benefits to society from recreational fishing than these based on the economic value of the caught fish.

- It is interesting to note that it would be possible to increase the aquaculture (the fish cage farming of rainbow trout) from about 10,000 t/year today with 30,000 t/year without jeopardizing the environmental goal that the Secchi depth in the Gulf of Finland (GF) should be lower than or equal to 7 m. This assumes that the fish cage farming is done in a sustainable manner so that the wild fish used as fodder for the fish cage production would not harm the wild fish stocks and so that the feed conversion rate must not be higher 1.2. It also means that the fish cage production is mainly localized to:

(1) The northern part of the Baltic Proper (around Åland, in the Finnish Archipelago Sea and in the Roslagen area in Sweden) where it would be 15,000 t/year.
(2) The Bothnian Sea (BS) – 10,000 t/year, and
(3) To the Bothnian Bay (BB) with 5,000 t/year higher than today.

A production of 30,000 t of rainbow trout would correspond to the creation of more than 7,000 full time jobs in the Baltic Sea region.

Some important messages from Table 5.17 include:

- No action is an alternative with a high societal costs since the willingness-to-pay for environmental measures in the Baltic Sea is high and since valuable fish production could be optimized compared with present conditions.
- The HELCOM strategy is associated with a high societal cost since its measures would be excessively expensive and this strategy has thus far also been difficult to implement.
- The "optimal" strategy in Table 5.17 involves effective and cost-effective measures against eutrophication, primarily in urban sewage treatment. It could be implemented at a relatively low cost and still bring the trophic state in the Baltic Sea back to how it was 100 years ago. The nutrient abatement costs of this strategy could be borne solely by the substantial benefits related to improved water quality.

The "optimal" strategy advocated in this work gives:

- 8,730 t/year less phosphorus inputs to the Baltic Sea at a cost of about 367 million euro/year.
- A Secchi depth of about 8.4 m in the Bothnian Bay, 8.0 m in the Bothnian Sea, 7.0 m in the Gulf of Finland 6.0 m in the Gulf of Riga and 10.7 m in the Baltic Proper.
- Reduced cyanobacterial biomass by a factor of 10 in the Baltic Proper.
- A new situation very close to steady-state will be reached about 20 years after measures have been implemented.
- A yearly fishing quota for predatory fish corresponding to 30% of total modeled predatory fishing and this would ensure that fish production is not harmed.
- Increased fish cage farming by 30 kt/year.
- Decreased value of the total fish production in the Baltic Sea by about 250 million euro/year.

Ongoing adjustments of the fishing rate and fishing quota should be made to changes in abiotic factors (water temperature, phosphorus loading, salt-water intrusions and oxygen conditions) regulating the fish production potential of the Baltic Sea system.

It should be stressed that the fish production potential of the Baltic Sea system is particularly sensitive to the frequency and magnitude of saltwater intrusions from the Kattegat and to temperature changes in surface waters.

Epilogue

This is the third and last book in our "Baltic Sea trilogy". The first book focused on tools and criteria for sustainable coastal ecosystem management for smaller coastal areas in the Baltic Sea than those discussed in this book. The second book dealt with the eutrophication in the Baltic Sea and remedial measures to mitigate eutrophication. That book studied the same five major sub-basins in the Baltic Sea as this book, the Baltic Proper, the Bothnian Sea, the Bothnian Bay, the Gulf of Finland and the Gulf of Riga. This book has a focus on methods to determine the fish production potential of the entire Baltic Sea system. We also present a new approach to determine fish quota based on holistic ecosystem modeling. This book motivates and presents a management plan for the Baltic Sea including a cost-benefit analysis.

To develop scientifically warranted programs of conservation, management and remediation is a great challenge. In this situation, quantitative models are essential to predict, to guide assessment and to direct intervention. We would like to regard the CoastWeb-model presented in this book as a new complimentary and general tool to set fish quota based on the fish production potential of a given system such as the Baltic Sea. It is also an approach to handle "trade-offs" and test working hypotheses concerning aquatic foodweb processes and interactions. The fact that the CoastWeb-model, in spite of its breadth and complexity , may be driven by relatively few readily accessible variables, and that it is based on a general production unit which may be repeated for different groups of functional organisms, gives, as we see it, a certain robustness and attractiveness to the model and provides a framework for its practical usefulness and predictive power, which are essential components in models for aquatic management. The minimalistic approach has been essential to us, and it is interesting to note that so much information about complicated ecosystem interactions can be obtained from a model based on this structure.

In ecosystem sciences, there is a need to optimize the model size, i.e., to create a balance between simplification and complication. The predictive power of any model depends on the number of driving variables, the uncertainty of the values used for the driving variables and uncertainties related to the model structure. The practical use of any model depends on the accessibility of the driving variables. Without equations there are few or no possibilities to gain the scientific insights that go with predictive power, and small or no possibilities to make, e.g., meaningful simulations of consequences of remedial measures. Without equations there is, we would

L. Håkanson et al., *The Fish Production Potential of the Baltic Sea*, Environmental Science and Engineering, DOI 10.1007/978-3-642-11562-2, argue contrary to many persons' opinions, very little understanding and knowledge on interactions in aquatic ecosystems, since such systems at most scales are very complex, where "everything depends on everything else". So, mathematical models are fundamental tools to gain understanding about ecosystem interactions. But the CoastWeb-model presented in this book has not been, and no models of this kind could be, deduced only from logical reasoning, like models of physical phenomena. "Logics" in ecology is in the eye of the beholder and explanations at one scale generally focus on processes at the next lower scale, and so on down to the level of the atom. Evidently, ecosystems are much more complex than the physical, chemical or geological parts making up the whole. There may be different approaches to handle, e.g., algorithms for biouptake, consumption rates and distribution coefficients in natural ecosystems. The biggest challenge is to find the best of these alternatives, i.e., the alternatives that are mechanistically most reasonable, the ones which provide the best predictive power in the widest possible domain from the fewest and most readily available driving variables when tested against empirical data. Reliable empirical data are needed at many steps in the derivation of ecosystem models. The ultimate limitation does not, we would argue, lie in the mathematics of modeling but in the access to empirical data and in the knowledge of ecosystem processes that only empirical data can provide.

Chapter 2 gave basic information on the conditions in the Baltic Sea, e.g., on the morphometry including the criteria to define the limit for the surface-water layer from the theoretical wave base using sedimentological criteria. That chapter also presented the water fluxes among and within the sub-basins and between the vertical water layers. These water fluxes are important for the quantification of all fluxes of salt, nutrients and SPM regulating all monthly concentrations in all twelve water layers in the Baltic Sea and also for the transport of planktonic organisms among the sub-basins. We have presented the dynamic mass-balance model (CoastMab) for suspended particulate matter for the entire Baltic Sea. Chapter 2 also gave approaches to predict chlorophyll-a concentrations and Secchi depths from dynamically modeled values of phosphorus, SPM and salinity and monthly light conditions. These approaches are of fundamental importance in the CoastWeb-modeling because the foodweb model is driven by chlorophyll-a concentrations and the Secchi depth regulates the depth of the photic layer. The water fluxes determined from the CoastMab-model for salinity are used throughout the CoastWeb-model. We have demonstrated that the CoastMab-model for phosphorus, which prior to this work has been validated for many independent aquatic systems and been demonstrated to predict very well, also predicts TP-concentrations in the Baltic Sea very well. When modeled values are compared to empirical annual data the coefficient of determination is 0.98 and the slope is close to 1 (0.96), and this is better than when the empirical data from the sub-basins in the Baltic Sea are split into two files and regressed against one another. We have shown how the model predicts TP-concentrations in water and sediments, and also the target bioindicators. In fact, the inherent uncertainties in the available empirical data used to run and test the model set the limit to the predictive power of the model. Chapter 2 also gave comprehensive compilations of how the conditions in the Baltic Sea have changed during the last 50 years. These trend analyses provide a framework to understand the present situation and future developments. Chapter 2 also gave results from extensive literature surveys related to the key fish species and groups of functional organisms and what the organisms eat. This information is important not just to test predictions using the CoastWeb-model but this literature review also disclosed where important knowledge and systematically collected data are missing.

The basic structure of the foodweb model (CoastWeb) was given in Chap. 3, which also gave a short comparison between this modeling approach and the Ecopath/Ecosim approach. This was done to highlight the specific features of the CoastWeb-approach and to stress that it actually provides a new dimension to understand and quantitatively simulate the factors regulating the fish production potential of coastal systems. CoastWeb is based on general principles and processes that apply for most aquatic systems. Simulations in Chap. 3 demonstrate that zoobenthos is an important food for prey fish in the Baltic Sea and that threats to the production of zoobenthos (e.g., low oxygen conditions) would be serious to the fish production in the system.

Chapter 4 gave all sub-models for all functional groups in detail and comparisons between dynamically modeled values and empirically-based values (norm-values) for all functional groups in three scenarios. It should be stressed that there is generally good correspondence between dynamically modeled biomasses or production values and the norm-values. We have also compared modeled biomasses for prey and predatory fish to empirical data compiled from our literature review. Given the fact that our modeled values apply for entire defined basins and provide monthly mean values and that most of the existing data emanate from individual sites, the correspondence is good and within order-of-magnitude ranges. In Chap. 4, we gave three scenarios, which were meant to provide gradients to illustrate the range of the model and how modeled values correspond the norm-values in such wider domains. (1) The TP-inflow to the Baltic Proper was reduced in steps from present-day conditions to very oligotrophic conditions. (2) The salt-water inflow from Kattegat was changed in steps to study how salinity variations would likely influence the system. (3). We gave a temperature scenario to illustrate how possible future water temperature increases might influence the system. Those studies demonstrate that reductions in tributary loading of phosphorus and increased salt-water intrusions would create an oligotrophication of the Baltic Sea system. This is easy to understand and state but here we have demonstrated this by quantitative data and explained the underlying processes, which is far more demanding and scientifically valuable than simple qualitative statements. Increased water temperatures will create higher nutrient concentrations, an eutrophication of the entire system, especially in the Bothnian Bay and the Bothnian Sea, with the highest land uplift and the largest ice-cover.

Thousand-year-old sediments influence the Baltic Sea ecosystem today. When the old bottom areas rise after being depressed by the glacial ice, they will be influenced by the waves, which will resuspend the sediments. The land uplift in the Baltic Sea varies from about 9 mm/year in the northern part of the Bothnian Bay to zero in the south-central part of the Baltic Proper. It has been shown that land uplift may contribute with large amounts of, e.g., phosphorus, nitrogen and suspended particulate matter to the Baltic Sea system. Land uplift influences the entire system in many profound ways, and this has been demonstrated in this work.

Chapter 4 also introduced the dynamic model to predict two fundamental functional categories of fish in the Baltic Sea, prey and predatory fish. The fish model is meant to account for all important factors regulating the production of fish in a general way. The basic aim of the model is that it should capture typical functional and structural patterns in all Baltic Sea basins. It accounts in a relatively simple manner for many complicated processes, like fishing (by birds, animals and man) and fish migration to and from basins. Food choices are handled by distribution coefficients regulating how much of the different available food sources a given organism would consume. Beside these distribution coefficients, and the way the food choices are structured (the food choice panel), fundamental concepts in the fish model are: (1) metabolic efficiency ratios, which express how much of the food consumed by the predator that will increase the biomass of the predator and how much that will be lost by respiration and faeces, (2) actual consumption rates, which are defined from the ratio between the actual biomass of the predator and the normal biomass of the predator, and the normal consumption rates, which are related to the turnover time of the predator. We have demonstrated that the CoastWeb-model gives predictions which agree well with the values given by the empirical norms, and also expected and requested divergences from these regression lines when they do not provide sufficient resolution.

In Chap. 4, it was also shown how the modeled proportions of predatory fish to the total fish biomass vary among the basins and seasonally and one can note that this ratio on average varies around 0.05–0.07 for the default period (1997–2006), and this should be compared to the empirically-based mean value of 0.06 (given in Fig. 2.46). The empirically-based biomass of predatory fish in the Baltic Sea around the year 2000 should (see Chap. 2) be about 100 kt ww, and our modeling gives 104 kt ww for the entire Baltic Sea

The total biomass of prey fish has fluctuated very much indeed during the last 4 decades. Typically, the annual biomasses for the dominating species of prey fish should vary around 2,500 and 5,000 kt ww. This is also what this modeling shows (see Chap. 5).

It should be stressed that the model presented in this book (CoastWeb) can provide good predictions in all Baltic Sea basins without basin-specific tuning and without taking nitrogen concentrations into account.

The main aim of the CoastWeb-model is to quantitatively describe typical, characteristic foodweb interactions so that production, biomasses and predation can be determined for the functional groups of organisms included in the model. Note that the model predicts functional groups, not species. There are many simplifications in the CoastWeb-model. They are necessary for several reasons, (1) to keep the model as small as possible (it is still quite extensive), (2) to keep the driving variable as few and as accessible as possible (otherwise few people are likely use the model), (3) to be able to critically test the model using empirical data or empirical regressions. The idea has not been to include everything but to focus on the key functional groups of organisms and on key abiotic/biotic relationships.

In Chap. 5, we gave several scenarios to illustrate the practical use of the CoastWeb-approach. In the last section of Chap. 5, we have put the results from these scenarios together and presented an holistic management plan for the Baltic Sea including a cost-benefit analysis. We have shown how the CoastWeb-modeling may be a useful tool in contexts of settling fish quota adjusted to changing environmental conditions, the influences of invasions of jellyfish and the cultivation of mussels and the expected consequences for the fish production potential after reducing the anthropogenic nutrient loading. So, the final compilation in Chap. 5 combines the responces of nutrient reductions, the connected consequences for the fisheries and a cost-benefit analysis related to different remedial strategies. Chapter 5 also includes a scenario where we have accounted for aquaculture and studied the effects of increasing fish farming of rainbow trout in the Baltic Sea and discussed how large fish farms could be recommended in the different basins.

- In the "optimal" scenario we reduce 8,730 t/year of phosphorus and no nitrogen. The annual costs for this would be about 367 million euro if this is done in a cost-effective manner. We argue that most of this would go to the building of sewage treatment plants in the Baltic countries and Poland. The costs to reduce 15,016 t/year of TP and 133,170 t of nitrogen according to the HELCOM strategy would be 3,100 million euro/year. That is 2,733 million euro/year higher than our strategy!
- According to the HELCOM strategy, one would create a situation where the Baltic Sea would be more oligotrophic than knowledgeable Baltic Sea managers should ask for. The total fish production would be much lower than today and the concentrations of organic toxins (such as PCBs, dioxins, etc.) in fish would likely be higher. The value of the fish caught via professional and recreational fishing would be about 1,130 million euro/year, as compared to 1,880 million euro/year today and 1,620 million euro/year in our "optimal" scenario. However, looking at the willingness-to-pay, one should note that this value is about 1,000 million euro/year and this value should be added to the "optimal" scenario because this is the sum that people would be ready to pay to do something about the present conditions in the Baltic Sea.
- The target variable for the "optimal" scenario is that the Secchi depth in the Gulf of Finland could again return to 7 m, which is was 100 years ago. This concrete goal is met in the optimal scenario, but not in the HELCOM scenario, where the Secchi depth would be as high as about 8.5 m, compared to about 4 m today.
- The economic value of the prey and predatory fish is evidently quite uncertain and varies among the Baltic States and with time depending on supply and demand. We have used a general value of 5 euro/kg for predatory fish and 0.62 euro/kg for prey fish and 6 euro/kg for the cultivated rainbow trout from fish cage farming. This gives for the "optimal" scenario a total value of the cultivated fish (30,000 t/year) of 180 million euro/year. The added value related to small-scale coastal and recreational fishing is about 190 million euro/year in the "optimal"

scenario, as compared to about 250 million euro/year today. So, the value of the recreational fishing is substantial.
- It is interesting to note that it would be possible to increase the aquaculture (the fish cage farming of rainbow trout) from about 10,000 t/year today with 30,000 t/year without jeopardizing the environmental goal that the Secchi depth in the Gulf of Finland (GF) should be 7 m. This assumes that the fish cage production is mainly, (1) in the northern part of the Baltic Proper (around Åland, in the Finnish Archipelago Sea and in the Roslagen area in Sweden), 15,000 t/year, (2) in the Bothnian Sea, 10,000 t/year and (3) in the Bothnian Bay, 5,000 t/year. Our strategy for the fish cage production would create more than 7,000 new jobs. Important presuppositions for this scenario are, (1) that a significant fraction (at least 20%) of the food eaten by the cultivated fish should come from wild fish caught in the Baltic Sea, such as sprat, i.e., from commercially less attractive and more abundant prey fish that could be caught without negative consequences for the Baltic Sea foodweb and fisheries and (2) that this fish farming would not jeopardize other natural fish stocks which supply feed (pellets) to the cultivated fish.

In our management plan for the Baltic Proper, we calculate changes in three key bioindicators, Secchi depth (as a measure of water clarity and the depth of the photic zone), chlorophyll-a concentrations (as a measure of phytoplankton production and biomass) and concentration of cyanobacteria (as a measure of harmful algae), and also biomasses of prey and predatory fish.

We have motivated why remedial actions should not focus on nitrogen and there are four main reasons:

1. It is not possible to provide scientifically relevant predictions of how the Baltic Sea system would respond to costly reductions in nitrogen loading since there are several major uncertainties related to the quantification of (a) nitrogen fixation, (b) wet and dry deposition of nitrogen, (c) the algorithm regulating the particulate fraction for nitrogen and hence also (d) sedimentation of particulate nitrogen and (e) denitrification.
2. Nitrogen reductions in the Baltic Sea are likely to favor the blooming of harmful algae (cyanobacteria), and such events should be avoided.
3. There are no validated mass-balance models for nitrogen which have been blind tested for independent coastal systems and been demonstrated to yield good predictive power.
4. In spite of the fact that costly measures have been implemented to reduce nitrogen transport from agriculture, urban areas (e.g., from water purification plants) and industries, the nitrogen concentrations in the surface-water in the main basins in the Baltic Sea, the Baltic Proper, have remain almost constant for 30 years.

The general phosphorus model presented in this book can provide good predictions of chlorophyll-a concentrations in all Baltic Sea basins without basin-specific tuning and without taking nitrogen concentrations into account. These findings fundamentally contradict the popular "vicious circle theory" which asserts that phosphorus diffusion from deep sediments is driven by nitrogen limited diatoms. Phosphorus rather than nitrogen seems to limit the long-term (growing season period) primary production in the Baltic Sea.

We have also discussed why oxygenation of the deep-water layer in the Baltic Proper, and chemical treatment of the sediments to reduce the diffusion of phosphorus from the sediments below the halocline at 75 m should be largely useless. This is because the sediment concentrations of TP in the deep-water layer in the Baltic Proper are low and close to the TP-concentration in older clay, e.g., of glacial origin. The main reason for the low sediment concentration of TP is that it is well documented that more than 90% of the phosphorus in the deep-water layer appears in dissolved forms, so the sedimentation of phosphorus is by necessity limited.

For the future, we believe that research in the following areas would help to reduce the uncertainties addressed in this book.

- The limited empirical knowledge regarding most biomasses for most functional groups of organisms and dominating species in the Baltic Sea basins is a problem which can only be handled by collecting more and better data from the system in a systematic manner.
- Today, there are uncertainties concerning the values used for the distribution coefficients on the food choice panels, e.g., regulating the consumption by prey fish of herbivorous zooplankton relative to predatory zooplankton. It would be valuable to get more information on that issue.
- Better information is also needed on consumption rates and metabolic efficiency ratios for the functional groups in the Baltic Sea system. How much of a given prey biomass is actually consumed by the predator per time unit? Are there other more relevant approaches to estimate consumption rates quantifying the fraction of the prey in the system being lost per time unit from grazing by the given predator?
- There are also uncertainties regarding structural foodweb changes at trophic boundaries. What is actually happening to foodweb structures, species composition and biomasses at trophic boundaries? Sudden changes may take place in the production and abundance of key functional organisms when there are changes in abiotic state variables, such as phosphorus, salinity and water temperature.
- The CoastWeb-model is based on many new structures and approaches, and it would be interesting to try to expand this type of modeling, e.g., to larger water bodies, such the Great Lakes of America, the Caspian Sea, the Black Sea, Norwegian fjords and Lake Ladoga. It would be a great challenge, with many possibilities for the management of such large water bodies, to try to adopt the modeling principles presented in this work, e.g., to set quota for fisheries also in such important systems. Evidently, large water bodies may have to be differentiated into functional parts, just like we have done for the Baltic Sea, and such divisions may not be related to geographical or national boundaries. So, to paraphrase Winston Churchill, "This is not the end, not even the beginning of the end, but it is the end of the beginning".

Thus, the CoastWeb-model should be regarded as a tool for sustainable Baltic Sea management and science. The scenarios in Chap. 5 have been included to stress and illustrate this point.

> What makes a free thinker is not his beliefs, but the way in which he holds them. If he holds them because his elders told him they were true when he was young, or if he holds them because if he did not he would be unhappy, his thought is not free; but if he holds them because, after careful thought, he finds a balance in their favour, then his thought is free, however odd his conclusions may seem – Bertrand Russell.

A
Appendices

Table A.1 A compilation of the differential equations for the dynamic SPM-model (CoastMab) using data for the Gulf of Riga (GR) to exemplify the calculation routines. Abbreviations and dimensions are given in Table 2.11

Surface water (SW)

$M_{SPMSWGR}(t) = M_{SPMSWGR}(t - dt) + (F_{SWSPMBPGR} + F_{xSPMDWSWGR} + F_{SPMETSWGR} + F_{SPMtribGR} - F_{SPMSWGRBP} - F_{SPMSWDWGR} - F_{SPMSWETGR} - F_{xSPMSWDWGR} - F_{SPMminSWGR})dt$

$M_{SPMSWGR}(t)$ = Mass (amount of SPM) in the SW-compartment at time t (g)
$F_{SWSPMBPGR}$ = Flow into the SW-compartment from the Baltic Proper (BP; g/month); see below
$F_{xSPMDWSWGR}$ = Flow from deep water to surface water (upward mixing; g/month)
$F_{SPMETSWGR}$ = Flow (resuspension) from ET-areas to the SW-compartment (g/month)
$F_{SPMtribGR}$ = Flow into the SW-compartment from tributaries (g/month)
$F_{SPMSWGRBP}$ = Flow from the SW-compartment and out to the Baltic Proper (g/month)
$F_{SPMSWDWGR}$ = Flow (sedimentation) from the SW-compartment to deep-water (DW) compartment (g/month)
$F_{SPMSWETGR}$ = Flow (sedimentation) from the SW-compartment to ET-areas (g/month)
$F_{xSPMSWDWGR}$ = Flow from surface water to deep water (downward mixing; g/month)
$F_{SPMminSWGR}$ = Flow (mineralization) from the SW-compartment (g/month)

ET-areas (ET)

$M_{SPMETGR}(t) = M_{SPMETGR}(t - dt) + (F_{SPMSWETGR} + F_{SPMLUGR} - F_{SPMETDWGR} - F_{SPMETSWGR} - F_{SPMminETGR})\cdot dt$

$M_{SPMETGR}(t)$ = Mass (amount of SPM) in the ET-compartment at time t (g)
$F_{SPMLUGR}$ = Flow into the SW-compartment from land uplift (g/month)
$F_{SPMETDWGR}$ = Flow (resuspension) from ET-areas to the DW-compartment (g/month)
$F_{SPMminETGR}$ = Flow (mineralization) from the ET-areas (g/month)

Deep water (DW)

$M_{SPMDWGR}(t) = M_{SPMDWGR}(t - dt) + (F_{SPMSWDWGR} + F_{SPMETDWSWGR} + F_{xSPMSWDWGR} + F_{SPMDWBPGR} - F_{xSPMDWSWGR} - F_{SPMDWADWGR} - F_{SPMDWGRBP} - F_{SPMminDWGR})dt$

$M_{SPMDWGR}(t)$ = Mass (amount of SPM) in the DW-compartment at time t (g)
$F_{SPMDWBPGR}$ = Flow into the DW-compartment from the Baltic Proper (g/month)
$F_{SPMDWADWGR}$ = Flow (sedimentation) from the DW-compartment to A-areas (ADW; g/month)
$F_{SPMDWGRBP}$ = Flow from the DW-compartment and out to the Baltic Proper (g/month)
$F_{SPMminDWGR}$ = Flow (mineralization) from the DW-compartment (g/month)

Table A.1 (continued)

A-areas (ADW)

$M_{SPMADWGR}(t) = M_{SPMAGR}(t - dt) + (F_{SPMDWADWGR} - F_{BurSPMGR} - F_{SPMminADWGR})dt$

$M_{SPMADWGR}(t)$ = Mass (amount of SPM) in the ADW-compartment at time t (g)
$F_{BurSPMGR\ P}$ = Flow (burial) from the ADW-compartment (g/month)
$F_{SPMminADWGR}$ = Flow (mineralization) from the ADW-compartment (g/month)

Algorithms for fluxes

Inflow:
$F_{SPMtribGR} = F_{TPtribGR} \cdot Y_{TribGR}$
$F_{SWSPMBPGR} = Q_{SWBPGR} \cdot SPM_{SWBP}$
$F_{SPMDWBPGR} = Q_{MWBPGR} \cdot SPM_{MWBP}$

Outflow
$F_{SPMSWGRBP} = Q_{SWGRBP} \cdot SPM_{SWGR}$
$F_{SPMDWGRBP} = Q_{DWGRBP} \cdot SPM_{DWGR}$

Production (mass of SPM from primary production from the CoastWeb-model)
$M_{SPMprodGR} = (M_{BPGR} + M_{PHGR} + M_{ZHGR}) \cdot 1{,}000$

Sedimentation
$F_{SPMSWETGR} =$
$M_{SPMSWGR} \cdot (v_{SWGR}/D_{SWGR}) \cdot ET_{GR} \cdot (1 - DC_{ResSPMSWGR}) + Y_{ResGR} \cdot DC_{ResSPMSWGR})$
$F_{SPMSWDWGR} =$
$M_{SPMSWGR} \cdot (v_{SWGR}/D_{SWGR}) \cdot (1 - ET_{GR}) \cdot (1 \cdot (1 - DC_{ResSPMSWGR}) + Y_{ResGR} \cdot DC_{ResSPMSWGR})$
$F_{SPMDWADWGR} =$
$M_{SPMDWGR} \cdot Y_{TGR} \cdot (v_{DWGR}/D_{DWGR}) \cdot ((1 - DC_{ResSPMDWGR}) + Y_{ResGR} \cdot DC_{ResSPMDWGR})$

Burial
$F_{BurSPMGR} = M_{SPMAGR} \cdot (1/(Age_{ADWGR})) \cdot Y_{LU}$

Resuspension
$F_{SPMETSWGR} = M_{SPMETGR} \cdot R_{ResGR} \cdot (1 - Vd_{GR}/3)$
$F_{SPMETDWGR} = M_{SPMETGR} \cdot R_{ResGR} \cdot (Vd_{GR}/3) \cdot Y_{LU}$
$F_{SPMLUGR} = F_{TPLUGR} \cdot Y_{TribGR}$

Mixing
$F_{xSPMSWDWGR} = M_{SPMSWGR} \cdot R_{MixSWDWGR}$
$F_{xSPMDWSWGR} = M_{SPMDWGR} \cdot R_{MixSWDWGR} \cdot V_{SWGR}/V_{DWGR}$

Mineralization
$F_{SPMminSWGR} = M_{SPMSWGR} \cdot R_{MinSWGR}$
$F_{SPMminDWGR} = M_{SPMDWGR} \cdot R_{MinDWGR}$
$F_{SPMminETGR} = M_{SPMETGR} \cdot R_{Minsed}$

Other model variables and algorithms
$Area_{GR} = 16{,}700 \cdot 10^6$ (km^2)
$Area_{EGR} = 7{,}810 \cdot 10^6/(Area_{GR} - Area_{belowDwbGR})$ (km^2)
$A_{areasGR} = Area_{GR} \cdot (1 - ET_{GR})$ (dim. less)
$Area_{AboveDwbGR} = Area_{GR} - Area_{BelowDwbGR}$ (km^2)
$Area_{BelowDwbGR} = 3{,}500 \cdot 10^6$ (km^2)
$Area_{UpfilftedperyrGR} = (Area_{GR} - Area_{BelowDwbGR}) \cdot 0.001 \cdot LR_{mm/monthGR} \cdot 12/D_{wbGR}$ (km^2)
$Age_{ETGR} = (12/Strat_{GR})$ (months)
$Age_{ADWGR} = 12 \cdot 10/(Sed_{cm/ryrGR})$ (months)

Table A.1 (continued)

$\mathrm{Amp_{Rmig}} = 0.2$ (dim. less)

$\mathrm{Amp_{Trib}} = 0.5$ (dim. less)

$\mathrm{C_{BPGR}} = 1{,}000 \cdot \mathrm{M_{BPGR}/V_{SWGR}}$ (conc. bacterioplankton; t/m^3)

$\mathrm{Chl_{GR}} = \mathrm{Y_{DayLGR} \cdot TP_{SWGR} \cdot (DF_{SWGR}/1) \cdot Y_{4GR} \cdot Y_{TempChl}}$ [μg/l]

$\mathrm{DC_{ResSPMDWGR}} = \mathrm{F_{SPMETDWSWGR}/(F_{SPMDWBPGR}+F_{SPMETDWSWGR}+F_{SPMSWDWGR}+F_{xSPMSWDWGR})}$ (dim. less)

$\mathrm{DC_{ResSPMSWGR}} = \mathrm{(F_{SPMETSWGR})/(F_{SPMETSWGR}+F_{SPMPPGR}+F_{SWSPMBPGR}+F_{SPMtribGR}+F_{xSPMDWSWGR})}$ (dim. less)

$\mathrm{D_{DWGR}} = \mathrm{(D_{MaxGR}-D_{wbGR})/2}$ (m)

$\mathrm{DF_{SWGR}} = 1-\mathrm{PF_{SWGR}}$ (dim. less)

$\mathrm{D_{MaxGR}} = 56$ (m)

$\mathrm{D_{wbGR}} = (45.7 \cdot (\mathrm{Area_{GR}} \cdot 10^{-6})^{0.5}/(21.4+(\mathrm{Area_{GR}} \cdot 10^{-6})^{0.5}))$ (m)

$\mathrm{D_{SWGR}} = \mathrm{D_{wbGR}}/2$ (m)

$\mathrm{d_{GR}} = 100 \cdot 2.6/(100+(\mathrm{W_{GR}}+\mathrm{IG_{GR}} \cdot (1-\mathrm{W_{GR}}/100)) \cdot (2.6-1))$ (g/cm^3)

$\mathrm{DWT_{GR}}$ = (1.00, 4.29), (2.00, 3.50), (3.00, 3.32), (4.00, 3.26), (5.00, 3.41), (6.00, 3.68), (7.00, 3.76), (8.00, 3.83), (9.00, 3.91), (10.0, 4.02), (11.0, 4.73), (12.0, 5.39) (°C)

$\mathrm{ET_{GR}} = \mathrm{(Area_{GR}-Area_{BelowDwbGR})/Area_{GR}}$ (dim. less)

$\mathrm{F_{TPLUGR}} = (10^6) \cdot \mathrm{LU_{GR}}/12$ (g TP/month)

$\mathrm{F_{TPtribGR}} = (((202+582+335))/12) \cdot 10^6 \cdot \mathrm{Y_{QGR}}$ (g/month)

Ice limit = 0.9 (°C)

$\mathrm{Lat_{GR}} = 57.7$ (°N)

$\mathrm{LR_{mm/monthGR}} = 0.625/12$ (mm/month)

$\mathrm{LU_{GR}} = 12 \cdot (\mathrm{Area_{AboveDwbGR}}+\mathrm{Area_{UpfilftedperyrGR}}) \cdot 0.001 \cdot \mathrm{LR_{mm/monthGR}} \cdot ((1-(\mathrm{W_{GR}}-15)/100)) \cdot (\mathrm{d_{GR}}+0.2) \cdot ((\mathrm{TP_{Clay}} \cdot \mathrm{Area_{EGR}}+(1-\mathrm{Area_{EGR}}) \cdot \mathrm{TP_{AsedGR}})) \cdot 1{,}000 \cdot 10^{-6}$ (μg TP/year)

$\mathrm{NBM_{BPGR}} = \mathrm{Y_{SPMGR}} \cdot 0.001 \cdot \mathrm{V_{SWGR}} \cdot 10^{(0.973 \cdot (0.27 \cdot \log(\mathrm{ChlGR})+0.19)-0.438)}$ (kg ww)

$\mathrm{PF_{SWGR}} = (\mathrm{Y_{PFGR}}+(\mathrm{M_{TPBioSGR}}/(\mathrm{M_{TPSWGR}}+\mathrm{M_{TPBioSGR}}))$ (dim. less)

$\mathrm{Q_{DWGRBP}} = \mathrm{Q_{GRBP}}-\mathrm{Q_{SWGRBP}}$ (m^3/month)

$\mathrm{Q_{MWBPGR}} = \mathrm{QB_{PGR}} \cdot (1-\mathrm{DC_{SWDWGR}})$ (m^3/month)

$\mathrm{Q_{SWBPGR}} = \mathrm{Q_{BPGR}} \cdot \mathrm{DC_{SWDWGR}}$ (m^3/month)

$\mathrm{Q_{SWGRBP}} = \mathrm{Q_{SWBPGR}}+\mathrm{Q_{TribGR}}+(\mathrm{Q_{PrecGR}}-\mathrm{Q_{EvaGR}})$ (m^3/month)

$\mathrm{Ref_{Temp}} = 9$ (°C)

$\mathrm{R_{MinDWGR}} = \mathrm{R_{minGR}} \cdot (\mathrm{DWT_{GR}}/\mathrm{Ref_{Temp}})^{1.2}$ (1/month)

$\mathrm{R_{MinGR}} = (\mathrm{M_{BPGR}}/\mathrm{NBM_{BPGR}}) \cdot 0.01 \cdot (0.99/\mathrm{ET_{GR}})$ (1/month)

$\mathrm{R_{Minsed}} = 0.01 \cdot (30/\mathrm{Y_{LU}})$

$\mathrm{R_{MinSWGR}} = \mathrm{R_{minGR}} \cdot (\mathrm{SWT_{GR}}/\mathrm{Ref_{Temp}})^{1.2}$ (1/month)

$\mathrm{R_{MixSWDWGR}}$ = (if $\mathrm{Sal_{DWGR}} > \mathrm{Sal_{SWGR}}$ then $\mathrm{R_{mixdefGR}} \cdot (1/(1+\mathrm{Sal_{DWGR}}-\mathrm{Sal_{SWGR}}))^2$ else $\mathrm{R_{MixdefGR}}$) (1/month)

$\mathrm{R_{MixdefGR}} = \mathrm{Strat_{GR}} \cdot \mathrm{ET_{GR}}/12$ (1/month)

$\mathrm{R_{ResGR}}$ = if $\mathrm{SWT_{GR}} <$ Ice limit then $(\mathrm{SWT_{GR}}+ 0.2) \cdot 1/\mathrm{Age_{ETGR}}$ else $1/\mathrm{Age_{ETGR}}$ (1/month)

$\mathrm{Seasnorm_{Latma}}$ = −1.000, −1.000, −1.000, −1.000, 2.170, 2.510, 0.630, 0.240, 0.050, −0.030, −0.660, −0.920 (dim. Less)

$\mathrm{Seasnorm_{Latmin}}$ = 1.040, 1.370, 0.560, 0.380, −0.290, −0.230, −0.620, −0.710, −0.790, −0.740, −0.280, 0.320 (dim. less)

$\mathrm{Seasnorm_{Qmax}}$ = −0.710, −0.480, −0.170, −0.170, 0.620, 1.740, 0.520, 0.090, −0.160, −0.200, −0.630, −0.440 (dim. less)

$\mathrm{Seasnorm_{Qmin}}$ = 0.580, 0.810, 0.840, 1.580, −0.100, −1.000, −1.000, −1.000, −0.820, −0.560, 0.110, 0.540 (dim. less)

$\mathrm{Strat_{GR}}$ = if $\mathrm{ABS(SWT_{GR}-DWT_{GR})} < 4$ then $1+1/1/(1+\mathrm{ABS(SWT_{GR}-DWT_{GR})})$ else $1/\mathrm{ABS(SWT_{GR}-DWT_{GR})}$ (dim. less)

Table A.1 (continued)

$Sec_{GR} = Y_{QSec}\cdot GR\cdot(10^{(-(zGR+0.5)\cdot(\log(SPMSWGR)/1+0.3)/2+zGR)})$ (m)
$Sed_{DWGR} = 10^2\cdot F_{SPMDWAGR}/(A_{AreasGR}\cdot 30)$ (μg/cm²·d)
SWT_{GR} = (1.00, 2.00), (2.00, 1.13), (3.00, 0.755), (4.00, 1.76), (5.00, 3.27), (6.00, 4.16), (7.00, 5.11), (8.00, 6.10), (9.00, 12.6), (10.0, 11.1), (11.0, 7.02), (12.0, 7.02) (°C)
$SPM_{DWGR} = M_{SPMDWGR}/V_{DWGR}$ (mg/l)
$SPM_{SWGR} = (M_{SPMSWGR}+ M_{SPMprodGR})/V_{SWGR}$ (mg/l)
$Temp_{CritGR}$ = if SWT_{GR} < Ice_limit then Ice_limit/(Ice_limit+SWT_{GR}) else Y_{DRGR}
TP_{Clay} = 0.36 (mg/g dw)
$TP_{AsedGR} = M_{TPADWGR}/((10^3)\cdot V_{AsedGR}\cdot d_{GR}\cdot(1-W_{GR}/100))$ (mg/g dw)
$TP_{SWGR} = 1{,}000\cdot(M_{TPSWGR}+M_{TPBioSGR})/V_{SWGR}$ (μg/l)
TP_{Nat}/TP_{TotGR} = 0.18 = 202/(202+582+335) (dim. less)
$V_{AsedGR} = Area_{belowDwbGR}\cdot 10\cdot 0.01\cdot(Vd_{DWGR})/3$ (m³)
$Vd_{DWGR} = 3\cdot D_{MVDWGR}/(D_{MaxGR}-D_{wbGR})$ (dim. less)
$Vd_{GR} = 3\cdot D_{MVGR}/D_{MaxGR}$ (dim. less)
$v_{DWGR} = v_{Def}\cdot Y_{SalDWGR}\cdot Y_{SPMDWGR}\cdot Y_{LU}\cdot Temp_{CritGR}$ (m/month)
$v_{SWGR} = v_{def}\cdot Y_{SPMSWGR}\cdot Y_{SalSWGR}\cdot Temp_{critGR}$ (m/month)
v_{Def} = 6 m/month
Y_{DRGR} = if DR_{GR} < 0.26 then DR_{GR}/0.26 else 0.26/DR_{GR}
$V_{DWGR} = 18\cdot 10^9$ (m³)
$V_{SWGR} = 392\cdot 10^9$ (m³)
W_{GR} = 75 (% ww)
Y_{DayLGR} = HDL/12 (dim. less)
Y_{DR} = If DR < 0.26 then 1 else 0.26/DR (calculates how changes in DR and turbulence influence sedimentation) (dim. less)
Y_{DW} = If T_{DW} < 7 (days) then Y_{DW} = 1 else $Y_{DW} = (T_{DW}/7)^{0.5}$ (calculates how changes in T and turbulence influence deep-water sedimentation)
Y_{LU} = if $((F_{TotTPinGR}+F_{TPLUGR})/F_{TotTPinGR})\cdot(0.76/Vd_{GR})$< 1 then 1 else $((F_{TotTPinGR}+F_{TPLUGR})/F_{TotTPinGR})\cdot(0.76/Vd_{GR})$ (dim. less)
$Y_{QGR} = 1+0.526\cdot((Lat_{GR}-35)^{2.18}/35^{2.18}\cdot Seasnorm_{Latmax}+(1-(Lat_{GR}35)^{2.18}/35^{2.18})\cdot Seasnorm_{Latmin})+ 0.265\cdot((Q_{empGR}/(60\cdot 60\cdot 24\cdot 365))^{0.22}/ 5{,}000^{0.22}\cdot Seasnorm_{Qmax}+(1-(Q_{empGR}/(60\cdot 60\cdot 24\cdot 365))^{0.22}/5{,}000^{0.22})\cdot Seasnorm_{Qmin})$ (dim. Less)
$Y_{QSec} = (Q_{MWBPGR}+Q_{SWBPGR})/(Q_{TribGR}+Q_{MWBPGR}+Q_{SWBPGR})$ (dim. less)
$Y_{PFGR} = F_{TPETSWGR}/(F_{TPprecGR}+F_{TPETSWGR}+F_{TPSWBPGR}+F_{TPtribGR}+F_{dTPDWSWGR}+F_{xTPDWSWGR})$ (dim. less)
$Y_{ResGR} = (Age_{ETGR}+1)^{0.5}\cdot Y_{LU}$ (calculates how much faster resuspended sediments settle out) (dim. less)
$Y_{SalSW} = (1 + 1\cdot(Sal_{SWGR}/1-1) = 1\cdot Sal/1$ (calculates how changes in salinities > 1 psu influence sedimentation) (dim. less)
$Y_{SalDWGR} = (1+1\cdot(Sal_{SWGR}/1-1))$ (dim. less)
$Y_{SPMDWGR} = 1+0.75\cdot(SPM_{DWGR}/50-1)$ (dim. less)
$Y_{SPMSWGR} = 1+0.75\cdot(SPM_{SWGR}/50-1)$ (dim. less)
Y_{TGR} = if $T_{DWGR}/30 < 7$ then 1 else $((T_{DWGR}/30)/7)^{0.5}$ (dim. less)
$Y_{TribGR} = (1{,}000/2)\cdot(1-0.5\cdot(TP_{nat}/TP_{totGR}/0.5-1))$ (dim. less)
Y_{4GR} = if Y_{3GR} < 0.012 then 0.012 else Y_{3GR} (dim. less)
Y_{3GR} = if Sal_{SWGR} > 40 then $(0.06-0.1\cdot(Sal_{SWGR}/40-1))$ else Y_{2GR} (dim. less)
Y_{2GR} = if Sal_{SWGR} < 12.5 then Y_{1GR} else $(0.28-0.1\cdot(Sal_{SWGR}/12.5-1))$ (dim. less)
Y_{1GR} = if Sal_{SWGR} < 2.5 then $(0.20-0.2\cdot(Sal_{SWGR}/12.5-1))$ else $(0.20+0.02\cdot(Sal_{SWGR}/2.5-1))$ (dim. less)
$Y_{TempChl}$ = if SWT_{GR} > 4 then 1 else $(SWT_{GR}+0.1)/4$ (dim. less)
$z_{GR} = (10^{\wedge(0.15\cdot\log(1+SalSWGR)+0.3)-1})$ (dim. less)

Table A.2 Compilation of monthly SPM-fluxes in the Baltic Sea based on the mean monthly values. Values in kt SPM (= 10^9 g) per month. Abbreviations and dimensions are given in Table 2.11

Month	1	2	3	4	5	6	7	8	9	10	11	12	MV
burADWBP	1,647	1,864	2,041	2,257	2,528	2,509	2,135	1,798	1,571	1,321	1,182	1,382	1,853
burADWGF	13	38	26	6	6	6	6	6	5	5	4	5	11
burAMWBP	1,460	1,470	1,619	1,570	1,138	745	683	692	735	877	1,344	1,452	1,149
burAMWGF	38	99	74	20	16	16	16	17	19	18	17	17	31
burBB	557	606	769	797	914	907	818	740	651	476	466	543	687
burBS	880	952	1,209	1,256	1,446	1,438	1,301	1,181	1,042	762	744	863	1,090
burGR	7	11	9	6	6	6	6	7	7	6	7	7	7
DWABB	2,006	1,000	526	570	421	657	383	299	386	911	1,185	1,101	787
DWABS	1,104	1,171	1,455	1,430	1,552	1,577	1,467	1,331	1,173	878	892	1,069	1,258
DWADWBP	1,646	1,841	1,954	2,243	2,707	2,636	2,101	1,733	1,492	1,243	1,125	1,370	1,841
DWADWGF	22	64	42	10	10	10	10	9	9	8	7	8	17
DWAGR	13	21	16	12	11	11	11	13	12	11	12	13	13
DWBBBS	41	16	6	5	15	25	24	21	20	29	40	47	24
DWBPBS	4	5	5	5	5	3	3	3	2	2	3	4	4
DWBPGF	15	16	18	18	17	14	12	10	9	8	10	12	13
DWBPGR	5	6	6	6	5	4	4	4	3	3	4	5	5
DWBSBB	4	4	4	4	3	0	0	0	0	3	7	5	3
DWGFBP	15	14	13	15	15	16	17	16	15	14	13	14	15
DWGRBP	18	19	20	19	18	19	21	20	16	13	16	18	18
DWKABP	40	40	40	40	40	40	40	40	40	40	40	40	40
ETDWBB	753	252	330	518	1,166	145	101	138	788	1,272	1,185	1,139	649
ETDWBS	1,107	1,068	1,060	1,269	823	120	92	100	147	1,014	1,947	1,192	828
ETDWGR	23	23	25	27	29	24	23	8	3	12	27	25	21
ETMWBP	2,493	2,331	2,601	2,079	993	192	157	169	251	1,096	2,790	2,836	1,499
ETMWGF	29	12	30	51	56	49	50	33	5	25	41	34	35
ETSWBB	428	140	180	266	566	72	51	70	402	665	643	639	344
ETSWBP	3,446	3,187	3,443	2,842	1,455	279	214	225	328	1,419	3,663	3,878	2,032
ETSWBS	944	893	869	980	603	90	71	77	113	802	1,591	1,007	670

Table A.2 (continued)

Month	1	2	3	4	5	6	7	8	9	10	11	12	MV
ETSWGF	56	23	56	98	107	92	95	63	10	47	78	64	66
ETSWGR	27	26	28	32	35	28	25	9	3	13	30	28	24
LUBB	1,032	1,030	1,029	1,031	1,033	1,030	1,024	1,020	1,019	1,023	1,026	1,029	1,027
LUBP	4,507	4,532	4,549	4,559	4,547	4,485	4,423	4,364	4,306	4,263	4,362	4,455	4,446
LUBS	1,879	1,880	1,881	1,883	1,884	1,883	1,881	1,879	1,875	1,873	1,875	1,876	1,879
LUGF	294	294	293	293	294	294	294	294	294	294	294	294	294
LUGR	119	119	119	119	119	119	119	119	119	119	119	119	119
minADWBB	82	42	22	28	24	36	17	12	15	33	44	44	33
minADWBP	106	118	121	144	187	179	133	108	91	75	69	88	118
minADWBS	73	76	92	86	88	91	87	79	70	53	56	70	77
minADWGF	7	19	13	3	3	3	3	3	3	2	2	2	5
minADWGR	4	6	4	3	3	3	3	3	3	3	3	4	3
minAMWBP	108	106	109	114	95	60	48	47	49	57	89	104	82
minAMWGF	0.1	0.1	0.1	0.0	0.1	0.1	0.1	0.2	0.2	0.3	0.2	0.1	0.1
minDWBB	1	0.004	0.001	0.000	0.113	1	1	1	2	7	10	2	2
minDWBP	42	42	52	78	114	63	61	58	35	23	33	34	53
minDWBS	14	13	12	10	29	30	15	18	8	6	22	15	16
minDWGF	0.1	0.03	0.03	0.1	0.3	0.3	0.2	0.4	0.4	0.3	0.3	0.2	0.2
minDWGR	0.1	0.1	0.1	0.1	0.2	0.2	0.2	0.3	0.3	0.2	0.3	0.1	0.2
minETBB	240	286	314	375	419	376	342	343	340	287	253	242	318
minETBP	1,986	1,846	1,689	1,653	1,791	1,926	1,998	2,136	2,266	2,362	2,268	2,142	2,005
minETBS	717	674	640	590	530	606	705	780	846	899	820	751	713
minETGF	277	347	411	383	333	299	278	268	280	292	283	275	311
minETGR	119	117	121	118	114	106	101	103	113	120	121	121	114
minMWBP	19	15	18	26	36	20	21	22	14	10	21	18	20
minMWGF	1.5	0.3	0.3	0.8	2.2	2	2	4	5	5	5	2	3
minSWBB	0.8	0.000	0.000	0.001	0.8	10	16	19	12	13	18	2	8
minSWBP	63	47	69	165	445	459	714	744	364	151	142	64	286
minSWBS	3	2	2	3	51	78	53	61	22	8	15	3	25

Table A.2 (continued)

Month	1	2	3	4	5	6	7	8	9	10	11	12	MV
minSWGF	1.20	0.02	0.09	1.44	8	11	10	30	52	41	25	3	15
minSWGR	0.64	0.23	0.33	1.32	3	4	5	14	25	15	9	1	7
MWAMWBP	1,665	1,657	1,766	1,773	1,383	891	764	758	794	937	1,453	1,634	1,290
MWAMWGF	72	190	141	38	31	31	31	33	35	34	32	33	59
MWBPBS	4	5	5	5	5	3	3	3	2	2	3	4	4
MWBPGF	8	8	9	9	7	6	5	5	4	4	6	7	7
MWBPGR	5	6	6	6	5	4	4	4	3	3	4	5	5
MWDWBP	2,406	2,396	2,553	2,563	1,999	1,287	1,104	1,096	1,147	1,354	2,100	2,362	1,864
MWDWGF	20	54	40	11	9	9	9	9	10	10	9	9	17
MWGFBP	24	24	23	22	21	22	24	26	25	25	25	24	24
MWGRBP	18	19	20	19	18	19	21	20	16	13	16	18	18
MWKABP	70	70	70	70	70	70	70	70	70	70	70	70	70
prodBB	111	53	40	56	68	486	420	259	375	287	326	288	231
prodBP	4,287	3,834	3,915	4,808	8,105	9,425	3,783	8,453	8,099	7,639	4,510	4,053	5,909
prodBS	376	467	516	540	1,175	3,480	1,130	1,921	2,049	1,827	913	721	1,260
prodGF	467	226	133	520	1,440	1,479	650	927	833	690	557	503	702
prodGR	240	203	173	340	643	653	313	410	398	405	261	223	355
SWBBBS	41	16	6	5	15	25	24	21	20	29	40	47	24
SWBPBS	147	156	165	150	121	131	147	141	127	115	123	136	138
SWBPGF	153	163	172	173	166	167	164	152	135	120	126	140	153
SWBPGR	16	17	19	19	17	18	19	18	16	14	14	15	17
SWBPKA	156	167	166	215	286	252	177	144	122	103	107	135	169
SWBSBB	21	22	23	16	7	10	15	14	12	11	17	21	16
SWBSBP	112	118	120	116	127	117	96	80	65	58	85	106	100
SWDWBB	535	365	150	113	52	69	91	102	89	109	145	157	165
SWDWBS	506	535	715	534	572	520	446	394	358	290	357	483	476
SWDWGR	9	17	12	9	9	9	10	10	10	11	11	10	11
SWETBB	920	629	258	195	90	119	156	175	154	188	249	271	284
SWETBP	1,180	1,258	1,325	1,346	1,115	891	942	943	976	994	1,002	1,079	1,088

Table A.2 (continued)

Month	1	2	3	4	5	6	7	8	9	10	11	12	MV
SWETBS	353	373	499	372	399	362	311	275	250	202	249	337	332
SWETGF	161	400	256	59	57	64	70	72	66	68	68	66	117
SWETGR	35	64	47	34	33	35	38	38	37	40	41	37	40
SWGFBP	258	225	190	202	244	278	286	289	285	275	266	262	255
SWGRBP	32	32	31	33	37	38	36	35	35	34	33	33	34
SWKABP	319	319	319	319	319	319	319	319	319	319	319	319	319
SWMWBP	1,663	1,773	1,868	1,897	1,572	1,256	1,328	1,330	1,377	1,402	1,412	1,522	1,533
SWMWGF	94	233	149	35	33	38	41	42	38	40	40	38	68
tribBB	69	71	67	119	192	164	103	84	76	65	57	63	94
tribBP	696	695	619	808	1,136	988	634	519	474	445	480	642	678
tribBS	73	74	69	107	164	141	89	73	66	58	56	67	86
tribGF	216	217	196	280	412	356	227	186	169	153	155	199	230
tribGR	51	51	46	56	72	61	39	33	31	31	35	47	46
xDWMWBP	42	45	57	53	27	4	2	2	2	7	24	36	25
xDWMWGF	2	1	1	2	3	3	3	2	0	1	2	2	2
xDWSWBB	164	53	19	25	75	11	4	3	29	113	166	206	72
xDWSWBS	86	93	99	122	82	9	5	4	4	26	84	77	58
xDWSWGR	17	19	21	27	40	37	32	10	1	4	13	15	20
xMWDWBP	22	23	28	26	12	2	1	1	1	4	14	21	13
xMWDWGF	3	3	3	3	4	4	4	3	0	2	3	3	3
xMWSWBP	60	63	78	71	34	4	3	2	2	10	37	55	35
xMWSWGF	69	67	64	70	89	86	100	73	11	45	80	69	68.5
xSWDWBB	121	46	17	18	33	6	4	4	20	57	90	127	45
xSWDWBS	44	47	48	55	38	4	2	2	2	11	42	40	28
xSWDWGR	13	14	15	19	28	28	25	8	1	5	11	12	15
xSWMWBP	112	120	146	134	70	11	7	6	7	24	73	102	68
xSWMWGF	63	54	48	55	78	82	98	70	10	42	73	63	61

Table A.3 Basic equations for the CoastWeb-model for a system with two vertical compartments (here the Bothnian Bay, BB). Abbreviations and dimensions are given in Table 2.11

Bacterioplankton

$BM_{BPBB}(t) = BM_{BPBB}(t - dt) + (IPR_{BPBB} + MIG_{InBPBB} - CON_{BPZHBB} - EL_{BPBB} - MIG_{OutBPBB})dt$

$IPR_{BPBB} = R_{PRBP} \cdot (SPM_{SWBB}/1{,}000) \cdot V_{SWBB} \cdot Y_{SWTBB}{}^{1}$
$MIG_{InBPBB} = R_{MigPHBB} \cdot NBM_{BPBS}$
$CON_{BPZHBB} = BM_{BPBB} \cdot CR_{ZHBB}$
$EL_{BPBB} = BM_{BPBB}/T_{BP}$
$MIG_{OutBPBB} = R_{MigPHBB} \cdot BM_{BPBB}$

Benthic algae

$BM_{BABB}(t) = BM_{BABB}(t - dt) + (IPR_{BABB} - EL_{BABB} - CON_{BAZBBB} - ER_{BABB})dt$

$IPR_{BABB} = R_{IPRBABB} \cdot Area_{SecBB} \cdot (2 \cdot Sec_{BB}) \cdot Y_{TPBB} \cdot (HDL/12) \cdot Y_{SWTBB}{}^{1}$
$EL_{BABB} = BM_{BABB}/T_{BA}$
$CON_{BAZBBB} = BM_{BABB} \cdot CR_{BAZBBB}$
$ER_{BABB} = BM_{BABB} \cdot R_{ErBB}$

Jellyfish

$BM_{JEBB}(t) = BM_{JEBB}(t - dt) + (IPR_{ZHJEBB} + IPR_{ZPJEBB} + MIG_{InJEBB} - EL_{JEBB} - MIG_{OutJEBB})dt$

$IPR_{ZHJEBB} = Y_{SalJEBB} \cdot DC_{ZPtoPHBP} \cdot (1 - DC_{ZPZHJE}) \cdot CON_{ZHJEBB} \cdot MER_{ZP} \cdot Y_{SWTBB}{}^{0.5}$
$IPR_{ZPJEBB} = Y_{SalJEBB} \cdot (DC_{ZPtoPHBP}) \cdot DC_{ZPZHJE} \cdot CON_{ZPJEBB} \cdot MER_{ZP}$
$MIG_{InJEBB} = R_{MigPYBB} \cdot NBM_{JEBB} \cdot Y_{SalJEBB}$
$EL_{JEBB} = BM_{JEBB}/T_{JE}$
$MIG_{OutJEBB} = R_{MigPYBB} \cdot BM_{JEBB}$

Macrophytes

$BM_{MABB}(t) = BM_{MABB}(t - dt) + (IPR_{MABB} - CON_{MAZBBB} - EL_{MABB} - Er_{MABB})dt$

$IPR_{MABB} = R_{PRMA} \cdot Area_{BB} \cdot MA_{CovBB} \cdot 0.01 \cdot (HDL/12) \cdot Y_{SWTBB}{}^{1}$
$CON_{MAZBBB} = BM_{MABB} \cdot CB_{ZBvsMABB} \cdot 0.001$
$EL_{MABB} = BM_{MABB}/T_{MA}$
$Er_{MABB} = BM_{MABB} \cdot R_{erBB}$

Predatory fish

$BM_{PDBB}(t) = BM_{PDBB}(t - dt) + (IPR_{PDBB} + MIG_{InPDBB} - FISH_{PDBB} - EL_{PDBB} - MIG_{OutPDBB})dt$

$IPR_{PDBB} = BM_{ERPD} \cdot F_{PYPDBB} \cdot Y_{SWTBB}{}^{0.25}$
$MIG_{InPDBB} = R_{MigPDBB} \cdot NBM_{PDBB}$
$FISH_{PDBB} = 2 \cdot BM_{PDBB} \cdot R_{FishBB}$
$EL_{PDBB} = BM_{PDBB} \cdot 1/T_{PD}$
$MIG_{OutPDBB} = \text{if } BM_{PDBB}/NBM_{PDBB} > 1 \text{ then } (R_{migPDBB}) \cdot (BM_{PDBB}) \text{ else } 0.5 \cdot (R_{MigPDBB}) \cdot (BM_{PDBB})$

Phytoplankton

$BM_{PHBB}(t) = BM_{PHBB}(t - dt) + (IPR_{PHBB} + MIG_{InPHBB} - ELP_{HBB} - CON_{PHZHBB} - MIG_{OutPHBB}) \cdot dt$

$IPR_{PHBB} = PrimP_{BB}$
$MIG_{InPHBB} = R_{MigPHBB} \cdot NBM_{PHBS}$

Table A.3 (continued)

$EL_{PHBB} = BM_{PHBB} \cdot 1/T_{PH}$
$CON_{PHZHBB} = BM_{PHBB} \cdot CR_{ZHBB}$
$MIG_{OutPHBB} = R_{MigPHBB} \cdot BM_{PHBB}$

Prey fish
$BM_{PYBB}(t) = BM_{PYBB}(t - dt) + (IPR_{ZHPYBB} + IPR_{ZPPYBB} + IPR_{ZBPYBB} + MIG_{InPYBB} - CON_{PYPDBB} - EL_{PYBB} - FISH_{PYBB} - MIG_{OutPYBB})dt$

$IPR_{ZHPYBB} = DC_{ZPZBBB} \cdot (1 - DC_{ZPZH}) \cdot MER_{PY} \cdot CON_{ZHPYBB} \cdot Y_{SWTBB}{}^{0.25}$
$IPR_{ZPPYBB} = DC_{ZPZBBB} \cdot DC_{ZPZH} \cdot CON_{ZPPYBB} \cdot MER_{PY} \cdot Y_{SWTBB}{}^{0.25}$
$IPR_{ZBPYBB} = F_{ZBPYBB} \cdot MER_{PY} \cdot (1 - DC_{ZPZBBB}) \cdot Y_{SWTBB}{}^{0.25}$
$MIG_{InPYBB} =$ if $BM_{PYBB}/NBM_{PYBB} > 1$ then $0.5 \cdot R_{MigPYBB} \cdot NBM_{PYBS}$ else $R_{MigPYBB} \cdot NBM_{PYBS}$
$CON_{PYPDBB} = BM_{PYBB} \cdot CR_{PYBB}$
$EL_{PYBB} = BM_{PYBB} \cdot 1/T_{PY}$
$FISH_{PYBB} = BM_{PYBB} \cdot R_{FishBB}$
$MIG_{OutPYBB} =$ if $BM_{PYBB}/N_{BMPYBB} > 1$ then $R_{migPYBB} \cdot BM_{PYBB}$ else $0.5 \cdot R_{MigPYBB} \cdot BM_{PYBB}$

Zoobenthos
$BM_{ZBBB}(t) = BM_{ZBBB}(t - dt) + (IPR_{MAZBBB} + IPR_{BAZBBB} + IPR_{SedZBBB} - CON_{ZBPYBB} - EL_{ZBBB})dt$

$IPR_{MAZBBB} = (1 - DC_{BAMA}) \cdot CON_{MAZBBB} \cdot MER_{MA} \cdot Y_{SWTBB}{}^{0.25}$
$IPR_{BAZBBB} = DC_{BAMA} \cdot MER_{BA} \cdot CON_{BAZBGBB} \cdot Y_{SWTBB}{}^{0.25}$
$IPR_{SedZBBB} = M_{SedBB} \cdot DC_{SedBA} \cdot NCR_{ZBBB} \cdot (MER_{SedBB} \cdot (ET_{BB} + (1 - ET_{BB}) \cdot Y_{Eh1BB} \cdot Y_{EhBB})) \cdot Y_{SWTBB}{}^{0.25}$
$CON_{ZBPYBB} = BM_{ZBBB} \cdot CR_{PYBB}$
$EL_{ZBBB} = BM_{ZBBB} \cdot 1/T_{ZB}$

Herbivorous zooplankton
$BM_{ZHBB}(t) = BM_{ZHBB}(t - dt) + (IPR_{PHZHBB} + IPR_{BPZHBB} + MIG_{InZHBB} - EL_{ZHBB} - CON_{ZHPYBB} - CON_{ZHZPBB} - CON_{ZHJEBB} - MIG_{OutZHBB})dt$

$IPR_{PHZHBB} = DC_{PHBPBP} \cdot CON_{PHZHBB} \cdot MER_{PHZH} \cdot Y_{SWTBB}{}^{0.5}$
$IPR_{BPZHBB} = (1 - DC_{PHBPBP}) \cdot CON_{BPZHBB} \cdot MER_{BPZH} \cdot Y_{SWTBB}{}^{0.5}$
$MIG_{InZHBB} = R_{MigZPBB} \cdot NBM_{ZHBS}$
$EL_{ZHBB} = BM_{ZHBB} \cdot 1/T_{ZH}$
$CON_{ZHPYBB} = BM_{ZHBB} \cdot CR_{PYBB}$
$CON_{ZHZPBB} = BM_{ZHBB} \cdot CR_{ZPBB}$
$CON_{ZHJEBB} =$ if $BM_{JEBB} > 1$ then $BM_{ZHBB} \cdot CR_{JEBB} \cdot R_{ProdJE}$ else 0
$MIG_{OutZHBB} = R_{MigZPBB} \cdot BM_{ZHBB}$

Predatory zooplankton
$BM_{ZPBB}(t) = BM_{ZPBB}(t - dt) + (IPR_{ZPBB} + MIG_{InZPBB} - CON_{ZPPYBB} - EL_{ZPBB} - CON_{ZPJEBB} - MIG_{OutZPBB})dt$

$IPR_{ZPBB} = CON_{ZHZPBB} \cdot MER_{ZP} \cdot Y_{SWTBB}{}^{0.25}$
$MIG_{InZPBB} = R_{MigZPBB} \cdot NBM_{ZPBS}$
$CON_{ZPPYBB} = BM_{ZPBB} \cdot CR_{PYBB}$
$EL_{ZPBB} = BM_{ZPBB} \cdot 1/T_{ZP}$
$CON_{ZPJEBB} = BM_{ZPBB} \cdot C_{RJEBB} \cdot R_{ProdJE}$
$MIG_{OutZPBB} = R_{MigZPBB} \cdot BM_{ZPBB}$
Model variables
$Area_{BB} = 36{,}260 \cdot 10^6$

Table A.3 (continued)

$Area_{2SecBB} = Area_{BB} - Area_{2SecBB} \cdot 10^9$ (=littoral fraction above 2 Secchi depths)
$C_{BABB} = 1{,}000 \cdot BM_{BABB}/Area_{BB}$
$C_{BPBB} = 1{,}000 \cdot BM_{BPBB}/V_{SWBB}$
Chl_{ModBB} = Modeled concentration of chlorophyll-a in BB (μg/L)
$C_{MABB} = 1{,}000 \cdot BM_{MABB}/Area_{BB}$
$C_{PDBB} = 1{,}000 \cdot BM_{PDBB}/V_{SWBB}$
$C_{PHBB} = 1{,}000 \cdot BM_{PHBB}/V_{SWBB}$
$C_{PYSWBB} = 1{,}000 \cdot BM_{PYBB}/V_{SWBB}$
$CR_{BAZBBB} = (NCR_{ZBBB} + NCR_{ZBBB} \cdot (BM_{ZBBB}/NBM_{ZBBB} - 1))$
$CR_{ZBMABB} = (NCR_{ZBBB} + NCR_{ZBBB} \cdot (BM_{ZBBB}/NBM_{ZBBB} - 1))$
$CR_{JEBB} = (NCR_{JEBB} + NCR_{JEBB} \cdot (BM_{JEBB}/NBM_{JEBB} - 1))$
$CR_{PDBB} = Y_{Fish} \cdot (NCR_{PDBB} + NCR_{PDBB} \cdot (BM_{PDBB}/NBM_{PDBB} - 1))$
$CR_{PYBB} = (NCR_{PYBB} + NCR_{PYBB} \cdot (BM_{PYBB}/(NBM_{PYBB}) - 1))$
$CR_{ZHBB} = (NCR_{ZHBB} + NCR_{ZHBB} \cdot (BM_{ZHBB}/NBM_{ZHRBB} - 1))$
$CR_{ZPBB} = (NCR_{ZPBB} + NCR_{ZPBB} \cdot (BM_{ZPBB}/NBM_{ZPBB} - 1))$
$C_{ZBBB} = 1{,}000 \cdot BM_{ZBBB}/Area_{BB}$
$C_{ZHBB} = 1{,}000 \cdot BM_{ZHBB}/V_{SWBB}$
$C_{ZPBB} = 1{,}000 \cdot BM_{ZPBB}/V_{SWBB}$
$DC_{BAMA} = (1 - DC_{SedBA}) \cdot 0.75$
$DC_{SedBA} = 0.75$
$DC_{PDBBBS} = Q_{SWBSBB}/(Q_{SWBSBB} + Q_{SWBSBP})$
$DC_{PHBP} = 0.5$
DC_{PYPDBB} = if $(TP_{SWBB}/(TP_{SWBB}+22))^{0.4} < 0.9$ then 0.9 else $(TP_{SWBB}/(TP_{SWBB}+22))^{0.4}$
$DC_{PYPDBB2}$ = if $DC_{PYPDBB} > 0.99$ then 0.99 else DC_{PYPDBB}
$DC_{PHBPBP} = 0.5$
$DC_{ZHZP} = 0.8$
$DC_{ZPPHBP} = 0.75$
$DC_{ZPZB} = 0.65$
DC_{ZPZBBB} = if $DC_{ZPZB} > 0.9$ then 0.9 else DC_{ZPZB}
$DC_{ZPZH} = 0.2$
$DC_{ZPZHJE} = 0.5$
ET_{BB} = ET-areas in BB from the CoastMab-model
$F_{FishPDBB} = BM_{PDBB} \cdot R_{fishBB}$
$Y_{SalSec} = (Sal_{SWBB}/12)$
$IG = 12$
$Lat_{BB} = 64$
$MA_{percentage} = 0.01$
$MA_{CovBB} = Litfrac_{BB} \cdot MA_{Percentage}$
$MER_{BA} = 0.15$
$MER_{BPZH} = 0.24$
$MER_{SedBB} = MER_{BA} \cdot 0.25$
$MER_{MA} = 0.15$
$MER_{PD} = 0.25$
$MER_{PHZH} = 0.24$
$MER_{PY} = 0.16$
$MER_{ZP} = 0.32 \cdot 1$
$M_{SedBB} = ((12-5)/12) \cdot (IG/100) \cdot (1/(1-W/100)) \cdot (M_{SPMABB} + M_{SPMETBB})$
$NBM_{MABB} = NPR_{MABB} \cdot T_{MA}$
$NBM_{BABB} = NPR_{BABB} \cdot T_{BA}$
$NBM_{BPBB} = 0.001 \cdot V_{SWBB} \cdot 10^{(0.973 \cdot (0.27 \cdot \log(ChlModBB) + 0.19) - 0.438)}$

Table A.3 (continued)

$NBM_{FishBB} = Y_{ChlBB} \cdot 10^{-6} \cdot ((Area_{BB} \cdot 590 \cdot TP_{SWBB}{}^{0.71}))$
$NBM_{JEBB} = NBM_{ZPBB} \cdot 4 \cdot Y_{SalJEBB}$
$NBM_{PDBB} = (1 - DC_{PYPDGR2}) \cdot SMTH(NBM_{FishBB}, T_{PD}, NBM_{FishBB})$
$NBM_{PHBB} = Y_{ChlBB} \cdot (10^{-6}) \cdot (1{,}500 - V_{SecBB}) \cdot (10^9) \cdot (30 \cdot TP_{SWBB}{}^{1.4})$
$NBM_{PYBB} = (DC_{PYPDBB}) \cdot SMTH(NBM_{FishBB}, T_{PY}, NBM_{FishBB})$
$NBM_{ZBBB} = Y_{ChlBB} \cdot (10^{-6}) \cdot 810 \cdot (TP_{SWBB}{}^{0.71}) \cdot Area_{BB}$
$NBM_{ZHRBB} = Y_{ChlBB} \cdot (DC_{ZHZP}) \cdot 10^{-6} \cdot (V_{SWBB}) \cdot 38 \cdot TP_{SWBB}{}^{0.64}$
$NBM_{ZPBB} = Y_{ChlBB} \cdot (1 - DC_{ZHZP}) \cdot 10^{-6} \cdot V_{SWBB} \cdot 38 \cdot TP_{SWBB}{}^{0.64}$
$NCR_{JEBB} = N_{JE}/T_{JE}$
$NCR_{PDBB} = 1/T_{PD}$
$NCR_{PYBB} = N_{PY} \cdot (NCR_{ZPBB} \cdot 0.15 + NCR_{PDBB} \cdot 0.85)$
$NCR_{ZBBB} = N_{ZB}/T_{ZB}$
$NCR_{ZHBB} = N_{ZH}/T_{ZH}$
$NCR_{ZPBB} = 1/T_{ZP}$
$N_{JE} = 2$
$NPR_{BABB} = 0.63 \cdot (A_{2Sec}/A) \cdot PR_{PHBB}$
$NPR_{MABB} = 0.001 \cdot Area_{BB} \cdot 1/52 \cdot 10^{(2.472+1.028 \cdot \log(MACovBB) - 0.516 \cdot 90/(90-LatBB))}$
$NPR_{ZHBB} = 0.148 \cdot F_{IPRPHBB}{}^{0.86}$
$NPR_{ZPBB} = 0.0759 \cdot F_{IPRPHBB}{}^{0.84}$
$N_{PY} = 2$
$N_{ZB} = 2$
$N_{ZH} = 2$
$MIG_{OutPDBS} =$ if $BM_{PDBS}/NBM_{PDBS} > 1$ then $(R_{MigPDBS}) \cdot (BM_{PDBS})$ else $0.5 \cdot (R_{MigPDBS}) \cdot (BM_{PDBS})$
$MIG_{OutPYBS} =$ if $BM_{PYBS}/NBM_{PYBS} > 1$ then $(R_{MigPYBS}) \cdot (BM_{PYBS})$ else $0.5 \cdot (R_{MigPYBS}) \cdot (BM_{PYBS})$
$PR_{BABB} = BM_{BABB}/T_{BA}$
$PR_{BPBB} = BM_{BPBB}/T_{BP}$
$PrimP_{BB} =$ if $Sec_{BB} > 1$ then
$((10^{-6}) \cdot ((2.13 \cdot Chl_{modBB}{}^{0.25+0.25})^4) \cdot (1/0.45) \cdot (1/0.2) \cdot 30.42 \cdot (1{,}500 - V_{SecBB}) \cdot 10^9)$ else
$((10^{-6}) \cdot ((2.13 \cdot Chl_{modBB}{}^{0.25+0.25})^4) \cdot (1/0.45) \cdot (1/0.2) \cdot 30.42 \cdot Area_{BB} \cdot (2 \cdot Sec_{BB})^2))$
$PR_{JEBB} = BM_{JEBB}/T_{JE}$
$PR_{MABB} = BM_{MABB}/T_{MA}$
$R_{ProdJE} = 8.5$
$PR_{PDBB} = BM_{PDBB}/T_{PD}$
$PR_{PHBB} = BM_{PHBB}/T_{PH}$
$PR_{PYBB} = BM_{PYBB}/T_{PY}$
$PR_{ZBBB} = BM_{ZBBB}/T_{ZB}$
$PR_{ZHBB} = BM_{ZHBB}/T_{ZH}$
$PR_{ZPBB} = BM_{ZPBB}/T_{ZP}$
$Q_{SWBSBB} = (Q_{evaBB} + Q_{SWBBB}S) - (Q_{tribBB} + Q_{precBB} + Q_{DWBSBB})$
$Q_{SWBSBP} = (1{,}055 \cdot 10^9)/12$
$R_{ErBB} =$ if $(0.1186 - 0.1338 \cdot \log(MA_{covBB}) + 0.0769 \cdot Vd_{BB}) < 0.1$ then 0.1 else $(0.1186 - 0.1338 \cdot \log(MA_{covBB}) + 0.0769 \cdot Vd_{BB})$
$R_{FishBB} = (BM_{PDBB}/NBM_{PDBB}) \cdot Y_{Arearef/BB} \cdot R_{fish/12}$
$R_{Fish} = 0.5$
$R_{IPRBABB} = 0.01$
$R_{Migconst} = 0.1$
$R_{MigPDBB} = R_{Migconst} \cdot Y_{SeasonBB}/T_{SWBB}$
$R_{MigPHBB} = 1/T_{SWBB}$
$R_{MigPYBB} = R_{MigPDBB}$

Table A.3 (continued)

$R_{MigZPBB} = 1/T_{SWBB}$
$R_{PRBP} = 12$
$RP_{RMA} = 0.025 \cdot (30.42/7)$
Sal_{SWBB} = Modeled salinity in SW-layer in BB
SecBB = Modeled Secchi depth in BB
Sec_{LakeBB} = if $(10^{(-(1+0.5)\cdot(\log(SPMSWBB)+0.3)/2+1)}) > Sec_{BB}$ then Sec_{BB} else $(10^{(-(1+0.5)\cdot(\log(SPMSWBB)+0.3)/2+1)})$
$Sed_{ABBcmyr}$ = Modeled salinity in sedimentation in A-areas in BB in cm/year
$smth_{BB}$ = if $Lat_{BB} > 63$ then 1 else $(63 - Lat_{BB})$
SPM_{SWBB} = Modeled SPM-concentration in SW-layer in BB
$T_{BA} = 4/30.42$
$T_{BP} = 2.8/30.42$
$Thres_{SalJE} = 10$
$T_{JE} = 120/30.42$
$T_{MA} = 300/30.42$
TP_{SWBB} = Modeled TP-concentration in SW-layer in BB
$T_{PY} = 2 \cdot 450/30.42$
$T_{PY} = 300/30.42$
$T_{ZB} = 128/30.42$
$T_{ZP} = 11/30.42$
Vd_{BB} = Form factor in BB
$V_{SWBB} = 1{,}067 \cdot 10^9$
$W = 75$
$Y_{Arearef/BB} = (10^{12}/Area_{BB})^{0.5}$
$Y_{AreasecBP} = (Area_{2SeclakeBP}/Area_{2SecBP})$
$Y_{ChlBB} = Chl_{modBB}/Chl_{modlakeBB}$
$Y_{ChlZBBB} = 1/Y_{area2SecBP}$
Y_{Eh1BB} = if $Sed_{ABBcmyr} > 0.75$ (cm/year) then $Y_{Eh1BB} = 0$ else $Y_{Eh1BB} = 1$
Y_{EhBB} = if $Sed_{ABBcmyr} < 0.075$ (cm/year) then $Y_{EhBB} = (1 - 1 \cdot (SedADWGR\_cm/year/0.075 - 1))$ else $Y_{EhBB} = 1$
Y_{FishBB} = if $Y_{fish1BS} < 0.2$ then 0.2 else $Y_{fish1BS}$
$Y_{Fish1BS}$ = if $TP_{SWBS} < 30$ then $(1 - 2.5) \cdot (NBM_{PYTPBB}/NBM_{refPYBB} - 1)$ else $(1 - 0.4) \cdot (NBM_{PYTPBB}/NBM_{refPYBB} - 1)$
$Y_{SWTBB} = SWT/9$
$NBM_{PYTPBB} = 10^{-6} \cdot (Area_{BB} \cdot 590 \cdot TP_{SWBB}{}^{0.71})$
$NBM_{refPYBB} = 10^{-6} \cdot (Area_{BB} \cdot 590 \cdot 30^{0.71})$
Y_{LU} = Modeled "clay factor" from CoastMab
$Y_{SalJEBB}$ = if $Sal_{SWBB} < Thres_{SalJE}$ then 0 else 1
$Y_{SalSWBB} = (1 + 1 \cdot (Sal_{SWBB}/1 - 1))$
$Y_{SalSecBB} = Sal_{SW}/12$
$Y_{SeasonBB}$ = if $(Y_{SeasonBB} - Y_{Season1BB}) \geq 0$ then $((Y_{SeasonBB} + Y_{Season1BB})/2) \cdot (Lat_{BB}/63)$ else $((Y_{Season1BB} + Y_{SeasonBB})/2) \cdot (63/Lat_{BB})$
If Lat > 63°N then AV = 1 else AV = (63-Lat); AV is an averaging function used in the smoothing function $Y_{Season1BB} = SMTH(Y_{SeasonBB}, AV, 1)$
$Y_{SecZPBB}$ = if $Sec_{BB} < 2$ then $(1 + 1 \cdot (Sec_{BB}/2 - 1))$ else 1
$Y_{TPBB} = (1 + 0.75 \cdot (TP_{SWBB}/10 - 1))$

References

Abrahamsson, O. and Håkanson, L., 1998. Modelling seasonal flow variability of European rivers. Ecol. Modell., 114:49–58.
Ackefors, H. and Olburs, C., 1996. Swedish Aquaculture Policy – a nightmare for the industry? Aquacult. Eur., 21:6–13.
Aertbjerg, G. (ed.), 2001. Eutrophication in Europe's coastal waters. European Environment Agency, Topic Report 7/2002, Copenhagen, 86p.
Aho, T., Ask, L., Axenrot, T., Bergström, L., Degerman, E., Edsman, L., Engwall, Y., Florin, A.-B., Gustavsson, T., Hemmingsson, M., Jonsson, A., Königsson, S., Lunneryd, S.-G., Nordwall, F., Nyberg, P., Paulrud, A., Persson, P., Sandström, A., Sjöstrand, B., Sköld, M., Sundelöf, A., Svedäng, H., Ulmestrand, M., Westerberg, H., Öresland, V. and Andersson, L., 2008. Fiskbestånd och miljö i hav och sötvatten. Fiskeriverket, Göteborg.
Aizaki, M., Otsuki, A., Fukushima, T., Hosomi, M. and Muraoka, K., 1981. Application of Carlson's trophic state index and other parameters. Verh. Int. Verein. Limnol., 21:675–681.
Alanärä, A., 2000. Optimizing feeding in fish farm,in (in Swedish, Optimering av utfodring vid kommersiell fisskodling). Vattenbruksinstitutionen, SLU, Rapport 21, Umeå, 32p.
Al-Hamdani, Z. and Reker, J. (eds.), 2007. Towards marine landscapes in the Baltic Sea. BALANCE interim report #10. Accessible at http://balance-eu.org/. ISBN: 978-87-7871-203-5.
Alimov, A.F., 1982. Productivity of invertebrate community of macrobenthos in continental water bodies of the Soviet Union (Review). Hydrobiol. J., 18:7–18 (in Russian).
Alimov, A.F., 1989. Introduction to production hydrobiology. Nauka, Leningrad (in Russian).
Ambio, 1990. Marine eutrophication, Ambio, 19:101–176, Special issue.
Ambio, 2000. Eutrophication and contaminants in the aquatic environment. Ambio, 4–5:183–290.
Anderson, G.C. and Zeutschel, R.P., 1970. Release of dissolved organic matter by marine phytoplankton in coastal and offshore areas of the Northeast Pacific Ocean. Limnol. Oceanogr., 15:402–407.
Andersson, M.H., Gullström, M., Asplund, M.E. and Öhman, M.C., 2007. Importance of using multiple sampling methodologies for estimating of fish community composition in offshore wind power construction areas of the Baltic Sea. Ambio, 36:634–636.
Anonymous, 2000. The state of the world fisheries and aquaculture 2000. FAO, Fisheries Department, Rome. ISBN 92-5-104492-9.
Anonymous, 2002. Report of the Workshop Course on Fish Assessment Techniques. ICES CM 2002/D:02.
Appelgren, K. and Mattila, J., 2005. Variation in vegetation communities in shallow bays of the northern Baltic Sea. Aquat. Bot., 83:1–13.
Arifin, Z. and Bendell-Young, L.I., 2001. Cost of selective feeding by the blue mussel (Mytilus trossulus) as measured by respiration and ammonia excretion rates. J. Exp. Mar. Biol. Ecol., 260:259–269.
Aro, E., 1989. A review of fish migration patterns in the Baltic. Rapp. P.-v. Réun. Cons. Int. Explor. Mer., 190:72–96.

Aro, E., 2000. The spatial and temporal distribution patterns of cod (*Gadus morhua callarias* L.) in the Baltic Sea and their dependence on environmental variability – implications for fishery management. Thesis. Finnish Game and Fisheries Research Institute.

Arrhenius, F. and Hansson, S., 1993. Food consumption of larval, young and adult herring and sprat in the Baltic Sea. Mar. Ecol. Prog. Ser., 96:125–137.

Axenrot, T. and Hansson, S., 2004. Seasonal dynamics in pelagic fish abundance in a Baltic Sea coastal area. Estuar. Coastal Shelf Sci., 60:541–547.

Azevedo, P.A., Cho, C.Y., Leeson, S. and Bureau, D.P., 1998. Effects of feeding level and water temperature on growth, nutrient and energy utilization and waste outputs of rainbow trout (Oncorhynchus mykiss). Aquat. Living Resour., 11:227–238.

BACC, 2008. Assessment of climate change for the Baltic Sea Basin. Springer, Heidelberg, 474p.

Baines, S.B. and Pace, M.L., 1991. The production of dissolved organic matter by phytoplankton and its importance to bacteria: patterns across marine and freshwater systems. Limnol. Oceanogr., 36:1078–1090.

Balistrieri, L.S. and Murray, J.W, 1983. Metal-solid interactions in the marine environments: estimating apparent equilibrium binding constants. Geochim. Cosmochim. Acta, 47:1091–1098.

Balistrieri, L.S., Murray, J.W. and Paul, B., 1992. The biogeochemical cycling of trace metals in the water column of Lake Sammamish, Washington: response to seasonally anoxic conditions. Limnol. Oceanogr., 37:510–528.

Balls, P.W., 1988. The control of trace metal concentrations in coastal seawater through partition onto particulate matter. Neth. J. Sea Res., 22:213–218.

Balls, P.W., 1989. The partition of trace metals between dissolved and particulate phases in European coastal waters: a compilation of field data and comparison with laboratory studies. Neth. J. Sea Res., 23:7–14.

BALTEX, 2006. Baltex Phase II 2003–2012. Science framework and implementation strategy. International BALTEX Secr., Publication, No. 34, pp.92, GKSS, Geestacht, Germany.

Beach Erosion Board, 1972. Waves in inland reservoirs. Tech. Mem. 132, Beach Erosion Corps of Engineers, Washington, DC.

Beeton, A.M. and Edmondson, W.T., 1972. The eutrophication problem. J. Fish. Res. Can., 29:673–682.

Benoit, G., 1995. Evidence of the particle concentration effect for lead and other metals in fresh waters based on ultraclean technique analyses. Geochim. Cosmochim. Acta, 59:2677–2687.

Benoit, G., Oktaymarshall, S.D., Cantu, A., Hood, E.M., Coleman, C.H., Corapcioglu, M.O. and Santschi, P.H., 1994. Partitioning of Cu, Pb, Ag, Zn, Fe, Al and Mn between filter-retained particles, colloids, and solution in 6 Texas estuaries. Mar. Chem., 45:307–336.

Bergström, S. and Carlsson, B., 1993. Hydrology of the Baltic basin. Swedish Meteorological and Hydrological Institute, Reports Hydrology No. 7, 32p.

Bergström, S. and Carlsson, B., 1994. River runoff to the Baltic Sea: 1950–1990. Ambio, 23: 280–287.

Bergström, L., Djodjic, F., Kirchmann, H., Nilsson, I. and Ulén, B., 2007. Phosphorus from farmland to water – status, flows and preventive measures in a Nordic perspective. Report Food 21 no. 4/2007. Swedish Agricultural University, Uppsala, 65p.

Bernardi Aubry, F., Acri, F., Bastianini, M., Pugnetti, A. and Socal, 2006. Picophytoplankton contribution to phytoplankton community structure in the Gulf of Venice (NW Adriatic Sea). Int. Rev. Hydrobiol., 91:51–70.

Bird, D.F. and Kalff, J., 1984. Empirical relationships between bacterial abundance and chlorophyll concentration in fresh and marine waters. Can. J. Fish. Aquat. Sci., 41:1015–1023.

Blanc, M., Banarescu, P., Gaudet, J.-L. and Hureau, J.C., 1971. European inland water fish: a multilingual catalouge. Fishing News (Books), London, 170p.

Blaxter, J.H.S. and Hunter, J.R., 1982. The biology of the clupeoid fishes. Adv. Mar. Biol., 20: 1–223.

Blazhchishin, A.I., 1984. Balance of settling matter in the Gdansk Basin of the Baltic Sea. Litologiya i Poleznye Iskopayemye, 5:67–76 (in Russian).

Blazhchishin, A.I. and Shuyskiy, Y.D., 1973. Inflow of terrigenic matter to the Baltic Sea. Litologiya i Poleznye Iskopayemye, 3:141–145 (in Russian).

Blazka, P., Backiel, T. and Taub, F.B., 1980. Ratio of primary and fish productions. The functioning of the freshwater ecosystems. Cambridge University Press, Cambridge, pp. 398–401.

Blindow, I., 2000. Distribution of charophytes along the Swedish coast in relation to salinity and eutrophication. Int. Rev. Hydrobiol. 85:707–717.

Bloesch, J. and Burns, N.M., 1980. A critical review of sedimentation trap technique. Schweiz. Z. Hydrol., 42:15–55.

Bloesch, J. and Uehlinger, U., 1986. Horizontal sedimentation differences in a eutrophic Swiss lake. Limnol. Oceanogr., 31:1094–1109.

Blomqvist, S., Gunnars, A. and Elmgren, R., 2004. Why the limiting nutrient differs between temperate coastal seas and freshwater lakes: a matter of salt. Limnol. Oceanogr., 49:2236–2241.

Blomqvist, S. and Larsson, U., 1994. Detrimental elements as tracers of settling resuspended particulate matter in a coastal area of the Baltic Sea. Limnol. Oceanogr., 49:880–896.

Blomqvist, S. and Rydin, E., 2009. How phosphorus sequestration in Baltic Sea bottom sediments can be improved. Swedish EPA Report 5914, Stockholm, 111p (in Swedish).

Boesch et al., 2008. Eutrophication of the seas along Sweden's west coast. Naturvårdsverket, Report 5898, Stockholm, 78p.

Bonsdorff, E., 2006. Zoobenthic diversity-gradients in the Baltic Sea: continuous post-glacial succession in a stressed ecosystem. J. Exp. Mar. Biol. Ecol., 330:383–391.

Boyle, J.F. and Birks, H.J.B., 1999. Predicting heavy metal concentrations in the surface sediments of Norwegian headwater lakes from atmospheric deposition: an application of a simple sediment-water partitioning model. Water Air Soil Pollut., 114:27–51.

Boulion, V.V., 1997. Regularities of the plankton primary production and their importance for the control and prediction of a trophic condition in water ecosystems. Biol. Inner Waters, 1:13–22 (in Russian).

Boulion, V.V. and Håkanson, L., 2003. A new general model to predict biomass and production of bacterioplankton in lakes. Ecol. Modell., 160:91–114.

Boulion, V.V. and Paveljeva, E.B., 1998. Relationship between bacterial abundance and chlorophyll concentration in plankton of freshwaters. Microbiology, 67:261–266 (in Russian).

Boulion, V.V. and Winberg G.G., 1981. A ratio between the primary production and fish productivity in water bodies. Bases of Study of Freshwater Ecosystems. Leningrad, pp. 5–10 (in Russian).

Brinkhurst, R.O., 1974. The benthos of lakes. Macmillan, London, 189p.

Brittain, J.E., 1998. European fish species of radioecological importance. ENEA, Rome, RT/AMB/98/4, pp. 19–21.

Brittain, J.E. and Brabrand, Å., 2001. Fish movement in rivers, lakes and estuaries in relation to contamination by radionuclides. Freshwater Ecology & Inland Fisheries Laboratory (LFI), Natural History Museums, University of Oslo, Oslo, Norway.

Brodeur, R.D., Sugisaki, H. and Hunt, G.L. Jr., 2002. Increases in jellyfish biomass in the Bering Sea: implications for the ecosystem. Mar. Ecol. Prog. Ser., 233:89–103.

Bryhn, A., Håkanson, L. and Eklund, J., 2008a. Variabilities and uncertainties in key coastal water variables as a basis for understanding changes and obtaining predictive power in modelling. Vatten 64:259–272.

BSAP, 2007. HELCOM Baltic Sea Action Plan – HELCOM Ministerial Meeting, Krakow, Poland, 15 November, 2007, 101p. (see HELCOM website).

Burban, P.-Y., Lick, W. and Lick, J., 1989. The flocculation of fine-grained sediments in estuarine waters. J. Geophys. Res., 94:8223–8330.

Burban, P.-Y., Xu, Y.-J., McNeiel, J. and Lick, W., 1990. Settling speeds of flocs in fresh water and seawater. J. Geophys. Res., 95:18213–18220.

Burbridge, P., Hendrick, V., Roth, E. and Rosenthal, H., 2001. Social and economic policy issues relevant to marine aquaculture. J. Appl. Ichthyol., 17:194–206.

Busch, W.-D.N. and Sly, P.G., 1992. The development of an aquatic habitat classification system for lakes. CRC Press, Boca Raton, FL.

Carlson, R.E., 1977. A trophic state index for lakes. Limnol. Oceanogr., 22:361–369.

Carroll, J. and Harms, I.H., 1999. Uncertainty analysis of partition coefficients in a radionuclide transport model. Water Res., 33:2617–2626.

Carstensen, J., Conley, J., Andersen, J.H. and Aertebjerg, G., 2006. Coastal eutrophication and trend reversal: a Danish case study. Limnol. Oceanogr., 51:398–408.

Casini, M., Cardinale, M. and Arrhenius, F., 2004. Feeding preferences of herring (*Clupea harengus*) and sprat (*Sprattus sprattus*) in the southern Baltic Sea. ICES J. Mar. Sci., 61:1267–1277.

Casini, M., Cadinale, M. and Hjelm, J., 2006. Inter-annual variation in herring, *Clupea harengus*, and sprat, *Sprattus sprattus*, condition in the central Baltic Sea: what gives the tune? Oikos 112:638–650.

Cato, I., 1977. Recent sedimentological and geochemical conditions and pollution problems in two marine areas in south-western Sweden. Striae Vol. 6, Societas Upsaliensis pro Geologia Quaternaria, Uppsala, 158p.

Chabot, D., Dutil, J.-D. and Couturier, C., 2001. Impact of chronic hypoxia on food ingestion, growth and condition of Atlantic cod, Gadus morhua. ICES CM 2001/V:05.

Chow-Fraser, P. and Trew, D.O., 1994. A test of hypothesis to explain the sigmoidal relationship between total phosphorus and chlorophyll a concentrations in Canadian lakes. Can. J. Fish. Aquat. Sci., 51:2052–2065.

Christiansen, C., Gertz, F., Laima, M.J.C., Lund-Hansen, L.C., Vang, T. and Jürgensen, C., 1997. Nutrient (P, N) dynamics in the southwestern Kattegat, Scandinavia: sedimentation and resuspension effects. Environ. Geol., 29:66–77.

Christensen, V., Walters, C.J. and Pauly, D., 2000. Ecopath with ecosim: a user's guide. Fisheries Centre, University of British Columbia, Vancouver, 130p.

Cole, J.J., Likens, G.E. and Strayer, D.L., 1982. Photosynethetically produced dissolved organic carbon: an important carbon source for planktonic bacteria. Limnol. Oceanogr., 27: 1080–1090.

Conan, P., Turley, C., Stutt, E., Pujo-Pay, M. and Van Wambeke, F., 1999. Relationship between phytoplankton efficiency and proportion of bacterial production to primary production in the Mediterranean Sea. Aquat. Microbial. Ecol., 17:131–144.

Corten, A., 1996. The widening gap between fisheries biology and fisheries management in the European Union. Fish. Res., 27:1–15.

Cowan, J.H. Jr., Houde, E.D. and Rose, K.A., 1996. Size-dependent vulnerability of marine fish larvae to predation: an individual-based numerical experiment. ICES J. Mar. Sci., 53:23–37.

Cummings, K.W., 1973. Trophic relations in aquatic insects. Ann. Rev. Entomol., 18:183–206.

Currie, D.J., 1990. Large-scale variability and interactions among phytoplankton, bacterioplankton, and phosphorus. Limnol. Oceanogr., 35:1437–1455.

Davenport, J., Smith, R.J.J.W., Packer, M., 2000. Mussels *Mytilus edulis*: significant consumers and destroyers of mesozooplankton. Mar. Ecol. Prog. Ser., 198:131–137.

Derenbach, J.B., Williams, P.J. and Le, B., 1974. Autotrophic and bacterial production: fractionation of plankton populations by differential filtration of samples from the English Cannel. Mar. Biol., 25:263–269.

De Robertis, A., et al., 2003. Differential effects of turbidity on prey consumption of piscivorous and planktivorous fish. Can. J. Fish. Aquat. Sci., 60:1517–1526.

Dillon, P.J. and Rigler, F.H., 1974. The phosphorus-chlorophyll relationship in lakes. Limnol. Oceanogr., 19:767–773.

Dippner, J.W., Hänninen, J., Kuosa, H. and Vourinen, I., 2001. The influence of climate variability on zooplankton abundance in the Northern Baltic Archipelago Sea (SW Finland). ICES J. Mar. Sci., 58:569–578.

Duarte, C.M., Conley, D., Carstensen, J. and Sánchez-Camacho, M., 2009. Return to Neverland: shifting baselines affect eutrophication restoration targets. Estuar. Coast., 32:29–36.

Eckhéll, J., Jonsson, P., Meili, M. and Carman, R., 2000. Storm influence on the accumulation and lamination of sediments in deep areas of the northwestern Baltic proper. Ambio, 29: 238–245.

Edler, L., 1979. Phytoplankton succession in the Baltic Sea. Acta Bot. Fenn. 110:75–78.

Ehrenberg, S.Z., Hansson, S. and Elmgren, R., 2005. Sublittoral abundance and food consumption of Baltic gobies. J. Fish. Biol., 67:1083–1093.

Eilola, K., Meier, H.E.M. and Almroth, E., 2009. On the dynamics of oxygen, phosphorus and cyanobacteria in the Baltic Sea; a model study. J. Mar. Syst., 75:163–184.

Elmgren, R., 1984. Trophic dynamics in the enclosed, brackish Baltic Sea. Rapp. P.-v. Cons. Int. Explor. Mer., 183:152–169.

Elmgren, R., 1989a. Trophic dynamics in the enclosed brackish Baltic Sea. Rapp. P.-v. Cons. Int. Explor. Mer., 183:152–169.

Elmgren, R., 1989b. Man's impact on the ecosystem of the Baltic Sea: energy flows today and at the turn of the century. Ambio, 18:326–332.

Emelyanov, E.M., 1986. Geochemistry of particulate matter and sediments in the Gdansk Basin and sedimentation processes. In: Emelyanov, E.M. and Lukashin, V.N. (eds.), Geochemistry of the sedimentation process in the Baltic Sea. Nauka, Moscow, pp. 57–115 (in Russian).

Emelyanov, E.M., 1988. Biogenic sedimentation in the Baltic Sea and its consequences. In: Winterhalter, B. (ed.), The Baltic Sea, Geological Survey of Finland, Special Paper 6: 127–135.

Emelyanov, E.M., 2001. Biogenic components and elements in sediments of the Central Baltic and their sedimentation. Mar. Geol., 172:23–41.

Enell, M., 1994. Environmental impact of nutrients from Nordic fish farming. Water Sci. Tech., 31:61–71.

Enin, U.I., Gröger, J. and Hammer, C., 2004. Species and length composition of fish in the southwestern Baltic Sea. J. Appl. Ichthyol., 20:369–375.

EPA, 1994. Eutrophication of soil, fresh water and the sea. National Swedish Environmental Protection Agency, Report 4244, 207p.

Erel, Y. and Stolper, E.M., 1993. Modeling of rare-earth element partitioning between particles and solution in aquatic environments. Geochim. Cosmochim. Acta, 57:513–518.

European Communities, 2006. Facts and figures on the CFP. Office for Official Publications of the European Communities, Luxembourg, 43p.

Evans, M.S., Arts, M.T. and Robarts, R.D., 1996. Algal productivity, algal biomass, and zooplankton biomass in a phosphorus-rich, saline lake: deviations from regression model predictions. Can. J. Fish. Aquat. Sci., 53:1048–1060.

FAO, 2000. The state of the world fisheries and aquaculture 2000. Food and Agriculture Organisation. Fisheries Department, Rome. ISBN 92-5-104492-9.

FAO, 2001. FAO Yearbook. Fishery statistics. Capture production 2001, Vol. 92/1. FAO, Quebec.

Finni, T., Kononen, K., Olsonen, R. and Wallström, K., 2001. The history of cyanobacterial blooms in the Baltic Sea. Ambio, 30:172–178.

Fischer, H. and Matthäus, W., 1996. The importance of the Drogden Sill in the sound for major Baltic inflows. J. Mar. Syst. 9:137–157.

Fiskeriverket, 2008. See: http://www.fiskeriverket.se/

Fiskeriverket, 2009. See: http://www.fiskeriverket.se/vanstermeny/statistikochdatabaser/provfiskeisotkustvatten/provfiskevidkusten/kustdatabas

Flinkman, J., Vuorinen, I. and Aro, E., 1992. Planktivorous Baltic herring (*Clupea harengus*) prey selectively on reproducing copepods and cladocerans. Can. J. Fish. Aquat. Sci., 49:73–77.

Fogg, G.E., 1966. The extracellular products of algae. Oceanogr. Mar. Biol., 4:195–212.

Froese, R. and Pauly, D., 2008. Accessible at http://FishBase.org

FRP, 1978. The sea; natural conditions and use (in Swedish, Havet; naturförhållanden och utnyttjande). Fysisk riksplanering (FRP), Bostadsdepartementet, Nr 7, 303p.

Fursenko, M.V. and Kuzmitskaja, N.K., 1975. Microbiological researches. Biological productivity of north lakes, 1. Proc. Zool. Inst. Acad. Sci. USSR, 57:53–76 (in Russian).

Gallego, E., Brittain, J.E., Håkanson, L., Heling, R., Hofman, D. and Monte, L., 2004. MOIRA: a computerised decision support system for the management of radionuclide contaminated freshwater ecosystems. Radioprotection, 98:83–102 (ISSN-0874-7016).

Goecker, M.E. and Kåll, S.E., 2003. Grazing preferences of marine isopods and amphipods on three prominent algal species of the Baltic Sea. J. Sea Res. 50:309–314.

Gorokhova, E., Aladin, N. and Dumont, H., 2000. Further expansion of the genus Cercopagis (*Crustacea, Branchiopoda, Onychopoda*) in the Baltic Sea, with notes on the taxa present and their ecology. Hydrobiologia, 429:207–218.

Gorokhova, E., Lehtiniemi, M., Viitasalo-Frösen, S. and Haddock, S.H.D., 2009. Molecular evidence for the occurrence of ctenophore *Mertensia ovum* in the northern Baltic Sea and implications for the status of the *Mnemiopsis leidyi* invasion. Limnol. Ocenogr., 54:2025–2033.

Gren, I.-M., 2001. International versus national actions against nitrogen pollution of the Baltic Sea. Environ. Resour. Econ., 20:41–59.

Gren, I.-M., Elofsson, K., 2008. Costs and benefits from nutrient reductions to the Baltic Sea. Report 5877. Swedish EPA, Stockholm, 67p.

Gustafsson, Ö. and Gschwend, P.M., 1997. Aquatic colloids: concepts, definitions and current challenges. Limnol. Oceanogr., 42:519–528.

Gustafsson, B.G. and Stigebrandt, A., 2007. Dynamics of nutrients and oxygen/hydrogen sulfide in the Baltic Sea deep water. J. Geophys. Res., 112(G02023), 12p.

Gutelmakher, B.L., 1986. Metabolism of plankton as a single whole. Nauka, Leningrad (in Russian).

Gyllenhammar, A. and Håkanson, L., 2005. Environmental consequence analyses of fish farms emissions related to different scales and exemplified by data from the Baltic – a review. Mar. Environ. Res. 60:211–243.

Gyllenhammar, A., Håkanson, L. and Lehtinen, K.-J., 2008. A mesocosm fish farming experiment and its implications for reducing environmental effects on a regional scale. Aquacult. Eng., 38:117–126.

Hairston, N.G. and Hairston, N.G., 1993. Cause-effect relationships in energy flow, trophic structure, and interspecific interactions. Am. Nat., 142:379–411.

Hajdu, S., Larsson, U., Andersson, A. and Huseby, S., 2007. Sommarens växtplanktonsamhälle har förändras. Havet, Naturvårdsverket, Stockholm, pp. 47–50.

Håkanson, L., 1977. The influence of wind, fetch, and water depth on the distribution of sediments in Lake Vänern, Sweden. Can. J. Earth Sci., 14:397–412.

Håkanson, L., 1982. Lake bottom dynamics and morphometry – the dynamic ratio. Water Resour. Res., 18:1444–1450.

Håkanson, L., 1984. Sediment sampling in different aquatic environments: statistical aspects. Water Resour. Res., 20:41–46.

Håkanson, L., 1991. Physical geography of the Baltic. The Baltic University. Session 1. Uppsala University, Uppsala, ISBN 91-506-0876-2.

Håkanson, L., 1995. Fish farming and environmental effects in lakes – new results motivate new criteria for evaluations (in Swedish, Fiskodling och miljöeffekter i sjöar – nya resultat motiverar nya bedömningsunderlag). Vatten 51, Bloms Boktryckeri, Lund, 103p.

Håkanson, L., 1999. Water pollution – methods and criteria to rank, model and remediate chemical threats to aquatic ecosystems. Backhuys Publishers, Leiden, 299p.

Håkanson, L., 2000. Modelling radiocesium in lakes and coastal areas – new approaches for ecosystem modellers. A textbook with Internet support. Kluwer Academic Publishers, Dordrecht, 215p.

Håkanson, L., 2002. The Baltic – present status, what it should be and how to get there. Accessible at http://www.sedimentology.geo.uu.se/BALTWEB_OH.pdf

Håkanson, L., 2003a. Consequences and correctives related to acidification, liming and mercury in lake fish – a case study for Lake Huljesjön, Sweden, using the LakeWeb-model. Environ. Modell. Assess., 8:275–283.

Håkanson, L., 2003b. Quantifying burial, the transport of matter from the lake biosphere to the geosphere. Int. Rev. Hydrobiol., 88:539–560.

Håkanson, L., 2004. Lakes – form and function. The Blackburn Press, Caldwell, NJ, 201p.
Håkanson, L., 2006. Suspended particulate matter in lakes, rivers and marine systems. The Blackburn Press, Caldwell, NJ, 331p.
Håkanson, L., 2008. Factors and criteria to quantify coastal area sensitivity/vulnerability to eutrophication. Presentation of a sensitivity index based on morphometrical parameters. Int. Rev. Hydrobiol., 3:372–388.
Håkanson, L., 2009. Modeling nutrient fluxes to, within and from the Kattegat to find an optimal cost-efficient Swedish remedial strategy. Uppsala University, Geotryckeriet, Uppsala, 122p, ISBN 978-91-633-4819-8.
Håkanson, L., Blenckner, T. and Malmaeus, J.M., 2004. New, general methods to define the depth separating surface water from deep water, outflow and internal loading for mass-balance models for lakes. Ecol. Modell., 175:339–352.
Håkanson, L. and Boulion, V.V., 2001a. Regularities in primary production, Secchi depth and fish yield and a new system to define trophic and humic state indices for aquatic ecosystems. Int. Rev. Hydrobiol., 86:23–62.
Håkanson, L. and Boulion, V.V., 2001b. A practical approach to predict the duration of the growing season for European lakes. Ecol. Modell., 140:235–245.
Håkanson, L. and Boulion, V., 2002a. The Lake Foodweb – modelling predation and abiotic/biotic interactions. Backhuys Publishers, Leiden, 344p.
Håkanson, L. and Boulion, V.V., 2002b. Empirical and dynamical models to predict the cover, biomass and production of macrophytes in lakes. Ecol. Modell., 151:213–243.
Håkanson, L. and Boulion, V.V., 2003a. Modelling production and biomasses of herbivorous and predatory zooplankton in lakes. Ecol. Modell., 161:1–33.
Håkanson, L. and Boulion, V.V., 2003b. A model to predict how individual factors influence Secchi depth variations among and within lakes. Int. Rev. Hydrobiol., 88:212–232.
Håkanson, L. and Boulion, V.V., 2003c. A general dynamic model to predict biomass and production of phytoplankton in lakes. Ecol. Modell., 165:285–301.
Håkanson, L. and Boulion, V.V., 2003d. Modelling production and biomasses of zoobenthos in lakes. Aqua. Ecol., 37:277–306.
Håkanson, L. and Boulion, V., 2004a. Modelling production and biomasses of prey and predatory fish in lakes. Hydrobiologia, 511:125–150.
Håkanson, L. and Boulion, V., 2004b. Empirical and dynamical models of production and biomasses of benthic algae in lakes. Hydrobiologia, 522:75–97.
Håkanson, L. and Bryhn, A.C., 2008a. Tools and criteria for sustainable coastal ecosystem management – with examples from the Baltic Sea and other aquatic systems. Springer, Heidelberg, 300p.
Håkanson, L. and Bryhn, A.C., 2008b. Eutrophication in the Baltic Sea – present situation, nutrient transport processes, remedial strategies. Springer, Berlin/Heidelberg, 261p.
Håkanson, L. and Bryhn, A.C., 2008c. Goals and remedial strategies for water quality and wildlife management in a coastal lagoon – a case-study of Ringkobing Fjord, Denmark. J. Environ. Manag., 86:498–519.
Håkanson, L. and Bryhn, A.C., 2008d. A dynamic mass-balance model for phosphorus in lakes with a focus on criteria for applicability and boundary conditions. Water Air Soil Pollut., 187:119–147.
Håkanson, L. and Bryhn, A.C., 2008e. Modeling the foodweb in coastal areas – a case study of Ringkobing Fjord, Denmark. Ecol. Res., 23:421–444.
Håkanson, L., Bryhn, A.C. and Hytteborn, J.K., 2007. On the issue of limiting nutrient and predictions of cyanobacteria in aquatic systems. Sci. Total Environ., 379:89–108.
Håkanson, L., Bryhn, A.C. and Hytteborn, J.A., 2009. A new general approach to quantify nitrogen fixation exemplified for the Baltic Proper. Open Mar. Biol. J., 3:36–48.
Håkanson, L. and Eckhéll, J., 2005. Suspended particulate matter (SPM) in the Baltic – new empirical data and models. Ecol. Modell., 189:130–150.
Håkanson, L. and Eklund, J.M., 2007. A dynamic mass-balance model for phosphorus fluxes and concentrations in coastal areas. Ecol Res., 22:296–320.

Håkanson, L., Ervik, A., Mäkinen, T. and Möller, B., 1988. Basic concepts concerning assessments of environmental effects of marine fish farms. Nordic Council of Ministers, NORD88:90, Copenhagen, 103p.

Håkanson, L. and Gyllenhammar, A., 2005. Setting fish quotas based on holistic ecosystem modelling including environmental factors and foodweb interactions – a new approach. Aquat. Ecol., 39:325–351.

Håkanson, L., Gyllenhammar, A. and Karlsson, M., 2002. The Baltic – present status, what it should be and how to get there (in Swedish, Östersjön – hur läget är, hur det borde vara och hur man kommer dit!). LUVA 02-01, Geotryckeriet, Uppsala, 110p.

Håkanson, L. and Jansson, M., 1983. Principles of lake Sedimentology. Springer, Berlin, 316p.

Håkanson, L., Kvarnäs, H. and Karlsson, B., 1986. Coastal morphometry as regulator of water exchange – a Swedish example. Estuar. Coast. Shelf Sci., 23:1–15.

Håkanson, L. and Lindgren, D., 2008a. A protocol (computer codes, equations and motivations) for a general operational foodweb model for coastal areas which can be run by readily accessible driving variables from maps and monitoring programs incorporating the eutrophication model discussed in WP5. Thresholds Project, 6th Framework Programme, EU, 89p.

Håkanson, L. and Lindgren, D., 2008b. On regime shifts and budgets for nutrients in the open Baltic Proper – evaluations based on extensive data between 1974 and 2005. J. Coast. Res., 24:246–260.

Håkanson, L. and Lindgren, D., 2009. Test and application of a general process-based dynamic coastal mass-balance for contaminants using data for radionuclides in the Dnieper-Bug estuary. Sci. Total Environ. 407:899–916.

Håkanson, L., Ostapenia, A. and Boulion, V.V., 2003a. A mass-balance model for phosphorus accounting for biouptake and retention in biota. Freshw. Biol., 48:928–950.

Håkanson, L., Ostapenia, A., Parparov, A., Hambright, K.D. and Boulion, V.V., 2003b. Management criteria for lake ecosystems applied to case studies of changes in nutrient loading and climate change. Lakes Reserv. Res. Manag., 8:141–155.

Håkanson, L. and Peters, R.H., 1995. Predictive limnology. Methods for predictive modelling. SPB Academic Publishing, Amsterdam, 464p.

Håkanson, L. and Rosenberg, R., 1985. Practical coastal ecology (Praktisk kustekologi). SNV PM 1987, Solna, 110p. (in Swedish).

Hansson, M., 2006. Cyanobakterieblomningar i Östersjön, resultat från satellitövervakning 1997–2005 (Blooms of cyanobacteria in the Baltic Sea, results from satellite monitoring 1997–2005). SMHI Reports Oceanography No 82, Norrköping (in Swedish).

Hansson, S. and Rudstam, L.G., 1990. Eutrophication and the Baltic fish communities. Ambio, 19:123–125.

Hårding, K.C. and Härkönen, T.J., 1999. Development in the Baltic grey seal (*Halichoerus grypus*) and ringed seal (*Phoca hispida*) populations during the 20th century. Ambio, 28: 619–627.

Harrison, W.G., Azam, F., Renger, E.H. and Eppley, R.W., 1977. Some experiments of phosphate assimilation by coastal marine plankton. Mar. Biol., 40:9–18.

Harden-Jones, F.R., 1968. Fish migration. Edward Arnold Publishers, London, 325 pp.

Harvey, C.J., Cox, S.P., Essington, T.E., Hansson, S. and Kitchell, J.F., 2003. An ecosystem model of food web and fisheries interactions in the Baltic Sea. ICES J. Mar. Sci., 60: 939–950.

Hatcher, A., 2000. Subsidies for European fishing fleets: the European Community's structural policy for fisheries 1971–1999. Mar. Policy, 24:129–140.

Hawley, N., Robbins, J.A. and Eadie, B.J., 1986. The partition of beryllium in fresh water. Geochim. Cosmochim. Acta, 50:1127–1131.

Heibo, E. and Karås, P., 2005. The coastal fish community in the Forsmark area SW Bothnian Sea. Svensk kärnbränslehantering AB, P-05-148.

Heikinheimo, O., 2008. Average salinity as an index for environmental forcing on cod recruitment in the Baltic Sea. Boreal Environ. Res., 13:457–464.

HELCOM, 1986. Water balance of the Baltic Sea. Baltic Sea Environment Proceedings 16, HELCOM, Helsinki, Finland.
HELCOM, 1987. First Baltic Sea pollution load compilation. Baltic Sea Environment Proceedings 20, HELCOM, Helsinki, 57p.
HELCOM, 1990. Second periodic assessment of the state of the marine environment of the Baltic Sea, 1984–1988; Background Document. Baltic Sea Environment Proceedings, 35B, 432p.
HELCOM, 1993. Second Baltic Sea pollution load compilation. Baltic Sea Environment Proceedings 45, HELCOM, Helsinki, 162p.
HELCOM, 1998. Third Baltic Sea pollution load compilation. Baltic Sea Environment Proceedings 70, HELCOM, Helsinki, 134p.
HELCOM, 2000. Baltic Sea Environment Proceedings 100.
HELCOM, 2002. Environment of the Baltic Sea area 1994–1998. Baltic Sea Environment Proceedings 82 B., 215p. Helsinki Commission, Helsinki, Finland. Accessible at www.helcom.fi
HELCOM, 2003. The Baltic Marine Environment 1999–2002. Baltic Sea Environment Proceedings 87, Helsinki Commission, Helsinki, Finland. Accessible at www.helcom.fi
HELCOM, 2005. Nutrient pollution to the Baltic Sea in 2000. Baltic Sea Environment Proceedings 100, HELCOM, Helsinki, 24p.
HELCOM, 2007a. Waterborne inputs of nitrogen and phosphorus. Indicator Fact Sheets 2007. Accessible at http://www.helcom.fi/environment2/ifs/ifs2007/en_GB/nutrient_load/ (accessed 27 February, 2009).
HELCOM, 2007b. Towards a Baltic Sea unaffected by eutrophication. HELCOM, Helsinki, 35p.
HELCOM, 2007c. HELCOM Baltic Sea Action Plan – HELCOM Ministerial Meeting, Krakow, Poland, 15 November, 2007, 101p. (see HELCOM website).
HELCOM, 2008a. Total and regional runoff to the Baltic Sea. Indicator Fact Sheets 2008. Accessible at http://www.helcom.fi/environment2/ifs/ifs2008/en_GB/Runoff/ (accessed 27 February, 2009).
HELCOM, 2008b. Phytoplankton biomass and species succession in the Gulf of Finland, Northern Baltic Proper and Southern Baltic Sea in 2008. Indicator Fact Sheets 2008. Accessible at http://www.helcom.fi/environment2/ifs/ifs2008/en_GB/Phyto_biomass/ (accessed 2 November, 2009)
HELCOM, 2008c. Eutrophication in the Baltic Sea. Baltic Sea Environment Proceedings No. 115B.
HELCOM and NEFCO, 2007. Economic analysis of the BSAP with focus on eutrophication. HELCOM, Helsinki, 112p.
Hinrichsen, H.-H., Möllmann, C., Voss, R., Köster, F.W. and Kornilovs, G., 2002. Biophysical modelling of larval Baltic cod (*Gadus morhua* L.) growth and survival. Can. J. Fish. Aquat. Sci., 59:1858–1873.
Hinrichsen, H.-H., St. John, M., Aro, E., Groenkjaer, P., Voss, R., 2001. Testing the larval drift hypothesis in the Baltic Sea: retention vs. dispersion due to influence of wind driven circulation. ICES J. Mar. Sci., 58:973–984.
Hjerne, O. and Hansson, S., 2002. The role of fish and fisheries in Baltic Sea nutrient dynamics. Limnol. Oceanogr., 47:1023–1032.
Holby, O. and Hall, O.L., 1991. Chemical fluxes and mass balances in a marine fish cage farm. II. Phosphorus. Mar. Ecol. Prog. Ser., 70:263–272.
Holmer, M., Black, K., Duarte, C.M., Marbà, N., Karakassis, I., (eds.), 2008. Aquaculture in the ecosyste. Springer, Berlin, 326p.
Holmlund, C.M. and Hammer, M., 1999. Ecosystem services generated by fish populations. Ecol. Econ., 29:253–268.
Honeyman, B.D. and Santschi, P.H., 1988. Metals in aquatic systems. Environ. Sci. Technol., 22:862–871.
Howarth, R.W., Marino, R. and Cole, J.J., 1988a. Nitrogen fixation in freshwater, estuarine, and marine ecosystems. 2. Biogeochemical controls. Limnol. Oceanogr., 33:688–701.

Howarth, R.W., Marino, R., Lane, J. and Cole, J.J., 1988b. Nitrogen fixation in freshwater, estuarine, and marine ecosystems. 1. Rates and importance. Limnol. Oceanogr., 33: 669–687.

Hüssy, K., St. John, M.A. and Böttcher, U., 1997. Food resource utilization by juvenile Baltic cod *Gadus morhua*: a mechanism potentially influencing recruitment success at the demersal stage? Mar. Ecol. Prog. Ser., 155:199–208.

Hutchings, J., 1996. Spatial and temporal variations in the density of northern cod and a review of hypotheses for the stock's collapse. Can. J. Fish. Aquat. Sci., 53:943–962.

Hutchinson, P. (ed.), 1997. Interactions between salmon culture and wild stocks of Atlantic solmon: the scientific and management issues. ICES J. Mar. Sci., 54:963–1227.

ICES, 2006a. HELCOM data. http://www.ices.dk/ocean/asp/helcom/helcom.asp?Mode=1

ICES, 2006b. ICES CM 2006/F. What plankton are fish really eating? Species and diets, availability and dependency. Accessible at http://www.ices.dk

ICES, 2007. Report of the ICES/HELCOM Working Group on Integrated Assessments of the Baltic Sea (WGIAB), 12–16 Match 2007, Hamburg, Germany. ICES CM 2007/BCC:04. 71 pp.

ICES, 2008a. ICES CM 2008/BCC:02. Accessible at http://www.ices.dk/products/CMdocs/CM-2008/BCC/sgprod08.pdf

ICES, 2008b. Report of the ICES Advisory Committee, 2008. ICES Advice, 2008. Book 8, 133 pp.

ICES, 2008c. Accessible at http://www.ices.dk/products/CMdocs/CM-2008/BCC/sgprod08.pdf

ICES Fisheries Statistics 1973–2000, Nominal Catch Statistics, STATLANT Programme, ICES, Copenhagen, Denmark, 2002. Accessible at http://www.ices.dk/fish/statlant.asp

IBSFC, 2003. Total allowable catches (TACs) established by the IBSFC for the respective years in thousand tonnes. Accessible at http://www.ibsfc.org

Ivanova, M.B., 1985. Production of planktonic crustaceans in fresh water. Nauka, Leningrad (in Russian).

Ivlev, V.S., 1961. Experimental ecology of the feeding of fishes. Translated from Russian by D. Scott. Yale University Press, New Haven, CT.

Jackson, J.B.C., Kirby, M.X., Berger, W.H., Bjorndal, K.A., Botsford, L.W., Bourque, B.J., Bradbury, R.H., Cooke, R., Erlandson, J., Estes, J.A., Hughes, T.P., Kidwell, S., Lange, C.B., Lenihan, H.S., Pandolfi, J.M., Peterson, C.H., Steneck, R.S., Tegner, M.J. and Warner, R.R., 2001. Historical overfishing and the recent collapse of coastal ecosystems. Science, 293:629–638.

Jacobsen, T.S., 1980. The Belt Project. Sea water exchange of the Baltic-measurements and methods. Report from the National Agency of Environmental Protection, Denmark. ISBN 87-503-3532-4, 106 pp.

Jarre-Teichmann, A., Wieland, K., MacKenzie, B., Hinrichsen, H.-H., Plikshs, M. and Aro, E., 2000. Stock-recruitment relationships for cod (*Gadus morhua callarias L.)* in the central Baltic Sea incorporating environmental variability. Arch. Fish. Mar. Res., 48:97–123.

Järvi, T. (coord.), 2002. Performance and ecological impacts of introduced and escaped fish: physiological and behavioural mechanisms – AQUAWILD. Final Report to: European Commission EC Contract No. FAIR CT 97-1957. National Board of Fisheries, Institute of Freshwater Research, Drottningholm, Sweden.

Javidpour, J., Sommer, U. and Shiganova, T., 2006. First record of *Menemiopsis leidyi* A. Agazziz 1985 in the Baltic Sea. Aquat. Invasions, 1: 299:302.

Jeppesen, E., Jensen, J.P., Kristinsen, P., Sondergaard, M., Mortensen, E., Sortkjaer, O. and Olrik, K., 1990. Fish manipulation as a lake restoration tool in shallow, eutrophic, temperate lakes: threshold levels, long-term stability and conclusions. Hydrobiologia, 200/201: 219–227.

Johansson, T., Håkanson, L., Borum, K. and Persson, J., 1998. Direct flows of phosphorus and suspended matter from a fish farm to the wild fish in Lake Southern Bullaren, Sweden. Aquacult. Eng., 17:111–137.

Jones, J.R. and Bachmann, R.W., 1975. Algal response to nutrient inputs in some Iowa lakes. Verh. Int. Ver. Limnol., 19:904–910.

Jonsson, P., 1992. Large-scale changes of contaminants in Baltic Sea sediments during the twentieth century. Thesis, Uppsala University, Sweden.
Jonsson, P, Carman, R. and Wulff, F., 1990. Laminated sediments in the Baltic – a tool for evaluating nutrient mass balances. Ambio, 19:152–158.
Jonsson, P., Persson, J. and Holmberg, P., 2003. Skärgårdens bottnar (The bottom areas of the archipelago). Swedish EPA, Report 5212, 112p.
Jönsson, A., 2005. Model studies of surface waves and sediment resuspension in the Baltic Sea. Dr thesis No 332, Linköping University, Linköping, 49p.
Jørgensen, S.E., 1998. An improved parameter estimation procedure in lake modelling. Lake Reserv. Res. Manag., 3:139–1432.
Jurvelius, J., Leinikki, J., Mamylov, V. and Pushkin, S., 1996. Stock assessment of pelagic three-spined stickleback (*Gasterosteus aculeatust*): a simultaneous up- and down-looking echo-sounding study. Fish. Res., 27:227–241.
Karås, P., 1999. Rekryteringsmiljöer för kustbeståndet av abborre, gädda och gös. Fiskeriverket Rapport 6:31–65.
Karagiannikos, A., 1996. Total Allowable Catch (TAC) and quota management system in the European Union.
Karjalainen, M., Engström-Öst, J., Korpinen, S., Peltonen, H., Pääkkönen, J-P., Rönkkönen, S., Suikkanen, S. and Viitasalo, M., 2007. Ecosystem consequences of cyanobacteria in the Northern Baltic Sea. Ambio, 36:195–202.
Karlson, A.M.L., Almqvist, G., Skora, K.E. and Appelberg, M., 2007. Indications of competition between non-indigenous round goby and native flounder in the Baltic Sea. ICES J. Mar. Sci., 64:479–486.
Karlson, K., Rosenberg, R. and Bonsdorff, E., 2002. Temporal and spatial large-scale effects of eutrophication and oxygen deficiency on benthic fauna in Scandinavian and Baltic waters – a review. Oceanogr. Mar. Biol. Annu. Rev., 40:427–489.
Karlsson, O.M., Jonsson, P.O., Lindgrejn, D., Lännergren, C., Mamaeus, J.M. and Stehn, A., 2009. Indications of recovery from hypoxia in the inner Stockholm archipelago (submitted).
Kasumyan, A.O. and Doving, K.B., 2003. Taste preferences in fishes. Fish Fish., 4:289–347.
Kautsky, L., 1991. Life in the Baltic Sea. The Baltic University, Uppsala University, Uppsala, 37p.
Kautsky, L. and Kautsky, H., 1989. Algal species diversity and dominance along gradients of stress and disturbance in marine environments. Vegetatio, 83:259–267.
Kautsky, L. and Kautsky, N., 2000. The Baltic Sea, including Bothnian Sea and Bothnian Bay. In: Sheppard, C. (ed.), Seas at the millennium: an environmental evaluation. Elsevier, Amsterdam, pp. 121–133.
Kitaev, S.P., 1984. Ecological principles of productivity of the lake of different natural zones. Nauka, Moscow (in Russian).
Kiirikki, M., Rantanen, P., Varjopuro, R. et al., 2003. Cost effective water protection in the Gulf of Finland. Focus on St. Petersburg. Finnish Environment Institute, Helsinki, 57p.
Knoechel, R. and Holtby, L.B., 1986. Cladoceran filtering rate: body length relationships for bacterial and large algal particles. Limnol. Oceanogr., 3:195–200.
Koelmans, A.A. and Lijklema, L., 1992. Sorption of 1,2,3,4-tetrachlorobenzene and cadmium to sediments and suspended solids in Lake Volkerak/Zoom. Water Res., 26:327–337.
Koelmans, A.A. and Radovanovic, H., 1998. Prediction of trace metal distribution coefficient (KD) for aerobic sediments. Water Sci. Technol., 37:71–78.
Köhler, J., 2006. Detergent phosphates: an EU policy assessment. J. Bus. Chem., 3:15–30.
Konopka, A. and Brock, T.D., 1978. Effect of temperature on blue-green algae (Cyanobacteria) in Lake Mendota. Appl. Environ. Microbiol., 36:572–576.
Kornilovs, G., Sidrevics, L. and Dippner, J.W., 2001. Fish and zooplankton interaction in the Central Baltic Sea. ICES J. Mar. Sci., 58:579–588.
Köster, F.W. and Möllmann, C., 2000a. Egg cannibalism in Baltic sprat (*Sprattus sprattus L.)*. Mar. Ecol. Prog. Ser., 196:269–277.

Köster, F.W. and Möllmann, C., 2000b. Trophodynamic control by clupeid predators on recruitment success in Baltic cod? ICES J. Mar. Sci., 57:310–323.
Köster, F.W., Möllmann, C., Hinrichsen, H.-H., Wieland, K., Tomkiewicz, J., Kraus, G. Voss, R., Makarchouk, A., MacKenzie, B.R., St. John, M.A., Schnack, D., Rohlf, N., Linkowski, T. and Beyer, J.E., 2005. Baltic cod recruitment – the impact of climate variability on key processes. ICES J. Mar. Sci., 62:1408–1425.
Köster, F.W., Möllmann, C., Neuenfeldt, S., Vinther, M., St. John, M.A., Tomkiewicz, J., Voss, R., Hinrichsen, H.-H., MacKenzie, B., Kraus, G. and Schnack, D., 2003a. Fish stock development in the central Baltic Sea (1974–1999) in relation to variability in the environment. ICES Mar. Sci. Symp., 219:294–306.
Köster, F.W., Hinrichsen, H.-H., Schnack, D., St. John, M.A., MacKenzie, B.R., Tomkiewicz, J., Möllmann, C., Kraus, G., Plikhs, M., Makarchouk, A. and Aro. E., 2003b. Recruitment of Baltic cod and sprat stocks: identification of critical life stages and incorporation of environmental variability into stock-recruitment relationships. Sci. Mar., 67(Suppl. 1):129–154.
Kotilainen, A.T., Hämäläinen, J.M.S. and Winterhalter, B., 2002. Reconstructing a continuous Holocene composite sedimentary record for the eastern Gotland Deep, Baltic Sea. Boreal Environ. Res. 7:1–12.
Kranck, K., 1973. Flocculation of suspended sediment in the sea. Nature, 246:348–350.
Kranck, K., 1979. Particle matter grain-size characteristics and flocculation in a partially mixed estuary. Sedimentology, 28:107–114.
Kuznetsov, S.I., 1970. Microflora of lakes and its geochemical activity. Nauka, Leningrad (in Russian).
Laine, A.O., 2003. Distribution of soft-bottom macrofauna in the deep open Baltic Sea in relation to environmental variability. Estuar. Coast. Shelf Sci., 57:87–97.
Lalonde, S. and Downing, A., 1991. Epiphyton biomass is related to lake trophic status, depth, and macrophyte architecture. Can. J. Fish. Aquat. Sci. 48:2285–2291.
Lappalainen, A., Shurukhin, A., Alekseev, G. and Rinne, J., 2000. Coastal-Fish Communities along the Northern Coast of the Gulf of Finland, Baltic Sea: responses to salinity and eutrophication. Int. Rev. Hydrobiol., 85:687–696.
Lappalainen, A., Westerbom, M. and Vesala, S., 2004. Blue mussels (*Mytilus edulis*) in the diet of roach (*Rutilus rutilus*) in outer archipelago areas of the western Gulf of Finland, Baltic Sea. Hydrobiologia, 514:87–92.
Larson, R.J., 1987. Daily ration and predation by medusae and ctenophores in Saanich inlet, B.C., Canada. Neth. J. Sea Res., 21:35–44.
Larson, U. and Hagström, A., 1979. Phytoplankton exudate release as an energy source for the growth of pelagic bacteria. Mar. Biol., 52:199–206.
Larsson, S. and Berglund, I., 2005. The effect of temperature on the energetic growth efficiency of Arctic charr (*Salvelinus alpinus* L.) from four Swedish populations. J. Therm. Biology, 30: 29–36.
Larsson, U., Elmgren, R. and Wulff, F., 1985. Eutrophication and the Baltic Sea: causes and consequences. Ambio, 14:9–14.
Larsson, U., Hajdu, S., Walve, J. and Elmgren, R., 2001. Baltic Sea nitrogen fixation estimated from the summer increase in upper mixed layer total nitrogen. Limnol. Oceanogr., 46:811–820.
Laznik, M., Stålnacke, P., Grimvall, A. and Wittgren, H.B., 1998. Riverine input of nutrients to the Gulf of Riga – temporal and spatial variation. J. Mar. Syst., 23:11–25.
Leclerc, M.J., Secretan, Y. and Boudreau, P., 2000. Integrated two-dimensional macrophyte-hydrodynamic modeling. J. Hydraulic Res. 38:163–172.
Lehtimäki, J. Moisander, P. Sivonen, K. and Kononen, K., 1997. Growth, nitrogen Fixation, and Nodularin production by two Baltic Sea Cyanbacteria. Appl. Environ. Microbiol., 63:1647–1656.
Lehtimäki, J., Sivonen, K., Luukainen, R. and Niemelä, S.I., 1994. The effects of incubation time, temperature, light, salinity, and phosphorus on growth and hepatotoxin production by *Nodularia* strains. Arch. Hydrobiol., 130:269–282.

Lehtiniemi, M. and Lindén, E., 2006 *Cercopagis pengoi* and *Mysis* spp. Alter their feeding rate and prey selection under predation risk of herring (*Clupea harengus membras*). Mar. Biol., 149:845–854.

Legović, T., 1987. A recent increase in Jellyfish populations: a predator-prey model and its implications. Ecol. Modell., 38:243–256.

Lehane, C. and Davenport, J., 2002. Ingestion of mesozooplankton by three species of bivalve: *Mytilus edulis*, *Cerastoderma edule* and *Aequipecten opercularis*. J. Mar. Biol. Assoc. UK, 82:3999/1–6.

Lemmin, U. and Imboden, D.M., 1987. Dynamics of bottom currents in a small lake. Limnol. Oceanogr., 32:62–75.

Leppäkoski, E., Gollasch, S., Gruszka, P. et al., 2002. The Baltic – a sea of invaders. Can. J. Fish. Aquat. Sci. 59:1175–1188.

Levinton, J.S., 2001. Marine biology: function, biodiversity, ecology, 2nd ed. Oxford University Press, New York, 515 pp.

Li, Y.-H., Burkhardt, L., Buchholtz, M.B., O'Hara, P. and Santschi, P.H., 1984. Partition of radiotracers between suspended particles and seawater. Geochim. Cosmochim. Acta, 48:2011–2019.

Lick, W., Lick, J. and Ziegler, C.K., 1992. Flocculation and its effect on the vertical transport of fine-grained sediments. Hydrobiologia, 235/236:1–16.

Lindahl, O., 2008. Musseloding för miljön – nu även i Östersjön. Havsutsikt, 3:4–5.

Lindquist, A., 2001. Catches of some "non-ICES" fish species in the Baltic. Fiskeriverket informerar 2001.

Lövdin, I., 2007. Tyst hav, Ordfront.

Lozán, J.L., Lampe, R., Mattäus, W., Rachor, E., Rumohr. H. and von Westernhagen, H. (eds.), 1996. Warnsignale aus der Ostsee. Pary, Buchverlag, Berlin, 385p.

Lucas, C.H. 2001. Reproduction and life history strategies of the common jellyfish, *Aurelia aurita*, in relation to its ambient environment. Hydrobiologia, 451:229–246.

Lucas, M.C. and Baras, E., 2001. Migration of freshwater fishes. Blackwell Science Ltd., Oxford.

Lučić, D., Njire, J., Morović, M., Precali, R., Fuks, D. and Bolotin, J., 2003. Microzooplankton in the open waters of the northern Adriatic Sea from 1990 to 1993: the importance of copepod nauplii densities. Helgol. Mar. Res., 57:73–81.

Lundström, K., et al. 2007. Estimation of grey seal (*Halichoerus grypus*) diet composition in the Baltic Sea. NAMMCO Sci. Pub., 6:177–196.

Mace, P.M, 2001. A new role for MSY in single-species and ecosystem approaches to fisheries stock assessment and management. Fish Fish., 2:2–32.

MacKenzie, B.R., Alheit, J., Conley, D.J., Holm, P. and Kinze, C.C., 2002. Ecological hypothesis for a historical reconstruction of upper trophic level biomass in the Baltic Sea and Skagerrak. Can. J. Fish. Aquat. Sci., 59:173–190.

MacKenzie, B.R., Hinrichsen, H-H., Plikshs, M., Wieland, K. and Zezera, A.S., 2000. Quantifying environmental heterogeneity: habitat size necessary for successful development of cod *Gadus morhua* eggs in the Baltic Sea. Mar. Ecol. Prog. Ser., 193:143–156.

MacKenzie, B.R. and Köster, F.W., 2004. Fish production and climate: sprat in the Baltic Sea. Ecology, 85:784–794.

Maitland, P.S., 1977. Freshwater fishes of Britain and Europe. Hamlyn, London, 256p.

Maitland, P.S., 1978. Biology of fresh waters. Blackie, Glasgow, 244p.

Mäkinen, T., 1991 (ed.), Marine aquaculture and environment, Vol. 22. Nordic Council of Minister, Nord, 126p.

Marino, R., Chan, F., Howarth, R.W., Pace, M.L. and Likens, G.E., 2006. Ecological constraints on planktonic nitrogen fixation in saline estuaries. I. Nitrogen and trophical controls. Mar. Ecol. Prog. Ser., 309:25–39.

Marker, A.F.H., 1965. Extracellular carbohydrate liberation in the flagellates *Isochrysis galbana* and *Prymnesium parvum*. J. Mar. Biol. Assoc. UK, 45:755–772.

Martinussen, M.B. and Båmstedt, U., 1999. Nutritional ecology of gelatinous planktonic predators. Digestion rate in relation to type and amount of prey. J. Exp. Mar. Biol. Ecol., 232:61–84.

Marumo, R. and Asaoka, O., 1974. Distribution of pelagic blue-green algae in the Northen Pacific Ocean. J. Oceanogr. Soc. Jpn, 30:77–85.

Matthiesen, H., Emeis, K.-C. and Jensen, B.T., 1998. Evidence for phosphate release from sediment in the Gotland Deep during oxic bottom water conditions. Meyniana, 50:175–190.

Mattila, J., Kankaanpää, H. and Ilus, E., 2006. Estimation of recent sediment accumulation rates in the Baltic Sea using artificial radionuclides 137Cs and 239,240Pb as time markers. Boreal Environ. Res., 11:95–107.

McDowall, R.M., 1988. Diadromy in fishes. Timber Press, Portland, OR, 308 pp.

Meier, H.E.M., Döscher, R. and Halkka, A., 2004. Simulated distributions of Baltic Sea-ice in warming climate and consequences for the winter habitat of the Baltic ringed seal. Ambio, 33:249–256.

Meijer, M.L., Dehaan, M.W., Breukelaar, A.W. and Buiteveld, H., 1990. Is reduction of benthivorous fish an important cause of high transparency following biomanipulation in shallow lakes? Hydrobiologia, 200/201:303–315.

Menshutkin, V.V., 1971. Mathematical modelling of populations and communities of aquatic animals. Leningrad Branch of Nauka Press, Leningrad (in Russian).

Mills, C.E., 1995. Medusae, siphonophores, and ctenophores as planktivorous predators in changing global ecosystems. ICES J. Mar. Sci., 52:575–581.

Moen, F.E. and Svensen, E., 2004. Marine fish and invertebrates. AquaPress, Essex, 608p.

Möller, P., Pihl, L. and Rosenberg, R., 1985. Fisk och bottendjur i grundområden i Bohuslän och Halland – en biologisk värdering. SNV PM 1911:7–95.

Möllmann, C., Kornilovs, G., Fetter, M. and Köster, F.W., 2004. Feeding ecology of central Baltic Sea herring and sprat. J. Fish Biol., 65:1563–1581.

Möllmann, C., Kornilovs, G. and Sidrevics. L., 2000. Long-term dynamics of main mesozooplankton species in the central Baltic Sea. J. Plankton Res., 22:2015–2038.

Möllmann, C. and Köster, F.W., 1999. Food consumption by clupeids in the Central Baltic: evidence for top-down control? ICES J. Mar. Sci. 56(Suppl.):100–113.

Möllmann, C. and Köster F.W., 2002. Population dynamics of calanoid copepods and the implications of their predation by clupeid fish in the Central Baltic Sea. J. Plankton Res. 24:959–977.

Monitor, 1988. Sweden's marine environment – ecosystem under pressure. Swedish Environmental Protection Agency, Stockholm, 207p.

Monakov, A.V., 1998. Feeding of freshwater invertebrates. Russian Academy of Sciences, Moscow (in Russian).

Monte, L. 1995. A simple formula to predict approximate initial contamination of lake water following a pulse deposition of radionuclide. Health Phys., 68(3):397–400.

Monte, L. 1996. Collective models in environmental science. Sci. Total Environ., 192:41–47.

Monte, L., Brittain, J.E., Håkanson, L. and Gallego, E. 1999. MOIRA models and methodologies for assessing the effectiveness of countermeasures in complex aquatic systems contaminated by radionuclides. ENEA, RT/AMP. 150 pp.

Monte, L., Håkanson, L. and Brittain, J., 1997. Prototype models for the MOIRA computerised system. ENEA, ISSN/1120-5555, Roma, 90p.

Monte, L., Van der Steen, J., Bergström, U., Gallego Diaz, E., Håkanson, L. and Brittain, J., 2000. The Project MOIRA A model-based computerised system for management support to identify optimal remedial straegies for restoring radionuclide contaminated aquatic ecosystens and drainage areas. Final Report. ENEA. ISSN/1120-5555, Rome, 89p.

Morgan, N.C., 1980. Transfer efficiency. The functioning of freshwater ecosystems. Cambridge University Press, Cambridge, pp. 329–334.

Muller, F.L.L., Tranter, M. and Balls, P.W., 1994. Distribution and transport of chemical constituents in the Clyde estuary. Estuar. Coast. Shelf Sci., 39:105–126.

Münsterhjelm, R., 1987. Flads and gloes in the archipelago. Geol. Surv. Finl. Spec. Pap., 2:55–61.

Myers, R.A. and Worm, B., 2005. Extinction, survival or recovery of large predatory fishes. Phil. Trans. R. Soc. B., 360:13–20.

Myrberg, K., 1998. Analysing and modelling the physical processes of the Gulf of Finland in the Baltic Sea. Monogr. Boreal Environ. Res. 10:50p.

NEFCO, 2007. Economic analysis of the BSAP with focus on eutrophication. HELCOM, Helsinki, 112p.

Neuenfeldt, S., Hinrichsen, H.-H., Nielsen, A. and Andersen, K.H., 2007. Reconstructing migrations of individual cod (*Gadus morhua L.*) in the Baltic Sea by using electronic data storage tags. Fish. Oceanogr., 16:526–535.

Nikolskiy, G.V., 1974. Fish ecology. Vysshaya Shkola, Moscow (in Russian).

Nilsson, J. Andersson, J., Karås, P. and Sandström, A., 2004. Recruitment failure and decreasing catches of perch (*Perca fluviatilis L.*) and pike (*Esox lucius L.)* in the coastal waters of southeast Sweden. Boreal Environ. Res., 9:295–306.

Nissling, A., Johansson, U., Jacobsson, M., 2006. Effects of salinity and temperature conditions on the reproductive success of turbot (Scophthalmus maximus) in the Baltic Sea. Fish. Res., 80:230–238.

Northcote, T.G., 1978. Migratory strategies and production in freshwater fishes. In: Gerking, S.D. (ed.), Ecology of freshwater fish production. Blackwell Scientific Publications, Oxford, pp. 326–359.

Odum, E., 1986. Ecology. Mir, Moscow (in Russian).

OECD, 1982. Eutrophication of waters. Monitoring, assessment and control. OECD, Paris, 154p.

Oglesby, R.T., 1977. Relationships of fish yield to lake phytoplankton standing crop, production, and morphoedaphic factors. J. Fish. Res. Board Can. 34:2271–2279.

Ojaveer, E., 2003. Fishes of Estonia. Estonian Academy Publishers, Tallinn.

Ojaveer, E. and Lehtonen, H., 2001. Fish stocks in the Baltic Sea: finite or infinite resource? Ambio, 30:217–221.

Olsen, Y., Jensen, A., Reinertsen, H., Borsheim, K.Y., Heldal, M. and Langeland, A., 1986. Dependence of the rate of release of phosphorus by zooplankton on P:C ratio in the food supply, as calculated by a recycling model. Limnol. Oceanogr., 31:34–44.

Omstedt, A. and Axell, L.B., 2003. Modeling the variations of sainity and temperature in the large Gulfs of the Baltic Sea. Cont. Shelf Res., 23:265–294.

Omstedt, A. and Chen, D., 2001. Influence of atmospheric circulation on the maximum ice extent in the Baltic Sea. J. Geophys. Res., 106:4493–4500.

Omstedt, A., Elken, J., Lehmann, A. and Piechura, J., 2004. Knowledge of the Baltic Sea Physics gained during the BALTEX and related programmes. Prog. Oceanogr., 63:1–28.

Omstedt, A. and Rutgersson, A., 2000. Closing the water and heat cycles of the Baltic Sea. Meteorol. Z., 9:57–64.

Ostapenia, A.P., 1989. Seston and detritus as structural and functional components of water ecosystems. Thesis for a Doctor's degree, Kiev (in Russian).

Ostmann, M., Suursaar, Ü. and Kullas, T., 2001. The oscillatory nature of the flows in the system of straits and small semienclosed basins of the Baltic Sea. Cont. Shelf Res., 21:1577–1603.

Ottosson, F. and Abrahamsson, O., 1998. Presentation and analysis of a model simulating epilimnetic and hypolimnetic temperatures in lakes. Ecol. Modell. 110:223–253.

Overbeck, J., 1972. Distribution pattern of phytoplankton and bacteria, microbial decomposition of organic matter and bacterial production in eutrophic, stratified lake. Productivity problems of freshwater. Warszawa-Krakow, 227–237.

Pace, M.L., 1993. Heterotrophic microbial processes. The trophic cascade in lakes. Cambridge University Press, Cambridge, pp. 252–277.

Pace, M.L., McManus, G.B. and Findlay, S.E.G., 1990. Plankton community structure determines the fate of bacterial production in a temperate lake. Limnol. Oceanogr., 35:795–808.

Paoli, A., Celussi, M., Valeri, A., Larato, C., Bussani, A., Fonda Umani, S., Vadrucci, M.R., Mazziotti, C. and Del Negro, P., 2007. Picocyanobacteria in Adriatic transitional environments. Estuar. Coast. Shelf Sci., 75:13–20.

Pearson, T.H. and Rosenberg, R., 1976. A comparative study on the effects on the marine environment of wastes from cellulose industries in Scotland and Sweden. Ambio, 5:77–79.

Peltonen, H., Vinni, M., Lappalainen, A. and Pönni, J. 2004. Spatial feeding patterns of herring (*Clupea harengus L.*), sprat (*sprattus sprattus L.*), and the three-spined stickleback (*Gasterosteus aculeatus L.*) in the Gulf of Finland, Baltic Sea. ICES J. Mar. Sci. 61:966–971.

Persson, J., 1999. On the role of morphometry in coastal ecosystem modelling and management. Dr. Thesis, Uppsala University, Uppsala.

Persson, J., Håkanson, L., Wallin, M., 1994. A geographical information system for coastal water planning based on chart information (in Swedish). Nordiska ministerrådet. TemaNord, Copenhagen, No. 667, 190p.

Persson, G. and Jansson, M., 1988. Phosphorus in freshwater ecosystems. Hydrobiologia, 170:340p.

Peters, R.H. 1986. The role of prediction in limnology. Limnol. Oceanogr., 31:1143–1159.

Peters, R.H., 1991. A critique for ecology. Cambridge University Press, Cambridge, 366p.

Pihl, L., Svensson, A., Moksnes, P.O. and Wennhage, H., 1999. Distribution of green algal mats throughout shallow soft bottoms of the Swedish archipelago in relation to nutrient loads and wave exposure. J. Sea Res. 41:281–294

Pohl, C. and Hennings, U., 1999. The effect of redox processes on the partitioning of Cd, Pb, Cu, and Mn between dissolved and particulate phases in the Baltic Sea. Mar. Chem., 65:41–53.

Porter, K.G., Paerl, H., Hodson, R., Pace, M., Priscu, J., Riemann, B., Scavia, D. and Stockner, J., 1988. Microbial interactions in lake food webs. Complex interactions in lake communities. Springer, New York, pp. 209–227.

Purcell, J.E. 2003. Predation on zooplankton by large jellyfish, *Aurelia labiata, Cyanea capillata* and *Aequorea aequorea*, in Prince William Sound, Alaska. Mar. Ecol. Prog. Ser., 246: 137–152.

Pustelnikov, O.S., 1977. Geochemical features of suspended matter in connection with recent processes in the Baltic Sea. Ambio, 5:157–162.

Quémerais, B., Cossa, D., Rondeau, B., Pham, T.T. and Fortin, B., 1998. Mercury distribution in relation to iron and manganese in the waters of St. Lawrence River. Sci. Total Environ., 213:193–201.

Rahm, L., Jönsson, A. and Wulff, F., 2000. Nitrogen fixation in the Baltic proper: an empirical study. J. Mar. Syst. 25:239–248.

Rai, H., 1978. Chlorophyll pigments in the Central Amazon lake ecosystems. Verh. Int. Verein. Limnol., 20:1192–1197.

Read, P. and Fernandez, T., 2003. Management of environmental impacts of marine aquaculture in Europe. Aquaculture 226:139–163.

Reila, H., 1979. Tursats Pärnu lahes. Abiks kalurile, 1:7–9.

Remane, A., 1934. Die Brackwasserfauna. Verhandlungen der Deutschen Zoologischen Gesellschaft 36:34–74.

Reynolds, C.S., 1987. Cyanobacteria water-blooms. Adv. Bot. Res., 13:67–143.

Reynolds, C.S. and Walsby, A.E., 1975. Water-blooms. Biol. Rev., 50:437–481.

Rodionova, N.V. and Panov, V.E., 2006. Establishment of the Ponto-Caspian predatory cladoceran Evadne anonyx in the eastern Gulf of Finland, Baltic Sea. Aquatic Invasions 1:7–12.

Rönkkönen, S., Ojaveer, E., Raid, T. and Viitasalo, M., 2004. Long-term changes in Baltic herring (Clupea harengus membras) growth in the Gulf of Finland. Can. J. Fish. Aquat. Sci., 61: 219–229.

Rosenberg, R., 1985. Eutrophication – the future marine coastal nuisance? Mar. Pollut. Bull., 16:227–231.

Rubio, V.C., Sánchez-Vázques, F.J., Madrid, J.A., 2005. Effects of salinity on food intake and macronutrient selection in European sea bass. Physiol. Behav., 85:333–339.

Ruohonen, K., Vielma, J. and Grove, D.J., 1998. Effects of feeding frequency on growth and food utilisation of rainbow trout (*Oncorhynchus mykiss*) fed low-fat herring or dry pellets. Aquaculture, 165:111–121.

Rutgersson, A., Omstedt, A. and Räisänen, J., 2002. Net precipitation over the Baltic Sea during present and future climate conditions. Clim. Res., 22:27–39.

Sadchikov, A.P. and Frenkel, O.A., 1990. Release of dissolved organic matter by living phytoplankton (methodical aspects). Hudrobiol. J., 26:84–87 (in Russian).
Sadchikov, A.P. and Makarov, A.A., 1997. Release of dissolved organic matter by living phytoplankton in three water bodies of different trophy. Hydrobiol. J., 33:104–107 (in Russian).
Salazkin, A.A. and Ogorodnikova, V.A.,1984. Methodical recommendations on sampling and treatment of materials at hydrobiological investigations of fresh waters. Nauka, Leningrad (in Russian).
Salomons, W. and Förstner, U., 1984. Metals in the hydrocycle. Springer, Heidelberg, 349p.
Salovius, S. and Kraufvelin, P., 2004. The filamentous green alga *Cladophora glolmerata* as a habitat for littoral macro-fauna in the northern Baltic Sea. Ophelia, 58:65–78.
Samuelsson, M., 1996. Interannual salinity variations in the Baltic Sea during the period 1954–1990. Cont. Shelf Res., 16:1463–1477.
Sandberg, J., Elmgren, R. and Wulff, F., 2000. Carbon flows in Baltic Sea food webs – a re-evaluation using a mass balance approach. J. Mar. Syst., 25:249–260.
Sandström, A., Eriksson, B.K., Karås, P., Isæus, M. and Schreiber, H., 2005. Boating activities influences the recruitment of near-shore fishes in a Baltic Sea archipelago area. Ambio, 34: 125–130.
Sandström, A. and Karås, P., 2002. Effects of eutrophication on young-of-the-year freshwater fish communities in coastal areas of the Baltic. Environ. Biol. Fish., 63:89–101.
Santschi, P.H. and Honeyman, B.D., 1991. Radioisotopes as tracers for the interactions between trace elements, colloids and particles in natural waters. In: Vernet, J.-P. (ed.), Heavy metals in the environment. Elsevier, Amsterdam.
Saulamo, K. and Lappalainen, J., 2007. Effects of abiotic factors on movements of pikeperch during pre-spawning and spawning season in a Baltic archipelago. Hydrobiologia, 579: 271–277.
Saulamo, K. and Neuman, E., 2002. Local management of Baltic fish stocks – significance of migrations. Fiskeriverket informerar, 9:1–18.
Savchuk, O.P., 2000. Studies of the assimilation capacity and effects of nutrient load reductions in the eastern Gulf of Finland with a biogeochemical model. Boreal Environ. Res., 5:147–163.
Savchuk, O.P., 2005. Resolving the Baltic Sea into seven subbasins: N and P budgets for 1991–1999. J. Mar. Syst., 56:1–15.
Savchuk, O.P. and Swaney, D.P., 2000. Water and Nutrient Budget of the Gulf of Riga. Accessible at http://data.ecology.su.se/mnode/Europe/Gulf%20of%20Riga/rigabud.htm (accessed April 19, 2006).
Savchuk, O.S., Wulff, F., Hille, S., Humborg, C. and Pollehne, F., 2008. The Baltic Sea a century ago – a reconstruction from model simulations, verified by observations. J. Mar. Syst., 74: 485–494.
Scheffer, M., 1998. Ecology of shallow lakes. Population and Community Biology Series, Vol. 22. Chapman and Hall, London, 384p.
Schernewski, G. and Neumann, T., 2005. The trophic state of the Baltic Sea a century ago: a model simulation study. J. Mar. Syst., 53:109–124.
Schmölcke, U., 2008. Holocene environmental changes and the seal (Phocidae) fauna of the Baltic Sea: coming, going and staying. Mammal Rev. 38:231–246.
Schneider, G. 1989. Estimation of food demands of *Aurelia Aurita* medusae populations in the Kiel Bight/Western Baltic. Ophelia, 31:17–27.
Schneider, G. and Behrends, G., 1994. Population dynamics and the trophic role of *Aurelia aurita* medusae in the Kiel Bight and western Baltic. J. Mar. Sci., 51:359–367.
Schindler, D.W., 1978. Factors regulating phytoplankton production and standing crop in the world's freshwaters. Limnol. Oceanogr., 23:478–486.
Schindler, D.W., Hecky, R.E., Findlay, D.L., Stainton, M.P., Parker, B.R., Patterson, M.J., Beaty, K.G., Lyng, M. and Kasian, S.E.M., 2008. Eutrophication of lakes cannot be controlled by reducing nitrogen input: results of a 37-year whole ecosystem experiment. PNAS, 105: 11254–11258.

Schou, J.S., Neye, S.T., Lundhede, T. et al., 2006. Modelling cost-efficient reductions of nutrient loads to the Baltic Sea. NERI Technical Report No. 592. NERI, Copenhagen, 69p.

Seifert, T., Tauber, F. and Kayser, B., 2001. A high resolution spherical grid topography of the Baltic Sea, 2nd ed. Baltic Sea Science Congress, Stockholm 25–29. November 2001. Poster #147. Accessible at http//www.io-warnemuende.de/iowtopo/

Seinä, A. and Palusuo, E., 1996. The classification of the maximum annual extent of ice cover in the Baltic Sea 1720–1995. In: MERI, Report 27. FIMR, Helsinki, 93p.

Sellner, K.G., 1997. Physiology, ecology and toxic properties of marine cyanobacteria blooms. Limnol. Oceanogr., 42:1089–1104.

Shafer, M.M., Overdier, J.T., Phillips, H., Webb, D., Sullivan, J.R. and Armstrong, D., 1999. Trace metal levels and partitioning in Wisconsin rivers. Water Air Soil Pollut., 110:273–311.

Skoptsov, V.G., Velichko, G.M. and Koljadin, S.A., 1983. Some questions of feeding and growth of Coregonus in conditions of commodity growing. In: Trophic relationships and their role in productivity of natural waters. Nauka, Leningrad, pp. 141–145 (in Russian).

SMHI, 2003. Djuphavsdata för havsområden 2003. Svenskt vattenarkiv, Nr 73, 69 sid.

SMHI, 2007. Baltic Sea ice climate. Accessible at http://www.smhi.se/cmp/jsp/polopoly.jsp?d=9048&a=25783&l=en (March 24, 2009).

SMHI and FIMR, 1982. Climatological ice atlas for the Baltic Sea, Kattegat, Skagerrak and Lake Vänern (1963–1979). Sjöfartsverkets tryckeri, Norrköping, 220p.

Smith, V.H., 1979. Nutrient dependence of primary productivity in lakes. Limnol. Oceanogr., 24:1051–1064.

Smith, V.H., 1985. Predictive models for the biomass of blue-green algae in lakes. Water Resour. Bull., 21:433–439.

Smith, V.H., 2003. Eutrophication of freshwater and coastal marine ecosystems: a global problem. Environ. Sci. Pollut. Res., 10:126–139.

Snickars, M., Sandström, A., Lappalainen, A., Mattila, J., Rosqvist, K. and Urho, L., 2009. Fish assemblages in coastal lagoons in land-uplift succession: the relative importance of local and regional environmental gradients. Estuar. Coast. Shelf Sci., 81:247–256.

SNV, 1993. Eutrophication of land, freshwater and the Sea (in Swedish, Eutrofiering av mark, sötvatten och hav). Swedish Environmental Protection Agency, Report 4134, Stockholm, 199p.

Söderberg, K. och Gårdmark, A., 2003. Kustfisk och fiske. Bottniska Viken 2003, pp. 19–21, ISSN 1400 2760.

Sogard, S.M. and Able, W., 1991. A comparison of eelgrass, sea lettuce macroalgae, and marsh creeks as habitats for epibenthic fishes and decapods. Estuarine Coastal Shelf Sci. 33:501–519.

Sondergaard, M., Riemann, B. and Jorgensen, N.O.G., 1985. Extracellular organic carbon (EOC) released by phytoplankton and bacterial production. Oikos, 45:323–332.

Sparholt, H., 1996. Causal correlation between recruitment and spawning stock size of central Baltic cod? ICES J. Mar. Sci., 53:771–779.

Sparholt, H., Aro, E. and Modin, J., 1991. The spatial distribution of cod (*Gadus morhua* L.) in the Baltic Sea. Dana, 9:45–56.

Stålnacke, P., Grimvall, A., Sundblad, K. and Tonderski, A., 1999. Estimation of riverine loads of nitrogen and phosphorus to the Baltic Sea, 1970–1993. Environ. Monit. Assess., 58: 173–200.

Stigebrandt, A., 2001. Physical oceanography of the Baltic Sea. In: Wulff, F., Rahm, L. and Larsson, P., (eds.), A systems analysis of the Baltic Sea. Springer, New York, pp. 19–74.

Stigebrandt, A. 2008. Så lindras Östersjöns övergödningssymptom. Webbutskrift från Göteborgs universitet, December 18, 2008.

Stigebrandt, A. and Gustafsson, B.G., 2007. Improvement of Baltic Proper water quality using large-scale ecological engineering. Ambio, 36:280–286.

Stockner, J.G. and Porter, K.G., 1988. Microbial food webs in freshwater planktonic ecosystems. In: Complex interactions in lake ecosystems. Springer, New York, pp. 69–83.

Stockner, J.G. and Shortreed, K.S., 1988. Algal picoplankton production and contribution to food webs in oligotrophic British Columbia lakes. Verh. Int. Ver. Limnol., 23:1–214.

Stone, L. and Weisburd, R.S.J., 1992. Positive feedback in aquatic ecosystems. Trends Ecol. Evol., 7:263–267.

Straskraba, M., 1980. The effect of physical variables on freshwater production: analyses based on models. In: The functioning of freshwater ecosystems. Cambridge University Press, Cambridge, pp. 13–84.

Subasinghe, R., 2004. Aquaculture in the global food supply. FAO, Rome.

Suchman, C.L. and Brodeur, R.D. 2005. Abundance and distribution of large medusae in surface waters of the northern California Current. Deep Sea Res. II, 52:51–72.

Sundblad, G., Härmä, M., Lappalainen, A., Urho, L. and Bergström, U., 2009. Transferability of predictive fish distribution models in two coastal systems. Estuar. Coast. Shelf Sci., in press.

Svensson, C., 2006. Ingenjörsgeologiska Exkursion, Sveriges jordarter. Accessible at http://connywww.tg.lth.se/Sv.jordarterdokument/DeglaciationiSvdokument/Landhjningen.html (September 14, 2006).

SVT, 1997. Ymparisto, 1997: 13. Kalavirrat – tietoa kalan tarjonnasta ja käytöstä, ISBN 951-776-138-4.

Swedish Environmental Advisory Council, 2005. A Strategy for Ending Eutrophication of Seas and Coasts. Swedish Government Official Reports, Environmental Advisory Council, Jo 1968: A, Stockholm, 59p.

Swedish EPA, 2008a. Ingen övergödning (No over-enrichment), revised version. Swedish EPA report 5840, Stockholm, 123p.

Swedish EPA, 2008b. Sweden's Commitment to the Baltic Sea Action Plan. Swedish Environmental Protection Agency, Report 5830. Swedish EPA, Stockholm, 150p (in Swedish).

Swingle, H.S., 1950. Relationships and dynamics of balanced and unbalanced fish populations. Bull. Agric. Exp. State AL, Polytech. Inst., 274:3–73.

Swingle, H.S., 1952. Farm pond investigations in Alabama. J. Wildlife Manag, 16:243–249.

Tessier, A., Carignan, R., Dubreuil, B. and Rapin, F., 1989. Partitioning of zinc between the water column and the oxic sediments in lakes. Geochim. Cosmochim. Acta, 53: 1511–1522.

Thomas, J.F., 1971. Release of dissolved organic matter from natural population of marine phytoplankton. Mar. Biol., 11:311–323.

Thurow, F., 1997. Estimation of the total fish biomass in the Baltic Sea during the 20th century. ICES J. Mar. Sci., 54:444–461.

Toivonen, A.-L., Roth, E., Navrud, S. et al., 2004. The economic value of recreational fisheries in Nordic countries. Fish. Manag. Ecol., 11:1–14.

Tonn, W.M., Magnusson, J.J., Rask, M. and Toivonen, J., 1990. Intercontinental comparison of small-scale fish assembalges. The balance between local and regional processes. Am. Nat., 136:345–375.

Tõnno, I., 2004. The impact of nitrogen and phosphorus concentration and N/P ratio on cyanobacterial dominance and N_2 fixation in some Estonian lakes. Doctoral thesis, Tartu University Press, Tartu.

Törnblom, E. and Rydin, E., 1998. Bacterial and phosphorus dynamics in profundal Lake Erken sediments following the deposition of diatoms: a laboratory study. Hydrobiologia, 364: 55–63.

Totti, C., Cangini, M., Ferrari, C. Kraus, R., Pompei, M., Pugnetti, A., Romagnoli, T., Vanucci, S. and Socal, G., 2005. Phytoplankton size-distribution and community structure in relation to mucilage occurrence in the northern Adriatic Sea. Sci. Total Environ., 353:204–217.

Tucker, J.M., 1999. Marine fish culture. Springer, Berlin, 760p.

Turesson, H., et al. 2002. Prey size selection in piscivorous pikeperch (*Stizostedion lucioperca*) includes active prey choice. Ecol. Freshw. Fish, 11:223–233.

Turner, A., 1996. Trace-metal partitioning in estuaries: importance of salinity and particle concentration. Mar. Chem., 54:27–39.

Turner, R.K., Georgiou, S., Gren, I.-M. et al., 1999. Managing nutrient fluxes and pollution in the Baltic: an interdisciplinary simulation study. Ecol. Econ., 30:333–352.

Turner, A., Millward, G.E., Bale, A.J. and Morris, A.W., 1993. Application of the Kd concept to the study of trace metal removal and desorption during estuarine mixing, Estuar. Coast. Shelf Sci., 36:1–13.

Uzars, D., Baranova, T. and Yula, E., 2000. Variation in environmental conditions, feeding and growth of cod in the Eastern Baltic. ICES CM 2000/Q10.

Uzars, D., Baranova, T., Yula, E., 2001. Interannual variation in gonadal maturation of cod in the Gotland Basin of the Baltic Sea: influence of environment and fish condition. ICES CM 2001/V:29.

Uzars, D. and Plikshs, M., 2000. Cod (*Gadus morhua L.*) cannibalism in the Central Baltic: interannual variability and influence of recruit abundance and distribution. ICES J. Mar. Sci., 57:324–329.

Van Densen, W.L.T. and McCay, B.J. 2007. Improving communication from managers to fishers in Europe and the US. ICES J. Mar. Sci., 64:811–817.

Vahtera, E., Conley, D.J., Gustafsson, B.G., Kuosa, H., Pitkänen, H., Savchuk, O.P., Tamminen, T., Viitasalo, M., Voss, M., Wasmund, N. and Wulff, F., 2007. Internal ecosystem feedbacks enhance nitrogen-fixing cyanobacteria blooms and complicate management in the Baltic Sea. Ambio, 36:186–193.

Vichi, M., Ruardij, P. and Baretta, J.W., 2004. Link or sink: a modelling interpretation of the open Baltic biogeochemistry. Biogeosci. Discuss., 1:79–100.

Viherluoto, M., 2001. Food selection and feeding behaviour of Baltic Sea mysid shrimps. Academic Dissertation. University of Helsinki, Helsinki.

Viherluoto, M. and Viitasalo, M., 2001. Temporal variability in functional responses and prey selectivity of pelagic mysid *Mysis mixta* in natural prey assemblages. Mar. Biol., 138:575–583.

Viitasalo, M., Flinkman, J. and Viherluoto, M., 2001. Zooplanktivory in the Baltic Sea: a comparison of prey selectivity by *Clupea harengus* and *Mysis mixta*, with reference to prey escape reactions. Mar. Ecol. Prog. Ser., 216:191–200.

Viitasalo, M., Vuorinen, I. and Saesmaa, S., 1995. Mesozooplankton dynamics in the northern Baltic Sea: implications of variations in hydrography and climate. J. Plankton Res., 17: 1857–1878.

Voipio, A. (ed.), 1981. The Baltic Sea. Elsevier Oceanographic Series, Elsevier, Amsterdam, 418p.

Vollenweider, R.A., 1968. The scientific basis of lake eutrophication, with particular reference to phosphorus and nitrogen as eutrophication factors. Tech. Rep. DAS/DSI/68.27, OECD, Paris, 159p.

Vollenweider, R.A., 1969. A manual on methods for measuring primary production in aquatic environments. IBP, Oxford, Handbook, 12, 213p.

Vollenweider, R.A., 1976. Advances in defining critical loading levels for phosphorus in lake eutrophication. Mem. Ist. Ital. Idrobiol., 33:53–83.

Vorobev, G.A., 1977. Landscape types of lake overgrowing. Natural Conditions and Resources of North European Part of the Soviet Union, Vologda, pp. 48–60 (in Russian).

Vuorinen, I., Hänninen, J., Viitasalo, M., Helminen, U. and Kuosa, H., 1998. Proportion of copepod biomass declines with decreasing salinity in the Baltic Sea. ICES J. Mar. Sci., 55:767–774.

Wahlström, E., 2000. Effects of nutrient loading on lake foodwebs – a review (in Swedish, Effekter av ökad närsaltsbelastning på födovävar i sjöar – översikt av befintligt kunskapsläge). Vattenbruksinstitutionen, SLU, Rapport 24, Umeå, 34p.

Waller, U. and Boettger, T., 2001. The response of young cod (Gadus morhua) to environmental gradients. ICES CM.

Wallin, M., Håkanson, L. and Persson, J., 1992. Load models for nutrients in coastal areas, especially from fish farms (in Swedish with English summary). Nordiska ministerrådet, 1992:502, Copenhagen, 207p.

Walters, C.J., Christersen, V. and Pauly, D. 1997. Structuring dynamic models of exploited ecosystems from trophic mass-balance assessments. Rev. Fish Biol. Fish., 7:139–172.

Walters, C.J., Christersen, V., Pauly, D. and Kitchell, J.F. 2000. Representing density dependent consequences of life history strategies in aquatic ecosystems: Ecosim II. Ecosystems, 3:70–83.

Warren, L.A. and Zimmerman, A.P., 1994. The influence of temperature and NaCl on cadmium, copper and zinc partitioning among suspended particulate and dissolved phases in an urban river. Water Res., 28:1921–1931.

Wasmund, N., 1997. Occurrence of cyanobacterial blooms in the Baltic Sea in relation to environmental conditions. Int. Rev. Gesamten Hydrobiol., 82:169–184.

Wasmund, N. and Uhlig, S., 2003. Phytoplankton trends in the Baltic Sea. ICES J. Mar. Sci., 60:177–186.

Wasmund, N., Voss, M. and Lochte, K., 2001. Evidence of nitrogen fixation by non-heterocystous cyanobacteria in the Baltic Sea and re-calculation of a budget of nitrogen fixation. Mar. Ecol. Prog. Ser., 214:1–14.

Watras, C.J., Back, R.C., Halvorsen, S., Hudson, R.J.M., Morrison, K.A. and Wente, S.P., 1998. Bioaccumulation of mercury in pelagic freshwater food webs. Sci. Total Environ., 219: 183–208.

Watras, C.J., Morrison, K.A. and Bloom, N.S., 1995a. Chemical correlates of Hg and methyl-Hg in northern Wisconsin lake waters under ice-cover. Water Air Soil Pollut., 84:253–267.

Watras, C.J., Morrison, K.A., Host, J.S. and Bloom, N.S., 1995b. Concentration of mercury species in relationship to other site specific factors in the surface waters of northern Wisconsin lakes. Limnol. Oceanogr., 40:556–565.

Watt, W.D., 1966. Release of dissolved organic matter from the cells of phytoplankton populations. Proc. R. Soc., 164:521–551.

Weber, W.J. Jr., McGinley, P.M. and Katz, L.E., 1991. Sorption phenomena in subsurface systems: concepts, models and effects on contaminant fate and transport. Water Res., 25: 499–528.

Westerbom, M., Kilpi, M. and Mustonen, O., 2002. Blue mussels, *Mytilus edulis*, at the edge of the range: population structure, growth and biomass along a salinity gradient in the north-eastern Baltic Sea. Mar. Biol. 140:991–999.

Westin, L. and Nissling, A., 1991. Effects of salinity on spermatozoa motility, percentage of fertilized eggs and egg development of Baltic cod (*Gadus morhua L.)* and its implications for cod stock fluctuations in the Baltic. Mar. Biol., 108:5–9.

Westlake, D.F., 1980. Primary production. The functioning of freshwater ecosystems. Cambridge University Press, pp. 141–246.

Wetzel, R.G., 1964. A comparative study of the primary productivity of higher aquatic plants, periphyton, and phytoplankton in a large, shallow lake. Int. Rev. Gesamten Hydrobiol., 49: 1–61.

Wetzel, R.G., 1983. Limnology. Saunders College Publishing, Philadelphia, 767p.

Weyhenmeyer, G.A., 1996. The significance of lake resuspension in lakes. Uppsala dissertations from the faculty of science and technology 225. Sweden. ISBN 91-554-3820-2.

Wieland, K., Jarre-Teichmann, A. and Horbowa. K., 2000. Changes in the timing of spawning of Baltic cod: possible causes and implications for recruitment. ICES J. Mar. Sci., 57: 452–464.

Winberg, G.G., 1985. Main features of production process in the Naroch lakes. Ecological System of Naroch Lakes, Minsk, pp. 269–284 (in Russian).

Winberg, G.G., Alimov, A.F., Umnov, A.A. and Norenko, D.S., 1986. Productivity and rational use of lakes of Eravno-Kharga system. Investigation of the relationship between food basis and fish productivity. Ackefors, Leningrad, pp. 212–219.

Winsor, P., Rodhe, J. and Omstedt, A., 2001. Baltic Sea ocean climate: an analysis of 100 yr of hydrographic data with focus on the freshwater budget. Clim. Res., 18:5–15.

Wolter, K., 1982. Bacterial incorporation of organic substances released by natural phytoplankton population. Mar. Ecol. Prog. Ser., 17:287–295.

Wotton, R.J., 1990. Ecology of teleost fishes. Chapman and Hall, New York, NY, 404 pp.

Wulff, F., 2006. The Baltic Sea in not like other systems (Östersjön är inte som andra hav). In: Johansson, B., (ed.), The Baltic Sea – threats and hope (Östersjön – hot och hopp), Formas Fokuserar, Stockholm, pp. 51–63.

Wulff, F., Rahm, L. and Larsson, P. (eds.), 2001. A systems analysis of the Baltic Sea. Springer, Berlin, Ecological Studies 148, 455p.

Yan, L., Stallard, R.F., Key, R.M. and Crerar, D.A., 1991. Trace metals and dissolved organic carbon in estuaries and offshore waters of New Jersey, USA. Geochim. Cosmochim. Acta, 55:3647–3656.

You, C.-F., Lee, T. and Li, Y.-H., 1989. The partition of be between soil and water. Chem. Geol., 77:105–118.

Youngson, A.F., Dosdat, A., Saroglia, M. and Jordan, W.C., 2001. Genetic interactions between marine finfish species in European aquaculture and wild conspecifics. J. Appl. Ichthyol., 17:153–162.

Zaitsev, Y. and Mamaev, V., 1997. Marine biological diversity in the Black Sea. A study of change and decline. United Nations Publications, New York, 208p.

Zehr, J.P., Jenkins, B.D., Short, S.M. and Steward, G.F., 2003. Nitrogenase gene diversity and microbial community structure: a cross-system comparison. Environ. Microbiol., 5:539–554.

Zhou, J.L., Fileman, T.W., Evans, S., Donkin, P., Readman, J.W., Mantoura, R.F.C. and Rowland, S., 1999. The partition of fluoranthene and pyrene between suspended particles and dissolved phase in the Humber Estuary: a study of the controlling factors. Sci. Total Environ., 244: 305–321.

Index

www.ingramcontent.com/pod-product-compliance
Ingram Content Group UK Ltd.
Pitfield, Milton Keynes, MK11 3LW, UK
UKHW020633260926
475829UK00003BA/65
9783642115615